Planning and Installing
Photovoltaic Systems
A guide for installers, architects and engineers
second edition

Planning and Installing

Photovoltaic Systems

A guide for installers, architects and engineers

second edition

DGS
Deutsche Gesellschaft für Sonnenenergie e.V.

earthscan

London • Sterling, VA

First published by Earthscan in the UK and USA in 2008
First edition published in 2005, reprinted 2006

Copyright © The German Energy Society (Deutsche Gesellshaft für Sonnenenergie (DGS LV Berlin BRB), 2008

All rights reserved

ISBN-13: 978-1-84407-442-6

Typeset by Mapset Ltd, Gateshead, UK
Printed and bound in Croatia by Zrinski
Cover design by Yvonne Booth

For a full list of publications please contact:

Earthscan
8–12 Camden High Street
London, NW1 0JH, UK
Tel: +44 (0)20 7387 8558
Fax: +44 (0)20 7387 8998
Email: earthinfo@earthscan.co.uk
Web: **www.earthscan.co.uk**

22883 Quicksilver Drive, Sterling, VA 20166-2012, USA

Earthscan publishes in association with the International Institute for Environment and Development

A catalogue record for this book is available from the British Library

Library of Congress Cataloging-in-Publication Data

Planning and installing photovoltaic systems : a guide for installers, architects, and engineers / Deutsche Gesellschaft für Sonnenenergie (DGS). – 2nd ed.
    p. cm.
  Includes bibliographical references and index.
  ISBN-13: 978-1-84407-442-6 (pbk. : alk. paper)
  ISBN-10: 1-84407-442-0 (pbk. : alk. paper)
 1. Building-integrated photovoltaic systems–Installation. I. Deutsche Gesellschaft für sonnenenergie.
  TK1087.P5813 2007
  621.31'244–dc22

2007034796

Printed on elemental chlorine-free paper.

This guide has been prepared as part of the GREENPro project co-funded by the European Commission. Also available in the series: *Planning and Installing Bioenergy Systems: A Guide for Installers, Architects and Engineers* 978-1-84407-132-6
Planning and Installing Solar Thermal Systems: A Guide for Installers, Architects and Engineers 978-1-84407-125-8

Neither the author nor the publisher make any warranty or representation, expressed or implied, with respect to the information contained in this publication, or assume any liability with respect to the use of, or damages resulting from, this information.

Contents

Foreword   *xi*

CHAPTER 1 Photovoltaic Basics

1.1	**PV array systems and PV applications**	**1**
1.1.1	Overview	1
1.1.2	Stand-alone systems	1
1.1.3	Grid-connected systems	5
1.2	**Solar radiation**	**8**
1.2.1	The sun as an energy source	8
1.2.2	Distribution of solar radiation	9
1.2.3	Direct and diffuse radiation	10
1.2.4	Angle definition	11
1.2.5	Solar altitude and solar spectrum	11
1.2.6	Ground reflection	13
1.2.7	How solar radiation is measured	14
1.2.8	Tracking PV arrays	15
1.3	**The photovoltaic effect and how solar cells work**	**19**
1.3.1	How a solar cell works	19
1.3.2	Design and functioning of a crystalline silicon solar cell	21
1.4	**Solar cell types**	**23**
1.4.1	Crystalline silicon	23
1.4.2	Mono-crystalline (single-crystal) silicon cells	24
1.4.3	Polycrystalline silicon cells	26
1.4.4	Ribbon-pulled silicon cells	27
1.4.5	Anti-reflective coating on crystalline silicon cells	30
1.4.6	Front contacts	31
1.4.7	Back contacts	33
1.4.8	High-performance cells	34
1.4.9	Thin-film cell technology	40
1.4.10	Amorphous silicon cells	42
1.4.11	Copper indium diselenide (CIS) cells	43
1.4.12	Cadmium telluride (CdTe) cells	44
1.4.13	Thin-film solar cells made from crystalline silicon	48
1.4.14	Concentrating systems	49
1.4.15	Hybrid cells: HIT solar cells	51
1.4.16	Comparison of solar cell types and trends	52
1.5	**Electrical properties of solar cells**	**53**
1.5.1	Equivalent circuit diagrams of solar cells	53
1.5.2	Cell parameters and solar cell characteristic I–V curves	56
1.5.3	Spectral sensitivity	60
1.5.4	Efficiency of solar cells and PV modules	62

CHAPTER 2 PV Modules and Other Components of Grid-Connected Systems

2.1	**PV modules**	**65**
2.1.1	Cell stringing	65
2.1.2	Cell encapsulation	67
2.1.3	Types of modules	72
2.1.4	Design options for PV modules	75
2.1.5	Module cable outlets and junction boxes	84

	2.1.6	Wiring symbols	84
	2.1.7	Characteristic *I–V* curves for modules	85
	2.1.8	Irradiance dependence and temperature characteristics	87
	2.1.9	Hot spots, bypass diodes and shading	89
	2.1.10	Electrical characteristics of thin-film modules	93
	2.1.11	Quality certification for modules	98
	2.1.12	Interconnection of PV modules	100
2.2		**PV array combiner/junction boxes, string diodes and fuses**	**101**
2.3		**Grid-connected inverters**	**103**
	2.3.1	Wiring symbol and method of operation	103
	2.3.2	Grid-controlled inverters	105
	2.3.3	Self-commutated inverters	106
	2.3.4	Characteristics, characteristic curves and properties of grid-connected inverters	109
	2.3.5	Grid-connected inverter types and construction sizes in various power classes	117
	2.3.6	Further developments in grid-connected inverter technology	118
2.4		**Cabling, wiring and connection systems**	**123**
	2.4.1	Module and string cables	123
	2.4.2	Connection systems	124
	2.4.3	DC main cable	126
	2.4.4	AC connection cable	126
2.5		**Direct current load switch (DC main switch)**	**126**
2.6		**AC switch disconnector**	**127**
	2.6.1	Miniature circuit breaker (MCBs)	127
	2.6.2	Earth leakage circuit breaker	127

CHAPTER 3 Site Surveys and Shading Analysis

3.1		**On-site visit and site survey**	**129**
3.2		**Consulting with the customer**	**130**
3.3		**Shadow types**	**131**
	3.3.1	Temporary shading	131
	3.3.2	Shading resulting from the location	133
	3.3.3	Shading resulting from the building	133
	3.3.4	Self-shading	134
	3.3.5	Direct shading	135
3.4		**Shading analysis**	**137**
	3.4.1	Using a site plan and sun path diagram	137
	3.4.2	Using a sun path diagram on acetate	137
3.5		**Shade analysis tools using software**	**139**
3.6		**Shading, PV-array configuration and system concept**	**141**
	3.6.1	Connection in series (string concept)	142
	3.6.2	Connection in parallel	142
	3.6.3	Comparison of connection concepts	142
3.7		**Shading with free-standing/rack-mounted PV arrays**	**144**
	3.7.1	Reducing the mutual shading losses of rack-mounted PV modules	145
3.8		**Checklists for building survey**	**146**
	3.8.1	PV system checklist	147
	3.8.2	PV generator, inverter and meter	148
	3.8.3	Lines and installation	149
	3.8.4	Other	149
	3.8.5	Shading checklist	149

CHAPTER 4 Planning and Sizing Grid-Connected Photovoltaic Systems

4.1	**System size and module choice**	**151**
4.2	**System concepts**	**152**
4.2.1	Central inverter concept	153
4.2.2	Sub-array and string inverter concept	155
4.2.3	Module inverter concept	156
4.3	**Inverter installation site**	**158**
4.4	**Sizing the inverter**	**159**
4.4.1	Choosing the number and power rating of inverters	159
4.4.2	Voltage selection	161
4.4.3	Determining the number of strings	161
4.4.4	Sizing using simulation programs	164
4.5	**Selecting and sizing cables for grid-tied PV systems**	**165**
4.5.1	Cable voltage ratings	166
4.5.2	Cable current carrying capacity	166
4.5.3	Minimizing the cable losses/voltage drops	167
4.5.4	Sizing the module and string cabling	167
4.5.5	Sizing the DC main cable	170
4.5.6	Sizing the AC connection cable	171
4.6	**Selection and sizing of the PV array combiner/junction box and the DC main disconnect/isolator switch**	**172**
4.7	**Lightning protection, earthing/grounding and surge protection**	**173**
4.7.1	Lightning protection – direct strikes	175
4.7.2	Indirect lightning effects and internal lightning protection	176
4.8	**Yield forecast**	**178**

CHAPTER 5 System Sizing, Design and Simulation Software

5.1	**Use of sizing, design and simulation programs**	**181**
5.2	**Checking the simulation results**	**182**
5.3	**Simulation of shading**	**183**
5.4	**Market overview and classification**	**183**
5.5	**Programme descriptions**	**184**
5.5.1	Calculation programs	184
5.5.2	Time-step simulation programs	185
5.5.3	Simulation systems	194
5.5.4	Supplementary programs and data sources	195
5.5.5	Design and service programs	197
5.5.6	Web-based simulation programs	198

CHAPTER 6 Mounting Systems and Building Integration

6.1	**Introduction**	**199**
6.2	**Roof Basics**	**200**
6.2.1	The roof's tasks	200
6.2.2	Roof shapes	200
6.2.3	Roof constructions	201
6.2.4	Roof skin	202
6.2.5	Sloping roof	206
6.2.6	Flat roof	207
6.3	**Sloping roofs**	**209**
6.3.1	On-roof systems	209
6.3.2	In-roof systems	221
6.4	**Flat roofs**	**224**
6.4.1	On-roof systems for flat roofs	224
6.4.2	Roof-integrated systems	234

6.5	**Façade basics**	**235**
6.5.1	External wall structure	235
6.5.2	Façade types	237
6.5.3	Façade structures and construction methods	238
6.5.4	Fastenings	242
6.5.5	Joints and joint sealing	243
6.6	**Photovoltaic façades**	**244**
6.6.1	Mounting modules on existing façades	245
6.6.2	Façades with integrated modules	246
6.7	**Glass roofs**	**257**
6.8	**Solar protection devices**	**263**
6.8.1	Module fixing	264
6.8.2	Fixed solar shading	264
6.8.3	Moveable solar shading	267
6.9	**Mounting systems for free-standing installations**	**270**

CHAPTER 7 Installing, Commissioning and Operating Grid-Connected Photovoltaic Systems

7.1	**General installation notes**	**273**
7.1.1	Notes on DC installation	273
7.1.2	Notes on module mounting	273
7.1.3	Notes on module interconnection	275
7.1.4	Notes on cable laying	275
7.2	**Example installation of a grid-connected PV system**	**276**
7.2.1	Preparation	276
7.2.2	System installation: Step by step	277
7.3	**Guarantee**	**283**
7.4	**Breakdowns, typical faults and maintenance for PV systems**	**285**
7.4.1	Maintenance	285
7.4.2	Maintenance and upkeep checklist	286
7.5	**Troubleshooting**	**287**
7.6	**Monitoring operating data and presentation**	**288**
7.6.1	Internet-based system evaluation	291
7.6.2	Web-based data transmission and evaluation	292
7.6.3	Presentation and visualization	292
7.7	**Long-term experience and quality**	**293**
7.7.1	Long-term behaviour of PV modules	293
7.7.2	Quality and reliability of inverters	294

CHAPTER 8 Stand-alone Photovoltaic Systems

8.1	**Introduction**	**295**
8.2	**Modules in stand-alone PV systems**	**298**
8.3	**Batteries in stand-alone PV systems**	**298**
8.3.1	How lead-acid batteries work: Construction and operating principles	298
8.3.2	Types and designs of lead-acid batteries	300
8.3.3	Operating behaviour and characteristics of lead-acid batteries	303
8.3.4	Ageing effects	**307**
8.3.5	Selection criteria	307
8.3.6	Battery safety and maintenance	307
8.3.7	Recycling	310
8.4	**Charge controllers**	**310**
8.4.1	Series controllers	311
8.4.2	Shunt controllers (parallel controllers)	312
8.4.3	Deep discharge protection	312
8.4.4	MPP charge controllers	313

8.5	**Stand-alone inverters**	**313**
8.5.1	Sine-wave inverters	315
8.5.2	'Modified sine-wave' inverters	315
8.5.3	Square-wave inverters	315
8.5.4	Application criteria for inverters in stand-alone systems	315
8.6	**Planning and designing stand-alone systems**	**316**
8.6.1	Direct coupling of PV array, battery and loads	316
8.7	**Measuring electricity consumption**	**317**
8.8	**Sizing the PV array**	**318**
8.8.1	Model for calculating the yield of a PV array	318
8.8.2	Cable, conversion and adjustment losses	320
8.8.3	Summary of the design outcome	321
8.8.4	Brief summary of the calculation method for designing a PV array, taking the example of the small holiday home	322
8.9	**Sizing of the cable cross sections**	**323**
8.9.1	Charge controller cable	325
8.10	**Battery sizing**	**325**
8.11	**Use of an inverter**	**326**
8.12	**Photovoltaics in decentral electricity grids/mini-grids**	**327**
8.12.1	DC-coupled systems	328
8.12.2	AC-coupled systems	329

CHAPTER 9 Economics and Environmental Issues

9.1	**Cost trends**	**331**
9.2	**Technological trends**	**332**
9.3	**Economic Assessment**	**333**
9.3.1	Power production costs	333
9.4	**Environmental impact**	**335**
9.4.1	Energy payback and harvest factor	335
9.4.2	Pollutants in the production process	336
9.4.3	Module recycling concepts	337

CHAPTER 10 Marketing and Promotion

10.1	**Marketing PV: The basics**	**341**
10.1.1	Customer orientation: The central theme	341
10.1.2	The iceberg principle	342
10.1.3	The pull concept	342
10.2	**Greater success through systematic marketing**	**343**
10.2.1	The benefits come first	343
10.2.2	The four pillars of the marketing concept	344
10.2.3	Range of marketing options	347
10.2.4	Six steps to the target	353
10.3	**A good sales talk is fun**	**358**
10.3.1	What does 'successful selling' mean?	358
10.3.2	Build a bridge	359
10.3.3	Find out the customer's requirements	360
10.3.4	Offer solutions	362
10.3.5	Achieve the result	362

Bibliography	**367**
Index	**371**

Foreword

Photovoltaics (PV), the technology which converts sunlight into electricity, is one of the fastest growing sectors of the renewable energy industry. It is already well established in many countries and looks set to become one of the key technologies of the 21st century. The market is being driven by concerns about carbon emissions, energy security and the rising price of fossil fuels.

This edition of *Planning and Installing Photovoltaic Systems* is a fully updated translation and adaptation of a book published by the German Solar Energy Society (DGS) – Germany is the country with the largest number of grid-tied PV systems in the world. It contains details of new products, particularly regarding the latest developments in module technology, and illustrations of advances in module mounting technology and building-integrated photovoltaics (BIPV). The chapters dealing with site surveys, system design and installation have also been updated. A new chapter on marketing PV has been added. While the book's emphasis is on grid-tied PV systems – the fastest growing sector of the industry and the type of system most relevant to industrial countries in Europe and North America where a sophisticated electrical distribution grid is in place – a chapter on stand-alone systems is also included.

This book is aimed at an international readership. For the most part, it follows German practice and regulations. These will be similar to current practices and regulations in other European Union countries but there will occasionally be differences, and in the United States and other English-speaking countries, practices, codes and regulations will definitely be different. Because of this, readers need to refer to national electrical and building codes and regulations and abide by them. Metering and selling solar-generated electricity onto the grid will also differ from country to country and it is important for installers to familiarize themselves with local arrangements.

Frank Jackson, Green Dragon Energy, Berlin
October 2007

1 Photovoltaic Basics

1.1 PV array systems and PV applications

1.1.1 Overview

Photovoltaic (PV) systems can be grouped into stand-alone systems and grid-connected systems. In stand-alone systems the solar energy yield is matched to the energy demand. Since the solar energy yield often does not coincide in time with the energy demand from the connected loads, additional storage systems (batteries) are generally used. If the PV system is supported by an additional power source – for example, a wind or diesel generator – this is known as a photovoltaic hybrid system.

In grid-connected systems the public electricity grid functions as an energy store. In Germany, most PV systems are connected to the grid. Because of the premium feed-in tariff for solar electricity in Germany, all of the energy they generate is fed into the public electricity grid. The forecast for the next 40 years is that photovoltaics may provide up to one third of the power supply in Germany.

While more and more grid-connected PV systems will be installed in Europe and North America in the coming years, in the long term it is expected that ever-increasing numbers of stand-alone systems will be installed, especially in developing countries. Small individual power supplies for homes – known as solar home systems – can provide power for lights, radio, television, or a refrigerator or a pump. And, increasingly, villages are gaining their own power supplies with an alternating current circuit and outputs in the two-digit kilowatt range.

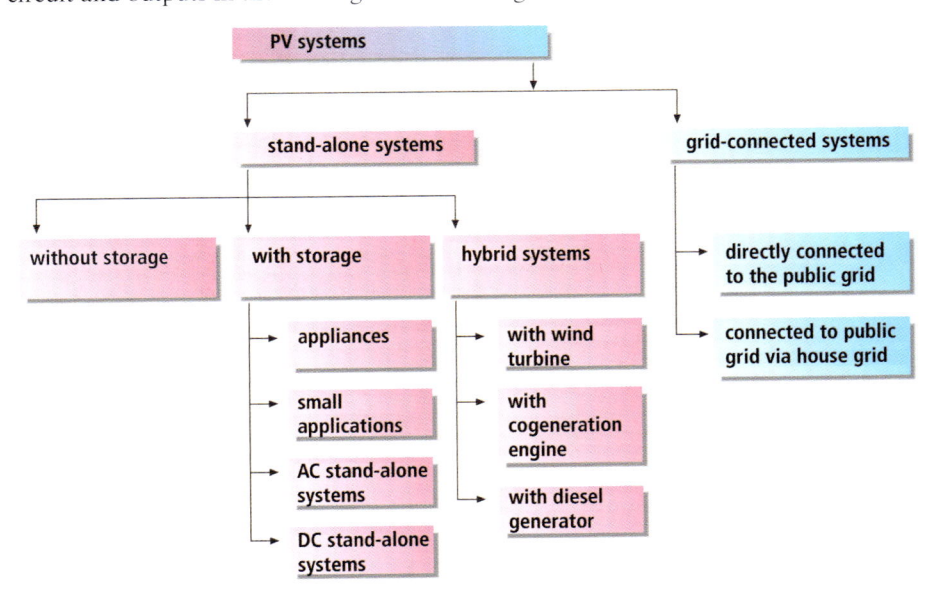

Figure 1.1 Types of PV systems

1.1.2 Stand-alone systems

The first cost-effective applications for photovoltaics were stand-alone systems. Wherever it was not possible to install an electricity supply from the mains utility grid, or where this was not cost-effective or desirable, stand-alone photovoltaic systems could be installed. The range of applications is constantly growing. There is great potential for using stand-alone systems in developing countries where vast areas are still frequently not supplied by an electrical grid. But technological innovations and new lower-cost production methods are opening up potential in industrialized countries as well.

Solar power is also on the advance when it comes to mini-applications: pocket calculators, clocks, battery chargers, flashlights, solar radios, etc., are well known examples of the successful use of solar cells in stand-alone applications.
Other typical applications for stand-alone systems:

- mobile systems on cars, camper vans, boats, etc.;
- remote mountain cabins, weekend and holiday homes and village electrification in developing countries;
- SOS telephones, parking ticket machines, traffic signals and observation systems, communication stations, buoys and similar applications that are remote from the grid;
- applications in gardening and landscaping;
- solar pump systems for drinking water and irrigation, solar water disinfection and desalination.

Figure 1.2 Milk frother
Source: Solarc

Figure 1.3 Garden light
Source: Solarwatt

Figure 1.4 Solar charger
Source: Solarc

PHOTOVOLTAIC BASICS 3

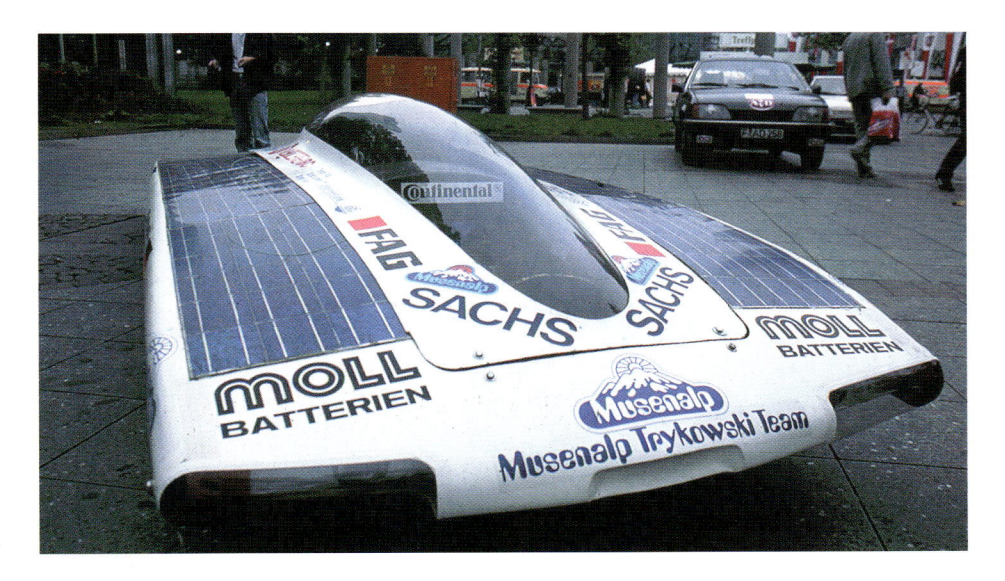

Figure 1.5 Solar car

*Figure 1.6
Mobile ice-cream stand with
solar freezer system
Source: Sepp Fiedler;
Solar Lifestyle GmbH*

*Figure 1.7
Solar boat
Source: D. A. Seebacher;
Aquawatt Yachtbau Company*

*Figure 1.8
Mountain cabin with small
stand-alone PV system
Source: Sonnenschein Company*

*Figure 1.9
The Rappenecker Hof Restaurant
(Black Forest, Germany) for day-trippers
obtains up to 70 per cent of its energy
from PV and wind power
Note: During the summer of 2003, the
system was modernized and a fuel cell
was added.*

Figure 1.10 Solar bus-stop lighting

*Figure 1.11 PV system to power buoys
Source: Sonnenschein Company*

Figure 1.12 PV flowers for a watering system in the Mauerpark, Prenzlauer Berg, Berlin, Germany

*Figure 1.13 Solar pump system for drinking water
Source: Siemens*

Stand-alone PV systems generally require an energy storage system because the energy generated is not usually (or infrequently) required at the same time as it is generated (i.e. solar energy is available during the day, but the lights in a stand-alone solar lighting system are used at night). Rechargeable batteries are used to store the electricity. However, with batteries, in order to protect them and achieve higher availability and a longer service life it is essential that a suitable charge controller is also used as a power management unit. Hence, a typical stand-alone system comprises the following main components:

1. PV module/s, usually connected in parallel or series-parallel;
2. charge controller;
3. battery or battery bank;
4. load(s);
5. inverter – in systems providing alternating current (AC) power.

Figure 1.14 Solar thermal desalination system; PV modules for the pump and control components make this compact system completely autonomous
Source: Fraunhofer ISE

Components and sizing of stand-alone PV systems are discussed in Chapter 8.

1.1.3 Grid-connected systems

A grid-connected PV system essentially comprises the following components:

1. PV modules/array (multiple PV modules connected in series or parallel with mounting frame);
2. PV array combiner/junction box (with protective equipment);
3. direct current (DC) cabling;
4. DC main disconnect/isolator switch;
5. inverter;
6. AC cabling;
7. meter cupboard with power distribution system, supply and feed meter, and electricity connection.

The individual components are described in detail in the section on '1.4 Solar cell types'. Figure 1.15 shows the typical layout of a grid-connected PV system.

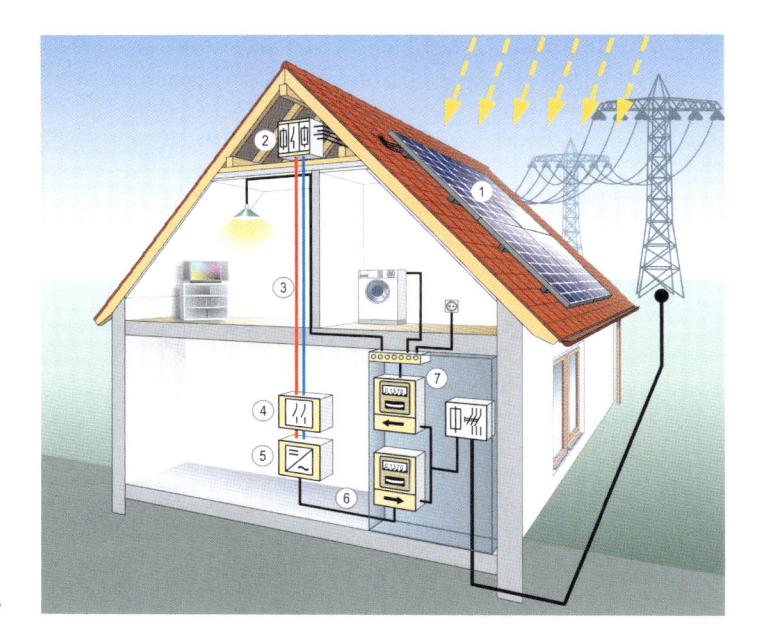

Figure 1.15 Principle of a grid-connected PV system
Source: C. Geyer/DGS LV Berlin BRB

Figure 1.16 Grid-connected PV system on the roof of a family house

Figure 1.17 Grid-connected PV system on the urban commercial estate Brockhill in Woking Borough, UK
Source: QJA-Services

PHOTOVOLTAIC BASICS 7

Figure 1.18: Grid-connected PV Cube at the Discovery Science Center in Santa Ana, California
Source: OJA-Services; copyright: Solar Design Associates Inc.

Figure 1.19 Power station towers at the Stadtwerke Duisburg, Germany, with glued-on PV modules
Source: Hoesch Contecna

Figure 1.20 Grid-connected PV system at a chicken farm

Figure 1.21 Grid-connected 100kW PV system on a noise barrier next to the A13 motorway near Domat/Ems in Switzerland
Source: TNC

Whereas the first PV system installations were mounted on the roofs of private family houses, PV systems are increasingly being installed on all kinds of buildings (e.g. apartment blocks, schools, and agricultural and industrial buildings). In addition, there is increasing use of other structures for photovoltaic systems (e.g. motorway noise barriers and train station platform roofs). There is now a great variety of design possibilities for integrating PV systems within buildings. These are discussed in detail in Chapter 7.

As well as this, energy utilities, operating companies and investment companies, in particular, are building large-scale grid-connected PV systems as ground-mounting systems.

Figure 1.22 5MW ground-mounting system at a former ash-settling basin near Espenhain, Leipzig, Germany

1.2 Solar radiation

1.2.1 The sun as an energy source

The sun supplies energy in the form of radiation, without which life on Earth could not exist. The energy is generated in the sun's core through the fusion of hydrogen atoms into helium. Part of the mass of the hydrogen is converted into energy. In other words, the sun is an enormous nuclear fusion reactor. Because the sun is such a long way from the Earth, only a tiny proportion (around two-millionths) of the sun's radiation reaches the Earth's surface. This works out at an amount of energy of 1×10^{18} kWh/a. Figure 1.23 compares this amount of energy to worldwide annual energy consumption and to fossil and nuclear energy resources. The energy sources that we primarily use in our industrial age are exhaustible. A supply shortage (from the technical and economic points of view) in easily extractable oil and natural gas reserves is anticipated in the first third of this century. Even if large new reserves were discovered, fossil fuels would still only last for a few more years.

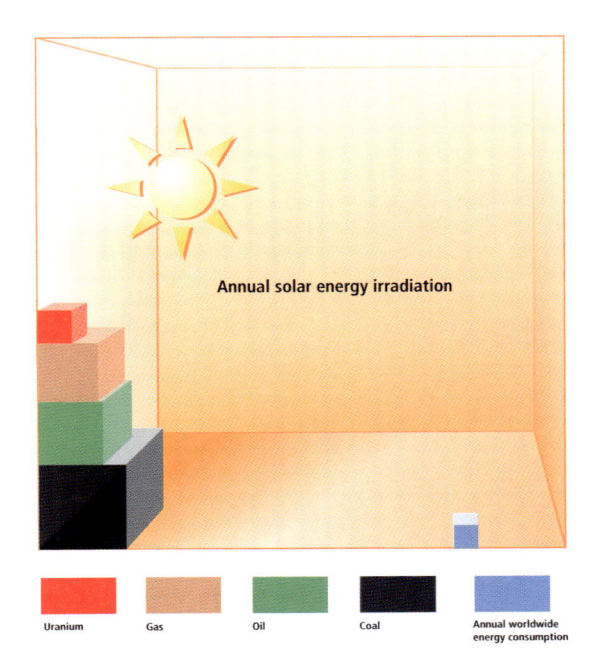

Figure 1.23 Energy content of annual solar radiation reaching the Earth's surface in comparison to worldwide energy consumption and fossil and nuclear energy resources
Source: BMWi (2000)

The amount of energy in the sunlight reaching the Earth's surface is equivalent to around 10,000 times the world's energy requirements. Consequently, only 0.01 per cent of the energy in sunlight would need to be harnessed to cover mankind's total energy needs.

1.2.2 Distribution of solar radiation

The intensity of solar radiation outside of the Earth's atmosphere depends upon the distance between the sun and the Earth. In the course of a year this varies between 1.47×10^8 km and 1.52×10^8 km. As a result, the irradiance E_0 fluctuates between $1325 W/m^2$ and $1412 W/m^2$. The average value is referred to as the solar constant:

solar constant: $E_0 = 1367 W/m^2$

This level of irradiance is not reached on the Earth's surface. The Earth's atmosphere reduces the insolation through reflection, absorption (by ozone, water vapour, oxygen and carbon dioxide) and scattering (caused by air molecules, dust particles or pollution). In good weather at noon, irradiance may reach $1000 W/m^2$ on the Earth's surface. This value is relatively independent of the location. The maximum insolation occurs on partly cloudy, sunny days. As a result of solar radiation reflecting off passing clouds, insolation can peak at up to $1400 W/m^2$ for short periods. If the energy content of solar radiation is added up over a year, this gives the annual global radiation in kWh/m^2. This value varies greatly depending upon the region, as shown in Figure 1.24.

Some regions at the equator reach values in excess of $2300 kWh/m^2$ per year, whereas Southern Europe receives maximum annual solar irradiance of $1700 kWh/m^2$ and Germany gets an average of $1040 kWh/m^2$. In Europe there are significant seasonal variations that are seen mainly in the difference between summer and winter insolation.

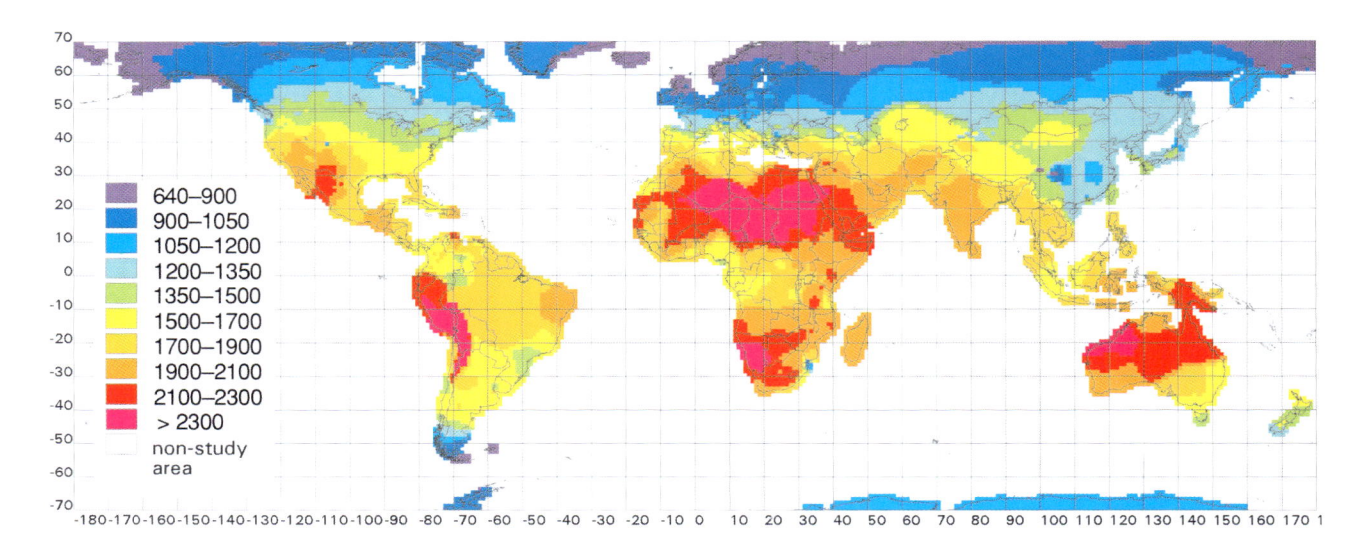

Figure 1.24 Worldwide distribution of annual solar irradiance in kWh/m² Source: METEONORM software by Meteotest

1.2.3 Direct and diffuse radiation

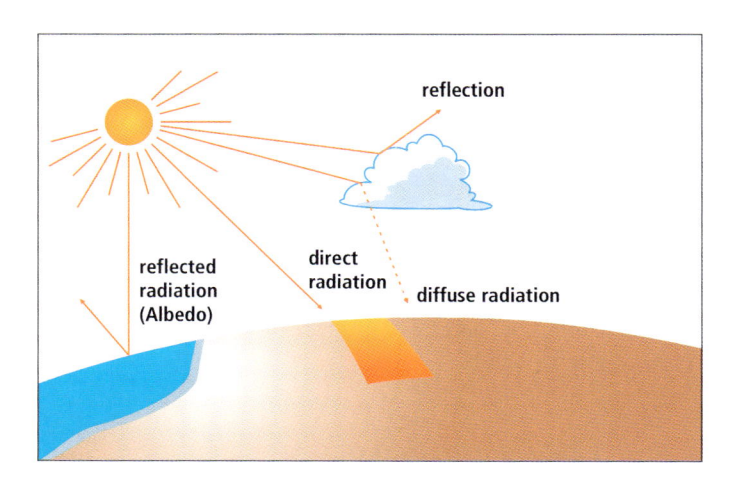

Figure 1.25 Sunlight as it passes through the atmosphere Source: V. Quaschning

Sunlight on the Earth's surface comprises a direct portion and a diffuse portion. The direct radiation comes from the direction of the sun and casts strong shadows of objects. By contrast, diffuse radiation, which is scattered from the dome of the sky, has no defined direction.

Depending upon the cloud conditions and the time of day (solar altitude), both the radiant power and the proportion of direct and diffuse radiation can vary greatly.

Figure 1.26 Global radiation and its components under different sky conditions

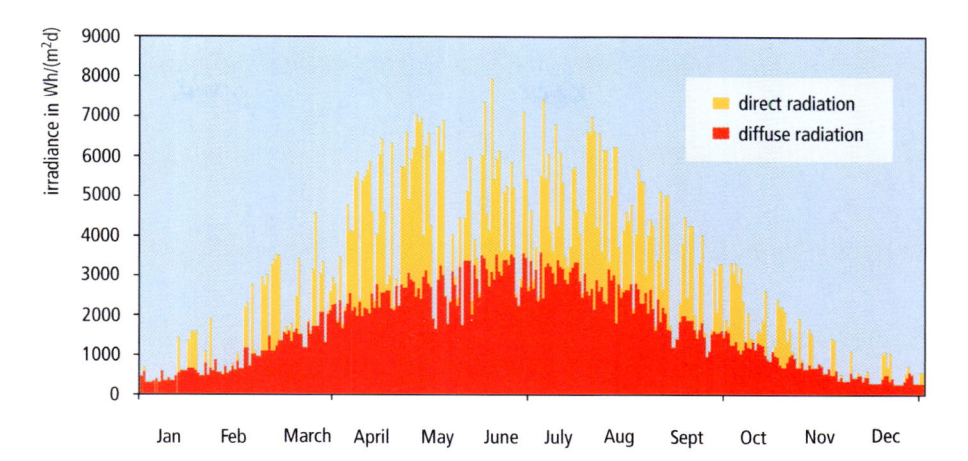

Figure 1.27
Typical development of daily totals of direct and diffuse radiation in Berlin

Figure 1.27 shows the proportion of direct and diffuse radiation in daily irradiance over the period of one year in Berlin. On clear days the direct radiation accounts for the greater part of the total radiation. On very cloudy days (especially in winter), the insolation is almost entirely diffuse. In Germany, the proportion of diffuse insolation is 60 per cent and direct radiation 40 per cent over the year.

1.2.4 Angle definition

Exact knowledge of the sun's path is important for calculating irradiance values and the yields of solar energy systems. The sun's altitude can be described at any location by the solar altitude and the solar azimuth.

When talking about solar energy systems, due south is generally given as $\alpha = 0°$. Angles to the east are indicated with a negative sign (east: $\alpha = -90°$). To the west, angles are given without a sign (or with a positive sign) (west: $\alpha = 90°$).

Figure 1.28
Defining angles in solar technology
Source: R. Haselhuhn/DGS LV Berlin BRB

1.2.5 Solar altitude and solar spectrum

The solar irradiance intensity depends, among other things, upon the solar elevation angle γ_S. This is measured from the horizontal. As the sun moves through the sky, the elevation angle changes during the day and also over the course of the year.

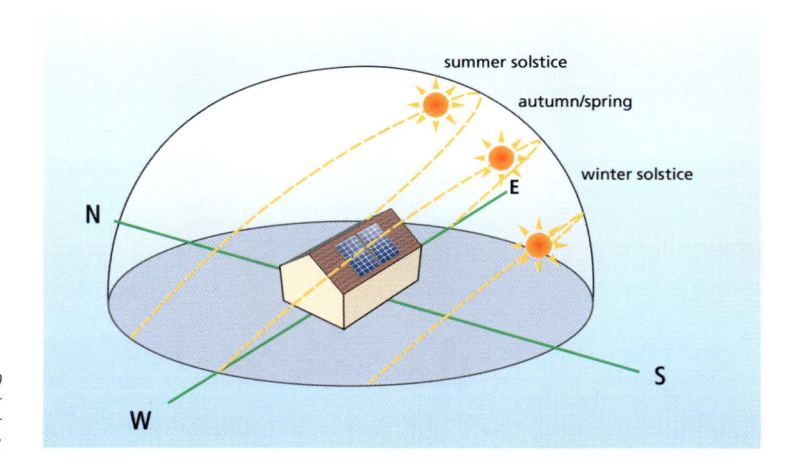

Figure 1.29
Path of the sun at particular times of the year
Source: V. Quaschning

When the solar altitude is perpendicular to the Earth, the sunlight takes the shortest path through the Earth's atmosphere. But if the sun is at a flatter angle, the path through the atmosphere is longer. This results in greater absorption and scattering of solar radiation and, hence, lower radiation intensity. The air mass factor (AM) specifies how many times the perpendicular thickness of the atmosphere the sunlight has to travel through the Earth's atmosphere. The relationship between solar altitude (height) γ_S and air mass is defined as follows:

$$AM = \frac{1}{\sin\gamma_S}$$

When the solar altitude is perpendicular ($\gamma_S = 90°$), AM = 1. This corresponds to the solar altitude at the equator at noon during the spring or autumn equinox.

Figure 1.30 shows the respective highest solar altitude on selected days in Berlin. The maximum solar elevation angle of $\gamma_S = 60.8°$ is attained on 21 June and corresponds to an air mass of 1.15. A maximum elevation angle of $\gamma_S = 14.1°$ and an air mass of 4 is reached on 22 December. For Europe, an air mass factor of 1.5 is used as the average annual value.

Solar radiation in space without the influence of the Earth's atmosphere is referred to as the AM 0 spectrum. When light passes through the Earth's atmosphere, the irradiance is reduced as a result of:

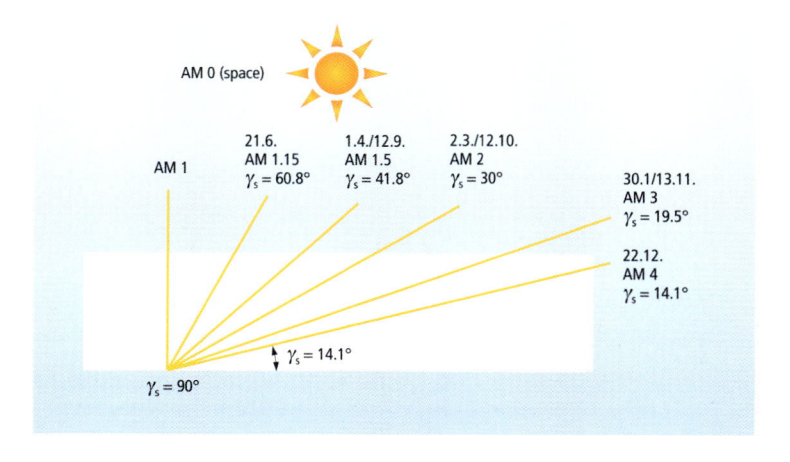

Figure 1.30
Solar altitude at noon over the course of the year in Berlin
Source: V. Quaschning

- reflection off the atmosphere;
- absorption by molecules in the atmosphere (O_3, H_2O, O_2, CO_2);
- Rayleigh scattering (molecular scattering);
- Mie scattering (scattering of dust particles and pollutants in the air).

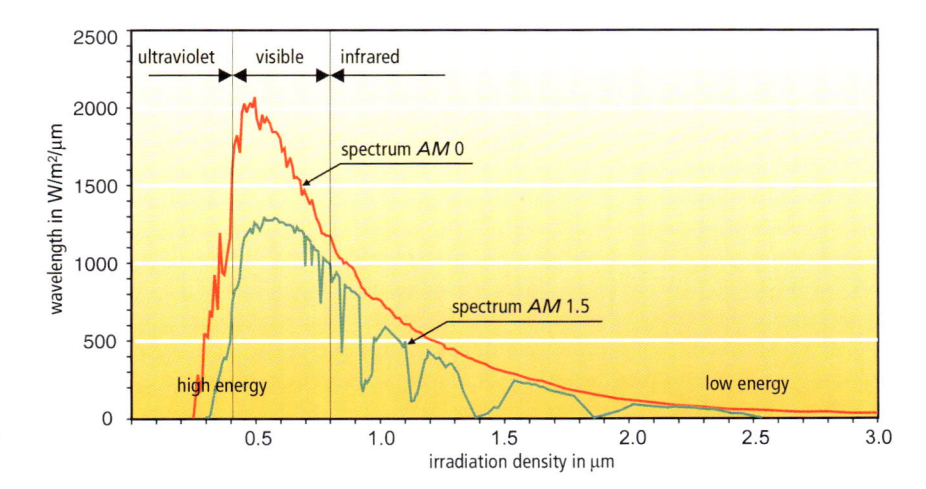

Figure 1.31
Solar spectrum AM 0 in space and AM 1.5 on the Earth at a solar altitude of 41.8°
Source: V. Quaschning

Table 1.1 shows the dependency of irradiance on the elevation angle γ_S. Absorption and Rayleigh scattering increase at lower solar altitudes. Scattering of pollution in the air (Mie scattering) is strongly location dependent. It is greatest in industrialized areas. Local weather effects such as clouds, rain and snowfall cause further weakening of irradiance.

Table 1.1
The dependence of irradiance on the angle of elevation γ_S

	AM	Absorption	Rayleigh scattering	Mie scattering	Overall reduction
90°	1.00	8.7%	9.4%	0–25.6%	17.3–38.5%
60°	1.15	9.2%	10.5%	0.7–29.5%	19.4–42.8%
30°	2.00	11.2%	16.3%	4.1–44.9%	28.8–59.1%
10°	5.76	16.2%	31.9%	15.4–74.3%	51.8–85.4%
5°	11.5	19.5%	42.5%	24.6–86.5%	65.1–93.8%

1.2.6 Ground reflection

When calculating irradiance on an inclined plane, the reflective component of the ground is included in the result. Depending upon the properties of the ground, an 'albedo' value is applied to take the reflectivity into account. This is required in some simulation programmes (e.g. SUNDI, PV*SOL and SolEm). The higher the albedo value, the higher the reflection of sunlight and, hence, the lighter the surrounding area and the greater the diffuse radiation. In general, an albedo value of 0.2 can be assumed. The albedo values for water apply to still water surfaces. Since water surfaces are always moving, waves are formed that reflect sunlight. Wolfgang Brösicke at the Fachhochschule für Technik und Wirtschaft (FHTW) Berlin calculated an albedo value of 0.51 for moving water surfaces and an angle of incidence to the sun of 60° (Brösicke, 1995). This value has since been confirmed by the increased yield of façade systems in front of water surfaces (Haselhuhn, 2004).

Table 1.2
Albedo values for different environments

Surface	Albedo	Surface	Albedo
Grass (July, August)	0.25	Asphalt	0.15
Lawn	0.18–0.23	Forests	0.05–0.18
Dry Grass	0.28–0.32	Heather and sandy areas	0.10–0.25
Untilled fields	0.26	Water surface ($\gamma_S > 45°C$)	0.05
Barren soil	0.17	Water surface ($\gamma_S > 30°C$)	0.08
Gravel	0.18	Water surface ($\gamma_S > 20°C$)	0.12
Clean concrete	0.30	Water surface ($\gamma_S > 10°C$)	0.22
Eroded concrete	0.20	Fresh layer of snow	0.80–0.90
Clean cement	0.55	Old layer of snow	0.45–0.70

1.2.7 How solar radiation is measured

Solar radiation is either measured directly using pyranometers or photovoltaic sensors, or indirectly by analysing satellite images. Pyranometers are high-precision sensors that measure solar radiation on a planar surface. They essentially comprise two hemispherical glass domes, a black metal plate as an absorber surface, the thermo-elements located below this, and a white metallic housing. Solar radiation falls through the hemispherical glass domes vertically onto the absorber surface, warming it up. Since the amount of warming directly depends upon the irradiance, the difference in temperature between it and the environment (or, more precisely, the white metallic housing) allows the irradiance to be calculated. The temperature difference is found via thermocouples that are wired in series. These deliver a voltage that is proportional to the difference in temperature. Using a voltmeter, it is then possible to work out the global radiation directly from the voltage and the calibration factor. If direct solar irradiance is screened out by fitting a shade ring, the diffuse radiation can be measured. Pyranometers achieve highly accurate measurements; but because they work on a thermal basis, they are somewhat slow to respond. As a result, rapid fluctuations in radiation, caused, for example, by a partially cloudy sky, are not captured satisfactorily. For longer measurement periods, measurement accuracy of 0.8 per cent is achieved on an annual average.

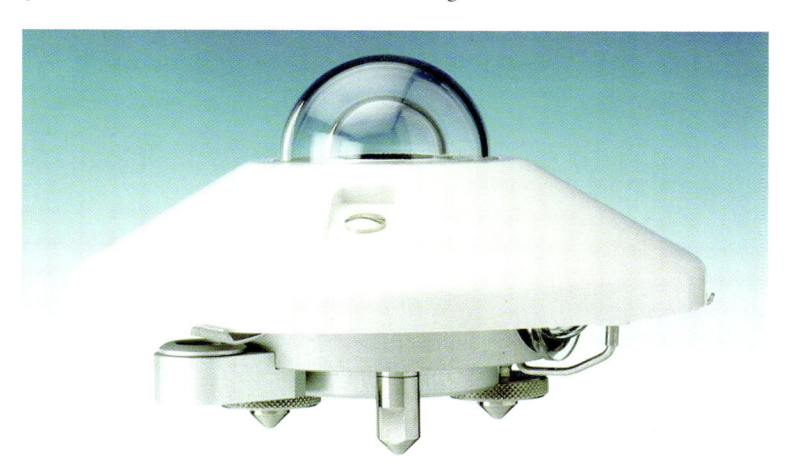

Figure 1.32
Pyranometer
Source: Kipp and Zonen
Lambrecht, Göttingen

Photovoltaic sensors cost significantly less than pyranometers. They generally use crystalline silicon sensors. A PV sensor consists of a solar cell that delivers a current proportional to the irradiance. However, because of the spectral sensitivity of these sensors, certain components of solar radiation are not measured accurately. A solar cell cannot measure long-wavelength infrared radiation. Depending upon the calibration and design of the sensor, measurement accuracy of 2 per cent to 5 per cent is achieved on the annual average. Accuracy of better than 4 per cent can be achieved through calibration and use of laminated temperature sensors for temperature compensation.

Figure 1.33
Photovoltaic sensor
Source: IKS

Figure 1.34
Irradiance meter with PV sensor
Source: Solarc

PV irradiance sensors are often used with larger PV arrays for monitoring the operation of the system. It is worth noting that using a sensor with the same cell technology (amorphous, mono-crystalline or polycrystalline silicon, cadmium telluride (CdTe) or copper indium diselenide (CIS)) increases accuracy and facilitates evaluation.

Data loggers in conjunction with evaluation units or state-of-the-art inverters can compare the measured solar radiation to the generated electrical power. This enables a diagnosis of how well the PV system is operating. A compact measuring unit with sensor, direct irradiance and temperature display, and optional data logger, is shown in Figure 1.34.

PYRANOMETER MANUFACTURERS AND PROVIDERS

Kipp & Zonen, Thies Clima, Uniklima Sensors, UMS.

MANUFACTURERS AND PROVIDERS OF PV SENSORS AND IRRADIANCE METERS WITH PV SENSORS

ESTI, IKS, Mencke & Tegtmeyer, NES, Solarc, Solarwatt, Tritec.

1.2.8 Tracking PV arrays

If a surface is moved to follow the sun, the energy yield increases. On days with high insolation and a large direct radiation component, a tracking system enables relatively large radiation gains to be achieved. In summer, a tracking system achieves around 50 per cent radiation gains on sunny days, and in winter, 300 per cent or more, compared to a horizontal surface.

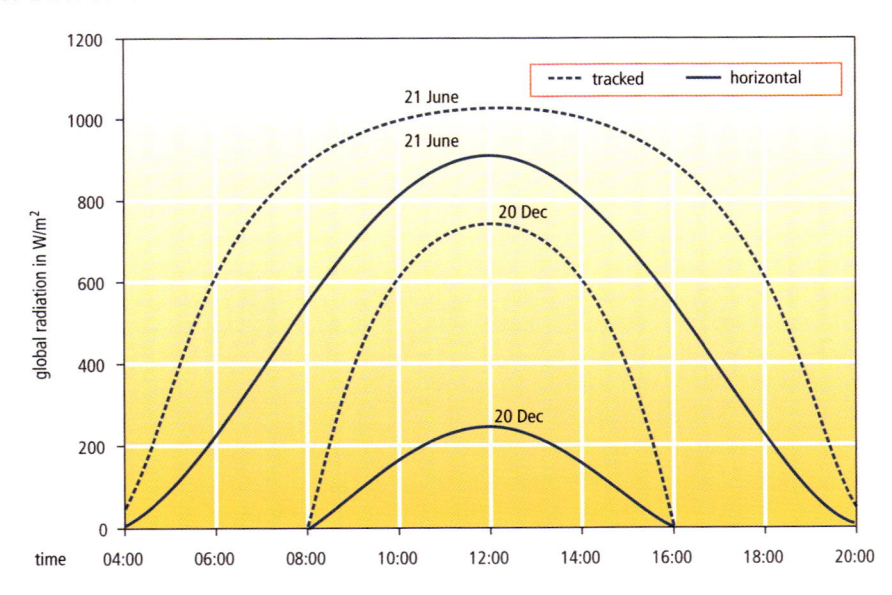

Figure 1.35
Differences in irradiance on horizontal and solar-tracking surfaces for cloud-free days and 50° latitude
Source: DIN82a

The vast majority of the energy gains when using a tracking system are achieved during summer. First, the absolute energy yield is higher than in winter; the proportion of cloudy days is also much higher in winter.

There are various types of tracker systems – PV systems that track the sun. One difference is between single-axis and dual-axis tracking. With dual-axis tracking the system always maintains the optimum alignment to the sun. Because dual-axis tracking is technically more complicated, single-axis tracking is often preferred. Here the system can either track the sun's daily path or its annual path. A system that tracks the annual path is relatively easy to implement. To do this, the tilt angle of the array needs to be adjusted at relatively large intervals of time (weeks or months). In some cases, this can be done manually.

Dual-axis tracking photovoltaic systems in Central Europe achieve an increased energy yield of approximately 30 per cent. Single-axis tracking provides an energy gain in the order of 20 per cent or so. In areas with higher irradiance, the energy gain is somewhat larger. Long-term tests with tracking systems at the Centre for Solar Energy and Hydrogen Research (Zentrum für Sonnenergie- und Wasserstoff-Forschung or ZSW) showed an average increased yield of 28 per cent for dual-axis tracking systems in Central Europe. At the Italian National Agency for New Technologies, Energy and the Environment (ENEA) solar test site in Monte Aquilone, Italy, an increased yield of 34 per cent was achieved. However, tracking systems are more complex to build, which also involves higher costs. They require a moving mounting system that can withstand high wind loads such as storms. The drive system can either use an electric motor or a thermo-hydraulic control system. Thermo-hydraulic systems work by exploiting the pressure difference resulting from fluids heating up.

If the tracking system fails, the PV array may become stuck in a poor-yield position, with the result that the energy yield will be severely reduced until the fault is repaired (Quaschning, 2000).

In the past, the higher energy yield in Central Europe did not generally compensate for the increased investment costs of a tracker system. As a result, tracking systems have not been widely used. However, more cost-effective single-axis tracking systems are now becoming available that can be economically viable under certain conditions. Where there is a good feed-in tariff for solar-generated electricity, these kinds of tracking systems can, in some circumstances, improve economic efficiency. Additional selling points for these systems, apart from the increased yield, are their optical effect (attractiveness) and the publicity effect that these systems generate.

Figure 1.36
Passive thermo-hydraulic tracking system
Source: Altec-Solartechnik

Figure 1.37
Solar tracker with electric motor system
Source: SOLON AG, Zwickert

Figure 1.38
Various tracking systems in use on the roof of the ufaFabrik in Berlin
Source: SOLON AG, Zwicker

PASSIVE TRACKING

The thermo-hydraulic system shown in Figure 1.36 has two tube tanks located at the sides of the PV array, over which two shading sheets are fitted. If the PV array surface is not aligned with the sun, the fluid in the tanks will be heated unevenly. The resulting pressure difference drives the fluid through a connecting pipe into the tube tank that has the lower temperature. The resulting shift in weight causes the PV array to rotate to face the sun. The principle of thermo-hydraulic systems means that their response can be sluggish, particularly in the morning. Because of this inertia in the system, the PV array will only move from the evening position to face the eastern morning sun after a lengthy period of sunshine (about one hour). The resultant loss means that the overall increased yield is reduced.

ACTIVE TRACKING

Inertial losses can be avoided if the modules are moved using an electronically controlled motor.

Astronomical: the electronic control system calculates the current position of the sun at the location and the tracking motor moves the modules perpendicular to the sun at preset time intervals using precise coordinates.

Sensor-controlled: rather than blindly aligning the modules with the astronomical position of the sun, a tracking system fitted with light sensors points the modules at the brightest point in the sky. Under a completely overcast sky, for example, the modules will be in a horizontal position. The motorized system shown in Figure 1.37 is activated via two small anti-parallel connected solar modules that are fitted opposite each other at right angles to the PV modules. When the array is directly aligned with the sun, the two solar modules receive the same insolation intensity. If the modules receive identical illumination, their voltages cancel each other out. If one of the modules is more brightly lit, a control voltage arises accordingly. This causes current to flow in one direction via a DC motor for as long as it takes for the voltage difference to be balanced out. Hence, the two solar control modules supply the DC motor simultaneously (Siegfriedt and Slickers, 2001).

Figure 1.39
Solar farm on the former Erlasee experimental vineyard near Arnstein, Germany: The planned 1500 independent SOLON Movers have a total power output capacity of 12MW
Source: SOLON AG, Paul Langrock

Large solar farms are increasingly being built with PV tracking systems. The high number of units means that on a per-unit basis, the additional costs and work involved in setting up the tracking system are greatly reduced compared to smaller-scale installations. In this case, either whole rows of modules are rotated on a shared axis, or multiple modules are mounted on a mast and moved as an array on one or two axes. The solar array of each SOLON Mover shown in Figure 1.39 is approximately 50m^2 in size and tracks the sun's astronomical position on two axes. In strong winds of gale force 8 and above, all Movers automatically move into a position that offers the least resistance to the wind and can even survive hurricanes without damage.

*Figure 1.40
PowerTracker system*

The PowerTracker from PowerLight is shown in Figure 1.40. Rows of modules with module power of up to 300kW are moved from east to west on a single axis using one motor. The module rows, which are arranged in parallel, are aligned in the north–south axis. The modules are secured to steel beams linked by square cross-members and pivot-mounted on multiple steel supports. Each cross-member is moved by a drive unit comprising a servo motor and microprocessor control unit at the end of the module rows. The system is controlled automatically, based on astronomical data. By automatically correcting the tilt angle, the backtracking feature prevents adjacent module rows from shading each other when the sun is low in the sky.

DEGERtrakers mounted on masts, as shown in Figure 1.41, can take a module surface area of up to 35m^2. The control system evaluates the reference values from two solar sensor cells to find the brightest point in the sky, activating the drive motor directly. A third cell on the rear is used to reset the system in the morning. Each mast is fitted with its own sensor and motor, which are powered via the solar modules (total power consumption is a maximum of 5W).

In the ATM tracking system up to 25 'towers', each with 20m^2 of PV module surface area, are mechanically linked to a central drive unit comprising motor, gear and astronomical controller.

TRACKING SYSTEM MANUFACTURERS

Altec Solartechnik, ATM Solar Solutions, Berger Solar, DEGERenergie, EGIS, Elektro-Spiegler, Lorentz, Mesatec, Pairan, PowerLight, RES, S & S, Solar-Trak, Solon AG, SPT SolarPower Tower GmbH, Traxle

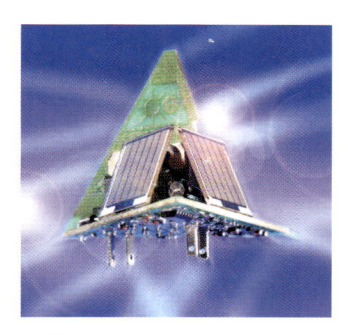

*Figure 1.41
DEGERconecter control system and
DEGERtraker tracking system
Source: DEGERenergie*

*Figure 1.42
Combined tracking of mechanically
linked ATM solar towers
Source: ATM Solar Solutions*

1.3 The photovoltaic effect and how solar cells work

The term photovoltaics means the direct conversion of light into electrical energy using solar cells. Semiconductor materials such as silicon, gallium arsenide, cadmium telluride or copper indium diselenide are used in these solar cells. The crystalline solar cell is the most commonly used variety. During 2006, these had a worldwide market share of 95 per cent.

1.3.1 How a solar cell works

The way in which solar cells work is shown below, taking crystalline silicon cells as an example. Highly pure silicon with a high crystal quality is needed to make solar cells. The silicon atoms form a stable crystal lattice. Each silicon atom has four bonding electrons (valence electrons) in its outer shell. To form a stable electron configuration, in each case in the crystal lattice two electrons of neighbouring atoms form an electron pair bond. By forming electron pair bonds with four neighbours, silicon achieves its stable noble gas configuration with eight outer electrons. An electron bond can be broken by the action of light or heat. The electron is then free to move and leaves a hole in the crystal lattice. This is known as intrinsic conductivity.

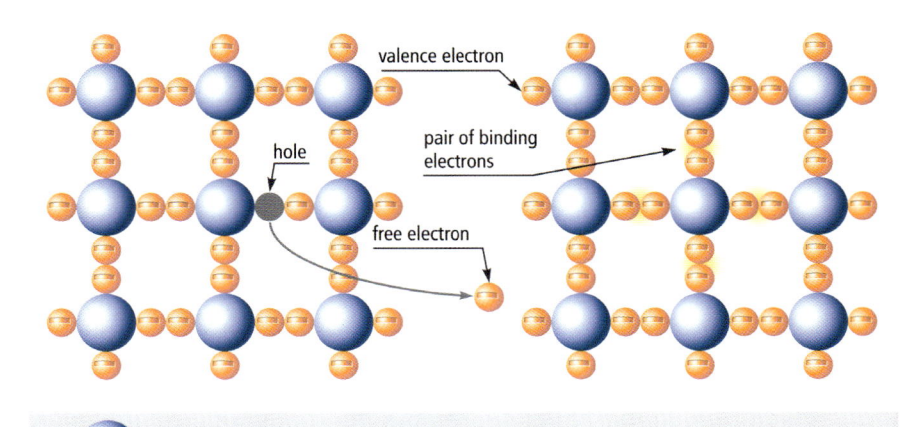

Figure 1.43
Crystalline structure of silicon and intrinsic conductivity
Source: V. Quaschning

Intrinsic conductivity cannot be used to generate electricity. So that the silicon material can be used to generate energy, impurities are deliberately introduced into the crystal lattice. These are known as doping atoms (see Figure 1.44). These atoms have one electron more (phosphorus) or one electron less (boron) than silicon in their outermost electron shell. Hence, the doping atoms result in 'impurity atoms' in the

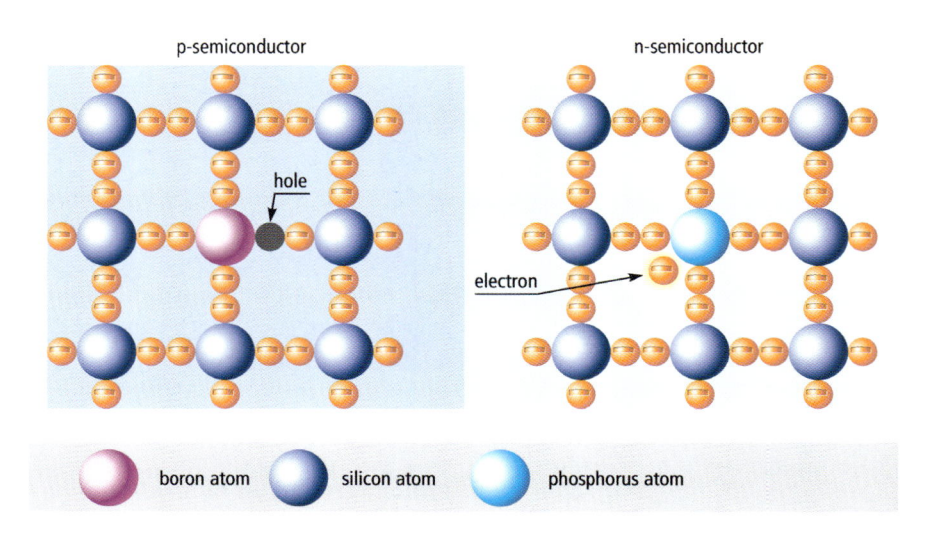

Figure 1.44
Extrinsic conduction in n- and p-doped silicon
Source: V. Quaschning

crystal lattice.

In the case of phosphorus doping (n-doped), there is a surplus electron for every phosphorus atom in the lattice. This electron can move freely in the crystal and hence transport an electric charge. With boron doping (p-doped), there is a hole (missing bonding electron) for every boron atom in the lattice. Electrons from neighbouring silicon atoms can fill this hole, creating a new hole somewhere else. The conduction method based on doping atoms is known as impurity conduction or extrinsic conduction. Considering the n- or p-doped material on its own, however, the free charges have no predetermined direction to their movement.

If n- and p-doped semiconductor layers are brought together, a p-n (positive-negative) junction is formed. At this junction, surplus electrons from the n-semiconductor diffuse into the p-semiconductor layer. This creates a region with few free charge carriers (see Figure 1.45). This region is known as the space charge region. Positively charged doping atoms remain in the n-region of the transition and negatively charged doping atoms remain in the p-region of the transition. An electrical field is created that is opposed to the movement of the charge carriers, with the result that diffusion does not continue indefinitely.

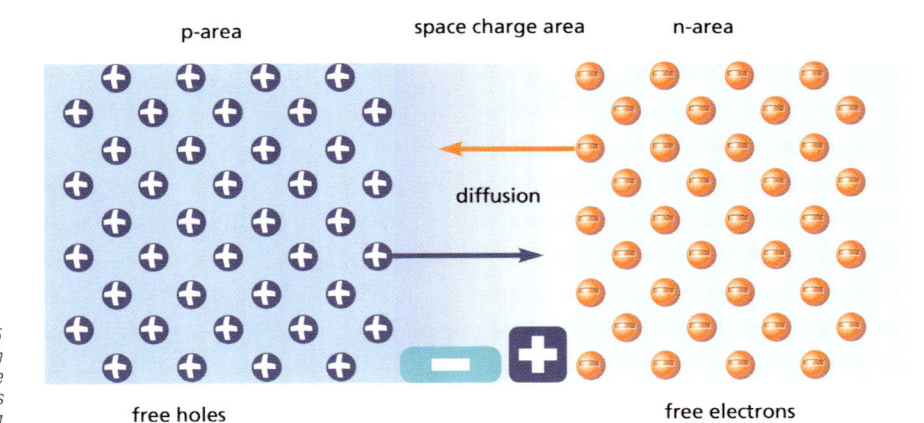

*Figure 1.45
Formation of a space charge region at the p-n junction through the diffusion of electrons and holes
Source: V. Quaschning*

If the p-n-semiconductor (solar cell) is now exposed to light, photons are absorbed by the electrons. This input of energy breaks electron bonds. The released electrons are pulled through the electrical field into the n-region. The holes that are formed migrate in the opposite direction, into the p-region. This process, as a whole, is called the photovoltaic effect. The diffusion of charge carriers to the electrical contacts causes a voltage to be present at the solar cell. In an unloaded state, the open circuit voltage OCV arises at the solar cell. If the electrical circuit is closed, a current flows.

Some electrons do not reach the contacts and recombine instead. Recombination refers to the bonding of a free electron to an atom lacking an outer electron (hole). Diffusion length here is the average distance that an electron covers in the crystal lattice during its lifetime until it meets an atom with a missing electron and bonds with it. Here, free charge carriers are lost and can no longer contribute to generating electricity. The diffusion length depends upon the number of impurity atoms in the crystal and must be large enough so that a sufficient number of charge carriers reach the contacts. The diffusion length depends upon the material. With one crystal impurity atom (doping) to 10 billion silicon atoms, this distance is 0.5mm. This corresponds to roughly twice the cell thickness. In the space charge region, there is a high probability of successful charge separation (electrons and holes) without recombination. Outside of the space charge region, the probability of recombination increases with the distance from the space charge region.

1.3.2 Design and functioning of a crystalline silicon solar cell

The classic crystalline silicon solar cell comprises two differently doped silicon layers. The layer that faces the sun's light is negatively doped with phosphorus. The layer below it is positively doped with boron. At the boundary layer, an electrical field is produced that leads to the separation of the charges (electrons and holes) released by the sunlight. In order to be able to take power from the solar cell, metallic contacts need to be fitted on the front and back of the cell. Screen printing is normally used for this purpose. On the back of the solar cell it is possible to apply a contact layer over the whole surface using an aluminium or silver paste. The front, by contrast, must let as much light through as possible. Here, the contacts are usually applied in the form of a thin grid or a tree structure. Sputtering or vapour depositing a thin film (anti-reflective coating) of silicon nitride or titanium oxide onto the front face of the solar cell reduces light reflection.

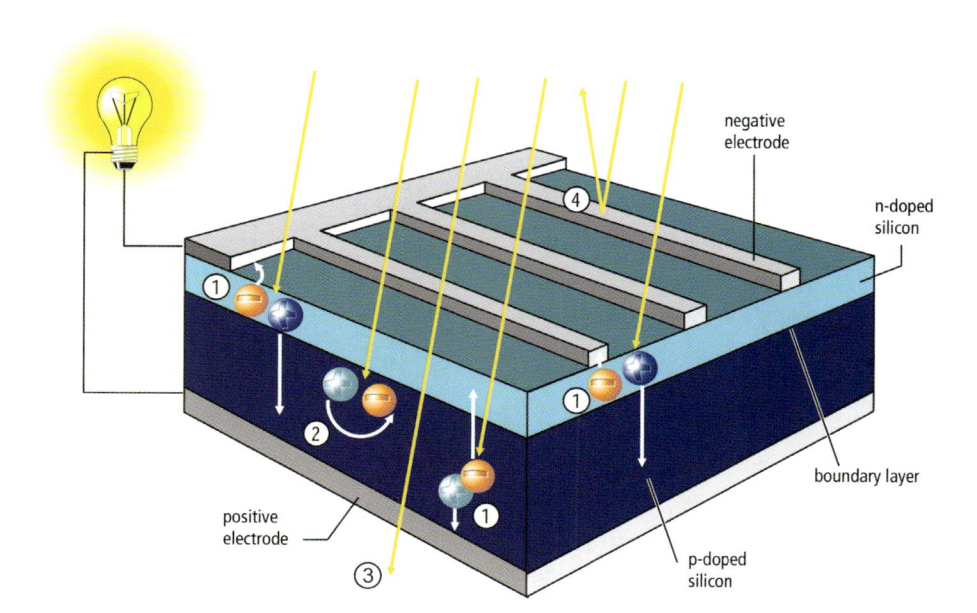

Figure 1.46
Design and functioning of a crystalline silicon solar cell
Notes:
1 charge separation;
2 recombination;
3 unused photon energy (e.g. transmission);
4 reflection and shading caused by front contacts.
Source: V. Quaschning

As described above, when light falls on the solar cell, charge carriers separate and if a load (in Figure 1.46, a light bulb) is connected, current flows. Losses occur at the solar cell due to recombination, reflection and shading caused by the front contacts. In addition, a large component of the long and short wavelength radiation energy cannot be used. As an example of this, the transmission losses are shown in Figure 1.46. A further portion of the unused energy is absorbed and converted into heat. Using the example of a crystalline silicon solar cell, the individual loss components are shown in the following 'energy balance sheet'.

Energy balance of a crystalline solar cell:

100 per cent irradiated solar energy;
- 3 per cent reflection and shading caused by front contacts;
- 23 per cent too low photon energy in long wavelength radiation;
- 32 per cent too high photon energy in short wavelength radiation;
- 8.5 per cent recombination losses;
- 20 per cent potential difference in the cell, particularly in the space charge region;
- 0.5 per cent series resistance (ohmic losses);

  = 13 per cent utilizable electrical energy.

1.4 Solar cell types

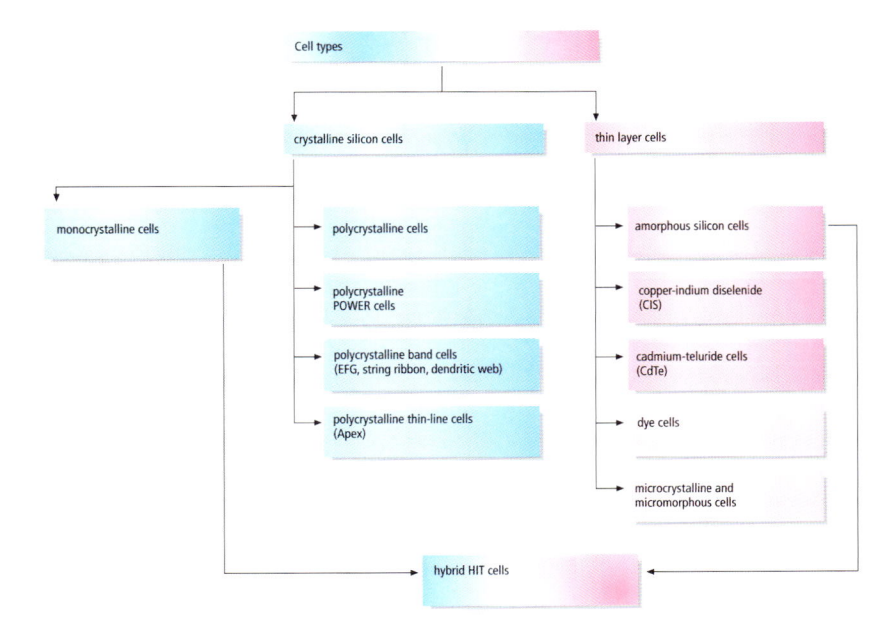

Figure 1.47
Types of solar cells
Source: D. Wunderlich/DGS LV Berlin BRB

1.4.1 Crystalline silicon

The most important material in crystalline solar cells is silicon. After oxygen, this is the second most abundant element on Earth and, hence, is available in almost unlimited quantities. It is present not in a pure form, but in chemical compounds, with oxygen in the form of quartz or sand. The undesired oxygen has to be first separated out of the silicon dioxide. To do this, quartz sand is heated together with carbon powder, coke and charcoal in an electric arc furnace to a temperature of 1800°C to 1900°C. This produces carbon monoxide and what is known as metallurgical silicon, which is about 98 per cent pure. But 2 per cent impurity in silicon is still much too high for electronics applications. Only billionths of a per cent are acceptable for photovoltaics, which falls to ten times less for the semiconductor industry (electronic grade silicon). The raw silicon is therefore purified further in chemical processes. Silicon is finely ground up and reacted with gaseous hydrogen chloride (hydrochloric acid) to form hydrogen and trichlorsilane – a liquid that boils at 31°C. In iterative stages, this liquid is distilled until the level of impurities falls to the required level. The current industry standard is a chemical vapour deposition process known as the Siemens process, which is used to extract ultra-pure silicon from trichlorsilane and hydrogen. The two gases are blown into a reactor where thin rods of high-purity silicon are located, heated to between 1000°C and 1200°C. Silicon from the trichlorsilane is deposited onto these rods. The silicon formed in this process is polycrystalline and is known as polysilicon. The rods grow in diameter to between 10cm and 15cm. These are broken up into chunks and used as a source material for mono-crystalline or polycrystalline silicon wafers, which are then turned into solar cells.

Because the purity requirements for silicon used in manufacturing solar cells are not as high as for electronic grade silicon, the solar industry primarily uses waste products from the semiconductor industry. Since 1998, however, there has not been enough silicon waste to cover the rapid growth in demand. The shortfall has mostly been made up using ultra-pure silicon, but which, in some cases, is of a slightly lower quality. Over the same period, processes have been developed that now make it possible to produce silicon with the quality required for solar cells (solar grade silicon and solar silicon), but involving less cost, time and energy expenditure.

Some manufacturers of solar silicon use fluidized bed reactors. Tiny particles of silicon are introduced into the reactor. Trichlorsilane or silane is blown into the

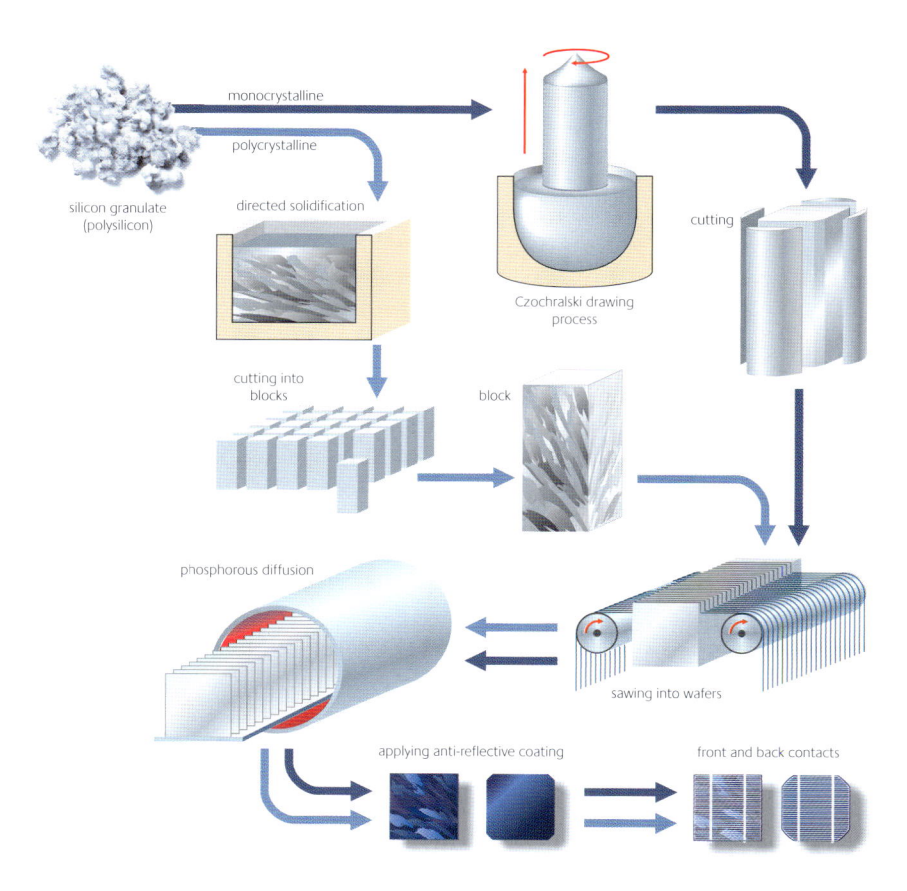

*Figure 1.48
Manufacturing mono-crystalline
and polycrystalline solar cells
from polysilicon*

reactor together with hydrogen. At 1000°C for trichlorsilane or 700°C for silane, the silicon from these materials is deposited onto the particles, which become larger and larger until they are so heavy that they fall to the bottom of the reactor and can be removed as silicon granulate. The tube reactor process, in contrast, is similar to the Siemens process. But instead of the rods, it uses a hollow silicon cylinder that only has to be heated to 800°C since silane is used as the source material. In the vapour to liquid deposition (VLD) process developed in Japan, silicon from gaseous trichlorsilane, which is introduced into a reactor together with hydrogen, is deposited onto a graphite tube heated to 1500°C. The silicon, which liquefies at between 1410°C and 1420°C, drops onto the reactor bottom where it solidifies into granulate. Other processes use silicon tetrachloride, which is reduced with zinc, or start directly with metallurgical silicon, which is converted into pure silicon by refinement, with plasma torches, or through the reduction of silicon carbide. The first large-scale production is set to begin in 2007.

1.4.2 Mono-crystalline (single-crystal) silicon cells

FABRICATION

The Czochralski process (crucible drawing process) has become established in the production of single-crystal silicon for terrestrial applications. In this process, the polycrystalline starting material (polysilicon) is melted in a quartz crucible at around 1420°C. A seed crystal with a defined orientation is dipped into the silicon melt and slowly drawn upwards out of the melt. During this process the crystal grows into a cylindrical mono-crystal up to 30cm in diameter and several metres in length. These cylindrical mono-crystals are cut to form semi-round or square bars, which are then cut with wire saws into slices (wafers) with a thickness of around 0.3mm. When cutting the mono-crystals and sawing the wafers, a large percentage of the silicon is lost as sawdust and needs to be re-melted, as do the conical ends of the rods. The wafers are chemically wet cleaned in etching and rinsing baths to remove sawing residues and marks. This cleaning process etches away approximately 0.01mm of the wafer on both sides. Starting from the raw wafers that have already been p-doped with boron, the thin n-doped layer is created through phosphorus diffusion. Phosphorus gas is diffused into a diffusion furnace at temperatures of between 800°C and 900°C, and the upper surface is doped. The heart of the solar cell, the p-n junction, is created. After applying the anti-reflective (AR) coating, the current collector lines are printed on the front, while the contacts appear on the back, in a screen printing process. The contacts have to be baked to contact the front side through the anti-reflective coating. Finally, the solar cells are etched at the edges to create a clean division between the p-layer and n-layer and to prevent a short circuit at the sides:

- *Efficiency:* 15 per cent to 18 per cent (Czochralski silicon).
- *Form:* depending upon how much of the mono-crystal is sliced away, round, semi-round or square cells are created. Round cells are cheaper than semi-round or square cells since less material is wasted in their production. Despite this, they are rarely used in standard modules because when placed next to each other in a module, they do not employ the space efficiently. However, in special modules for building integration where partial transparency is desired, or for solar home systems, round cells are a perfectly viable alternative.
- Usual sizes: 10cm^2 × 10cm^2 (4 inch); 12.5cm^2 × 12.5cm^2 (5 inch); or 15cm^2 × 15cm^2 (6 inch); Ø: 12.5cm or 15cm.
- *Thickness:* 0.2mm to 0.3mm.
- *Appearance:* uniform.
- *Colour:* dark blue to black (with AR); grey (without AR).

CELL MANUFACTURERS

Astro Power, Bharat Electronics, BHEL, BP Solar, Canrom, CEL, CellSiCo, Deutsche Cell, Eurosolare, GE Energy, GPV, Helios, Humaei, Isofoton, Kaifeng Solar Cell Factory, Kwazar JSC, Maharishi, Matsushita Seiko, Microsolpower, Ningbo Solar Energy Power, Pentafour Solec Technology, Photowatt, RWE Schott Solar, Sharp, Shell Solar, Solartec, Solar Wind Europe, Solec, Solmecs, Solterra, Suntech, Sunways, Telekom-STV, Tianjin Jinneng Solar Cell, Viva Solar, Webel SL, Yunnan Semiconductor.

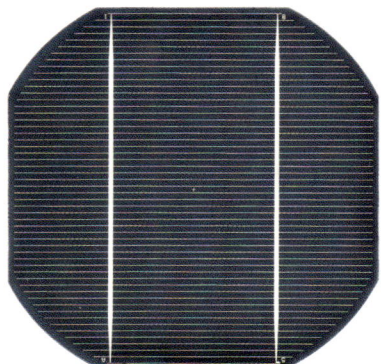

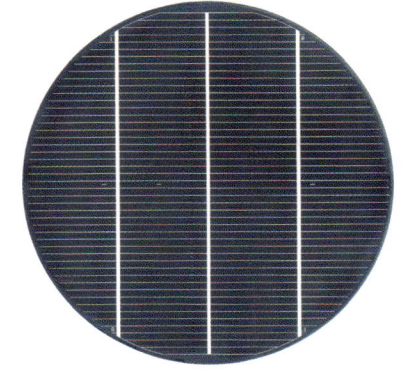

Figure 1.49
Square mono-crystalline cell
Source: Siemens Solar

Figure 1.50
Semi-round mono-crystalline cell
Source: Siemens Solar

Figure 1.51
Round mono-crystalline cell
Source: Siemens Solar

1.4.3 Polycrystalline silicon cells

FABRICATION

The silicon starting material is melted in a quartz crucible and cast into a cuboid form. Through controlled heating and cooling, the cast block cools evenly in one direction. The purpose of this directed solidification is to form large numbers of the largest possible homogeneous silicon crystals, with grain sizes from a few millimetres to several centimetres. The grain boundaries constitute crystal defects with an increased recombination risk and have an adverse effect on the efficiency of polycrystalline solar cells, which is somewhat lower than that of mono-crystalline cells. In the block casting method, large silicon blocks, or ingots, are created. The ingots are generally sawn into bars using a band saw, and then cut into wafers approximately 0.3mm thick using a wire saw. Sawing the wafers results in some of the silicon being lost as sawdust. After cleaning and phosphorus doping, the anti-reflective coating is applied. Finally, the contacts are printed and the edges are etched:

- *Efficiency:* 13 per cent to 16 per cent (with AR).
- *Form:* Square.
- *Usual sizes:* 10cm^2 × 10cm^2; 12.5cm^2 × 12.5cm^2; 15cm^2 × 15cm^2; 15.6cm^2 × 15.6cm^2; and 21cm^2 × 21cm^2 (4 inch; 5 inch; 6 inch; 6+ inch; and 8 inch).
- *Thickness:* 0.24mm to 0.3mm.
- *Appearance:* the block casting process forms crystals with different orientations. Because the light is reflected differently, the individual crystals can be clearly seen on the surface (frost pattern).
- *Colour:* blue (with AR); silver grey (without AR).

CELL MANUFACTURERS

Al-Afandi, BP Solar, Deutsche Cell, ErSol, Eurosolare, GPV, Kwazar JSC, Kyocera, Maharishi, Mitsubishi, Motech, Photovoltech, Photowatt, Q-Cells, RWE Schott Solar, Sharp, Shell Solar, Solar Power Industries, Solartec, Solterra, Suntech, Sunways, Tianjin Jinneng Solar Cell.

Figure 1.52
Cast polycrystalline silicon blocks
Source: Photowatt

Figure 1.53
Sawn polycrystalline silicon bars
Source: Photowatt

In polycrystalline cells there is a clear trend towards larger cells and, hence, more efficient module production, as well as higher module efficiency. Many manufacturers now offer 8 inch polycrystalline cells: the edge length is 8 inches (21cm). Larger cells will bring down the costs of cell and module production in future since fewer cells are needed per module. However, module manufacturers first need to adjust their production systems to accommodate the new sizes, and also develop new bypass diodes and junction boxes that are designed for the higher currents and diode temperatures. The system technology requirements are also higher (cables, inverters, etc.) since the systems have to handle higher currents.

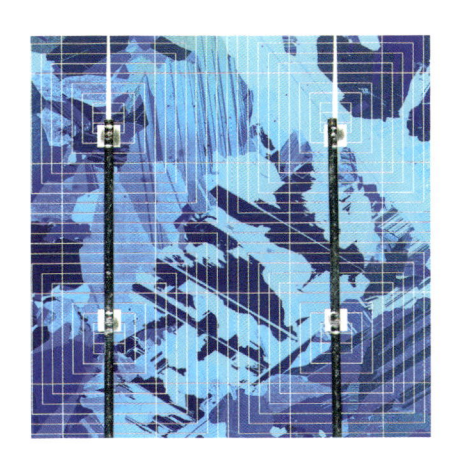

Figure 1.54
Polycrystalline wafer without anti-reflective coating
Source: Photowatt

Figure 1.55
Polycrystalline wafer with anti-reflective coating
Source: Photowatt

Figure 1.56
Polycrystalline cell with anti-reflective and contact grid lines
Source: Photowatt

Work is being carried out to make even thinner cells in the future. The main difficulty is printing the contacts since the paste has a different thermal expansion coefficient than silicon and the cells distort when the contacts are baked. With current technology, the limit should be around 0.1mm since polycrystalline wafers become increasingly unstable at the grain boundaries with decreasing thickness. Monocrystalline silicon, by contrast, is not so prone to breakage since wafers become flexible from about 0.08mm.

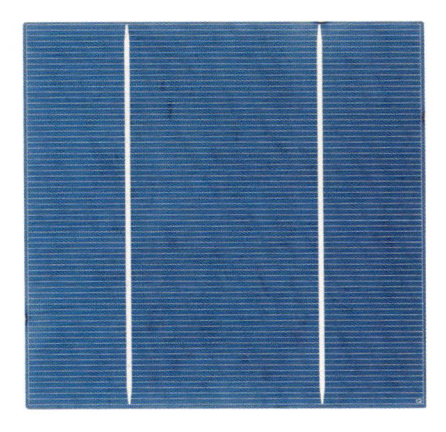

Figure 1.57
Six-inch and 8-inch cells compared in size
Source: Q-Cells

1.4.4 Ribbon-pulled silicon cells

With conventional methods of manufacturing crystalline silicon wafers, up to 40 per cent of the raw silicon is wasted as sawdust by the time the finished wafer is produced. The sawn wafers also require a thickness of around 0.3mm for mechanical reasons. To reduce the high material losses and increase material utilization, various ribbon-pulling processes have been developed. Here, films are pulled directly out of the silicon melt. The silicon ribbons already have the thickness of the future wafers. All that remains to be done, generally using lasers, is to cut the flat surfaces into pieces. This technological development has raised hopes that, in future, it will be possible to reduce the thickness of the silicon ribbons down as far as 0.1mm. Compared to wafer production using crucible pulling or block-casting methods, ribbon-pulling methods are more economical with energy and materials, and have a significant cost-reduction potential.

Three technologies have made it to production stage and are used in commercial solar cell production. In the Edge-Defined Film-Fed Growth (EFG) Technique and string ribbon processes, the pulled wafers consist of a silicon ribbon or strip. APex cells, in contrast, are polycrystalline thin-film solar cells on a cost-efficient substrate (Zimmermann, 2001).

POLYCRYSTALLINE EFG SILICON CELLS

Fabrication

The EFG process has been used in mass production applications for a number of years. An octagonal shaping die made from graphite is dipped into the silicon melt and pulled out. This creates octagonal tubes up to 6.5m in length, with sides 10cm or 12.5cm in length and an average wall thickness of 0.3mm. The finished wafers are cut from the eight sides. Material wastage in this process is less than 10 per cent. After phosphorus doping and application of the back contact layer, the wafers are provided with contact grid lines on the front and an anti-reflective coating. EFG silicon is polycrystalline but has very few grain boundaries and crystal defects. In appearance and electrical properties, therefore, the cells are more similar to mono-crystalline cells:

- *Efficiency:* 14 per cent.
- *Form:* Square.
- *Size:* 12.5cm^2 × 12.5cm^2.
- *Thickness:* average 0.24mm.
- *Appearance:* the EFG process produces long pulled crystals that can just be seen if you look closely. The cell surfaces are slightly uneven.
- *Colour:* blue (with AR).

Cell manufacturer

RWE Schott Solar.

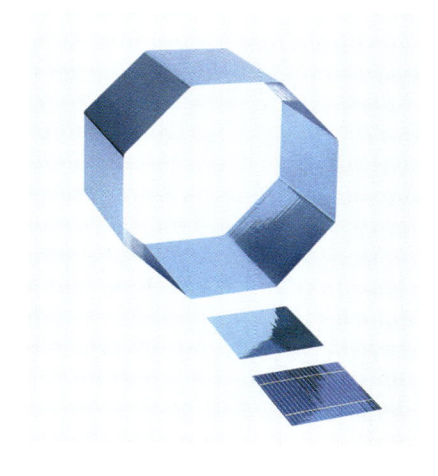

Figure 1.58
EFG ribbon-pulling machine
Source: RWE Schott Solar

Figure 1.59
Wafers are cut by laser from the octagonal tubes
Source: RWE Schott Solar

Figure 1.60
Square EFG cells
Source: RWE Schott Solar

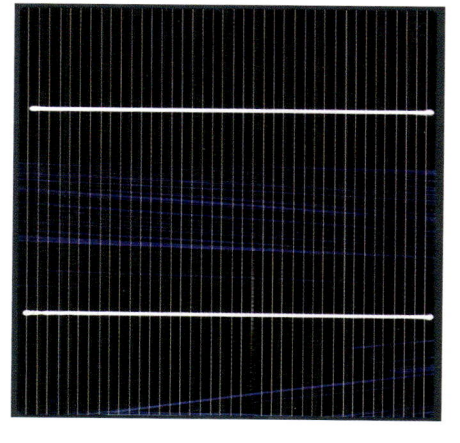

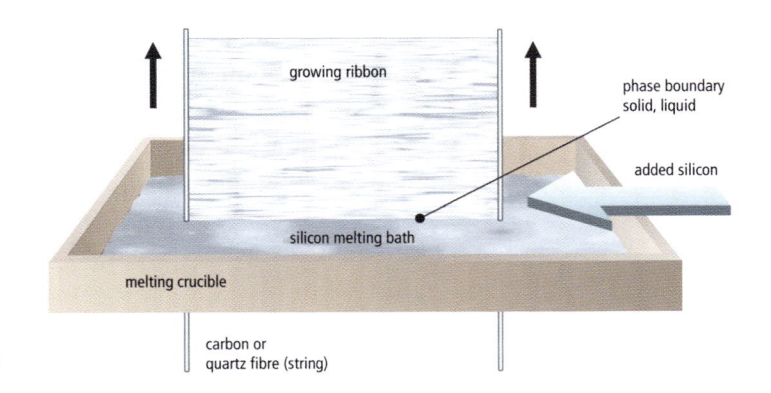

*Figure 1.61
Pulling a silicon ribbon in the string ribbon process
Source: Evergreen/Solarpraxis/DGS LV Berlin BRB*

POLYCRYSTALLINE STRING RIBBON SILICON CELLS

Fabrication

In the string ribbon process, two highly heated carbon or quartz fibres – the strings – are pulled vertically through a flat crucible of silicon melt. The liquid silicon forms a skin between the strings and crystallizes into an 8cm wide silicon strip – the ribbon. The pulling part of the process runs continuously: the strings are wound off rolls and raw silicon is continuously added to the crucible while the growing ribbon is cut into rectangular wafers at its finished end:

- *Efficiency:* 12 per cent to 13 per cent.
- *Form:* rectangular.
- *Size:* 8cm^2 × 15cm^2.
- *Thickness:* 0.3mm.
- *Appearance:* as for EFG.
- *Colour:* blue (with AR); silver grey (without AR).

Cell manufacturer

Evergreen Solar, EverQ.

*Figure 1.62
String ribbon solar cells during cell production
Source: Evergreen Solar*

POLYCRYSTALLINE APEX CELLS

Fabrication

APex cells represent the first application of a thin-film process with crystalline silicon that has made it to production stage. An electrically conductive ceramic substrate containing silicon replaces the thick silicon wafer and, in a horizontal continuous process, is given a covering of thin polycrystalline silicon of between 0.03mm and 0.1mm in thickness, which functions as a photovoltaic layer. Large-format solar cells are produced that have similar properties to conventional polycrystalline cells. High process temperatures in the region of 900°C to 1000°C are still required; but the small amounts of high-grade semiconductor material that are needed and the fast production speed promise cost advantages:

- *Efficiency:* 9.5 per cent.
- *Form:* square.
- *Size:* 20.8cm^2 × 20.8cm^2.
- *Thickness:* 0.03mm to 0.01mm + ceramic substrate.
- *Appearance:* as for polycrystalline solar cells, but smaller crystallites.
- *Colour:* blue (with AR); silver grey (without AR).

CELL MANUFACTURER
GE Energy.

*Figure 1.63
Production process for APex solar cells
Source: Sunset*

1.4.5 Anti-reflective coating on crystalline silicon cells

So that as much light as possible penetrates the cell, an anti-reflective coating of silicon nitride or titanium dioxide is applied. This ensures that as little light as possible is reflected off the surface of the cell and reduces reflection losses to a few per cent. Silicon nitride also has the effect of passivating any crystal defects on the surface. Passivation prevents charge carrier pairs from recombining.

*Figure 1.64
Colour palette of mono-crystalline cells
(efficiencies: 11.8 to 15.4 per cent)
Source: Solartec photo archive*

PHOTOVOLTAIC BASICS 31

Figure 1.65
Green polycrystalline cell with special anti-reflective coating (efficiency 11.8 per cent)
Source: RWE Schott Solar

Figure 1.66
Golden polycrystalline cell with special anti-reflective coating (efficiency 12 per cent)
Source: RWE Schott Solar

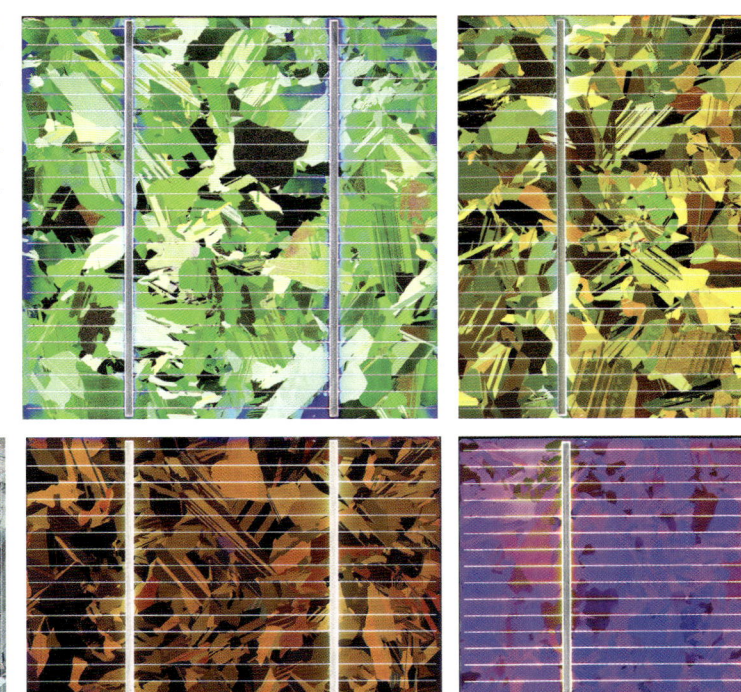

Figure 1.67
Silver polycrystalline cell without anti-reflective coating
Source: Ersol

Figure 1.68
Brown polycrystalline cell with special anti-reflective coating (efficiency 12.5 per cent)
Source: RWE Schott Solar

Figure 1.69
Violet polycrystalline cell with special anti-reflective coating (efficiency 13.2 per cent)
Source: RWE Schott Solar

This AR coating causes the originally grey crystalline wafers to take on a blue (polycrystalline cells) or dark blue to black (mono-crystalline cells) colouring. As well as this yield-optimizing anti-reflective coating, it is possible to create different colour tones by varying the coating thickness. The colours are caused by the reflection of a different part of the spectrum of light in each case. Currently, the colours green, gold, brown and violet can be produced. However, the optical effect comes at the price of lower cell efficiency. It is also possible to leave out the AR completely and have the wafers in their original silver grey (polycrystalline cells) or dark grey (mono-crystalline cells). Cells without AR are more frequently used for façade integration. They are simple to manufacture and the neutral colour tone is often desirable to architects. At the same time, it is accepted that up to 30 per cent of the sunlight will be reflected off the surface of the solar cell.

1.4.6 Front contacts

So that the cells can be integrated within an electrical circuit, metallic contacts are applied to both sides. A fine metal grid is used on the side facing the sun to keep the shaded area as small as possible. The front contacts are generally applied using a screen printing process. In this a silver paste is applied through a screen onto the

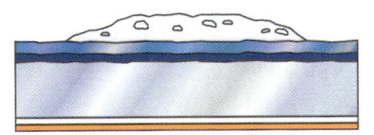

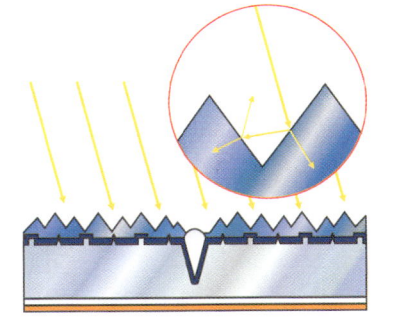

Figure 1.70
Comparison of screen printing method and Saturn technology: Creating front contacts and surface texture
Source: BP/Solarpraxis

silicon wafer. The individual lines (contact fingers) have a width of about 0.1mm to 0.2mm in this process. Two collector contact lines (busbars) about 1.5mm to 2.5mm thick run across the thin contact fingers. These busbars are later connected to the back contacts of the next cell in the string via a thin soldered copper strip. The contact fingers and busbars are sintered by firing at 800°C to 900°C and forced through the anti-reflective coating beneath them.

Special technologies have been developed for high-performance solar cells to improve the contact properties and minimize reflection on the cell surface. One example is known as the Saturn process. In this, the contact line is cut in using a laser. The width of the contact lines at 0.02mm is considerably reduced compared to screen printing. As a result, less of the cell surface is shaded, in turn allowing more lines to be cut into the solar cell. Since these laser-cut grooves can be filled with contact material, ohmic losses in conducting the charge carriers are reduced. In buried contact technology, instead of the V-shaped troughs, grooves with a square cross section are created and n-doped at an increased concentration (n^{++}) before the contact lines are applied.

Figure 1.71
Polycrystalline cell with screen-printed front contacts
Source: BP

Figure 1.72
Mono-crystalline cell with laser-formed front contacts
Source: BP

In addition, a textured surface consisting of tiny pyramids further reduces reflection losses. Light striking the surface is reflected and refracted repeatedly by the pyramid surfaces. This allows more light to penetrate the cell and be absorbed. This effect is known as light trapping. Depending upon the process and manufacturer, different surface structures or textures are etched into the cell (e.g. inverted pyramids).

Finer screen-printing masks that will create contact fingers just 0.03mm wide and buried contact processes for polycrystalline solar cells are due to be launched soon. In addition, the first cells with three busbars are being produced. These allow the greater currents from increasingly large polycrystalline cells to be transported with ever higher efficiency and low electrical losses.

Figure 1.73
Polycrystalline solar cells with three busbars
Source: Kyocera

Figure 1.74
Front contact lines along the grain boundaries of polycrystalline silicon
Source: AIAU

Figure 1.75
Decorative front contact pattern design (96.3 per cent efficiency compared to optimized front contacts)
Source: AIAU

Figure 1.76
Decorative front contact pattern design (98 per cent efficiency compared to optimized front contacts)
Source: AIAU

Figure 1.77
Designer module with diagonal power buses, commercially available from 2006 (module efficiency: 12 to 12.5 per cent)
Source: Powerquant Photovoltaik GmbH

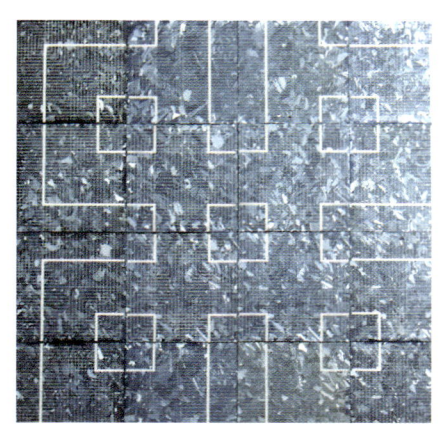

As part of the BIMODE international research project, the Atomic Institute of the Austrian Universities (AIAU) experimented with the design of the front contact lines to yield an additional optical effect. The front contact patterns that they designed showed efficiency losses of no more than 5 per cent compared to the optimized standard patterns. When they attempted to place the front contacts along the grain boundaries of polycrystalline silicon, cell efficiency increased. However, these front contact patterns were applied by hand, which is expensive. The research has now led to a commercial solar module made from special cells with diagonal power buses that can be arranged as desired in the module.

1.4.7 Back contacts

Unlike the front contacts, the metal contacts on the back of the cells can be applied across much more of the surface area. While these cannot be seen in standard modules with an opaque rear cover, they are visible in special modules for building integration that have a transparent rear cover and can be utilized as an additional design element. To optimize efficiency, a full-surface aluminium coating is printed on the back between the screen-printed contacts in point or strip form, which are 2.5mm to 6mm wide. When sintered, the aluminium coating becomes the back surface field (BSF), a strong p^+-doping that creates an additional electrical field. The BSF is intended to passivate crystal defects on the surface and reduce recombination of charge carriers on the back of the cell. Like an electrical mirror, it bounces the charge carriers back inside the cell.

Figure 1.78
Back point contacts and back surface field

Figure 1.79
Back strip contacts and back surface field

Figure 1.80
Back grid contact

Figure 1.81
Rays and circles back contact

1.4.8 High-performance cells

Using costly production processes, research laboratories have for many years been able to produce highly efficient crystalline silicon cells with efficiencies up to nearly 25 per cent. In these the electrical and optical losses are minimized. One example of the processes used to produce the high-grade wafers, which are used as a starting material, is the float-zone method. This process enables the production of mono-crystalline solar cells with greater purity and, hence, 1 per cent to 2 per cent higher efficiency. The source material required here is a highly pure polycrystalline silicon rod with a mono-crystalline silicon seed at its tip – but this material is expensive. This rod is lowered through an electromagnetic coil and melted in a ring shape using high-frequency fields, beginning at the mono-crystalline tip. Impurities are transported in the liquid zone to the last end of the rod to be heated. When it cools, a mono-crystalline structure forms with high purity and crystal quality in the whole rod.

To minimize recombination losses at the surfaces and contacts, the back can be passivated with silicon oxide or amorphous silicon. Point contacts on the back with a heavily doped local back surface field (p^{++}) also increase the efficiency. Finally, optical losses are minimized via textured surfaces and buried contacts with minimal cell shading. Here, microscopically small pyramid, ridge or groove-type structures are created on the cell surface using lasers or cutting tools. These function as light traps to prevent reflection.

Some of the production steps are very expensive and are therefore only used for laboratory cells. However, many processes have already transferred into mass production. BP Solar's Saturn cells with textured surface and buried contacts (see '1.4.6 Front contacts') have been commercially available for a number of years. Sanyo, with its heterojunction with intrinsic thin-layer (HIT) hybrid technology (see '1.4.13 Thin-film solar cells made from crystalline silicon'), also produces high-performance cells for the standard market. Cell concepts for back-contacted cells are new. Previously, these have only been used in concentrator applications.

BACK-CONTACTED SOLAR CELLS

In these cell, both the positive and the negative contacts are connected to the back of the cell. As well as avoiding shading by the front contacts on the side facing the sun, this also facilitates the subsequent creation of cell strings in a solar module and allows a uniform appearance with close cell spacing.

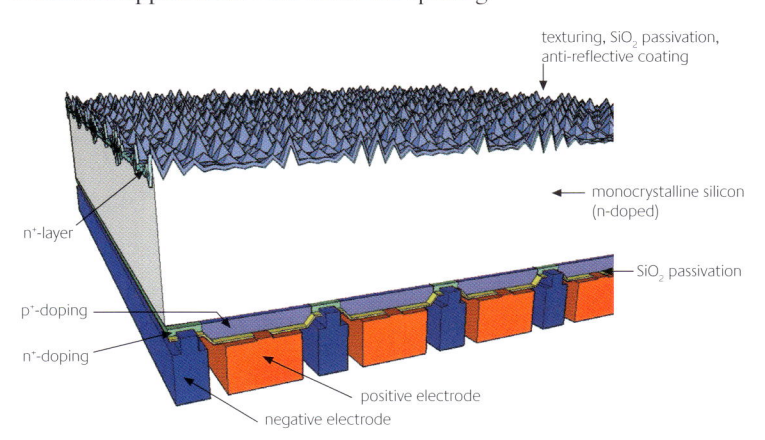

Figure 1.82
Structure of a back-contacted high-performance SunPower A-300 cell
Source: SunPower Corp

SunPower A-300 cells for commercial modules are a cost-optimized offshoot from cells for space travel that were developed for the National Aeronautics and Space Administration (NASA). The raw wafers are produced from mono-crystalline float-zone silicon (solar grade), which unlike conventional solar wafers is n-doped. The high silicon quality is necessary because the charge carriers have to diffuse through the entire cell thickness to reach the p-n junction on the underside. This is created through n- and p-doping in strips and is connected via metal fingers that also function as reflectors. Both surfaces are passivated with silicon oxide. To reduce recombination at the contacts, the electrodes are only connected with the p- and n-layer via point contact holes in the passivation layer.

These cells achieve efficiencies of up to 21.5 per cent in mass production and enable module efficiencies of up to 18.6 per cent. They have broad spectral sensitivity across a wide range of the solar spectrum. The slight n-doping on the front increases sensitivity in the short-wavelength blue range and provides good performance under low light conditions. Compared to standard crystalline cells, the cell voltage is somewhat higher:

- *Efficiency:* 20.8 per cent.
- *Form:* semi-round.
- *Size:* 12.5cm × 12.5cm nominal (5 inch).
- *Thickness:* 0.27mm.
- *Appearance:* uniform; no contact grid.
- *Colour:* velvet black.

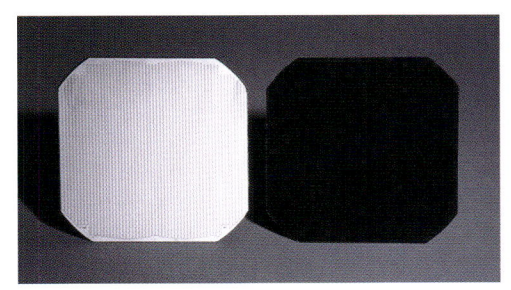

Figure 1.83
Back with contact lines and front of the A-300 cell
Source: SunPower Corp

Figure 1.84
Modules without front contacts with white- and anthracite-coloured backing film
Source: SunPower Corp

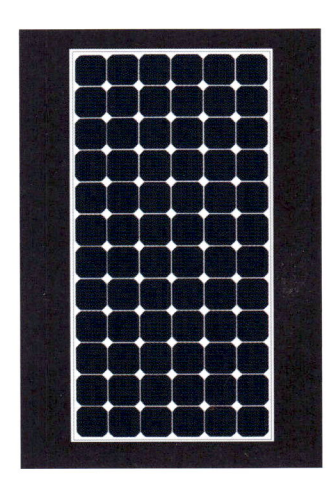

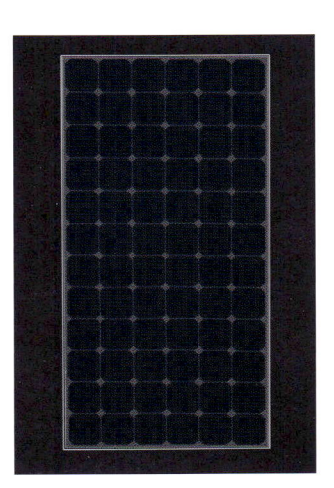

CELL MANUFACTURER
SunPower Corporation.

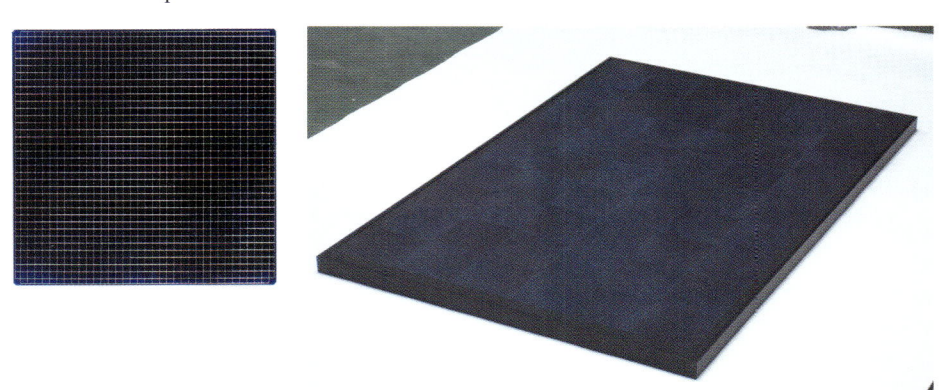

Figure 1.85
Maxis BC+ back-contacted cells
Source: Photovoltech

Figure 1.86
PV module with Maxis BC+ cells
Source: Photovoltech

The Maxis BC+ back-contacted cell from Photovoltech does have a thin contact grid on the front; but it is connected entirely via the busbars on the back. To achieve this, the wafers are perforated by laser along what later become the busbar lines. The cells are made from polycrystalline wafers, the surface of which is etched to form equally distributed pit points with no defined direction. This texturing makes the grain boundaries indistinct and the surface appears matt. After the anti-reflective coating is applied, the fine diamond-shaped contact grid is printed on. This fills the lasered holes with metal and creates the electrical connection to the conductor paths on the back. These then need to be separated using a laser so that there is no short circuit between the front and back. The front (negative contact) leads out from one side of these square cells via a pair of conductors. The back (positive contact) has three contact pairs on the other three sides so that the cells in the module can be wired in series even through a corner.

To date, cell efficiencies of up to 16.5 per cent have been achieved and a prototype module efficiency of 14.9 per cent. The uniform appearance coupled with high power density makes the modules particularly suitable for building integration. BC+ cells made from mono-crystalline wafers and with a 6-inch format are in the test phase:

- *Efficiency:* 15.4 per cent.
- *Form:* square.
- *Size:* 12.5cm × 12.5cm (5 inch).
- *Thickness:* 0.33mm.
- *Appearance:* matt and almost uniform with fine diamond-shaped contact grid. The polycrystalline structure is almost invisible.
- *Colour:* dark blue.

CELL MANUFACTURER
Photovoltech.

TRANSPARENT SOLAR CELLS

FABRICATION
Cell manufacturer Sunways offers two different transparent cells. The older polycrystalline version (formerly POWER cell) is made from polycrystalline wafers that then go through a mechanical structuring process. Using a fast-rotating roller, grooves are milled into the front and back of the silicon wafers. The direction of the grooves in the front and back are rotated 90° to each other. At the points where the grooves intersect, microscopically small holes appear. The solar cell lets light through at these points. The evenly spaced hole structure results in the cell being transparent. Depending upon the size of the holes this can vary between 0 per cent and 30 per cent. For technical reasons a small opaque border remains at the edge of the transparent cell. The cell can also be produced to be light sensitive on both sides.

In the newly developed Transparent Sunways Solar Cell, the mono-crystalline or, alternatively, polycrystalline wafer also receives an evenly spaced but coarser hole structure, produced with lasers. Depending upon the customer's requirements, the holes can be square, round or any other shape. The textured surface gives the cell an all-black appearance. In the module, only the cells' hole structure determines the transparency since the spaces between cells are blacked out by screen printing:

- *Efficiency:* 10 per cent (milled version) or 13.8 per cent (laser cut), each with 10 per cent transparency.
- *Form:* square.
- *Size:* $10cm^2 \times 10cm^2$; $12.5cm^2 \times 12.5cm^2$.
- *Thickness:* 0.3mm.
- *Appearance:* depends upon type with polycrystalline frost pattern appearance and gauze-like transparency, or with coarse hole pattern and special contact design.
- *Colour:* depends upon type, such as polycrystalline cells or black.

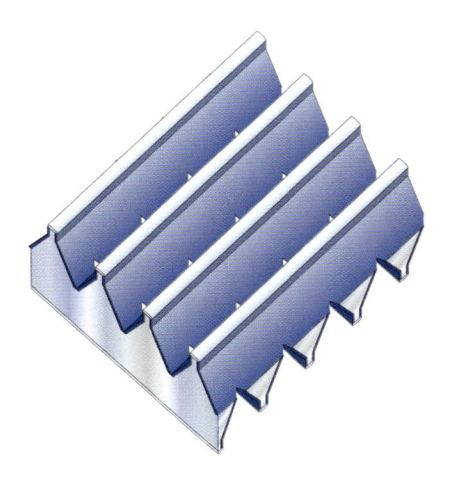

*Figure 1.87
Isometry of the Transparent Sunways solar cells with milled grooves
Source: Sunways*

*Figure 1.88
Milled polycrystalline cells with different anti-reflective coatings
Source: Sunways*

CELL MANUFACTURER
Sunways.

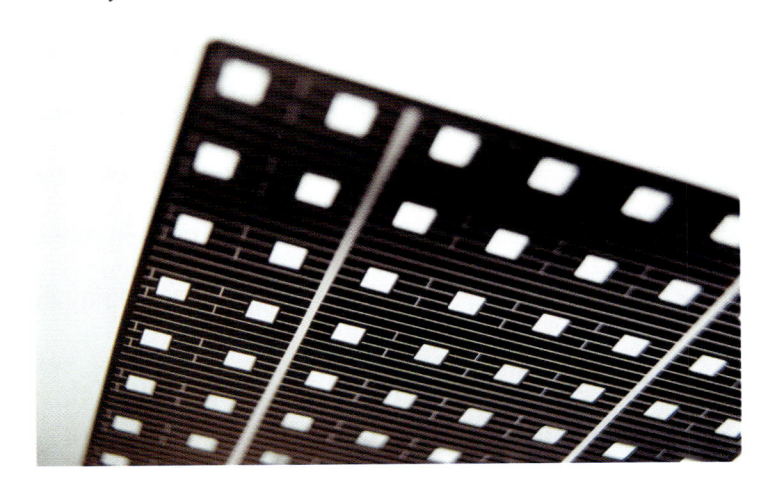

*Figure 1.89
New Transparent Sunways solar cells with lasered hole pattern
Source: Sunways*

NEW SOLAR CELL CONCEPTS

SPHERICAL SOLAR CELLS

During the early 1990s, Texas Instruments conducted research on spherical silicon solar cells but did not take them to the production stage. The Canadian firm Spheral Solar Power began the first commercial mass production of mono-crystalline spherical solar cells in 2004. Modules by Kyosemi Corporation of Japan are in the pilot stage. These spherical solar cells with a diameter of 0.7 m (Spheral) or 1mm to 1.2mm (Kyosemi) are made from p-doped silicon (e.g. from drops of liquid silicon formed into balls when they fall in a vacuum). The surface is n-doped and sometimes given an anti-reflective coating.

With Kyosemi, contacts are provided via two opposing electrodes made from silver and aluminium in the p-core and the n-shell. These are wired in series or in parallel using fine copper wires. The balls are embedded in a transparent synthetic resin and turned into transparent modules on a substrate plate of white synthetic resin, which reflects back incident light between the spheres.

Spheral spherical cells are connected between two superimposed aluminium sheets that are insulated with a thin plastic layer. The upper perforated aluminium sheet holds the spheres mechanically and connects the n-layer. The spheres are etched slightly from below until the p-doped core is exposed. This is passivated and electrically contacted with the lower aluminium sheet. Since all of the cell spheres are connected in parallel in this method, the cell sandwiches are cut into 15cm (6 inch) squares and these laminates are wired in series by ultrasound-welding the sheets.

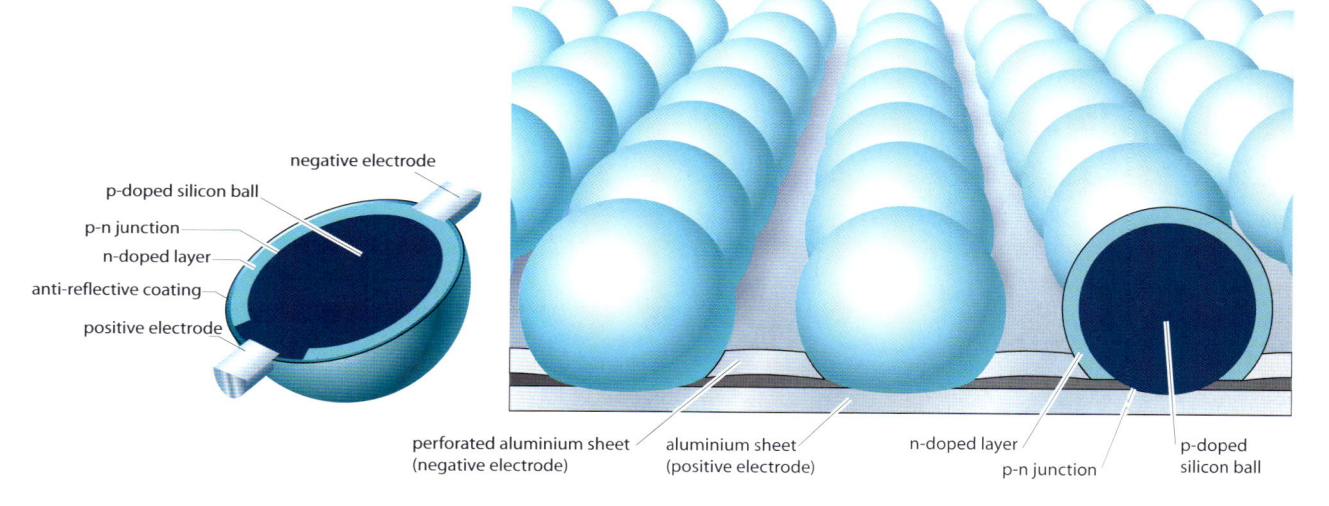

Figure 1.90
Various spherical cell concepts with similar names: (left) structure of the Spheral spherical cell by Kyosemi; (right) structure and electrical connection of the Spheral spherical cells by Spheral Solar Power
Source: Kyosemi/DGS LV Berlin BRB

Kyosemi has already demonstrated cells with approximately 12.5 per cent efficiency. The first large area modules from Spheral on aluminium trapezoidal sheet substrates achieve a shade under 9.5 per cent efficiency. Owing to their almost spherical p-n junction, spherical solar cells can optimally catch even light falling at an angle and, hence, utilize diffuse radiation better than flat wafers or thin-film cells. Since this is not taken into account in standard measurement procedures under standard test conditions (STC), the modules promise higher yields per kilowatt peak (kWp). Since spherical cells also enable the production of flexible modules, resilient spherical cell modules are particularly suited to mobile applications, such as boats and vehicle roofs, and for integration within roofing elements and corrugated roofs. The relatively simple manufacturing processes and the reduced silicon requirements compared to wafer production offer possible cost reductions for the future.

Figure 1.91
Macro shot of Spheral spherical solar cells
Source: Kyosemi

*Figure 1.92
Spheral spherical solar cell module
made from 6-inch units
Source: Kyosemi*

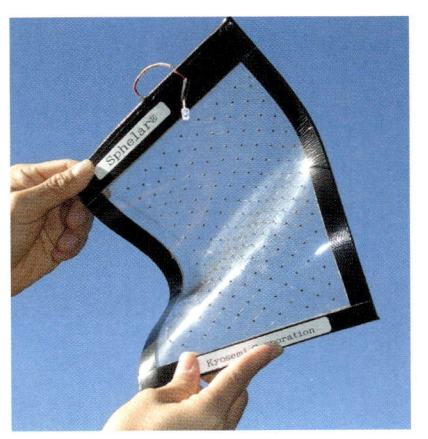

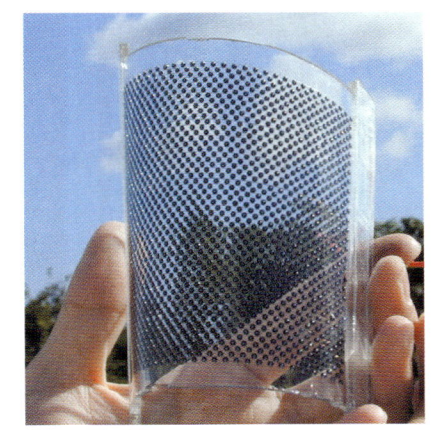

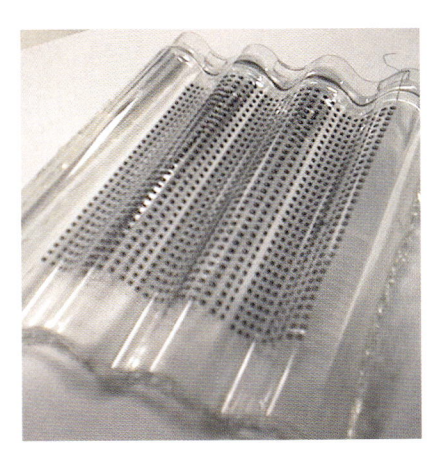

*Figure 1.93
Prototype Spheral modules:
Because they are flexible, transparent
and can be wired up in any desired way,
spherical cell modules are particularly
suited to building integration and
use in electronic devices
Source: Kyosemi*

SLIVER CELLS

Produced by Australian firm Origin, sliver cells are based on a 1mm thick monocrystalline wafer made from float-zone silicon. This is etched perpendicularly to the surface in order to produce multiple 0.05mm thin strips with equal-sized gaps between them. Once sliced up in this way, the wafers are n-doped with phosphorus on one side and p-doped with boron on the other. The gaps are textured, passivated and given an anti-reflective coating. Finally the strips, which are 10cm long, 1mm wide and 0.05mm thick, are detached from the wafer. In a module, they are arranged between two sheets of glass with the long doped edges facing each other. Arrangement and electrical connection of the several thousand cells per square metre are performed automatically. The module can be set up for any voltage (e.g. for 12V applications or higher voltages for grid-connected systems). The cells, which are absolutely symmetrical and, hence, active on both sides, are generally arranged allowing a generous spacing between them. This allows more light to be utilized with the same number of cells, which means higher performance. A diffuse reflector on the back of the module casts the light falling into the gaps back onto the module.

*Figure 1.94
Sliver cells as strip cells and modules
Source: C. Geyer/DGS Berlin BRB*

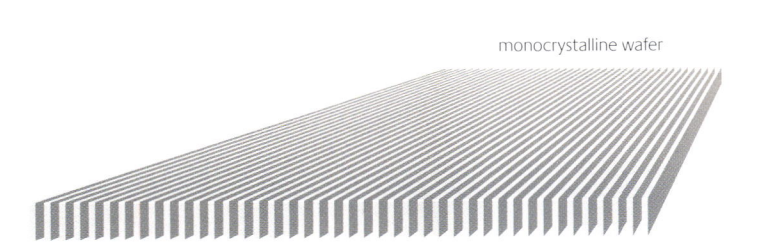

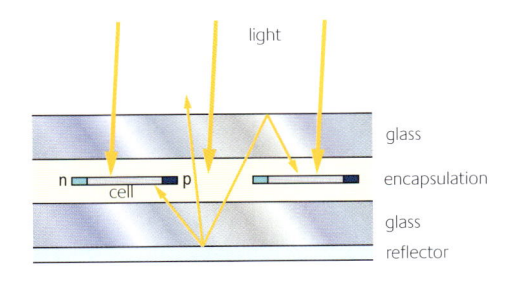

In this way the silicon requirements per megawatt (MW) are reduced to around one seventh of conventional wafer cells. Efficiencies of up to 19 per cent have been measured in laboratory cells. Test modules have achieved between 17.7 per cent and 13 per cent efficiency (depending upon the areas of the modules actually covered with cells). The first commercial glass modules will come onto the market with 40W and 10 per cent efficiency. Power outputs exceeding 100W are planned for the future. With their transparency and light sensitivity on both sides, these cells are particularly suited to partially transparent architectural applications or free-standing noise barriers. The flexible slivers also mean that flexible modules can be produced. However, in price terms the sliver modules will only be able to compete with conventional silicon modules when they achieve high production volumes.

Figure 1.95
Individual cell slivers and finished sliver cell module
Source: Origin

1.4.9 Thin-film cell technology

Since the 1990s there has been increased development of thin-film processes for manufacturing solar cells. In these, photoactive semiconductors are applied as thin layers to a low-cost substrate (in most cases, glass). The methods used include vapour deposition, sputter processes (cathode sputtering) and electrolytic baths. Amorphous silicon, copper indium diselenide (CIS) and cadmium telluride (CdTe) are used as semiconductor materials. Because of the high light absorption of these materials, layer thicknesses of less than 0.001mm are theoretically sufficient for converting sunlight. The materials are more tolerant to contamination by foreign atoms. Compared to manufacturing temperatures of up to 1500°C for crystalline silicon cells, thin-film cells require deposition temperatures of between only 200°C and 600°C. The lower material and energy consumption and the capability for highly automated production

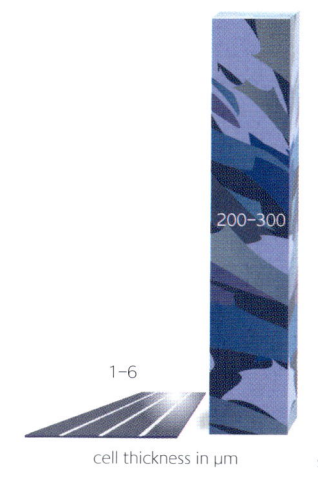

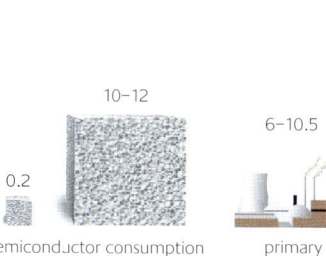

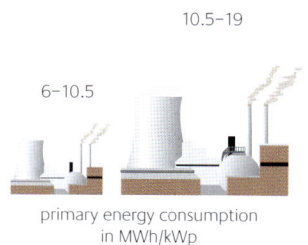

Figure 1.96
Comparison of cell thickness, material consumption and energy expenditure for thin-film cells (left) and crystalline silicon cells (right)
Source: manufacturers' information, Solarpraxis

with a large throughput offer considerable savings potentials when compared to crystalline silicon technology.

Thin-film cells are not restricted in their format to standard wafer sizes, as is the case with crystalline cells. Theoretically, the substrate can be cut to any size and coated with semiconductor material. However, because only cells of the same size can be connected in series for internal wiring, for practical purposes only rectangular formats are common.

A further distinguishing feature of thin-film cells that differentiates them from crystalline cells is the way in which they are connected together. While crystalline solar cells are soldered together from cell to cell (external interconnection), thin-film cells are interconnected monolithically during the coating and layering process. The cells are electrically separated and interconnected by means of structuring stages, in which each cell layer is cut into strip-like individual cells (see '2.1.1 Cell stringing' in Chapter 2). This creates thin transparent grooves between the individual cells. In order to achieve as great an energy yield as possible, these are made as thin as possible and are hardly visible to the naked eye. They can, however, be used as a design element and be deliberately widened. The wider the grooves between the cells, the greater the transparency. The semi-transparent optical effect can also be created by forming additional grooves perpendicular to the cell strips.

The electrical contact is created on the back with an opaque metal coating. On the front side facing the light this function is fulfilled by a highly transparent and conductive metal oxide layer called the transparent conductive oxide (TCO) layer. Typical TCO materials include zinc oxide (ZnO), tin oxide (SnO2) and indium tin oxide (ITO). The TCO layers are an important cost factor in thin-film cell production.

With thin-film technology, the terms cell and module as used with crystalline technology need to be supplemented by another term: the raw module. Here a cell consists of a long, narrow strip of semiconductor material on the substrate glass. The raw module describes the completely coated glass sheet with multiple cell strips connected in rows. When this is encapsulated with a laminating material (EVA) and protected with a second glass sheet, this is known as a module. The usual distinction between cell and module efficiency also does not transfer straightforwardly since thin-film cells are not produced or measured individually. Hence, in the scientific literature the efficiency is often related to the aperture area (i.e. the photovoltaically active surface without edge and frame).

Despite the relatively low efficiency, the energy yield can, under certain conditions, be quite considerable. The utilization of diffuse and low light is better with thin-film cells and there is a more favourable temperature coefficient (i.e. the decrease in performance at higher operating temperatures is less than with other technologies). Furthermore, because of their cell form (long narrow strips), thin-film cells are less sensitive to shading. Whereas a leaf on a crystalline module can completely cover a crystalline cell, with a thin-film module it might cover several cells at the same time, but only ever cover a small area of each respective cell (see '2.1.10 Electrical characteristics of thin-film modules' in Chapter 2).

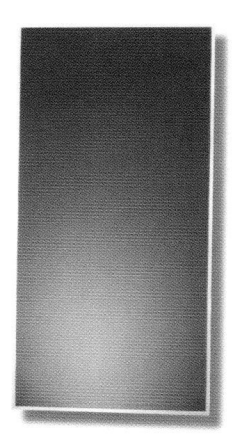

Figure 1.97
Typical strip appearance of a thin-film module, here made from CdTe
Source: First Solar

Figure 1.98
Semi-transparent thin-film modules made from amorphous silicon with additional separating steps, interior view

1.4.10 Amorphous silicon cells

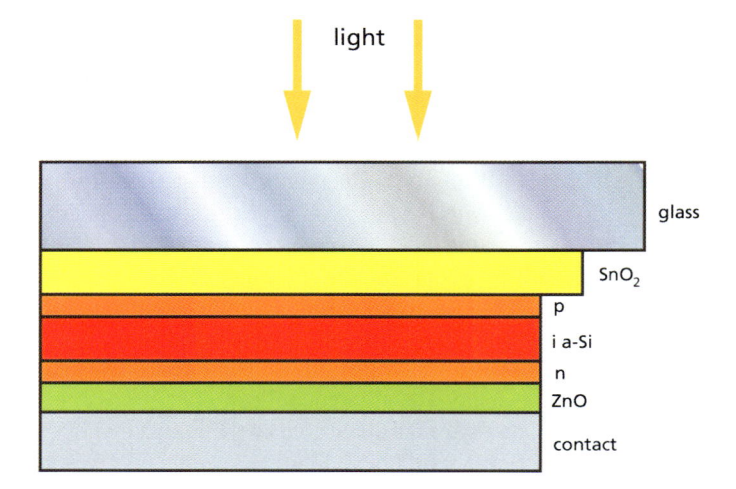

*Figure 1.99
Layered structure of an amorphous cell
Source: Solarpraxis*

FABRICATION

Amorphous (formless) silicon does not form a regular crystal structure, but an irregular network. As a result, open bonds occur that absorb hydrogen until saturation. This hydrogenated amorphous silicon (a-Si:H) is created in a plasma reactor by chemical gas-phase deposition of gaseous silane (SiH_4). Process temperatures of 200°C to 250°C are sufficient. The doping is carried out by mixing gases that contain the corresponding doping material (e.g. B2H6 for p-doping and PH3 for n-doping). Because of the very small diffusion length of doped a-Si:H, the free charge carriers in a direct p-n junction would not survive long enough to be able to contribute to generating electricity. Therefore an intrinsic (un-doped) i-layer is coated between the n- and p-doped layers, in which the lifetime of the charge carrier is substantially longer. This is where the light absorption and the charge generation occur. The p- and n-layers only create the electric field that separates the released charge carriers. For TCO front contacts, tin oxide (SnO_2), indium tin oxide (ITO) or zinc oxide (ZnO) are used. The lower TCO layer functions with the metallic back contact as a reflector. If the cells are deposited on the front side of the glass as in Figure 1.133, then this creates the characteristic p-i-n structure. Alternatively, they can also be deposited in a reverse sequence (n-i-p) on the back. This enables flexible solar modules to be created on non-transparent and lightweight substrate materials (e.g. on metal or plastic sheeting, which is ideal for integration within roof systems).

The disadvantage of amorphous cells is their low efficiency, which diminishes even during the first 6 to 12 months of operation owing to light-induced degradation (known as the Staebler-Wronski effect) before levelling off at a stable value, the nominal power rating. Often in amorphous cells, multiple pin structures are deposited on top of each other to make stacked solar cells. Tandem cells consist of two cell stacks; triple cells consist of three cell stacks. This allows higher efficiencies to be achieved since each part cell can be optimized for a different colour band of the solar spectrum – for example, by admixing germanium (a-SiGe). In addition, with stack cells, the ageing effect is reduced since the individual i-layers are thinner and are therefore less susceptible to light degradation:

- *Efficiency:* 5 per cent to 7 per cent module efficiency (stabilized condition).
- *Size:* standard modules, maximum $0.79m^2 \times 2.44m^2$; special modules, maximum $2m^2 \times 3m^2$.
- *Thickness:* 1mm to 3mm substrate material (non-hardened glass, metal, occasionally 0.05mm plastic), with approximately 0.001mm (1μm) coating, of which approximately 0.3μm amorphous silicon.
- *Appearance:* uniform appearance.
- *Colour:* reddish brown to blue or blue-violet.

*Figure 1.100
Flexible amorphous modules based on metallic foils that are glued onto trapezoidal sheet sections
Source: Hoesch Contecna; modules: United Solar*

*Figure 1.101
Trier-Birkenfeld University of Applied Sciences, Germany
Source: RWE Schott Solar*

CELL MANUFACTURERS

BP Solar, Canon, Dunasolar, ECD Ovonics, EPV, Free Energy Europe, Fuji Electric, ICP, Iowa Thin Film Technologies, Kaneka, MHI, RWE Schott Solar, Sanyo, Shenzhen Topray Solar, Sinonar, Solar Cells, Terra Solar, Tianjin Jinneng Solar Cell, United Solar Ovonic, VHF Technologies.

1.4.11 Copper indium diselenide (CIS) cells

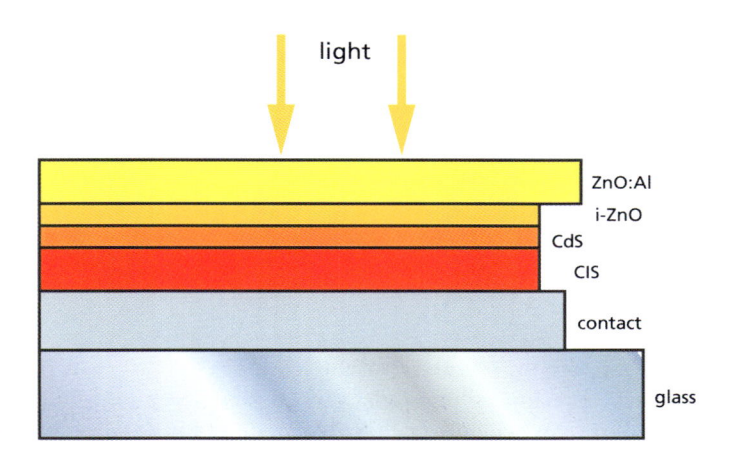

*Figure 1.102
Layered structure of a CIS cell
Source: Solarpraxis*

FABRICATION

The active semiconductor material in CIS solar cells is copper indium diselenide. The CIS compound is often also alloyed with gallium and/or sulphur. When fabricating the cells, the substrate glass is initially coated with a thin molybdenum layer as back contact using cathode sputtering. The p-type CIS absorber layer can be manufactured by simultaneously vaporizing the elements copper, indium and selenium in a vacuum chamber at temperatures of around 500°C to 600°C (manufacturer: Würth). Another possibility is to sputter the individual elements as individual layers at room temperature, and then to combine them to form CIS by briefly heating to 500°C (Shell Solar). Aluminium-doped zinc oxide (ZnO:Al) is used as the transparent front contact. This is n-conductive and is sputtered with an intrinsic zinc oxide (i-ZnO) intermediate layer. An n-type cadmium sulphide (CdS) buffer layer can be used to reduce losses caused by mismatching of the crystal lattices from the CIS and ZnO layers. This is deposited in a chemical bath. Unlike amorphous silicon, CIS solar cells are not susceptible to light-induced degradation. However, effective sealing must be ensured owing to the moisture sensitivity of the zinc oxide layer.

CIS modules are currently the most efficient of all thin-film technologies. With expansion to mass production volumes, significantly lower production costs than for crystalline silicon (Si) modules are expected. In Germany, owing to the negligible amounts of selenium and cadmium, CIS modules meet environmental regulations for domestic refuse dumps; but local regulations need to be checked. The manufacturer Sulfurcell Solartechnik replaces selenium with sulphur and has created a pilot product CIS cell using copper indium disulphide in a combined sputter and vapour deposition process. Research is under way into using indium sulphide buffer layers as a substitute for the CdS layer:

- *Efficiency:* 9 per cent to 11 per cent module efficiency.
- *Size:* standard modules, maximum 1.2m × 0.6m.
- *Thickness:* 2mm to 4mm substrate material (non-hardened glass) with 3μm to 4μm coating, of which approximately 1μm to 2μm CIS.
- *Appearance:* uniform appearance.
- *Colour:* dark grey to black.

CELL MANUFACTURERS

CIS Solartechnik GmbH, Daystar, EPV, Global Solar, Shell Solar, Showa Shell,

Figure 1.103
CIS modules based on copper indium disulphide
Source: Sulfurcell

Figure 1.104
Façade systems with CIS modules
Source: Würth Solar

Solarion, Sulfurcell, Würth Solar.

The manufacturer Solarion deposits CIS on ultra-thin polymer films and from these creates small-format flexible modules for aerospace and terrestrial applications in automotive engineering, communication and textiles. The first CIS cells from Daystar on flexible metallic foils were announced for the end of 2005. CIS Solartechnik is developing CIS strip cells on copper strips, which are designed to be placed next to each other in the future module and connected in series by shingling.

Figure 1.105
Glass-free CIS cells on plastic films
Source: Solarion

Figure 1.106
CIS cells on flexible metallic foils
Source: Daystar

1.4.12 Cadmium telluride (CdTe) cells

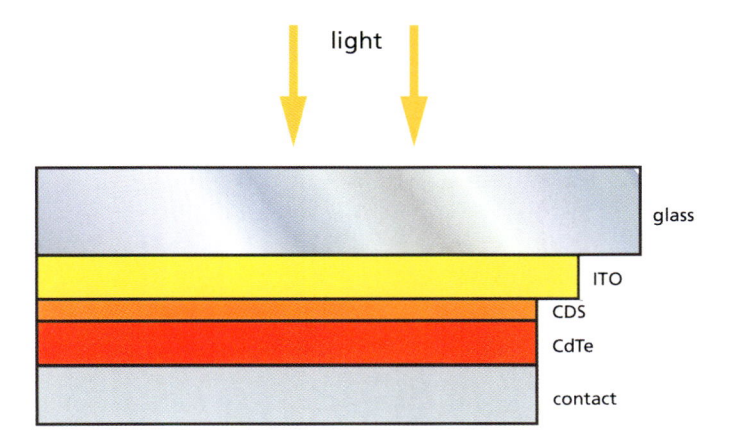

Figure 1.107
Layered structure of a CdTe cell
Source: Solarpraxis

FABRICATION

CdTe solar cells are manufactured on a substrate glass with a transparent TCO conductor layer usually made from indium tin oxide (ITO) as the front contact. This is initially coated with an n-type CdS window layer, which is as thin as possible, before being coated with the p-type CdTe absorber layer. The semiconductor layers are created using a simple vapour deposition process with low requirements for the vacuum used. The vapour source is heated to around 600°C. The substrate glass, which is somewhat cooler at 500°C, is held a short distance above it and 'steamed up' with the semiconductor materials. After the deposition, the CdS and CdTe layers are activated by tempering (deliberate heating up) in a chlorinated atmosphere at 400°C and re-crystallize to create a CdS–CdTe double layer. The metallic back contact is then applied using a sputter process. The back contact is a weak point in CdTe cells since it is responsible for ageing that may occur. Modern high-grade CdTe modules do not suffer any initial degradation.

CdTe technology has the lowest production costs among the current thin-film modules. Mass production on a larger scale may realize further high-cost saving potentials in the future. Market acceptance of the use of cadmium, which is a heavy metal, is the subject of intense discussion. But since cadmium occurs anyway as a waste product of zinc mining, its processing into harmless CdTe in solar modules can be seen as ecologically beneficial. CdTe is a non-toxic compound and is very stable. It only breaks down at temperatures in excess of 1000°C. There is no need to fear environmental and health risks even in a fire since the heavy metal would be encased in the glass, which melts at a much lower temperature. Manufacturers take back modules that have reached the end of their life and recycle them in an environmentally conscious way:

- *Efficiency:* 7 per cent to 8.5 per cent module efficiency.
- *Thickness:* 3mm substrate material (non-hardened glass) with 0.005mm coating.
- *Size:* standard module dimensions of 1.2m × 0.6m.
- *Appearance:* uniform.
- *Colour:* reflective dark green to black.

CELL MANUFACTURERS

Antec Solar Energy, First Solar.

*Figure 1.108
CdTe module
Source: Antec Solar Energy*

*Figure 1.109
On-roof system
Source: Antec Solar Energy*

*Figure 1.110
1.3MW ground-mounting system in Dimbach, Germany
Source: First Solar*

DYE-SENSITIZED NANO-CRYSTALLINE CELLS

A new kind of solar cell was announced by Swiss Professor Michael Grätzel in 1991 and could develop into an affordable alternative to silicon technology in future. The base material for the Grätzel cell is the semiconductor titanium dioxide (TiO_2). However, it does not function on the basis of a p-n junction in the semiconductor, but absorbs light in an organic dye similar to the way in which plants use chlorophyll to capture energy from sunlight by photosynthesis.

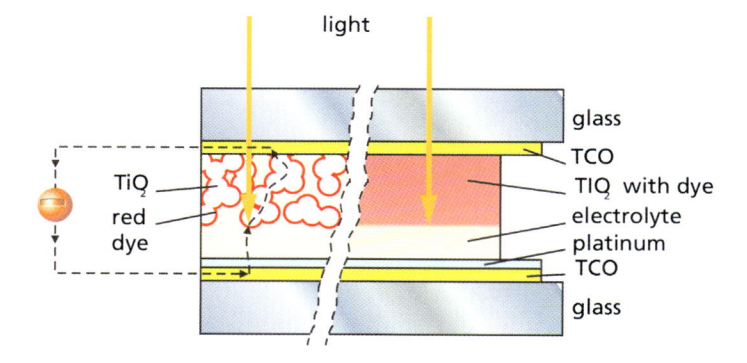

*Figure 1.111 Layered structure of a dye-sensitized cell (on the right)
Note: In reality, the TiO_2/dye and electrolyte layers are not as clearly separated from each other. The electrolyte completely permeates the porous semiconductor material (left side of the figure).
Source: Solarpraxis/DGS Berlin BRB*

Dye-sensitized solar cells are fundamentally different to conventional solar cells. A dyed titanium dioxide layer and a conducting saline solution as the electrolyte are located between two conducting and transparent electrodes (with TCO-coated glass sheets). Titanium dioxide is applied as a paste to the upper electrode using a screen printing process. At 450°C, the layer is cured to form a 10μm thick solid film. This creates a rough micro-porous structure consisting of particles that are 10μm to 30μm (0.00001mm to 0.00003mm) thick. The inner surface of this 'light sponge' is 1000 times greater than with a smooth film. Since the TiO_2 absorbs only ultraviolet light, the TiO_2 surface is given an ultra-thin coating of a ruthenium-based dye. The liquid electrolyte completely penetrates the porous layer and thus connects the dye electrically with the lower electrode.

If light hits the cell, the dye is excited and injects an electron into the titanium dioxide. The electron migrates through the TiO_2 particles to the upper electrode. It reaches the lower electrode via the outer electrical circuit. This transfers the electron with the help of the platinum catalyst to the electrolyte solution. The electrolyte transports the electron back to the dye and the cycle is completed.

The unique feature of the dye-sensitized cell concept is that the light absorption and charge carrier transport occur in different materials. The charges are generated by light absorption in the dye, while the TiO_2 semiconductor and the dissolved ions in the electrolyte are responsible for the transport. This has the advantage that recombination cannot occur even with contaminated semiconductor material. This means there is no need for clean room and vacuum technology during the production of the cells.

The materials used are non-toxic and are cheap to produce. Titanium dioxide is produced industrially in large amounts and is used, for example, in wall paints, toothpaste and paper. Expensive materials such as platinum and stable dyes are required only in very small amounts. There are still serious problems that have to be solved, however, before this technology can be mass produced, particularly with regard to long-term stability and the sealing of the bonding system. To improve the handling and to simplify the sealing, work is increasingly being done on thickening the liquid electrolyte to form a gel similar to that used in gel batteries.

To date, small laboratory cells have reached an efficiency of up to 12 per cent. Modules from the first limited batch produced by the Australian firm STI have an efficiency of around 5 per cent. In partnership with Swiss company Greatcell Solar, commercial modules with 8 per cent to 9 per cent efficiency are planned. Konarka Technologies, Inc, in the US, has begun pilot production of dye-sensitized cells for use in small appliances. The cells are printed on plastic films in a roll-to-roll process and are claimed to have 5 per cent efficiency. Peccell in Japan has also announced flexible dye-sensitized cells on a plastic substrate for indoor applications. Sony Europe is doing research into tandem cells with red and black dyes to improve the absorption spectrum.

The modest efficiencies under standard test conditions are balanced by comparatively high efficiency at low light intensities. Dye-sensitized cells have proven very tolerant to poor incident angles and shading. Unlike crystalline cell types, their efficiency actually increases at higher temperatures. They are therefore generally used for small appliances in interior spaces and for integration within buildings. Here the dye-sensitized cells offer exciting new design possibilities with their adjustable transparency and reddish-ochre tint; alternatively, grey-green can be selected depending upon the dye.

Figure 1.112
Prototype of a dye-sensitized cell module (50cm x 50cm)
Source: INAP Gelsenkirchen

Figure 1.113
First commercial module from a small production run
Source: STI, Australia

1.4.13 Thin-film solar cells made from crystalline silicon

Silicon is completely non-toxic and almost inexhaustible. Thin-film solar cells made from crystalline silicon are a promising alternative for the future. They benefit not only from the material advantages of crystalline silicon, but also from the manufacturing advantages of thin-film technology (cheap automated mass production with minimal material consumption). Research activities are progressing in two directions. High-temperature processes deposit high-quality silicon films on a cheap substrate (wafer equivalents, such as graphite, ceramic or silicon) at temperatures of 900°C to1100°C, creating large-grained structures similar to polycrystalline cells. With grain sizes up to 1mm, efficiency of up to 16 per cent is possible. This technology is already used in the production of APex cells. This cell type is classified as a crystalline cell since it is based on wafers.

In new experimental research, 20µm to 30µm thick silicon layers are deposited out of the gas phase (trichlorsilane) at temperatures exceeding 1000°C onto mono-crystalline wafers with a porous surface. After deposition of the contacts and application of the anti-reflective coating, a glass or plastic sheet is glued on. The holes in the surface of the porous wafer function like 'tear here' perforations, with the result that the mono-crystalline silicon thin film, which adheres to the glass plate, can be freed from the wafer like a peel-off film. After preparation, the wafer can be reused up to nine times. Efficiencies of up to 16.6 per cent have already been achieved in the laboratory with this method. However, the process is still years away from market.

MICROCRYSTALLINE AND MICROMORPHOUS SOLAR CELLS

The second type of low-temperature process is a 'conventional' thin-film technology with a deposition process as in amorphous silicon technology. Deposition at temperatures of between 200°C and 600°C produces silicon films with very fine-grained microcrystalline structures. The low temperatures enable the use of cheap substrates made of glass, metal or plastic. In order to be able to create layer thicknesses less than 10µm despite the poorer absorption capability of crystalline silicon, the light capturing must be optimized with light trap structures. For this purpose, the surfaces of the silicon and contact layers (TCO) are textured (see section 1.4.6 'Front contacts'). Microcrystalline cells have similar optical properties as crystalline wafer solar cells and achieve stable efficiencies of up to 8.5 per cent.

Better results can be achieved by combining amorphous silicon in tandem cells. Such tandem cells are described as micromorphous, a term derived from the words microcrystalline and amorphous. When combined, the two pin cells are able to use the solar spectrum better than they can individually because they then absorb the long-wave radiation as well. At the same time, they undergo only very slight light-induced degradation in contrast to purely amorphous cells. To date, maximum cell efficiencies of 12 per cent have been achieved. Commercial modules from Kaneka in Japan with module efficiencies of 9.1 per cent are available. These modules are often referred to as hybrid modules, but should not be confused with the HIT solar cells based on wafers that are mentioned below.

CRYSTALLINE SILICON ON GLASS (CSG) SOLAR CELLS

The crystalline silicon on glass (CSG) thin-film process was developed at the University of New South Wales in Australia. It was put into production in commercial modules in 2006 by CSG Solar AG, based in Saxony-Anhalt, Germany. The substrate glass is textured on both sides in a coating process. An anti-reflective coating of silicon nitride is deposited on one side, followed by amorphous silicon layers with 1.4µm overall thickness, and with n^+, p and p^+ doping. The silicon crystallizes in an oven at 600°C. Briefly heating to over 900°C anneals out any crystal defects. The now polycrystalline silicon layer is cut by laser into cell strips approximately 6mm wide and coated with an insulating synthetic resin. White pigments in the synthetic resin improve the reflection characteristics and the coupling of light into the silicon layer.

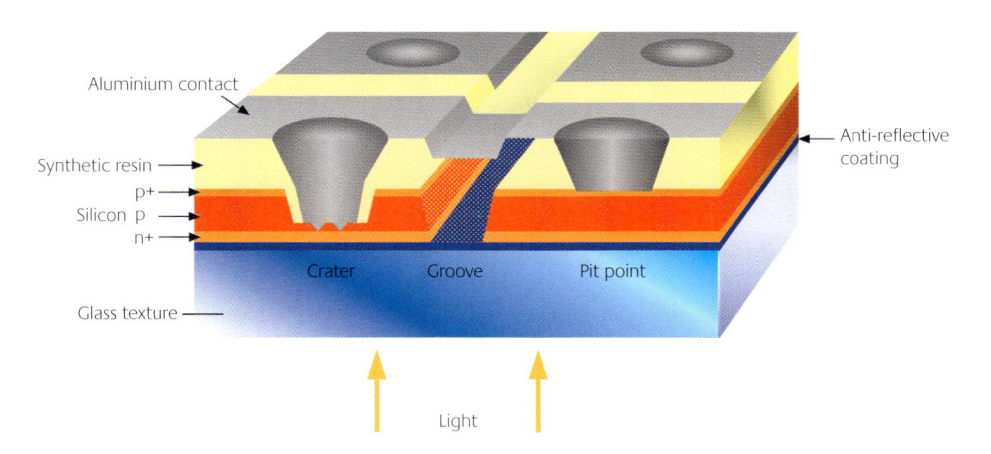

Figure 1.114
Layered structure of a CSG solar cell

The cells are contacted via crater and pit points, which are etched into the synthetic resin layer. Here an inkjet process is used to print an etching solution onto the cell at intervals. The pits to contact the p-layer are etched only briefly, while in the craters the p-layer also needs to be removed to contact with the thin n-layer. The walls of the craters are insulated with synthetic resin. The back contact is made of aluminium and is sputtered onto the entire surface, connecting the hole contacts to each other. Finally, the aluminium layer is divided into individual fields by laser so that only the desired cell regions are electrically connected to each other.

The first mass-produced modules will offer efficiency of just under 9 per cent. An increase to 12 per cent or 13 per cent is forecast for the next few years. This estimate increases for tandem cells to 16 per cent or 17 per cent, and for triple cells to 18 per cent or 19 per cent.

Figure 1.115
CSG prototype module
Source: CSG Solar AG

1.4.14 Concentrating systems

So-called III-V semiconductors, such as indium gallium arsenide (InGaAs), indium gallium phosphide (InGaPh) or germanium, which consist of elements of Group III and Group V in the periodic table, enable the production of highly efficient solar cells. In these, multiple solar cells made from different materials and optimized for different parts of the solar spectrum are stacked one above the other (known as multi-junction cells). But since these cells are extremely expensive, low-cost lenses are used to collect sunlight from a larger receiving area and concentrate it on small cells that are often only a few square millimetres in size and have efficiencies of more than 30 per cent with concentrated light. The current record efficiency is 39 per cent. By producing solar cells that comprise four or five layers, this could increase to efficiencies approaching 50 per cent in the future.

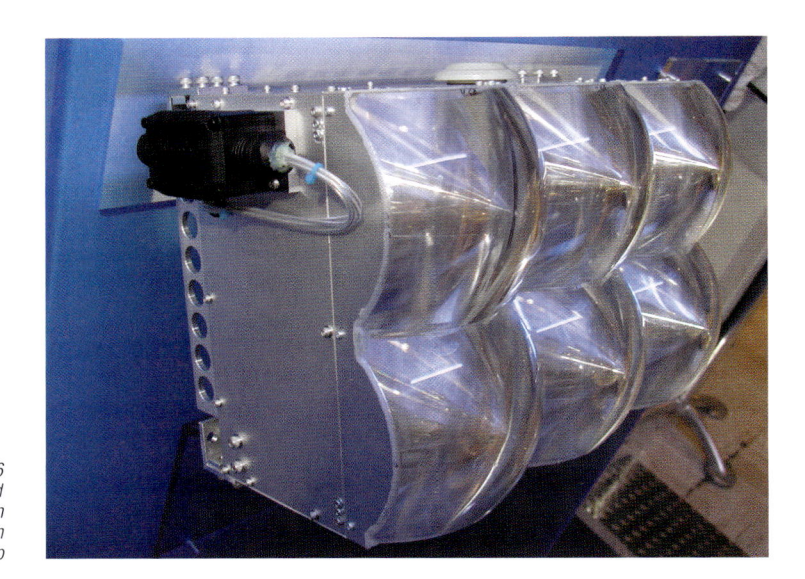

*Figure 1.116
Mini-module with anti-fog-coated Fresnel lenses, as well as ventilation opening and condensation drain
Source: Sharp*

In the modules, Fresnel lenses are generally used for optical concentration by a factor of around 500. Sufficient heat dissipation has to be ensured in the cells. Concentrator modules have to track the sun since only direct radiation can be concentrated. Hence, concentrator systems are mainly suitable for regions with a high proportion of direct radiation. Some companies have redoubled their activities with respect to this technology, announcing that the modules will be commercially available within a few years. Significant cost reduction is expected when used in Southern countries. Modules tested and measured on open ground have already achieved efficiencies of just under 27 per cent. However, these efficiencies are no longer determined under STC conditions. They are temperature compensated for 25°C; but they relate only to the direct radiation component. The aim is to develop market-ready concentrator modules that can resist high temperature conditions in the long term and maintain high efficiencies for many years.

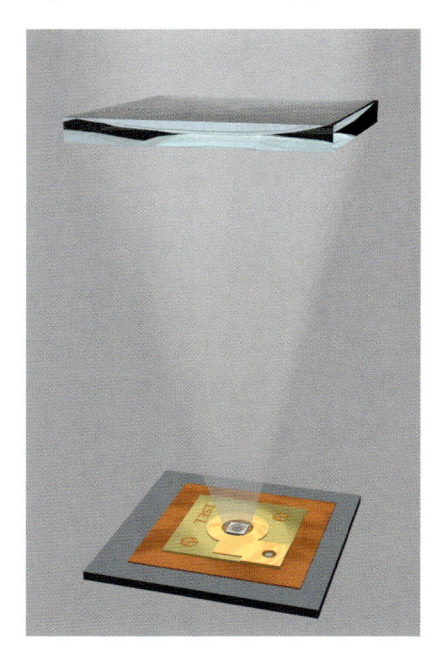

*Figure 1.117
Principle of concentration: A Fresnel lens with an area of 4cm x 4cm concentrates the light by a factor of around 500 onto a solar cell with a diameter of 2mm
Source: Concentrix Solar*

*Figure 1.118
Flatcon module prototype comprising 48 cell units in a glass box
Source: Concentrix Solar*

1.4.15 Hybrid cells: HIT solar cells

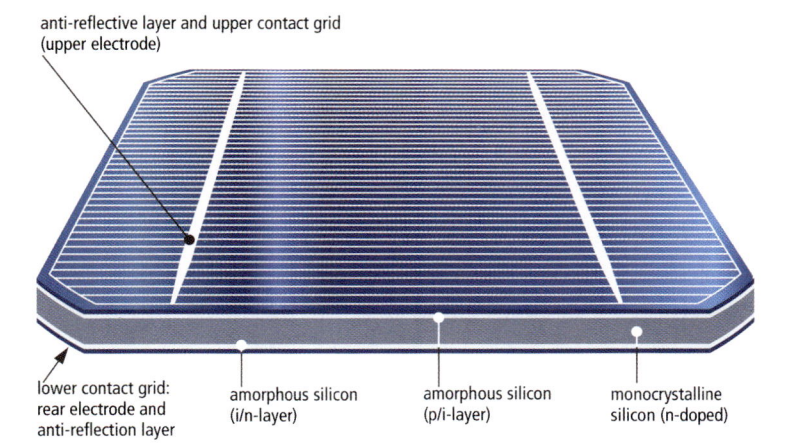

*Figure 1.119
Layered structure of the hybrid HIT solar cell
Source: Sanyo*

Labels: anti-reflective layer and upper contact grid (upper electrode); lower contact grid: rear electrode and anti-reflection layer; amorphous silicon (i/n-layer); amorphous silicon (p/i-layer); monocrystalline silicon (n-doped)

FABRICATION

The HIT solar cell is a combination of a crystalline and a thin-film solar cell. HIT (heterojunction with intrinsic thin layer) refers to the structure of these hybrid solar cells. This comprises crystalline and amorphous silicon that is bonded with an additional un-doped thin film (intrinsic thin layer). A mono-crystalline wafer forms the core of the HIT cell and is coated on both sides with a thin layer of amorphous silicon (a-Si). As intermediate layer, an ultra-thin undoped (intrinsic) i-layer made from amorphous silicon bonds the crystalline wafer with each amorphous silicon layer. A p-doped a-Si layer is deposited on the front side, which forms the p-n junction with the n-doped mono-crystalline wafer. Whereas in conventional silicon solar cells the same semiconductor material is doped differently in order to create a p-n junction, with HIT solar cells this occurs between the two structurally different semiconductors. This is known as a heterojunction. The amorphous p/i layer and the n-doped wafer create a pin structure as with amorphous thin-film cells. The back of the wafer is coated with highly n-doped amorphous silicon to prevent the free charge carriers recombining on the back electrode. On the cell surfaces, an anti-reflective coating and the wafer texture minimize reflection losses.

There is no deterioration of the efficiency as a result of light-induced degradation that is characteristic of amorphous thin-film cells. Compared to crystalline solar cells, the HIT cell is distinguished by greater energy yields at higher temperatures and utilization of a wider spectrum. Here, for each degree Celsius increase in temperature, the performance worsens by only 0.33 per cent compared with 0.45 per cent for crystalline silicon. The HIT cell saves energy and material in the cell manufacture. The required deposition temperature is just 200°C. This means that the wafers are exposed to a smaller thermal load and can be reduced in thickness by around 0.2mm. Operating results from HIT modules are now available and are worth noting. When operating a PV system with HIT modules, a 7 per cent additional annual yield was obtained in a direct comparison with polycrystalline modules:

- *Efficiency:* 18.5 per cent.
- *Form:* square (rounded corners).
- *Size:* 10.4cm × 10.4cm; 12.5cm × 12.5cm.
- *Thickness:* 0.2mm.
- *Appearance:* uniform.
- *Colour:* dark blue to almost black.

CELL MANUFACTURER

Sanyo.

1.4.16 Comparison of solar cell types and trends

For grid-connected solar systems, solar cells made from single-crystal and poly-crystal silicon are generally used. The lower efficiency of polycrystalline silicon (cast block or ribbon pulled) is balanced out by a price advantage in manufacturing costs. Modules made from amorphous silicon have thus far been used predominantly in leisure applications (small appliances, camping and boats) and in architecture (façade systems and semi-transparent glazing). Now that reservations concerning their stability and their ageing behaviour have proven unfounded in new long-term test results, amorphous modules are currently enjoying a renaissance and are becoming increasingly established in larger systems. CIS modules have the highest efficiencies among the thin-film modules that have reached the mass-production stage; they have now been used in various pilot projects. As a cost-effective thin-film alternative, CdTe modules are already in use in megawatt-class ground-mounting systems. Highly efficient thin-film solar cells made from what are known as III-V semiconductors such as gallium arsenide (GaAs) or germanium (Ge), which comprise elements from Groups III and V in the periodic table, are not competitive in terms of price and are therefore used only in space flight applications and concentrator systems, generally with additional III to V compounds such as GaInAs or GaInP. Efficiencies exceeding 30 per cent are, in principle, possible only with multi-layer cells. Hence, such Group III to V tandem and triple cells are interesting objects for research in the effort to set new world records for cell efficiency. Concentrator systems will soon be on the market for use in sunny regions of the world.

Among the organic solar cells, only dye-sensitized cells are currently market ready. These are an interesting and cost-effective alternative for the future. With their colouring and transparency they look set to create new accents, especially in building integration. In Australia, the first modules have been commercially produced in a small production run. At the top end of module efficiencies among commercially available modules, HIT hybrid modules have been pushed out of first place by mono-crystalline modules using SunPower high-performance cells. Modules with even higher efficiencies will soon be available here. The maximum values of the efficiency of solar cells and modules are summarized in Table 1.3. The average values for the modules available in the market will be lower.

Table 1.3
Maximum efficiencies in photovoltaics
Notes: a In a stabilized state.
b Measured with concentrated irradiance.
c Small production run.
Source: Fraunhofer ISE, University of Stuttgart, Quaschning, Photon 2/2000

Solar cell material	Cell efficiency η_z (laboratory) (%)	Cell efficiency η_z (production) (%)	Module efficiency η_M (series production) (%)
Monocrystalline silicon	24.7	21.5	16.9
Polycrystalline silicon	20.3	16.5	14.2
Ribbon silicon	19.7	14	13.1
Crystalline thin-film silicon	19.2	9.5	7.9
Amorphous silicon[a]	13.0	10.5	7.5
Micromorphous silicon[a]	12.0	10.7	9.1
CIS	19.5	14.0	11.0
Cadmium telluride	16.5	10.0	9.0
III-V semiconductor	39.0[b]	27.4	27.0
Dye-sensitized call	12.0	7.0	5.0[c]
Hybrid HIT solar cell	21	18.5	16.8

High efficiencies are a decisive criterion when space is limited and the aim is to install the greatest power output capacity per unit of area. Figure 1.120 shows a system with modules of the most widely used cell technologies. The individual arrays with the different modules each have the same power output capacity of approximately 1kWp.

Figure 1.120
PV system with modules using the various solar cell technologies at the University of Applied Sciences in Weihenstephan, Germany: (left to right) Polycrystalline, mono-crystalline, CIS, amorphous, CdTe; each approximately 1kWp
Source: Soltec

For the foreseeable future, mono-crystalline and polycrystalline silicon technology will dominate the market. Continuous progress has been made during recent years in implementing higher efficiencies in mass production. In addition, work is being carried out on producing larger and thinner silicon wafers. Theoretically, a wafer thickness of 50μm would be sufficient. However, wafers that are this thin are difficult to handle because they are so brittle; hence, they cannot be cut or printed with contacts. As a result, substantially thinner wafers are possible only if the silicon blocks are cut with lasers instead of wire saws in the future and new contacting technologies, such as lasered contact points, are implemented industrially. All processes in wafer, cell and module production need to be improved and adapted to the new wafer and cell thicknesses. As well as promising cost-reduction potentials, thinner wafers may also alleviate the supply shortages in solar silicon.

These shortages have opened up new market opportunities for thin-film modules. Significantly greater sales figures and better standardization of the production processes are needed to tap the long-heralded cost reduction potentials, which are only attainable with genuine mass production. Because of their lower efficiencies, thin-film modules will only be competitive through significantly lower prices and higher yields. At the moment, the higher planning and installation costs largely eat up any price advantages. It is now up to the newly installed sizeable systems to confirm the promising yields from some test and comparison systems based on the better poor-lighting and temperature characteristics.

Amorphous silicon modules dominate the thin-film market. Here, modules with greater power output (up to 100W) are now available that have already been used in large-scale systems and, owing to their better temperature characteristics, are of interest for substantial projects in Southern countries. With micromorphous technology, a new version is on the scene with higher and more stable efficiencies. As well as high efficiencies and modules with greater individual power output, with some thin-film technologies even thinner films using toxic materials and/or raw materials of limited availability are on the development agenda.

1.5 Electrical properties of solar cells

1.5.1 Equivalent circuit diagrams of solar cells

A solar cell consisting of p-doped and n-doped silicon material is, in principle, a large-scale silicon diode. Both have similar electrical properties. As an example, the characteristic curve of the BAY 45 silicon diode is shown in Figure 1.121. If a positive potential is present at the p-doped anode and a negative potential is present at the n-doped cathode, the diode is connected in forward-biased direction. The characteristic curve in the first quadrant applies. Starting from a particular voltage (the threshold voltage here is 0.7V), current flows. If the diode is connected in reverse-biased direction, current flow is prevented in this direction. The characteristic curve in the third quadrant applies. Only starting from a high breakdown voltage (here, 150V) does the diode become conductive. This can also lead to the destruction of the diode.

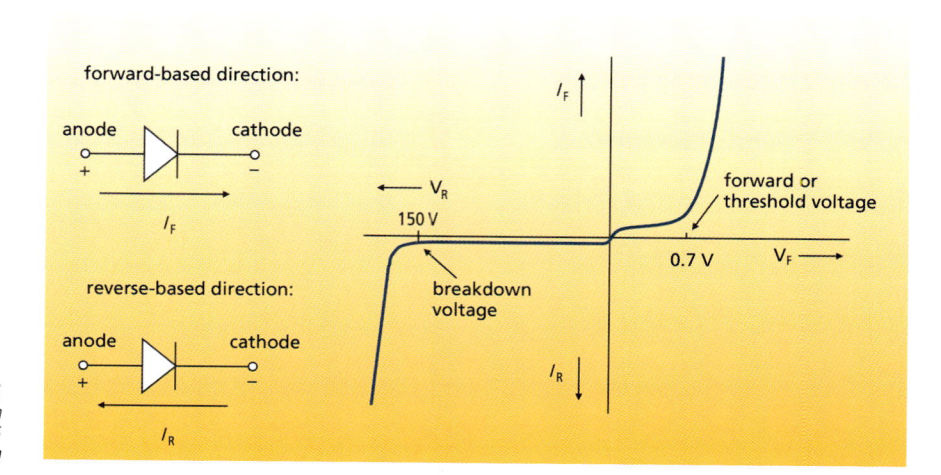

Figure 1.121
Current voltage curve for silicon diode BAY 45
Source: R. Haselhuhn

Table 1.4
The variables represented in an equivalent circuit diagram
Source: A. Wagner, R. Haselhuhn

Parameter	Formula sign	Unit
Voltages:		
Solar cell terminal voltage	V	V
Diode voltage	V_D	V
Temperature voltage	V_T	V
Currents:		
Solar cell terminal current	I	A
Diode current	I_D	A
Saturation current in diode reverse-biased direction	I_0	A
Photocurrent	I_{Ph}	A
Current through the parallel resistor	I_P	A
Diode factor	m	–
Coefficient of photocurrent	c_0	m²/V
Solar irradiance of cell	G	W/m²
Parallel resistance	R_P	Ω
Series resistance	R_S	Ω

The simplified equivalent circuit diagrams of the solar cells are considered in greater depth below.

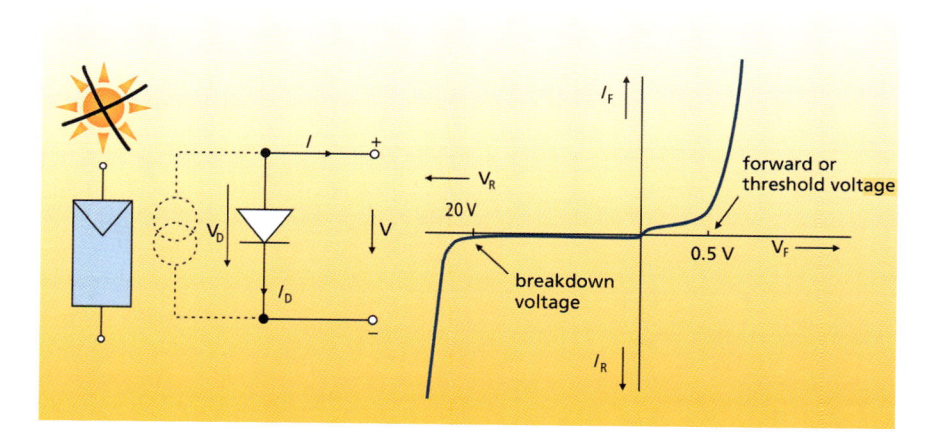

Figure 1.122
Dark equivalent circuit diagram and characteristic curve
Source: R. Haselhuhn

$$V = V_D$$

$$I = -I_D = -I_0 \times (e^{V/m \times VT} - 1)$$

An un-illuminated solar cell is described in the equivalent circuit diagram by a diode. Accordingly, the characteristic curve of a diode is also applicable. For a monocrystalline solar cell, one can assume a forward voltage of approximately 0.5V and a breakdown voltage of 12V to 50V (depending upon the quality and cell material).

PHOTOVOLTAIC BASICS

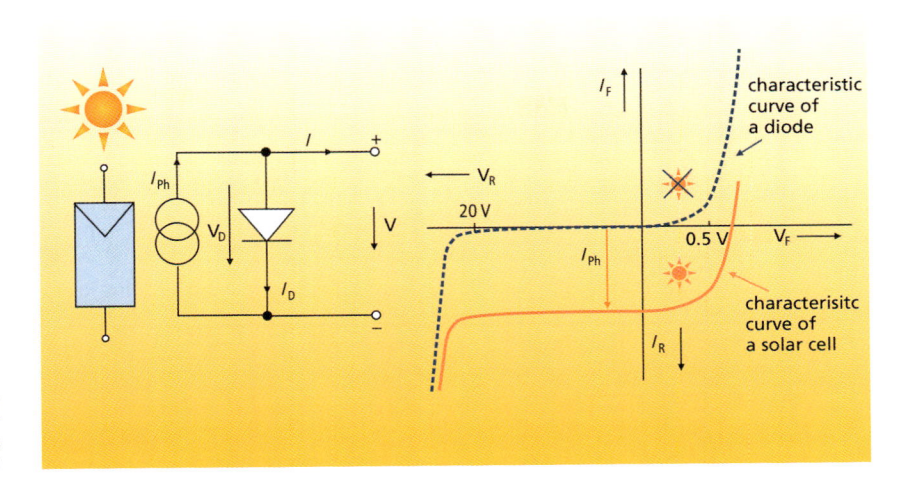

Figure 1.123
Illuminated equivalent circuit diagram and characteristic curve
Source: R. Haselhuhn

$V = V_D$

$I_{Ph} = c_0 \, x \, E$

$I = I_{Ph} - I_D$

When light hits the solar cell, the energy of the photons generates free charge carriers. An illuminated solar cell constitutes a parallel circuit of a power source and a diode. The power source produces the photoelectric current (photocurrent) I_{Ph}. The level of this current depends upon the irradiance. The diode characteristic curve is shifted by the magnitude of the photocurrent in the reverse-biased direction (into the fourth quadrant in Figure 1.124).

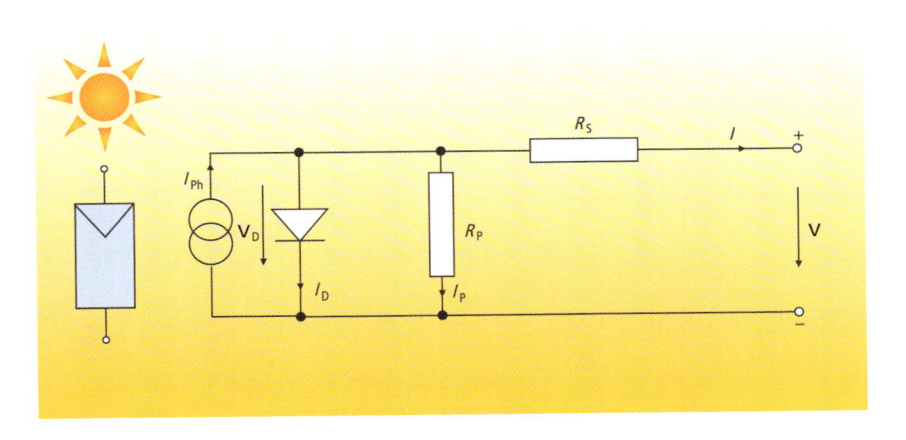

Figure 1.124
Extended equivalent circuit diagram
Source: R. Haselhuhn

$I = I_{Ph} - I_D - I_P$

$I_p = V_D/R_p = V + R_s x I/R_p$

This extended equivalent circuit diagram is termed a single-diode model of a solar cell and is used as a standard model in photovoltaics. In the solar cell, a voltage drop occurs as the charge carriers migrate from the semiconductor to the electrical contacts. This is described by the series resistor Rs, which is in the range of a few milliohms. In addition, what are known as leakage currents arise, which are described by the parallel resistor ($R_p >> 10\Omega$). Both resistors bring about a flattening of the solar cell characteristic curve. With the series resistor, it is possible to calculate current/voltage characteristic curves of solar cells at different irradiances and temperatures in accordance with the procedure of the DIN EN 60891/IEC 60891 standards.

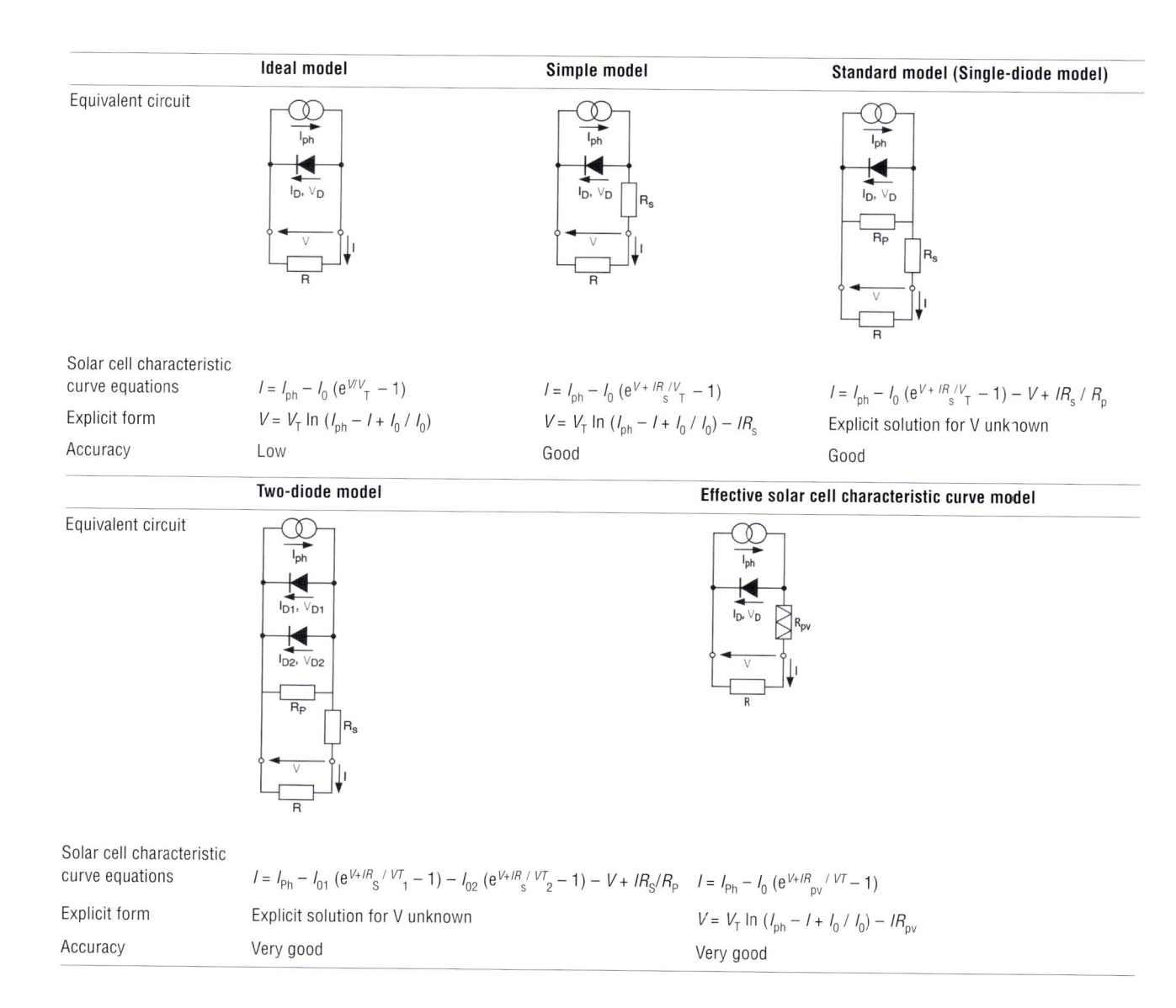

	Ideal model	Simple model	Standard model (Single-diode model)
Equivalent circuit			
Solar cell characteristic curve equations	$I = I_{ph} - I_0 (e^{V/V_T} - 1)$	$I = I_{ph} - I_0 (e^{V+IR_S/V_T} - 1)$	$I = I_{ph} - I_0 (e^{V+IR_S/V_T} - 1) - V + IR_S / R_p$
Explicit form	$V = V_T \ln(I_{ph} - I + I_0 / I_0)$	$V = V_T \ln(I_{ph} - I + I_0 / I_0) - IR_S$	Explicit solution for V unknown
Accuracy	Low	Good	Good

	Two-diode model	Effective solar cell characteristic curve model
Equivalent circuit		
Solar cell characteristic curve equations	$I = I_{Ph} - I_{01}(e^{V+IR_S/VT_1} - 1) - I_{02}(e^{V+IR_S/VT_2} - 1) - V + IR_S/R_p$	$I = I_{Ph} - I_0 (e^{V+IR_{pv}/VT} - 1)$
Explicit form	Explicit solution for V unknown	$V = V_T \ln(I_{ph} - I + I_0 / I_0) - IR_{pv}$
Accuracy	Very good	Very good

Table 1.5
Equivalent circuit diagrams for solar cells and their characteristic curve equations

1.5.2 Cell parameters and solar cell characteristic I–V curves

In the technical literature, frequently it is only the part of the current and voltage curve in which the solar cell produces current that is shown (fourth quadrant of the light characteristic curve in Figure 1.125). At the same time, the light characteristic curve is mirrored in the voltage axis. This part of the characteristic curve is then termed the solar cell characteristic curve.

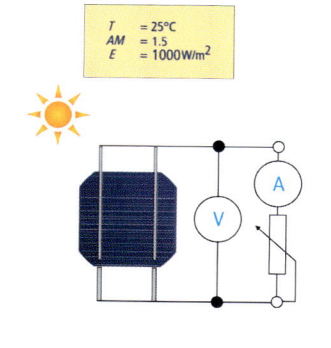

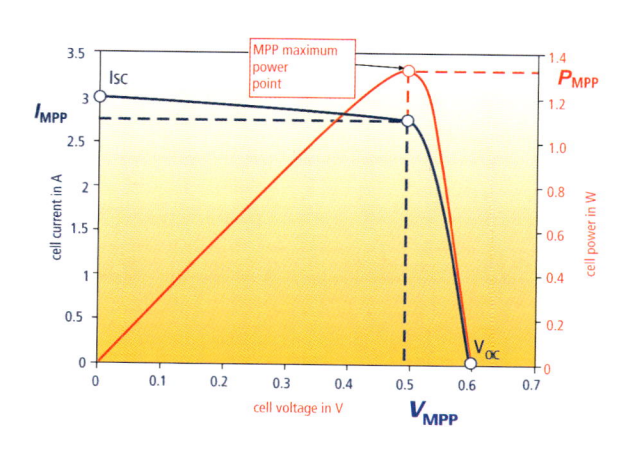

Figure 1.125
Current/voltage characteristic curve (I–V curve) for a crystalline silicon solar cell
Source: R. Haselhuhn

If light falls on an unloaded solar cell, a voltage of approx. 0.6V builds up. This can be measured as the open-circuit voltage V_{OC} at the two contacts. If the two contacts are short circuited via an ammeter, the short-circuit current I_{SC} can be calculated. In order to record a complete solar cell characteristic I–V curve, one requires a variable resistor (shunt), a voltmeter and an ammeter.

STANDARD TEST CONDITIONS (STC)

In order to be able to compare different cells or, indeed, PV modules with one another, uniform conditions are specified for determining the electrical data with which the solar cell characteristic I–V curve is then calculated. These standard test conditions, as they are known, relate to the IEC 60904/DIN EN 60904 standards:

1. vertical irradiance E of 1000 W/m^2;
2. cell temperature T of 25°C with a tolerance of ± 2°C;
3. defined light spectrum (spectral distribution of the solar reference irradiance according to IEC 60904-3) with an air mass $AM = 1.5$.

Basically, the I–V curve is characterized by the following three points:

1. The maximum power point (MPP) value is the point on the I–V curve at which the solar cell works with maximum power. For this point, the power P_{MPP}, the current I_{MPP} and voltage V_{MPP} are specified. This MPP power is given in units of peak watts (Wp).
2. The short-circuit current ISC is approximately 5 per cent to 15 per cent higher than the MPP current. With crystalline standard cells (10cm × 10cm) under STC, the short-circuit current I_{SC} is around the 3A mark.
3. The open-circuit voltage V_{OC} registers, with crystalline cells, approximately 0.5V to 0.6V, and for amorphous cells is approximately 0.6V to 0.9V.

The cell parameters and characteristic I–V curves of thin-film cells deviate from those of crystalline silicon cells – in some cases, very strongly. In amorphous cells, the MPP point is at 0.4V and, overall, the I–V curve is much flatter (see Figure 1.126). Owing to the lower efficiency, a lower current flows. To achieve the same power output as crystalline cells, a larger cell surface area is required. The less clearly marked MPP makes higher demands on the control technology in the inverters and MPP controllers.

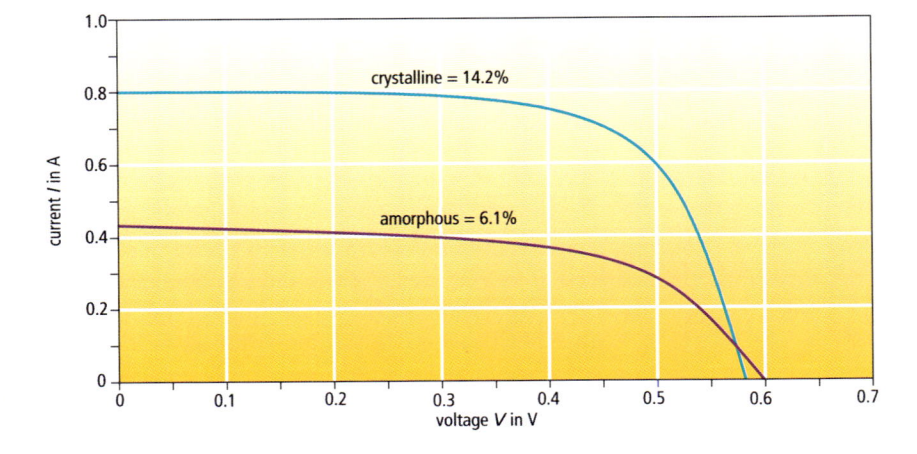

Figure 1.126
Comparison of current/voltage characteristic curves (I–V curves) of crystalline and amorphous silicon solar cells with an irradiance of 1000W/m^2 on 5cm × 5cm cell surface area and a temperature of 28°C
Source: R. Haselhuhn

The short-circuit current is linearly dependent upon the irradiance (i.e. if the irradiance doubles, the current also doubles). This is why a straight line results in the graph depicted in Figure 1.129. The open-circuit voltage V_{OC} stays relatively constant as the irradiance changes. Only when the irradiance falls below 100W/m^2 does the voltage break down. Mathematically, there is a logarithmic dependency between voltage and irradiance in crystalline solar cells.

ADDITIONAL SOLAR CELL MODELS

As well as the solar cell models introduced so far, still other models are used. Table 1.5 gives an overview of the most commonly used models with their equivalent circuit diagrams, the associated current and voltage equations and an assessment of the accuracy. To complete the equivalent circuit diagrams, a general load resistor R was added. The aim of all equivalent circuit diagrams and solar cell models is to describe the solar characteristic curve mathematically in sufficient quality. They help with theoretical understanding and form the basis for measurement and control devices in photovoltaics (e.g. maximum power point (MPP) controllers) or for simulation programmes (see Chapter 5). They make it possible to determine the points of maximum power under the changing operating conditions and in this way set the optimum operating point of the PV system. The starting point for this is to determine the gradient M of the characteristic curve (see Figure 1.127):

$$M = \frac{dV}{dI} = tan\varphi = \frac{\Delta V}{\Delta I}$$

The MPP point is found on the current/voltage characteristic curve at the point where the gradient M is 1 – hence, the angle of the gradient ω is 45°. Mathematically, the second derivation of the current/voltage function according to the voltage results in the power/voltage function. At the MPP point, the power is at its maximum. As a result, the gradient of the power/voltage characteristic curve equals 0 and the angle of gradient is also 0° (see also Figure 1.123, red curve).

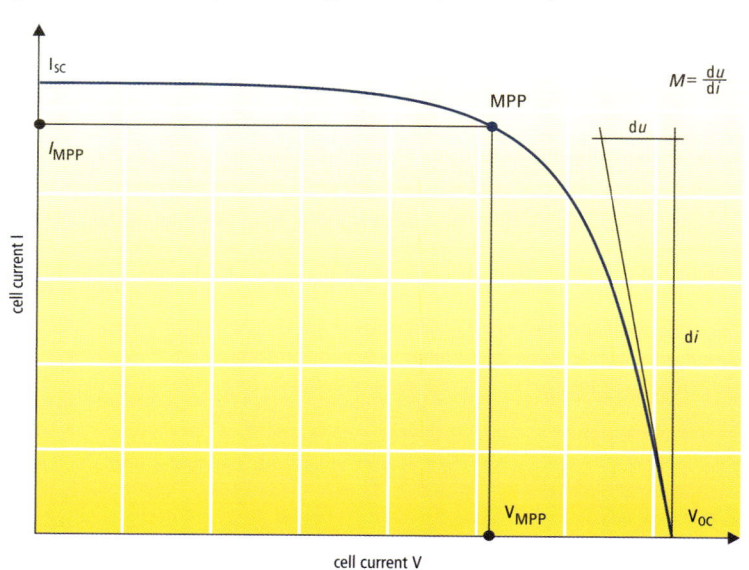

Figure 1.127
Gradient of the current/voltage characteristic curve of a solar cell
Source: R. Haselhuhn

The standard model is insufficiently accurate for various application fields. If a higher accuracy is required, the two-diode model or the effective solar cell model is used. In order to calculate with the two-diode model, six solar cell parameters must be known. An explicit (one-to-one) solution for the voltage in the standard model and in the two-diode model cannot, however, be calculated (Wagner, 2001).

THE EFFECTIVE SOLAR CELL MODEL

The effective solar cell model requires only four cell parameters to solve the current and voltage equations. This reduces the work for the calculation, but also for obtaining information on suitable module parameters. The special feature of using the effective solar cell model is that both resistors R_S and R_P of the standard model are combined into a fictive photovoltaic resistor R_{PV}. This photovoltaic resistor can take both positive and negative values. It is therefore not an ohmic resistor.

The four required cell parameters (R_{PV}, V_T, I0 and I_{ph}) can be calculated as follows from the gradient M and from the cell parameters of open-circuit voltage VOC, short-circuit current I_{SC}, MPP voltage V_{MPP} and MPP current I_{MPP}:

$$R_{pv} = -M\frac{I_{SC}}{I_{MPP}} + \frac{V_{MPP}}{I_{MPP}}\left(1 - \frac{I_{SC}}{I_{MPP}}\right)$$

$$V_T = -(M + R_{pv})I_{SC}$$

$$-\frac{V_{OC}}{V_T}$$

$$I_0 = I_{SC}\,e$$

$$I_{ph} = I_{SC}$$

The gradient M is required for the calculation. It is a function of the following parameters:

$$M = f(V_{OC}, I_{SC}, V_{MPP}, I_{MPP})$$

The following approximations to the characteristic curve can be derived with an accuracy of 1 per cent:

$$M = \frac{V_{OC}}{I_{SC}} + k_1\frac{I_{MPP}V_{MPP}}{I_K V_K} + k_2\frac{V_{MPP}}{V_{OC}} + k_3\frac{I_{MPP}}{I_{SC}} + k_4$$

with the equation constants:

$$k_1 = -5.411 \quad k_2 = 6.450 \quad k_3 = 3.417 \quad k_4 = -4.422$$

The equation constants were calculated using a numerical mathematical method (method of the least square (smallest squared) error). The cell and module parameters required for the calculation (V_{OC}, I_{SC}, V_{MPP} and I_{MPP}) can be gathered from the manufacturers' data sheets. From the gradient M, the four cell parameters named above are calculated. Using the equations for voltage and current from Table 1.5, all points on the solar characteristic curve can be calculated with good accuracy.

The effective solar cell model is the basis for the peak performance measuring device shown in Figure 1.128. This metering unit can calculate the nominal power (MPP power under standard test conditions) of PV modules under normal operating conditions. The accuracy of an on-site measurement of nominal power using this instrument on a PV module is specified at ± 5 per cent (Wagner, 1999).

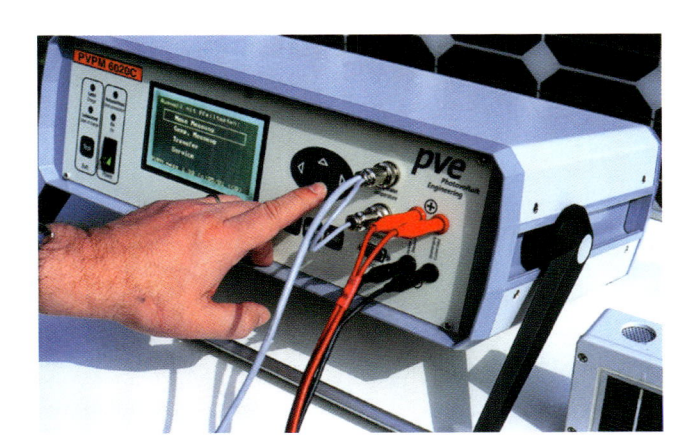

Figure 1.128
Peak power meter
Source: PV-Engineering

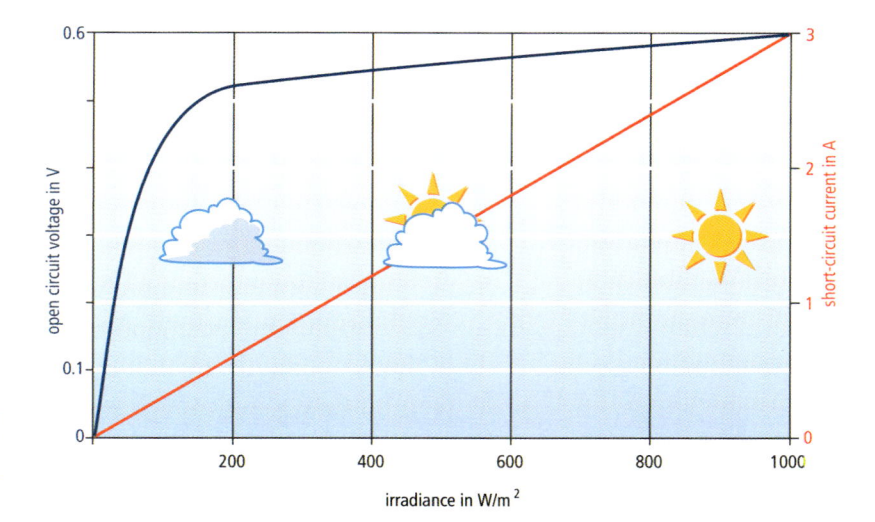

Figure 1.129
Open-circuit voltage VOC and short-circuit current ISC depending upon irradiance
Source: R. Haselhuhn

The filling factor describes the quality of solar cells. It is defined as a quotient of MPP and the theoretical maximum power that results as the product of short-circuit current I_{SC} and open-circuit voltage V_{OC}:

$$FF = \frac{V_{MPP} \times I_{MPP}}{V_{OC} \times I_{SC}} = \frac{P_{MPP}}{V_{OC} \times I_{SC}}$$

For crystalline solar cells, the filling factor is around 0.75 to 0.85, and for amorphous solar cells, around 0.5 to 0.7. As a graph, the filling factor can be determined as the relationship of area B to area A (see Figure 1.130).

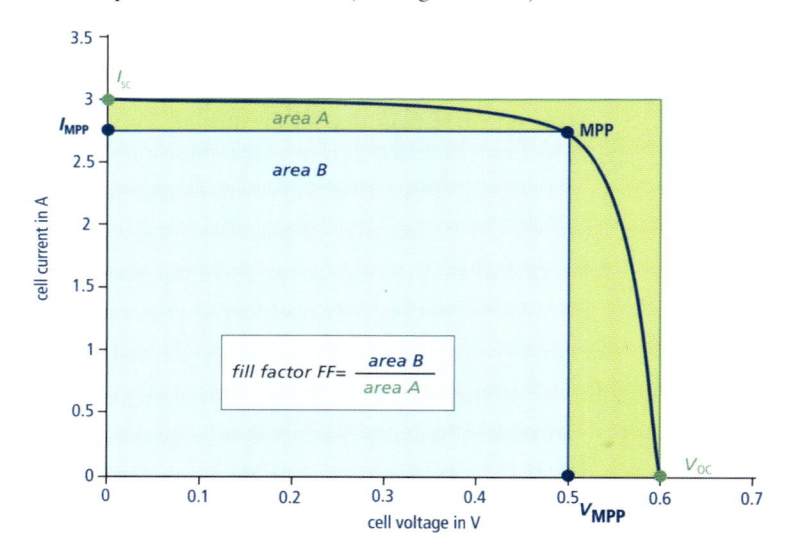

Figure 1.130
Filling factor of solar cells
Source: R. Haselhuhn

The most important solar cell parameters are listed in Table 1.5.
Since a PV module consists of solar cells connected together, the information in this chapter also applies to the following chapters in this volume.

1.5.3 Spectral sensitivity

Depending upon the materials and the technology used, solar cells are better or worse at converting the different colour bands of sunlight into electricity. The spectral sensitivity describes the wavelength range in which a cell works most efficiently and influences the efficiency under different irradiance conditions. Sunlight has the greatest energy in the visible light range between 400nm and 800nm.

While crystalline solar cells are particularly sensitive to long wavelength solar radiation, thin-film cells utilize the visible light better. Amorphous silicon cells can absorb short wavelength light optimally. In contrast, CdTe and CIS are better at absorbing medium wavelength light. A-300 mono-crystalline high-performance cells from SunPower utilize a very broad spectrum. Because their front doping is only slight, the exploitation of short wavelength radiation is increased, while the back passivation with silicon oxide helps to absorb the long wavelength range.

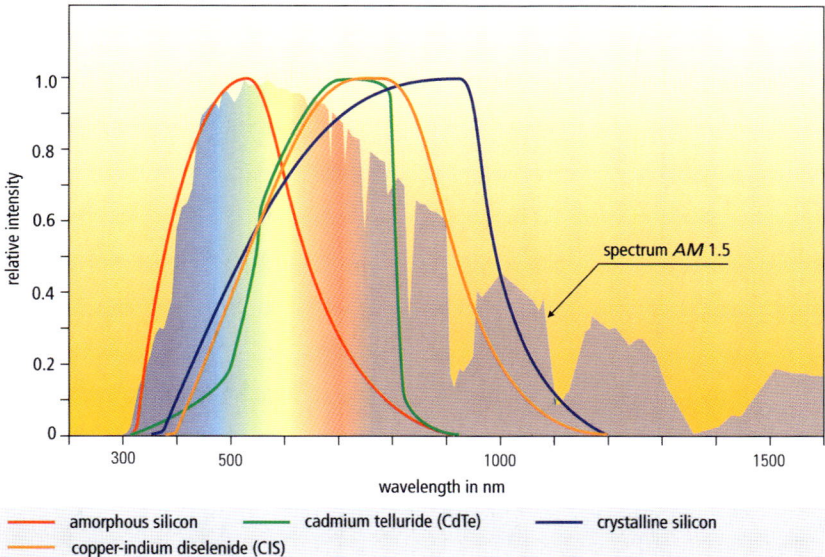

*Figure 1.131
Spectral sensitivity of different
solar cell types
Source: ISET Kassel; Mulligan, 2004*

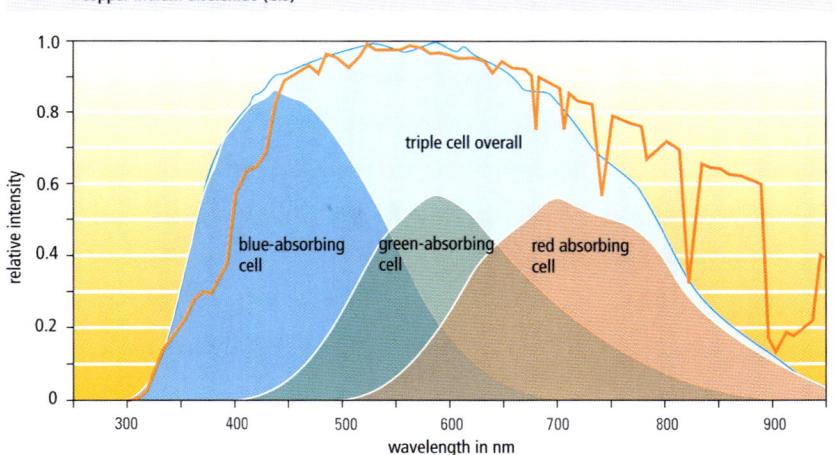

*Figure 1.132
Spectral sensitivity of an amorphous
triple solar cell and its individual
stacked cells
Source: Uni-Solar*

In stack cells, which are common mainly in amorphous thin-film technology, the individual cells arranged one on top of the other are optimized for different wavelength ranges (see Figure 1.132). Figure 1.133 shows the layered structure of a triple solar cell. Here, the top cell absorbs the blue light and allows the other components of the light to pass through. The green/yellow light is utilized by the middle cell; finally, the lower cell converts the red light. This division into different spectral zones enables the triple cell to achieve the greatest efficiency among the amorphous cells and, in addition, better utilizes low irradiance (see the section on '2.1.10 Electrical characteristics of thin-film modules' in Chapter 2).

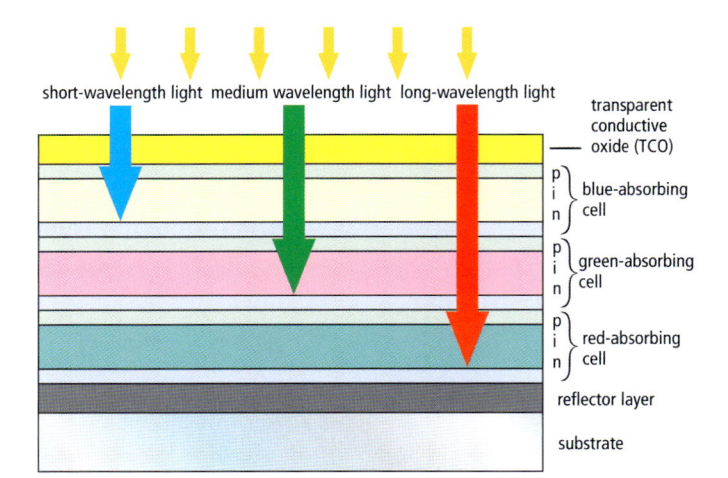

*Figure 1.133
Layered structure of a triple cell:
The three cell parts are sensitive
to different spectral ranges
Source: Unisolar, DGS LV Berlin BRB*

1.5.4 Efficiency of solar cells and PV modules

The efficiency η of solar cells is the result of the relationship between the power delivered by the solar cell and the power irradiated by the sun. Hence, it is calculated from the MPP P_{MPP}, the solar irradiance E and the area A of the solar cell as follows:

$$\eta = \frac{P_{MPP}}{A \cdot E} = \frac{FF \cdot V_{OC} \cdot I_{SC}}{E \cdot A}$$

In PV modules, the module surface area is used for A. On the data sheets, the efficiency is always specified under standard test conditions (STC):

$$\eta_n = \eta_{STC}$$

This yields the nominal efficiency of solar cells and modules:

$$\eta_n = \frac{P_{MPP(STC)}}{A \times 1000W/m^2}$$

The efficiency of solar cells depends upon irradiance and temperature. The efficiency at a particular irradiance or temperature is the result of the nominal efficiency minus the change in efficiency.

$$\eta = \eta_n - \Delta\eta$$

With the radiation factor s, the change in efficiency with irradiances deviating from STC can be calculated:

$$s = \frac{E}{1000W/m^2}$$

For example, $s = 0.5$ means the radiation factor is at half STC irradiance and, hence, irradiance is at 500W/m². The approximate change in efficiency with crystalline silicon cells results with constant temperature as follows:

$$\Delta\eta \approx -0.04 \times \eta_n \times \ln s$$

For example, with $s = 0.5$ and a solar cell efficiency under STC of 15.4 per cent, we get an efficiency 0.4 per cent lower than under STC. The efficiency with irradiance of 500W/m² in this case is 15.5 per cent. In contrast to this, amorphous triple cells at low irradiances achieve approximately 30 per cent greater efficiency than under STC (see Figure 2.76 in Chapter 2).

Moreover, the efficiency of crystalline solar cells falls with increasing temperature. Crystalline solar cells therefore reach their greatest efficiency at low temperatures. The temperature coefficients are material dependent. For the power temperature coefficient, a value of approximately –0.45 per cent per °C can be applied for crystalline silicon (see the section on '2.1.8 Irradiance dependence and temperature characteristics' in Chapter 2). The change in efficiency with constant irradiance is calculated by:

$$\Delta\eta \approx -0.045\% \times (25°C - T_{mod}) \times \eta_n$$

The temperature factor is, in addition, dependent upon the irradiance. With low irradiance, the power reduction as a result of temperature is not so high with crystalline cells. At 100W/m², it is still only –0.15 per cent. For amorphous cells, the power temperature coefficient actually rises at low irradiance (thus, for example, for amorphous cells up to +1.4 per cent per °C; see Figure 2.80 in Chapter 2).

SOLARWATT
Solar Technology which Convinces.

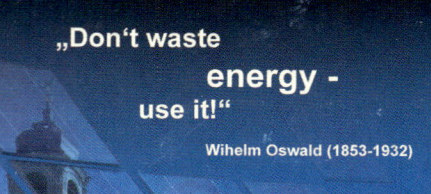

„Don't waste energy - use it!"

Wihelm Oswald (1853-1932)

SOLARWATT is a manufacturer of high-quality state-of-the-art solar modules according to certified quality and management systems. **SOLARWATT** is one of Germany's leading solar module manufacturers of mono-crystalline and polycrystalline silicon solar cells. The current production capacity amounts to 100 MWp.

SOLARWATT glass-foil solar modules combine the usual features of SOLARWATT products, such as high-quality, large energy yields and a very low power tolerance with easy installation. As framed standard modules glass-foil modules are mass produced. They are electronically and mechanically matched to all components of the solar plant.

SOLARWATT glass-glass solar modules display an exceptionally high level of mechanical stability. With 18% optical transparency, they are particulary suitable for use as warm facade or shading elements. The dimensions and level of optical transparency of the solar modules can be customised for individual solutions.

www.solarwatt.de

2 PV Modules and Other Components of Grid-Connected Systems

2.1 PV modules

2.1.1 Cell stringing

When manufacturing PV modules, the low individual power output of solar crystalline cells means that the solar cells are first connected in series (cell stringing). The front contacts of each cell are soldered with the back contacts of the next cell. Some manufacturers already use lead-free solder or lead-free connections. By connecting the negative pole (front) of each cell with the positive pole (back) of the following cell, the cells are connected in series. The individual cells are spaced several millimetres apart. Back-contacted solar cells are easy to connect together since strings are formed entirely on the back of the cells. This allows the cell spacing to be much smaller.

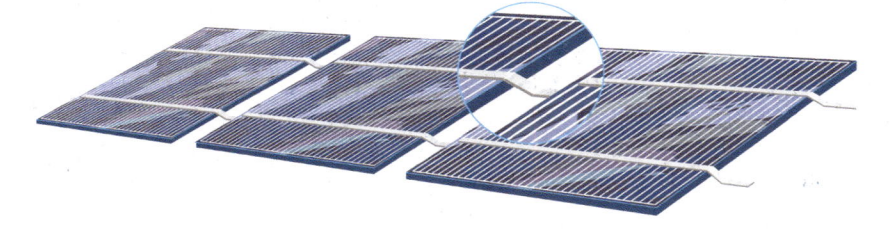

Figure 2.1
External series connection of crystalline solar cells
Source: Solarpraxis/DGS Berlin BRB

Figure 2.2
Automated cell stringing
Source: Scheuten Solar

Figure 2.3
Soldering the strings
Source: Solarwatt

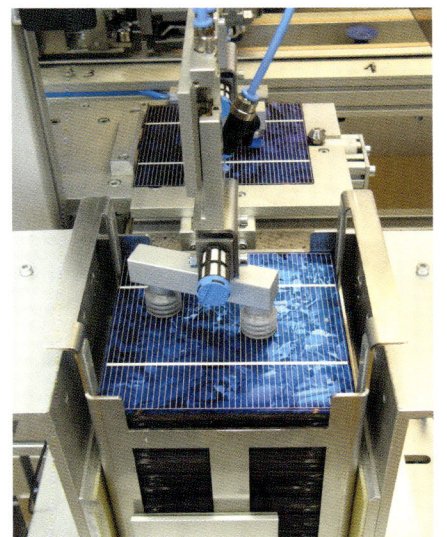

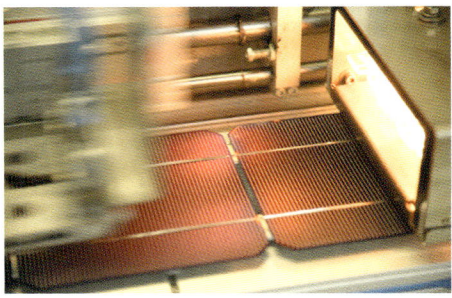

Figure 2.4
Detail from a stringing machine
Source: Solarwatt

Cell stringing machines are used in series production. The only time cells are still sometimes soldered by hand is in custom-made products. In most standard modules, strings of 36 or 72 cells are connected in series. The beginning and end of each string or up to three parallel strings are extended outwards for the later electrical connection.

INTEGRATED SERIES INTERCONNECTION WITH THIN-FILM CELLS

Crystalline cells are interconnected with one another in a special production stage, cell for cell. In contrast, the electrical connection of thin-film cells is an integral part of the cell fabrication and is achieved by cutting grooves in the individual layers. Here, the materials are cut into 0.5cm to 2cm wide cell strips, either by laser or mechanical scribing.

Figure 2.5 shows the fabrication sequence when the cells are deposited on the front side. The support material is then called the superstrate. This occurs with CdTe modules and most amorphous silicon modules. First of all, the transparent conductive oxide (TCO) layer is separated into parallel strips onto which the solar cell layer is deposited. The second cut is then made slightly offset but parallel to the first line. Finally, the back contact is applied and structured in a third stage. This creates an electrical connection from the back contact of one cell to the front contact of the next cell, with the individual cell strips being interconnected in series.

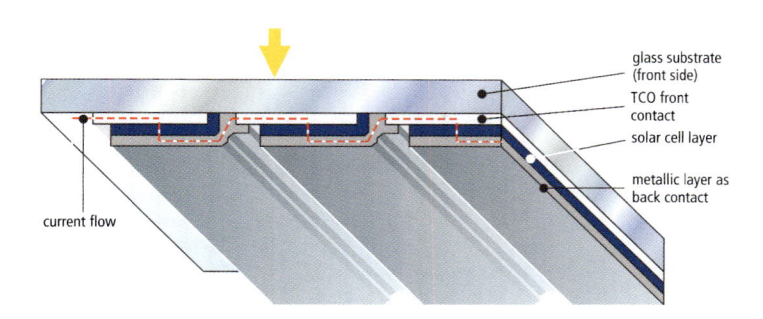

Figure 2.5
Integrated series connection of thin-film cells made from amorphous silicon or CdTe
Source: Solarpraxis/DGS Berlin BRB

CIS cells and amorphous cells based on flexible films are deposited in the reverse sequence, beginning with the back contact. The structuring occurs analogously.

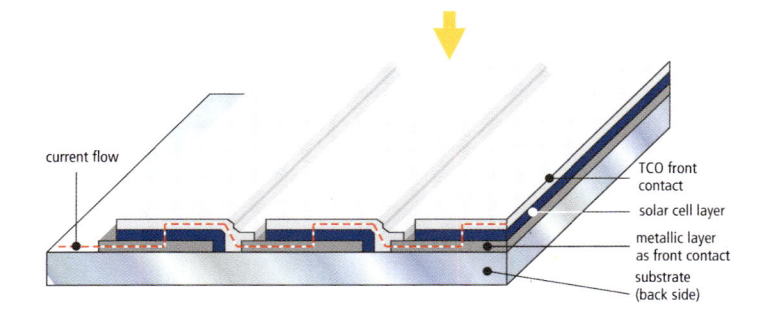

Figure 2.6
Integrated series connection of thin-film cells made from CIS or amorphous silicon
Source: Solarpraxis/DGS Berlin BRB

2.1.2 Cell encapsulation

To protect the cells against mechanical stress, weathering and humidity, the cell strings are embedded in a transparent bonding material that also isolates the cells electrically. For structural stabilization, the bonding system is applied to a substrate. In most cases, glass is used; but it is also possible to use acrylic plastic or Makrolon, metal or plastic sheeting. Depending upon the process, the solar cells can lie on, behind or between the substrate material. It is important that the covering on the light-sensitive side of the solar cell is made of highly transparent material to enable as much incident solar energy as possible to hit the solar cell. For this reason, white glass low in iron oxide is generally used as the front substrate, which allows up to 92 per cent of the light to penetrate. Because of its low iron content, white glass reflects less light and does not have the distinct green tint of conventional glass. The glass is pre-stressed to enable it to withstand the high thermal loading. This pre-stressed so-called white glass is also known as solar glass. The standard thickness of solar glass is 4mm. For larger modules and/or special constructional requirements, sheets up to 10mm thick are used. The remaining 8 per cent losses are caused mainly by reflection. Because of this, glass manufacturers have developed various types of anti-reflective glass for PV modules that let even more light into the solar cells. Here, it is only useful to treat the front side of the front glass since the similar refraction index of ethylene vinyl acetate (EVA) and glass means that there are no reflection losses on the rear of the pane. Four different kinds of anti-reflective glass for PV modules are available. These solar glasses receive a porous coating of silicon dioxide (SolGel) – for example, in an immersion bath. Other anti-reflective coatings consist of multiple sputtered layers of silicon dioxide and silicon nitride (PV-lite) or are formed in an etching process as a porous upper layer on the glass (Sunarc). The fourth variant is cast glasses with a pyramid-shaped, grooved or finely textured surface (Albarino ornamental glass). Anti-reflective glasses increase light transmission by up to 3 per cent. This results in an increase in module STC performance (light with a vertical angle of incidence) of 2 per cent to 3 per cent, and in the annual yields (depending upon the location) by 3 per cent to 5 per cent since greater performance increases are achieved when the sunlight hits the glass at an angle. To date, only a few module manufacturers use the new anti-reflective glasses because no long-term data are yet available with regard to the durability of the coatings and their greater susceptibility to soiling.

Four different types of encapsulation are used for embedding the solar cells: EVA encapsulation, polyvinyl butyral (PVB) encapsulation, Teflon encapsulation and resin encapsulation.

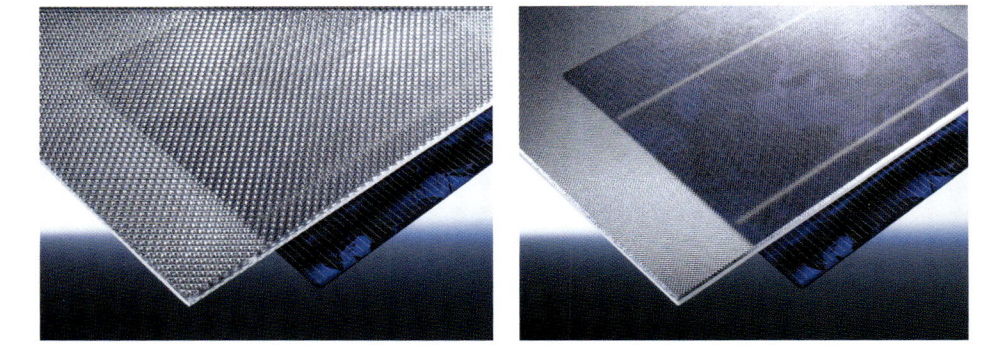

Figure 2.7
Anti-reflective glass: Ornamental glass Albarino P (left) with deep pyramidal surface texture and structured ornamental glass Albarino S (right)
Source: Saint-Gobain Glass

ENCAPSULATION WITH ETHYLENE VINYL ACETATE (EVA)

When encapsulating with EVA, the solar cells are laminated together in a vacuum chamber by applying negative and positive pressure at temperatures of up to 150°C (vacuum lamination process). The EVA melts during this process and surrounds the solar cells on all sides. The EVA requires UV-resistant weatherproofing on the front side. In most cases, this is a hardened glass sheet made of highly transparent white glass (solar glass). The backing in standard modules is generally a non-transparent (opaque) laminated film (see Figure 2.8), or more rarely a conventional sheet of hardened glass (see Figure 2.9). EVA encapsulation is mostly used for manufacturing standard and special modules. Module sizes of up to 2m × 3m can be fabricated. With large module sizes, it is possible that the cells begin to float during lamination, which makes it difficult to maintain equal gaps between the cells.

Using different substrates enables different lamination configurations to be fabricated.

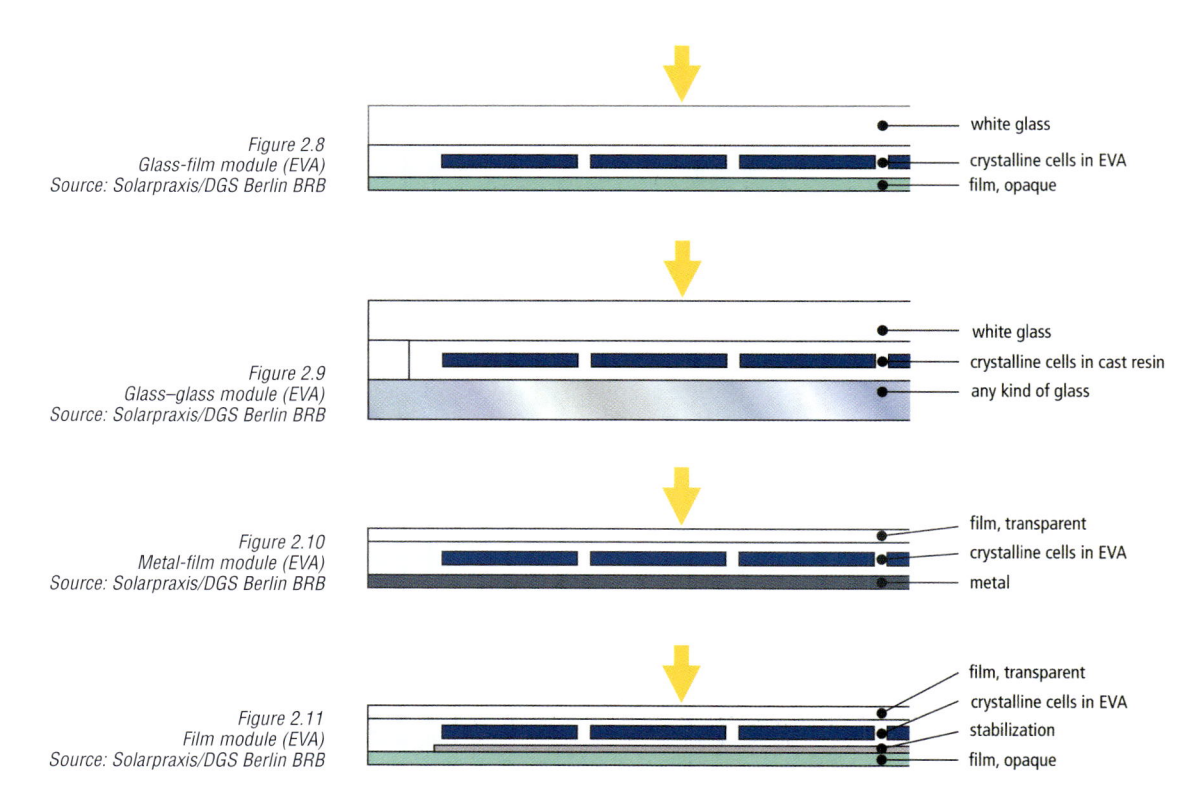

Figure 2.8
Glass-film module (EVA)
Source: Solarpraxis/DGS Berlin BRB

- white glass
- crystalline cells in EVA
- film, opaque

Figure 2.9
Glass–glass module (EVA)
Source: Solarpraxis/DGS Berlin BRB

- white glass
- crystalline cells in cast resin
- any kind of glass

Figure 2.10
Metal-film module (EVA)
Source: Solarpraxis/DGS Berlin BRB

- film, transparent
- crystalline cells in EVA
- metal

Figure 2.11
Film module (EVA)
Source: Solarpraxis/DGS Berlin BRB

- film, transparent
- crystalline cells in EVA
- stabilization
- film, opaque

ENCAPSULATION IN POLYVINYL BUTYRAL (PVB)

If PV modules are to be integrated within glass roofs or façades where the use of laminated safety glass (LSG) is required (e.g. in overhead glazing), PVB suggests itself as an encapsulation material. PVB has long been used in the glass industry as a sandwich layer in laminated safety glass. Amorphous PV glass–glass modules with PVB encapsulation have been on the market for some time. Now PVB laminates that are suitable for laminating crystalline modules are also available. As well as the conventional vacuum laminators used in module production, other laminating technologies from the glass industry can also be employed. Glass–glass modules with PVB have the same transparency as EVA and meet the requirements of IEC 61215 and 61646, and Protection Class II. Because of their integrated solar cells, PVB glass–glass modules cannot be considered the same as LSG; but they do meet the requirements for building approval (this is the case in Germany – national building codes and regulations need to be referred to).

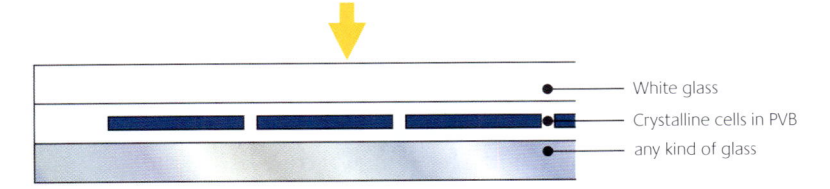

Figure 2.12
Glass–glass module (PVB)
Source: Solarpraxis/DGS Berlin BRB

— White glass
— Crystalline cells in PVB
— any kind of glass

ENCAPSULATION IN TEFLON

With Teflon encapsulation, the solar cells are enclosed in a special fluoropolymer (Teflon) in a process similar to the one described above. In contrast to EVA encapsulation, the encapsulated solar cells lie on a substrate plate and require no further covering on the front. Teflon is UV resistant, highly transparent, repels dirt, does not fade and has a reflection-free surface. With Teflon modules, the Teflon layer on the solar cells is only 0.5mm thick and is a very good conductor of heat. This therefore ensures that the cells are cooled regardless of the installation situation. As a substrate, a conventional hardened glass sheet can be used that meets the structural requirements or other opaque materials such as metal, slate, concrete and ceramics. The Teflon encapsulation is mostly used for small-scale special modules (solar roof tiles).

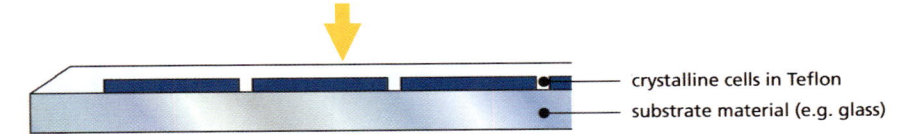

Figure 2.13
Teflon module
Source: Solarpraxis/DGS Berlin BRB

— crystalline cells in Teflon
— substrate material (e.g. glass)

ENCAPSULATION IN RESIN

Resin encapsulation uses a casting process. The solar cells are fixed between two glass sheets using adhesive pads. The sheets are bonded around the edges with a transparent spacer that is adhesive on both sides. The front sheet consists of a highly transparent, hardened white glass; the rear glass sheet consists of conventional hardened glass that meets structural requirements. The resulting cavity is filled with highly transparent casting resin, which then hardens. The type of resin determines whether or not UV light is needed for the hardening. This can play a role in designing the rear glass sheet. If a resin is used that requires UV light, it is not possible to apply tinted glass on the back since the missing UV light prevents the resin from hardening between the solar cells and the coloured sheet. With resin encapsulation, it is possible to fabricate module sizes up to 2.5m × 3.8m. The advantage of this type of encapsulation is the precise positioning of the cells, which is maintained during casting. This allows uniform cell gaps to be maintained even with large modules. Resin encapsulation is usually used with custom-made modules for integrating within buildings (façades, transparent roofs and solar shading devices). This also enables the

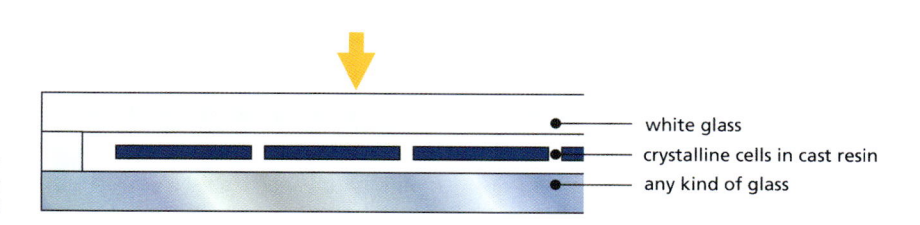

Figure 2.14
Glass–glass module (resin)
Source: Solarpraxis/DGS Berlin BRB

- white glass
- crystalline cells in cast resin
- any kind of glass

encapsulation of solar cells between two Makrolon sheets to form a glass-free module, which can also be curved if required.

The resin used for manufacturing modules is also used for producing sound-absorbing glazing. Thus, a module encapsulated with resin has sound attenuating properties from the outset.

SPECIAL FEATURES WHEN ENCAPSULATING THIN-FILM RAW MODULES

There are a few differences when encapsulating with crystalline cells. Thin-film raw modules already consist of a glass sheet (substrate sheet) that is coated with the semiconductor material. It is not possible to use hardened glass for these substrate sheets as the high temperature used for the semiconductor coating would destroy the glass strengthening. If the finished thin-film module is to fulfil special demands in terms of toughness and fracture behaviour (e.g. it is being used in a cold façade), the raw module must be additionally furnished with a sheet of toughened safety glass (TSG). A thin-film module with TSG therefore always consists of two glass sheets.

Due to the different coating technologies used in thin-film cell fabrication, the semiconductor material is either located below the superstrate or on the substrate sheet, depending upon whether the support glass is the front or back (see '2.1.1 Cell stringing). The position of the semiconductor material affects the possible structure of the module.

Amorphous and CdTe raw modules are deposited onto superstrates. This enables the substrate glass to also be used as the front weatherproofing. The back of these raw modules can be provided with a Tedlar film or a second glass sheet of any kind. EVA is employed as the standard encapsulation material. For use in building envelopes, PVB in combination with heat strengthened glass is usual for the back.

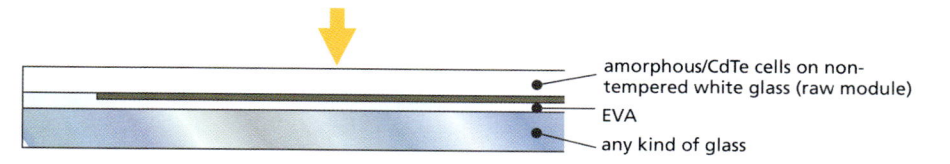

Figure 2.15
Glass–glass module
(amorphous/CdTe cells in EVA)
Source: Solarpraxis/DGS Berlin BRB

- amorphous/CdTe cells on non-tempered white glass (raw module)
- EVA
- any kind of glass

With CIS raw modules and amorphous raw modules made of coated films, the semiconductor coating is located on the substrate. The CIS module receives an additional glass sheet as the front weatherproofing – heat-strengthened or toughened safety glass can be used if required. CIS cells put particularly high demands on the moisture properties of the encapsulation material since any moisture penetration would result in cell degradation and therefore needs to be prevented.

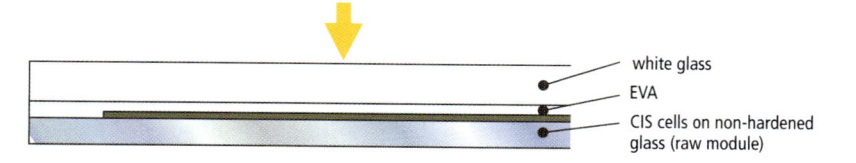

Figure 2.16
Glass–glass module (CIS cells in EVA)
Source: Solarpraxis/DGS Berlin BRB

- white glass
- EVA
- CIS cells on non-hardened glass (raw module)

Whereas prefabricated CIS modules always consist of two glass sheets, in some thin-film modules the glass sheets are replaced by plastic ones. The front of several small-format raw module units are embedded in an EVA-based fluoropolymer (see the previous section 'Encapsulation in teflon') and are stabilized on the back with lacquered steel sheeting.

When fabricating thin-film modules with resin encapsulation, neither the front nor back module glass can be used as substrate glass for the thin-film cells since these sheets are made of heat-strengthened glass (HSG). As with mono- and polycrystalline cells, the thin-film raw modules are positioned and cast between two module glass sheets. Such a glass–glass module therefore consists of three sheets and is thicker than a glass–glass module with resin-cast mono- or polycrystalline cells.

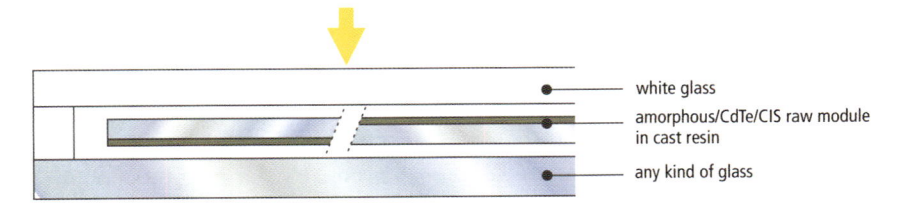

Figure 2.17
Glass–glass module (thin-film cells in resin)
Source: Solarpraxis/DGS Berlin BRB

NEW MODULE CONCEPTS

As an alternative to EVA encapsulation, materials are being tested that enable even simpler, faster and more cost-effective manufacturing processes. The first amorphous and crystalline module prototypes using silicon as an embedding material – which does not require any lamination process – have been demonstrated at trade fairs. The use of polyurethane (PU) is more advanced. This is a plastic that has been used for 20 years – for example, to coat car windscreens. A thermoplastic PU (TPU) film has been developed for encapsulating solar cells. It does not require a vacuum for lamination since it is processed in roll laminators. This continuous process enables a quadrupling of the production speed. The mechanical and optical properties of EVA and TPU are very similar. However, if modules are faulty, the TPU encapsulation can be opened again by heating it up. The first manufacturers of thin-film and special modules are already using the new material. The next obvious step in development would be to produce the frame and the rear cover from polyurethane as well, which would almost enable the modules to be moulded in one go. The front glass pane with the solar cells laminated onto it in EVA is placed in a mould and this is injected with a viscous PU elastomer. The material hardens in a few seconds and the complete module, including frame and mounting points, back and junction box, is complete. So far, test modules with a surface area of up to $1.5m^2$ have been produced to meet the requirements of IEC 61215. The market will decide whether these plastic modules with their solid frame become established.

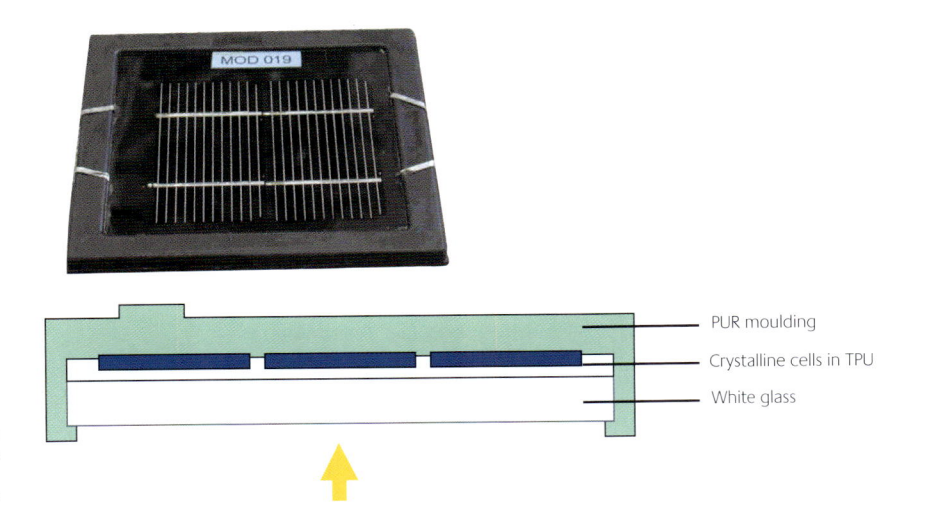

Figure 2.18
Module structure and prototype module with moulded PU frame and rear covering
Source: Bayer MaterialScience

The NICE module concept (new industrial solar cell encapsulation) has been developed using the same principle as insulated glass windows and is now set to go into full-scale production. The NICE concept eliminates both the EVA encapsulation material (or similar) and the soldering of the solar cells. A glue is applied to the rear glass that fixes the adjacently positioned overlapping contact strips and solar cells. For the external module connection, a connection leads out to the edge region. The front glass is positioned on a thermoplastic spacer (TPS), which supports the glass on edge strips of between 5mm and 10mm wide and simultaneously functions as a seal and adhesive. The two panes are pressed together and the remaining cavity is filled with inert gas. There is also a negative pressure in the cavity that holds the module sandwich together and provides a low-resistance electrical contact between the cells and the metal strips coated with solder. The first test modules meet the requirements of IEC 61215. Unlike conventional EVA encapsulation, the NICE modules are not subject to the risk of degradation resulting from ageing or yellowing encapsulating films. In addition, it is possible to repair the modules and they are easier to recycle. As well as improvement in the mechanical properties, work is also under way on temperature-resistant and water vapour-proof sealing materials.

2.1.3 Types of modules

Various module classifications are used commercially. The general term 'module' (or panel) is defined more precisely by highlighting the module's specific qualities.
Modules can be classified according to:

- Cell type:
  - mono-crystalline modules;
  - polycrystalline modules;
  - thin-film modules (amorphous, CdTe and CIS modules).
- Encapsulation material:
  - teflon modules;
  - PVB modules;
  - resin modules (the EVA classification module is not generally used).
- Encapsulation technology:
  - lamination (with EVA, PVB or teflon; see the following section on 'Laminates').
- Substrate:
  - film modules;
  - glass–film modules (or glass–Tedlar modules);
  - metal–film modules;
  - acrylic plastic modules;
  - glass–glass modules.
- Frame structure:
  - framed modules;
  - frameless modules.
- Construction-specific additional functions:
  - toughened safety glass (TSG) modules;
  - laminated safety glass (LSG) modules;
  - insulating glass modules;
  - insulating glass modules for overhead glazing;
  - stepped insulating glass modules;
  - laminated glass modules.

In addition, there are further distinctions, which will be explained in more detail in the following sections:

- standard modules;
- special modules;
- custom-made modules.

LAMINATES

Modules that are fabricated in laminating processes (EVA, PVB and teflon encapsulation) are also described as laminates. Depending upon the rest of the module structure, these are then known as film laminates, glass–film laminates or glass–glass laminates. If the description 'laminate' is used alone, then this refers to glass–film laminates. However, in photovoltaics, the term 'laminate' has become established as a general term for frameless modules.

STANDARD MODULES

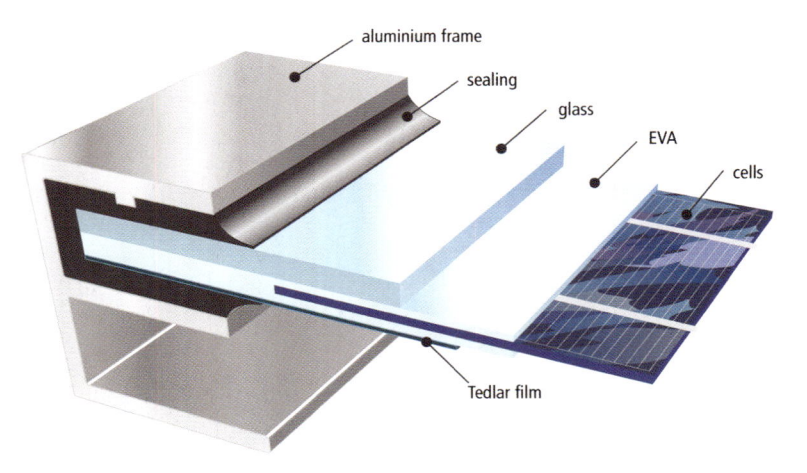

Figure 2.19
Frame of a standard module
Source: Pilkington Solar

Standard modules are modules manufactured with the aim of achieving maximum energy yields per square metre at as little cost as possible. These are mostly glass–film laminates encapsulated in EVA, which are offered with fixed dimensions and outputs. There are standard modules with and without aluminium frames. They are deployed where no special demands are made on the modules in terms of shape and size, either as rack-mounted systems or incorporated with special profile systems as part of building-integrated systems. A typical standard module consists of 36 to 216 cells and has a power output of 100Wp to 300Wp (crystalline cells). The cells are often arranged in four to eight consecutive rows, resulting in a rectangular module with dimensions of, for example, 1.6m × 0.8m. Material savings, simplified mounting, new system designs and, last but not least, aesthetic demands mean that today standard modules are available with nominal power ratings of up to 330Wp and dimensions of 2.15m × 1.25m.

Figure 2.20
Frameless standard module
Source: Solar-Fabrik

Figure 2.21
Framed standard modules
Source: MHH

SPECIAL MODULES

Special modules are modules that are mass produced for special purposes; special materials or a special frame may be necessary. Examples of special modules include all small-scale applications and lightweight modules used for solar vehicles, boats,

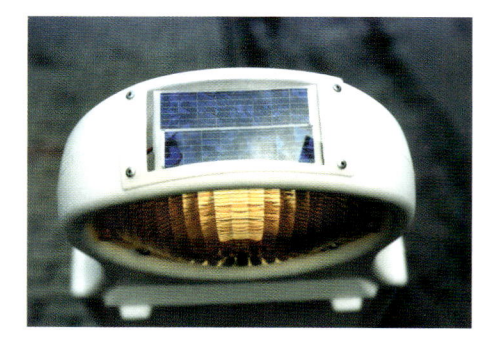

Figure 2.22
Amber warning lantern
Source: Solarwatt

Figure 2.23
Lightweight module
Source: Solarwatt

Figure 2.24
Solar roof tile Terra Piatta Solar
Source: Pfleiderer

camping and solar tiles. The latter require a frame, for example, that ensures that the roof is protected against rain and snow.

CUSTOM-MADE MODULES

Custom-made modules are modules that are specially fabricated for a specific location. This could be for a cold or warm façade, a glazed roof or a shading device. The location determines the module's structure, size and shape. The design possibilities for custom-made modules will be described in more detail in the next section.

Figure 2.25
Trapezoidal custom-made module in production
Source: Scheuten Solar

2.1.4 Design options for PV modules

Solar modules are deployed primarily in and on buildings. For this reason, solar modules should be considered not only as technical components for generating electricity, but as a versatile construction material that harmonizes with the building and, while complying with building regulations, can be integrated within its envelope. This requires modules with diverse appearances and functional qualities for different buildings and application areas. The PV market, meanwhile, offers a broad repertoire of design options to meet the demands made on solar modules as a construction product.

A large variety of standard and special modules are prefabricated for the market. With standard modules, designers can choose between various cell types, sizes and frame structures. Depending upon the application, there is also a large range of special modules. What these two types of modules have in common is that they are marketed as finished products that give designers a choice, but do not allow them to affect the appearance. In contrast, custom-made modules are fabricated to order. The two main components from which PV modules are formed – solar cells and glass (or film) – offer the designer a multitude of design possibilities. The encapsulation material plays a central role in whether they can be used in façades or overhead glazing. Apart from the choice of cells and their arrangement, the product attributes of different types of glazing can be combined with one another to create multifunctional modules that allow tailored-made solutions to meet the building's architectural requirements.

The design possibilities for custom-made modules include the:

- cell type;
- glass size;
- cell coverage;
- glass format;
- cell shape;
- glass type;
- cell contacting;
- encapsulation material;
- cell background.

The way in which these parameters are combined determines the modules' appearance. Thus, in conjunction with the module manufacturer, designers can create individual modules with different:

- colours (depending upon the cell type, cell background, cell contacting and glass type);
- transparency (see 'Cell arrangement and transparency');
- flexibility: curved modules with a minimum radius of 0.9m can be fabricated from crystalline solar cells by embedding the cells between curved sheets or curving finished modules; thin-film modules are permanently flexible and rollable when deposited on malleable substrates;
- grids;
- structural function (e.g. heat insulation and overhead glazing).

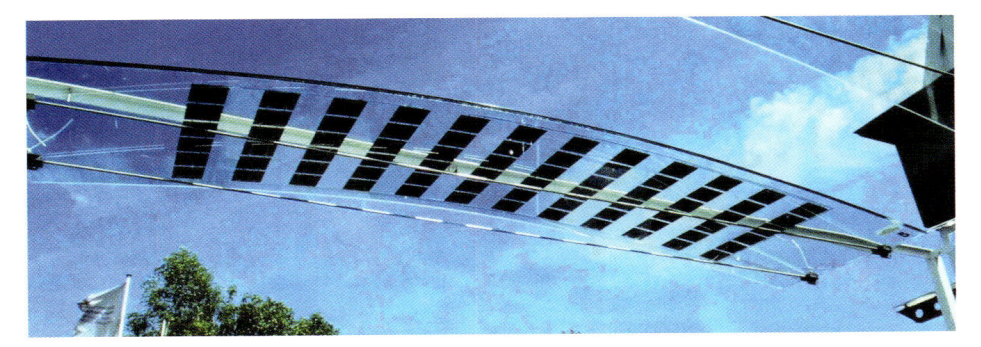

Figure 2.26
Curved glass custom module:
The cold curved glass–glass module
(resin) is held in shape via tie rods
Source: Saint Gobain

CELL TYPES

There is a large choice of cells on the market. What at first sight seems to be a uniform range reveals considerable diversity on closer examination. The crystalline cells differ according to the manufacturing processes in their:

- structure (homogenous, crystalline);
- shape (rectangular, square, semi-square, round);
- size (10cm × 10cm; 12.5cm × 12.5cm; 15cm × 15cm; 21cm × 21cm);
- colour (blue, black): by using special anti-reflective coatings, the colours magenta, gold, brown and green are also available and, if the anti-reflective coating is omitted, silver grey, as well.

Other variations are possible by using different patterns for the current collectors and connection contacts.

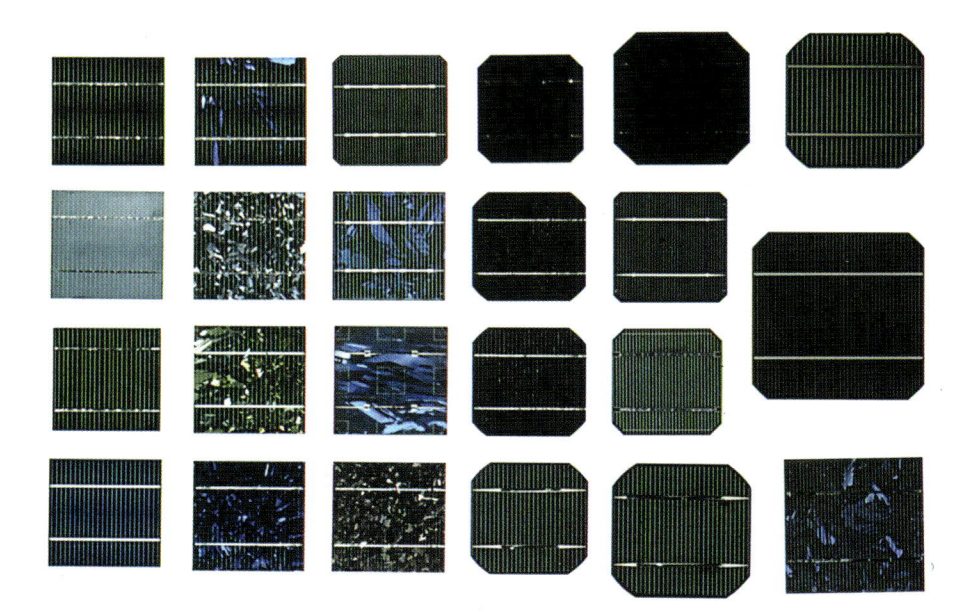

Figure 2.27
Various crystalline cells
Source: Scheuten Solar

Thin-film cells are not restricted by standard wafer sizes but only by the size of the substrate glass. This is freely selectable. Because the substrate is completely coated with solar cells, there is no grid of cells and current paths typical of crystalline modules. Instead, the solar surface is homogenous and of one colour. Only the dividing lines between the cell strips can be seen on closer examination. The colouring is determined by the cell material (a-Si, CIS or CdTe). Typical module formats are 1.2m × 0.6m or 1m × 0.6m, up to just under 2m × 1m (depending upon the cell manufacturer). To fabricate larger modules up to a maximum of 2m × 3m, the thin-film cells are divided into several separate arrays when coating and then combined in the module. Another feature of thin-film technology is the ability to manufacture flexible solar cells on malleable substrates.

A more detailed description of the various cell types and their appearance can be found in Chapter 1.

Figure 2.28
CIS solar cells
Source: Würth Solar, RWE Schott Solar, Antec Solar

Figure 2.29
Amorphous silicon cells
Source: Würth Solar, RWE Schott Solar, Antec Solar

Figure 2.30
CdTe solar cells
Source: Würth Solar, RWE Schott Solar, Antec Solar

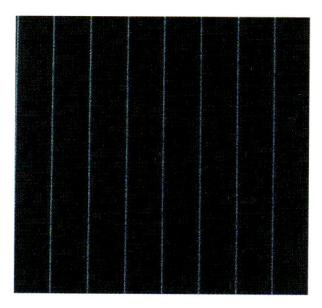

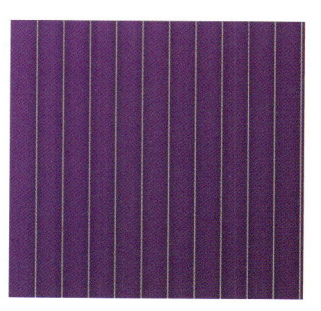

CELL ARRANGEMENT AND TRANSPARENCY

In glass–glass modules, considerable scope for design is also provided by the positioning of the cells and the possible interplay of structure, light and shadow. With mono- and polycrystalline modules, the distance of the cells to one another and to the edge of the glass can be freely defined. This enables specific control of transparency (in terms of light and energy) and shading. If a crystalline solar module is to be evenly transparent, special transparent solar cells are recommended.

Figure 2.31
Former Shell solar cell factory in Gelsenkirchen, Germany
Source: (modules) Scheuten Solar

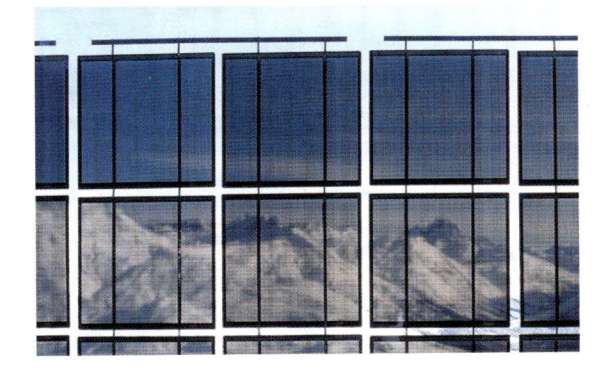

Figure 2.32
Glass–glass modules with transparent solar cells
Source: (cells) Sunways; (modules) Solarnova

In the glazing for the entrance hall of the Shell solar cell factory (see Figure 2.31), the solar cells have been arranged in stripes to emphasize the long, narrow bands of light.

The glazing in the two figures presented below shows how cells can be arranged to ensure desired shading of the spaces behind (see Figures 2.33 and 2.34). In the canopy of the town hall in Monthey, on the other hand, cell distances of 3cm to 4cm were chosen so as not to prevent too much light from entering the building (see Figure 2.34).

Figure 2.33
Effect of glass–glass modules on interior spaces, Solar Office Doxford International
Source: OJA-Services; photographer: Dennis Gilbert; modules: Kyocera

Figure 2.34
Town hall in Monthey, Switzerland
Source: (modules) Scheuten Solar

With thin-film modules, semi-transparency is also created by additional grooves perpendicular to the cell strips. Here, the solar cell material is removed in strips so that light can penetrate through the module at these points. This creates a finely checked pattern that gives the thin-film modules an even and neutrally coloured transparency. Alternatively, increasing the cell spacing results in strip-like transparency.

Figure 2.35
Amorphous thin-film modules with finely checked 10 per cent transparency
Source: (modules) RWE Schott Solar/Glaswerke Arnold

Figure 2.36
Jubilee Campus, Nottingham University, UK
Source: OJA-Services; copyright: Ove Arup and Partners; modules: BP

CELL SHAPES

Crystalline cells are square, semi-square or quadratic, and can therefore be easily arranged to provide good coverage in rectangular modules. If the modules deviate from a rectangular form because, for example, the façade design requires this, there are two possibilities in terms of the cell coverage along the diagonal module edges.

The cells are either stepped back along the diagonal edge or cut parallel to it. Because this reduces the current flow relative to the rest of the cells, cut cells cannot be connected to them in series. They remain electrically inactive. The visually homogenous appearance of such a module entails additional costs than that of the stepped variant.

Thin-film cells also behave very similarly. Computer-aided glass-cutting enables the creation of diagonal or round shapes. The active surface of the semiconductor is, however, always rectangular (i.e. within asymmetrical areas the power is also determined by the largest possible rectangular area). The area outside of the active rectangular area is not electrically active, but cannot be distinguished visually from the active area.

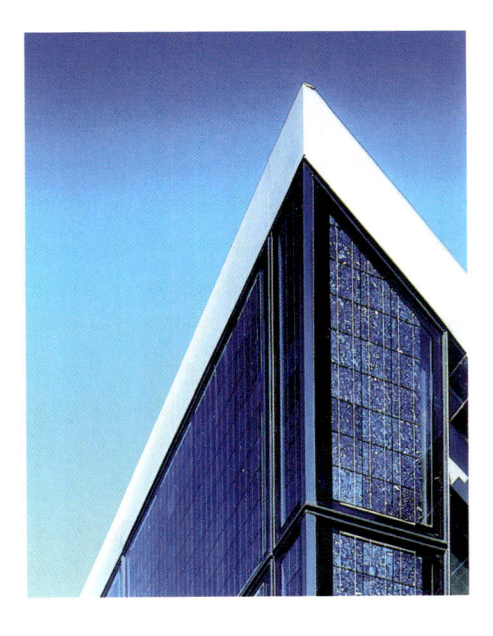

Figure 2.37
Bayerische Landesbank in Munich, Germany
Source: (modules with cut cells) Saint Gobain

Figure 2.38
Enecolo, The Hague, The Netherlands
Source: (modules with stepped cells) Saint Gobain

CELL BACKGROUND

Larger cell gaps provide interesting opportunities for designing the space between the cells. With glass–film laminates or purely film laminates, the rear (Tedlar) film can be coloured or transparent.

Figure 2.39
Glass–film modules with black front contacts and black Tedlar polyester film as backing material on this heritage protected church in Riethnordhausen near Erfurt, Germany
Source: (modules) Solarwatt

Figure 2.40
Semi-transparent sunshade louvres on the saw-tooth roof of the Jakob-Kaiser building in Berlin, Germany
Source: (glass–Tedlar modules) Solon

FRONT CONTACTS

With little expenditure it is possible to vary the pattern and colour of the contact lines (see the section on '1.4.6 Front contacts' in Chapter 1). Instead of the silver-coloured standard contacting, the conducting paths can be tinted to match the cell colour. Combined with a corresponding background colour, crystalline modules can also make a very homogenous visual impression (see Figure 2.39).

GLASS SIZE

Standard modules are available in a variety of formats and sizes. The preferred surface areas lie between 0.5m^2 and 2.5m^2 as these sizes are easiest to handle both in terms of mounting and electrically. For custom-made modules, module sizes up to 2.5m × 3.8m can be realized (module dimensions exceeding these are technically feasible, but are associated with exponential increases in costs). With custom-made modules, it is therefore possible to incorporate architectural grid requirements and clad geometrically complicated façades. The front and rear glass can be variously sized to create one- or multiple-sided stepped glazing.

Figure 2.41
PV installations at the Olympic Village, Sydney, Australia
Source: OJA-Services; copyright: MIRVAC Lend Lease; modules: BP Solar

GLASS FORMAT

All glass formats can generally be fabricated (triangles, trapeziums and curves).
However, it should be taken into consideration that, compared with rectangular custom-made modules, custom-made shapes entail a disproportionate increase in cost.

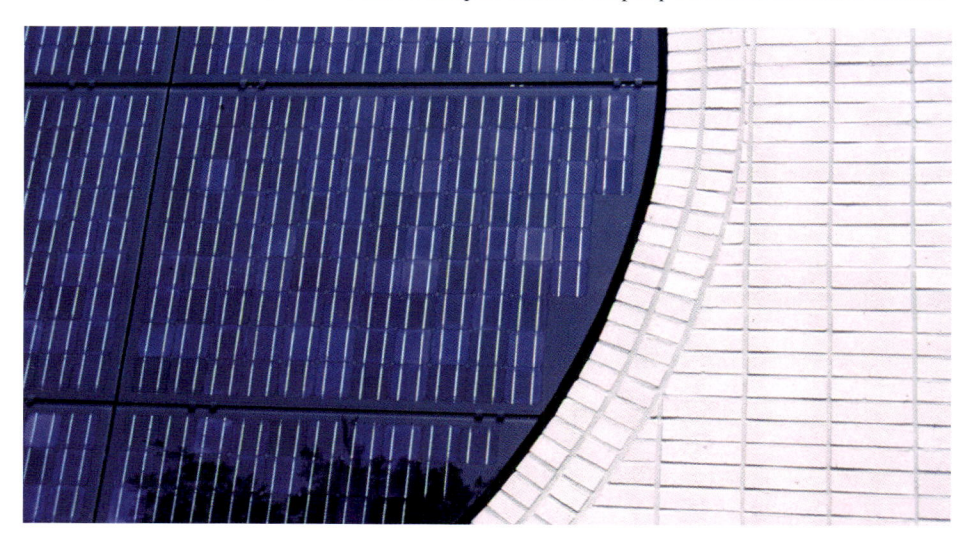

Figure 2.42
EnergieNed, Arnheim, The Netherlands
Source: (rounded laminates) Creaglas

GLASS TYPE: MULTIFUNCTIONAL MODULES FOR BUILDING ENVELOPES

The front covering must be of highly transparent material. Only the glass surface allows scope for design and can be executed, for example, as structural glass or with an 'orange skin' surface.

On the other hand, the rear covering can be utilized as a design element. Glass–glass modules can use almost all glass produced by the industry. For instance, photovoltaic modules can be changed visually and furnished to provide additional constructional features such as weatherproofing, solar protection, heat insulation, soundproofing, safety functions or electromagnetic shielding. They therefore meet all demands made on external building components. Possible glass types include, for example:

- body-tinted glass (see Figure 6.132 in Chapter 6);
- coloured coated glass, in which the ceramic colour is baked on the rear side;
- screen-printed glass, in which both the pattern and the colour can be chosen (see Figures 6.145 and 6.146 in Chapter 6);
- reflective glass, in which the rear side is coated with layers of reflecting metal oxide;
- solar protection glass, which is coated on the back with selectively reflecting metal oxide layers that reflect long-wave solar radiation; in contrast, visible light penetrates almost unhindered through the glass so that the interior of the building remains both bright and cool in summer (see Figure 6.137 in Chapter 6);
- sound-insulating glass, which has a high sound-insulating effect due to its structure (see Figure 6.170 in Chapter 6 and Figure 2.45);
- laminated safety glass which consists of two sheets of glass bonded together with PVB film (upon shattering, the bonding layer holds the glass fragments together and the sheet remains in the frame; as safety glass, LSG enables the use of modules in overhead areas – three sheets in total; see section '6.7 Glass roofs' in Chapter 6);
- insulating glass consisting of any desired inner and outer sheets, in which the space between them is filled with inert gas and which ensures the heat insulation of the building; the internal sheet can also be freely chosen;
- insulating glass for use in overhead areas (warm roof) with LSG as the inner sheet.

Figure 2.43
Body-tinted glass backing
Source: Saint Gobain

Figure 2.44
Schott administration building in Barcelona, Spain: Insulating glass structure with dyed inner pane
Source: RWE Schott Solar

Figure 2.45
Arena event hall in Treptow, Berlin, Germany: Heat and sound insulation glazing as module backing – inner pane: LSG; external pane: structured glass (structure facing into the inter-pane cavity)
Source (modules) Saint Gobain

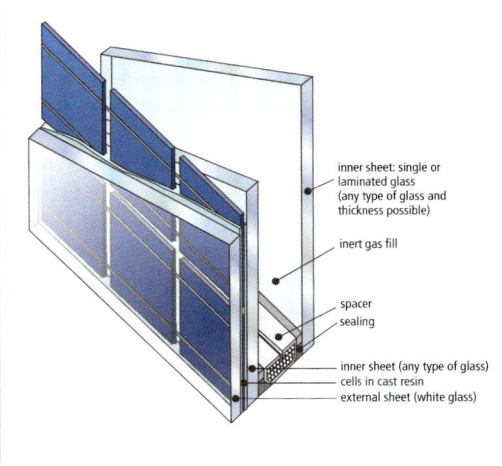

Figure 2.46
PV module incorporated within insulating glass
Source: Scheuten Solar

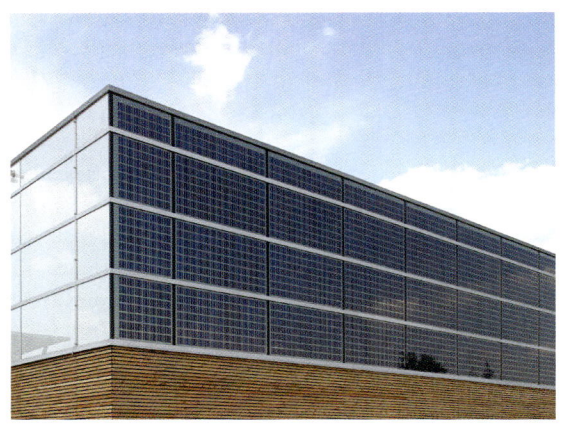

Figure 2.47
Multifunctional PV system in the south-facing façade of the sports hall in Burgweinting, Regensburg, Germany
Notes: The solar cells absorb most of the sunlight and function as a sunshade. The insulating glass features a heat-reflective coating on the outer pane (a heat barrier during summer). The inside pane is made from LSG to withstand ball impact, with two light-scattering PVB films (anti-glare). The cell spacing of approximately 20mm has been optimized for daylight illumination and thermal efficiency in the sports hall.
Source: Building Construction Authority of the City of Regensburg, Germany

CUSTOM-MADE MODULES FROM ACRYLIC PLASTIC OR MAKROLON

The use of transparent acrylic plastic or Makrolon opens up additional interesting design possibilities. In some applications, glass is not desired as a material because it is liable to shatter (e.g. maritime applications or curved modules for architectural design applications). With the correct tools, acrylic plastic can be very easily sawn, drilled, sanded and polished, enabling any shape to be created. It can be bent when cold and moulded when hot. The minimal cold bending radius for cell arrays of 10cm × 10cm is 350 times the thickness of the strongest acrylic plastic sheet. The heat moulding occurs outside the cellular array area. The rear acrylic plastic sheet can be coloured, screen printed or painted. The non-weathering plastic Makrolon has similar advantages. It is impact resistant and can be used in the form of solid sheets or multi-wall sheets.

Figure 2.48
International Motor Show (IAA) in Frankfurt, Germany, 1999
Source: (acrylic plastic modules) Sunovation

CUSTOM-MADE MODULES WITH INTEGRATED LIGHT-EMITTING DIODE (LED)

Figure 2.49
Solar flags with transparent crystalline solar cells and integrated LEDs in the archways of the walls of the heritage protected Castello Doria fortress in Porto Venere, Italy
Source: Sunways

PV modules with integrated light-emitting diodes (LEDs) offer new design possibilities with lighting effects for prestigious or architecturally sensitive applications. They simultaneously offer transparency during the day and illumination at night, for which they generate power during the day. The modules are also suitable for glazing at bus stops, which require illumination at night.

Figure 2.50
Prototype Lumiwall module with transparent micromorphous thin-film solar cells and integrated high-performance LEDs
Source: Sharp

NOISE PROTECTION MODULES

The Spanish module manufacturer Isofoton has developed a new kind of ceramic module specifically for use in noise barriers. Since conventional modules have a weight per unit area that is too low for sound insulation, a ceramic plate is used as the back substrate material. Between this ceramic plate and the transparent Tedlar front covering, the solar cells are laminated in EVA. This design enables average noise attenuation of 32dB. As a result, the first PV modules have been installed in a noise barrier project.

Figure 2.51
'PV soundless' noise barrier along the A 92 motorway near Freising, Germany, with sound-insulating ceramic modules in the top five rows
Source: Freisinger Stadtwerke Versorgungs-GmbH

2.1.5 Module cable outlets and junction boxes

To run the electrical supply cables for the cell strings from the embedding material to the outside, either a rear glass panel with holes is used or the rear film is penetrated. In these cases, a junction box is fixed to the entry point. The module junction boxes must have minimum protection to IP 54 and Protection Class II. When mounting, care should be taken to avoid any water penetration. This is prevented by using drip loops. Many modules are supplied complete with connecting leads and reverse polarity-proof, touch-proof plugs to make installation easier. The modules can then simply be plugged together without opening the module junction boxes. Another possibility for the cable outlets is to run the cables out along the glass edges. This option is used in custom-made modules where a junction box on the back would have a visually undesirable effect (e.g. in glass façades).

Figure 2.52
Cable exit on the back with junction box

Figure 2.53
Cable exit at the side with module lead plug connector system
Source: Scheuten Solar

Figure 2.54
Cable exit along glass edge

2.1.6 Wiring symbols

This wiring symbol is used for a:

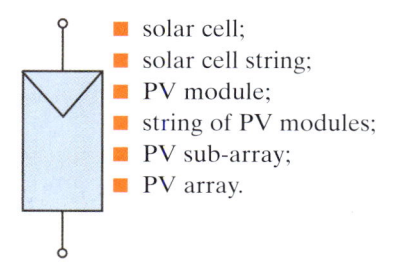

- solar cell;
- solar cell string;
- PV module;
- string of PV modules;
- PV sub-array;
- PV array.

2.1.7 Characteristic *I–V* curves for modules

As mentioned in section 2.1.1, several solar cells are interconnected in order to achieve greater power. Here, two types are possible: series and parallel cell interconnection. In PV modules, the solar cells are largely connected in series to create a higher voltage. Figure 2.55 demonstrates the change in the electrical parameters and characteristic *I–V* curve when three solar cells are connected in series. It can be seen that the cell voltages increase while the current remains constant.

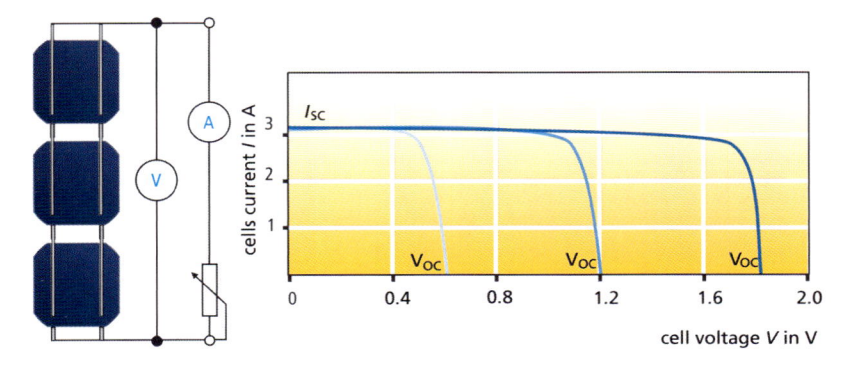

Figure 2.55
I–V curve (current/voltage curve) for three solar cells connected in series
Source: R. Haselhuhn

In the early days of photovoltaic electricity generation, the first application area for PV systems was as stand-alone systems. For stand-alone systems, standard 12V batteries were generally used, which were charged via the modules. For this reason, a voltage level of 17V was initially chosen for PV modules. This voltage is above the battery voltage to ensure optimum charging. With silicon solar cells, the voltage value of 17V is provided by a series connection of 36 to 40 solar cells. Since the PV market at that time concentrated on producing such PV modules, these modules were known as standard modules. Figure 2.56 illustrates the current-voltage characteristic curve (the *I–V* curve) and the power-voltage curve for a 50W module. The *I–V* curves result from connecting 36 solar cells in series.

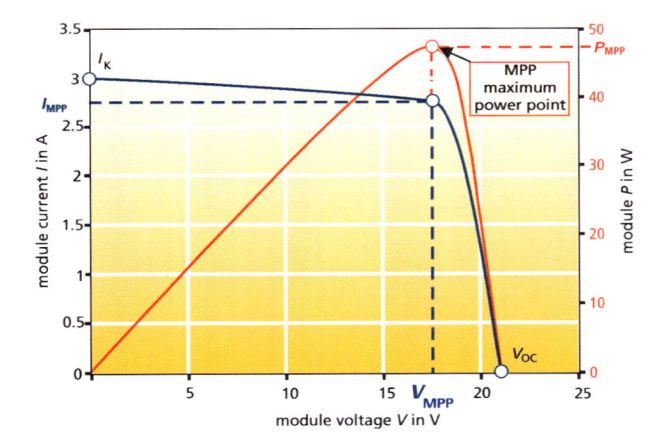

Figure 2.56
I–U curve for a mono-crystalline 50W module
Source: R. Haselhuhn

All modules that are produced in large volumes, usually with EVA encapsulation, have now come to be referred to as standard modules. These achieve power outputs of up to 300W. As well as connecting solar cells in series, multiple cell strings are often connected in parallel, particularly in modules with higher power output. Figure 2.57 shows the change in the characteristic *I–V* curve as a result of connecting three solar cells in parallel. Here, the voltage remains constant and the current increases. It is unusual to have PV modules in which only one cell is connected to another because of the low voltage. Several solar cells are always first connected in series to form a string, and this string is then connected in parallel to a module. With large PV modules, two or more solar cell strings, each with 36 solar cells, are often connected in parallel.

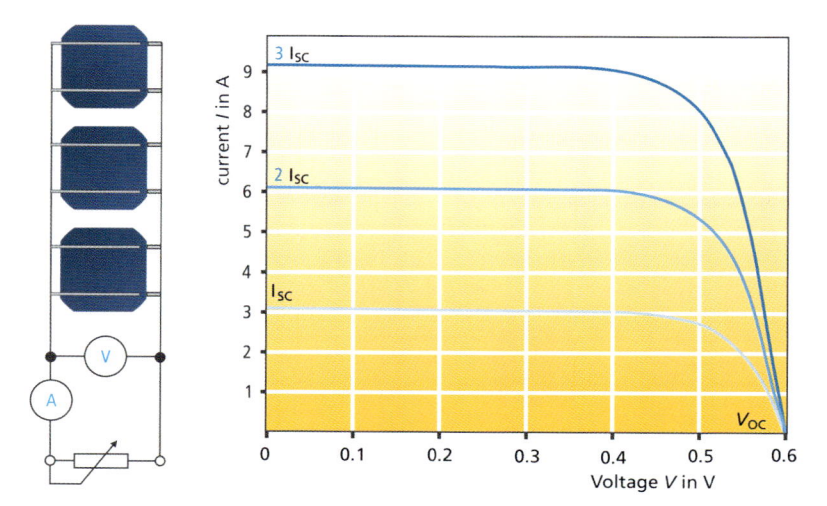

Figure 2.57
I–V curve for three solar cells connected in parallel
Source: R. Haselhuhn

The electrical parameters for PV modules are determined by the manufacturers at STC. The short-circuit current I_{SC}, the open-circuit voltage VOC and the maximum power rating P_{max} or P_{MPP} are specified for the solar modules with a tolerance up to ± 10 per cent.

In reality these conditions occur very rarely; however, if the sun shines with the specified intensity, then the cell temperature would be higher than 25°C. For this reason the nominal operating cell temperature (NOCT) is often specified as well. This cell temperature is determined for an irradiance level of 800 W/m², an ambient temperature of 20°C and a wind velocity of 1m/s. From this the temperature coefficients for current and voltage can be determined. Table 2.1 shows a typical data sheet for a monocrystalline PV module. Apart from the electrical data, information is also provided on the dimensions, weight, limit values for thermal and mechanical stress, and temperature dependence.

Table 2.1
Data sheet information taken in accordance with DIN EN 50380
Source: Shell Solar

Mono-crystalline solar module:	SQ 160-C
Electrical parameter symbol unit	
Behaviour under STC:	nominal power (MPP power under STC) P_n (also P_{max}) W_p 160
Nominal voltage:	V_{MPP} V 35.0
Nominal current:	I_{MPP} A 4.58
Short-circuit current:	I_{SC} A 43.5
Open-circuit voltage:	V_{OC} V 4.9
Module efficiency η %:	12.14
Behaviour at 800W/m², NOCT, AM 1.5:	MPP power P_{800} W_p 115
MPP voltage:	V_{MPP} V 32.0
MPP current:	I_{MPP} A 3.6
Short-circuit current:	I_{SC} A 40.0
Open-circuit voltage:	V_{OC} V 3.95
Efficiency reduction at 25°C and 200W/m²:	η % 8
Maximum permissible system voltage:	V_{max} V 1000
Thermal parameter	
NOCT °C 46	
Temperature coefficient of MPP power:	γ %/K −0.52
Temperature coefficient of MPP voltage:	$β_{MPP}$ mV/K −167
Temperature coefficient of short-circuit current:	α A/K 1.4
Temperature coefficient of open-circuit voltage:	β mV/K −161
Additional characteristics and properties	
Number of bypass diodes:	3
Height × width × thickness:	622mm × 814mm × 40mm
Weight:	7.2kg
Manufacturer's warranty ten years:	90% related to $P_{STC\ min}$
25 years:	80% related to $P_{STC\ min}$
Product warranty:	two years
Approvals and certificates:	IEC 61215, UL 1703, TÜV SK II, EN ISO 9001

Additional diagrams with V/I identification lines at different irradiation and temperatures provide greater transparency. In addition they are required for PV simulation programs (see Chapter 7). An informative data sheet provides important parameters for planning and creates trust among investors and installers. Unfortunately not all manufacturers give full information on the PV module as prescribed in Standard DIN EN 50330 'Data sheet and identification plate information of photovoltaic modules' – for instance, the electrical parameters with low-level irradiation and temperature coefficients and reverse current load capacity.

2.1.8 Irradiance dependence and temperature characteristics

PV systems very rarely operate under STC conditions. The electrical output and the I–V curves of PV modules depend upon temperatures and irradiance, so modules are usually only partly loaded during operation. During the course of a day the irradiance varies more than the temperature. The changes in irradiance affect the module current most of all since the current is directly dependent upon the irradiance. When irradiance drops by half, the electricity generated also reduces by half.

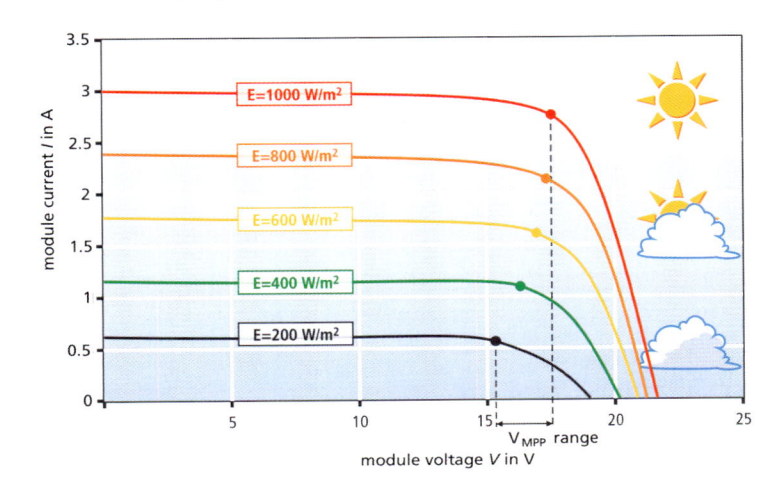

Figure 2.58
Module I–V curves for varying irradiance and constant temperature
Source: R. Haselhuhn

By contrast, the MPP voltage stays relatively constant with changing irradiance. Figure 2.58 shows that the maximum change in the MPP voltage resulting from changing irradiance is approximately 4V with a polycrystalline 150W standard module. But since in many PV systems a larger number of PV modules are connected in series, the fluctuation in the MPP voltage under changing irradiance can add up to more than 40V. With low irradiance values of just a few W/m², the voltage breaks down. Downstream inverters then operate in the fixed voltage range (i.e. at low irradiance) and the modules no longer operate in the MPP: the inverter's operating point no longer corresponds to the MPP.

The module voltage is affected most of all by the module temperature. The MPP voltage deviation from the STC value of a ventilated 150W module can amount to 10V during summer and more than +10V in winter. The change in voltage of the module determines the system voltage and therefore the design of the entire PV system. In particular, the increase in voltage at low temperatures should be taken into account. When several modules are connected in series, this can amount to more than 100V and perhaps exceed the voltage resistance of downstream devices. When sizing PV systems, therefore, particular attention must be paid to this situation. The current hardly changes with changes in module temperature. It increases slightly with increasing temperature.

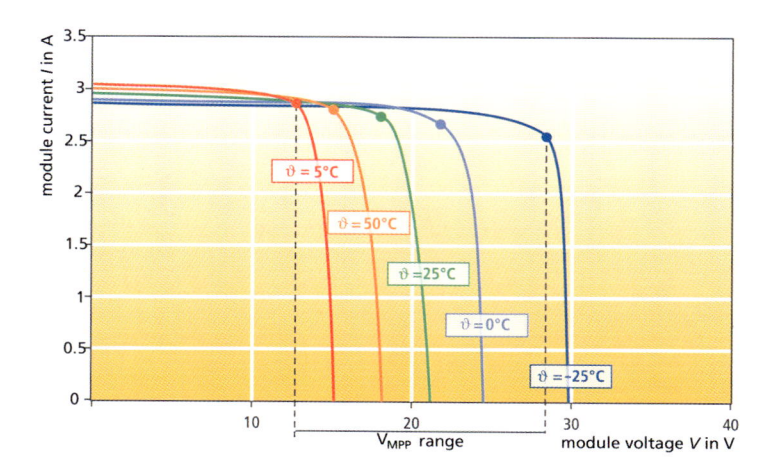

Figure 2.59
Module I–V curves at different module temperatures and with constant irradiance of 1000W/m²
Source: R. Haselhuhn

During summer, the power output of a module at high temperatures can be 35 per cent less than under STC, as depicted in Figure 2.60. In order to minimize this power loss, the PV modules should be able to dissipate heat easily (sufficient ventilation).

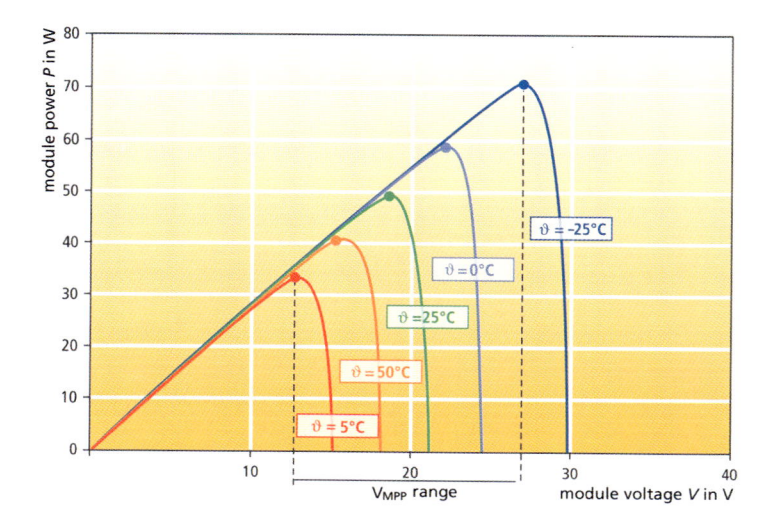

Figure 2.60
Module power at different module temperatures and with constant irradiance of 1000W/m²
Source: R. Haselhuhn

Apart from the nominal characteristics (STC), the temperature coefficients for the voltage and current change are also often specified on PV module data sheets as percentage, mV or mA per °C. These enable the electrical parameters to be calculated at any temperature. If the data sheet does not provide any information on the temperature coefficients, the graph for crystalline silicon modules depicted in Figure 2.61 can also be used to determine the parameters for temperature changes.

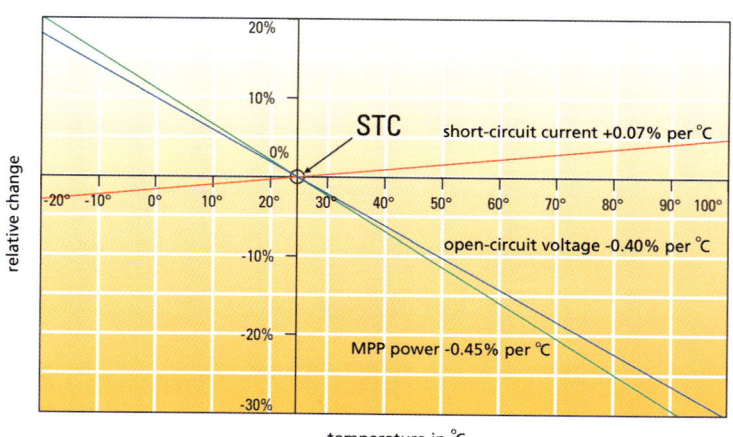

Figure 2.61
Temperature dependence of the electrical characteristics of PV modules
Source: R. Haselhuhn

The temperature coefficients for the open-circuit voltage and the short-circuit current are measured for the module certification to IEC 61215. The temperature coefficient for the MPP power is not usually measured, but determined using a calculation procedure to IEC 60891.

The coefficients presented in Table 2.2 can be specified for crystalline modules.

Table 2.2
Coefficients for crystalline modules
Source: manufacturers' specifications, measured data LEEE-TISO, TÜV and ECN

Typical temperature coefficient	Crystalline silicon modules	High-performance modules (HIT, SunPower)
Open-circuit voltage	−0.30 to −0.55%/°C	−0.25 to −0.29%/°C
Short-circuit current	+0.02 to +0.08%/°C	+0.02 to +0.04%/°C
MPP power (STC)	−0.37 to −0.52%/°C	−0.30 to −0.38%/°C

On a normal summer's day (in Germany or in a country with a similar climate) with an irradiance of 800W/m² and an ambient temperature of 20°C, a free-standing module has a normal operating temperature of around 42°C. The operating temperature is decisively influenced by the thermal ambient conditions that differ according to the type of installation and the mounting of the PV array. With a roof-integrated PV system, higher temperatures develop than with a well-ventilated system. In Figure 2.62 the temperature increases of a PV array relative to the ambient temperature when there are 1000W/m² at the module level are shown as red bars for the usual installation types. Of course, the temperature dependence of the power also affects the energy yield of the module. The reduced energy yields owing to module heating are depicted as blue bars. The stated values are the average values for Germany. The values can deviate from these guide values by approximately ±10 per cent for temperature increases and by ±30 per cent for yield reduction.

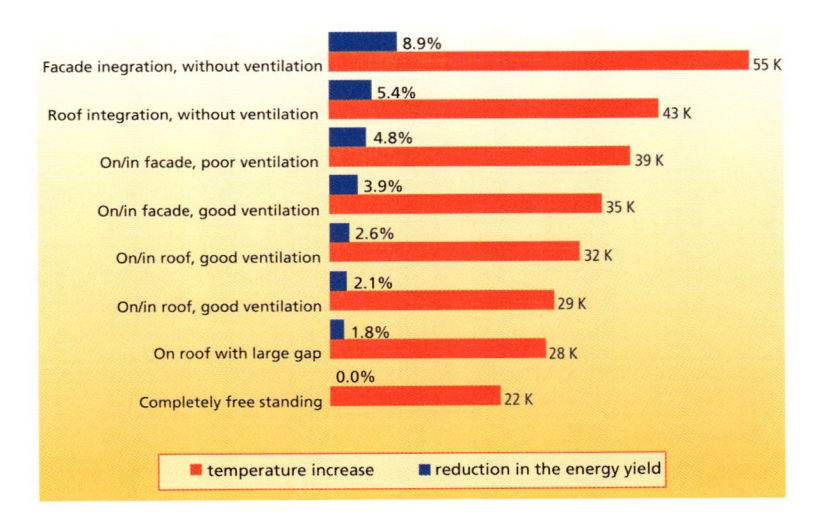

Figure 2.62
Temperature increase and reduction in the annual energy yield for various PV array installation methods
Source: ISE (1997)

Installation	Reduction in energy yield	Temperature increase
Facade inegration, without ventilation	8.9%	55 K
Roof integration, without ventilation	5.4%	43 K
On/in facade, poor ventilation	4.8%	39 K
On/in facade, good ventilation	3.9%	35 K
On/in roof, good ventilation	2.6%	32 K
On/in roof, good ventilation	2.1%	29 K
On roof with large gap	1.8%	28 K
Completely free standing	0.0%	22 K

In summary, it can be said that owing to the different irradiance and temperature conditions in Germany, a 2kWp system, for example, very rarely delivers a power of 2kW, but almost always has a current output that is considerably less.

2.1.9 Hot spots, bypass diodes and shading

Under certain operating conditions, a shaded solar cell can heat up to such an extent that the cell material is damaged and a so-called hot spot develops. This can happen, for instance, when relatively high reverse current flows through the unlit solar cell. A hot spot reduces the power of the solar cell only slightly provided that no contact strip is destroyed. However, the probability of cell failure and, hence, module failure increases each time the cell is shaded.

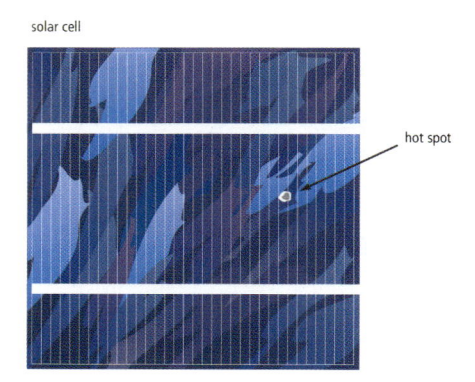

Figure 2.63
Solar cell with hot spot

First, let us consider the normal operating condition shown in Figure 2.64. A standard module with 36 cells is irradiated by the sun. The current generated in the solar cells is used by a load (resistance R).

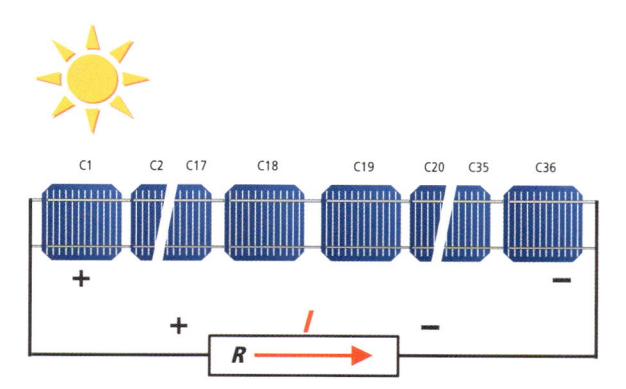

Figure 2.64
PV module with load
Source: R. Haselhuhn

If a leaf falls on the solar module so that a solar cell (C36 in Figure 2.65) is darkened, this solar cell becomes an electricity load. No more current is generated in this cell. Instead, it uses the current from the other cells: the direction of the voltage is reversed in the shaded cell. The current from the other illuminated solar cells is driven through the darkened cell. This current flow is then converted into heat. If there is a large enough current, this can lead to the hot spot effect already mentioned. The largest current that can flow is the short-circuit current. Short-circuit currents are, for example, a normal occurrence in stand-alone PV systems with short-circuit charge controllers.

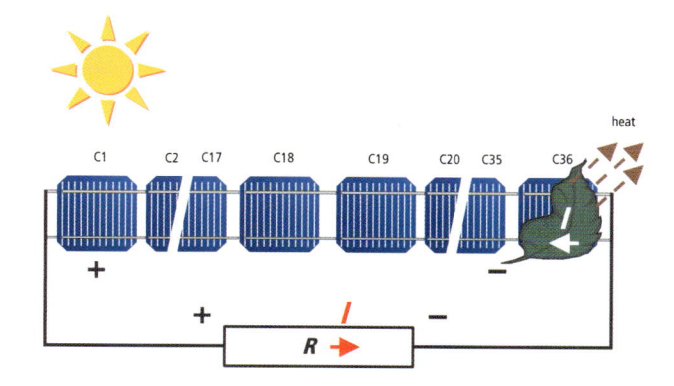

Figure 2.65
Shaded PV module without bypass diodes
Source: R. Haselhuhn

Eighteen to 20 cells can create a voltage of around 12V. As described in section '1.5.1 Equivalent circuit diagrams of solar cells' in Chapter 1, the breakdown voltage of a solar cell is between 12V to 50V. With this voltage it is possible for a reverse current to flow through the solar cells. To prevent a hot spot from developing, the current is diverted past the solar cells via a bypass. This bypass is provided via a bypass diode (see Figure 2.66). It prevents large voltages from building up across the solar cells in the reverse-biased direction. The largest shading tolerance would be attained if bypass diodes were connected across every cell. In practice, however, bypass diodes are usually connected for manufacturing reasons across 18 to 20 solar cells. Hence, modules with 36 to 40 cells have two bypass diodes, and modules with 72 solar cells have four bypass diodes.

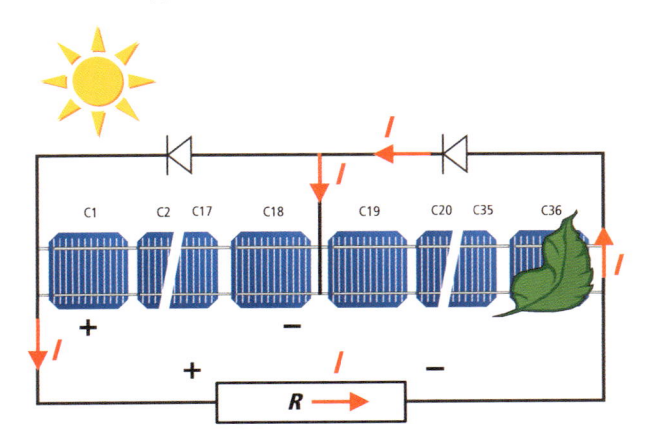

Figure 2.66
Shaded PV module with bypass diodes
Source: R. Haselhuhn

If a module lies in the shadow of an object (e.g. a chimney or antenna, as in Figure 2.67), the module's characteristic curve is changed by the bypass module. Without the bypass diode the entire current from the module would be determined by the shaded cell.

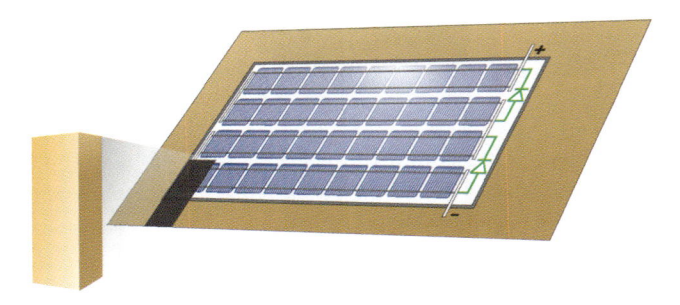

Figure 2.67
Shading of a cell in a 50W standard module with two bypass diodes
Source: V. Quaschning

In accordance with the lower irradiance on the cell, this would result in the red characteristic curve shown in Figure 2.68. The bypass diode ensures that the full current flows through at least the 18 non-shaded cells, which results in the green characteristic curve. It can also be seen that the MPP voltage drops by about half. The characteristic curves for shaded modules shown in Figure 2.68 apply for a standard module with 36 cells under STC conditions in which one cell is 75 per cent shaded.

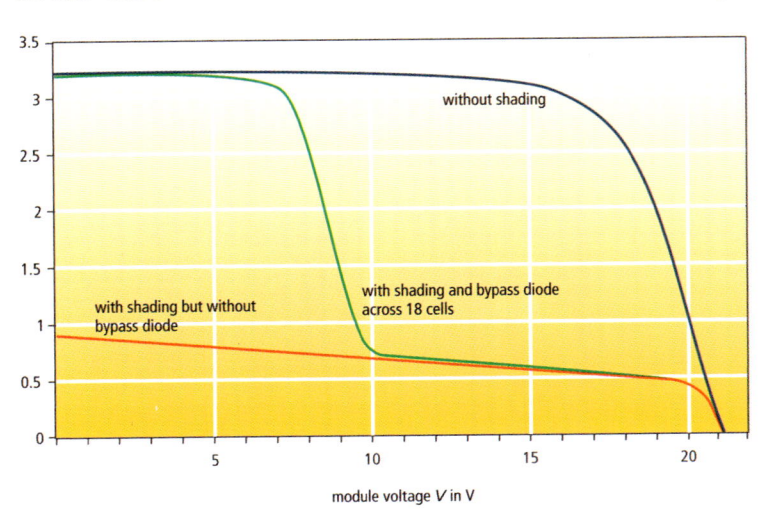

Figure 2.68
Module characteristic I–V curves with and without bypass diodes
Source: V. Quaschning

The bypass diodes are generally housed in the module junction box (see Figure 2.69). In some cases the module junction box is sealed with silicon.

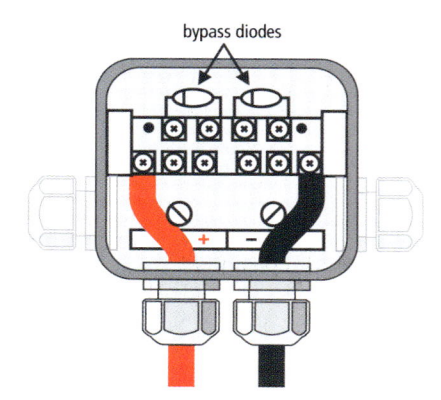

Figure 2.69
Module junction box with bypass diodes

Figure 2.70
Module junction box with one bypass diode
Source: Multi-Contact

Some manufacturers use laminating 'strip' bypass diodes in the modules. Here, more diodes are used in the module than are normally employed in a module junction box, which significantly increases shading tolerance. But in this case it is problematic to replace faulty bypass diodes. While the installation engineer can replace a faulty bypass diode that is not sealed into the module junction box, if it is sealed, only the manufacturer can replace it. However, faulty bypass diodes are relatively rare. They may be caused by insufficient heat dissipation combined with frequent module shading, or by connecting the module with the wrong polarity, or by surge voltages resulting from lightning strikes in the surrounding area. By using reverse polarity-proof module-connecting cables, ensuring sufficient heat dissipation in the junction box and the use of generously rated diodes, the probability of failure can be minimized.

Figure 2.71
Custom-made module for building integration with connection and bypass diodes at the edge of the module
Source: (module) Scheuten Solar; Astrid Schneider

*Figure 2.72
PV module with laminated
'strip' bypass diodes
Source: BP Solar*

2.1.10 Electrical characteristics of thin-film modules

The PV modules that are obtainable on the market generally behave similarly. Apart from their efficiencies, crystalline and thin-film modules differ in terms of their irradiance dependence, their temperature dependence, their spectral sensitivity and their shading tolerance. The lower efficiency of thin-film modules essentially only means that a larger area is required to achieve the desired power. The lower efficiency does not affect the specific energy yield per kilowatt of the thin-film modules. Unique in this context are amorphous silicon modules. The technologically determined effect of light degradation of the material (known as the Staebler Wronski effect) causes a reduction in efficiency during the first 6 to 12 months, which eventually levels out at a stable value. This value is specified by the manufacturers as the nominal power rating, which means that the modules are delivered with greater initial power than the nominal power rating. This must be taken into account when sizing system equipment, such as inverters. In addition to the initial degradation, reversible degradation occurs during operation in winter, but is reversed by the higher temperatures in summer (thermal regeneration). For this reason the efficiency of amorphous modules fluctuates either side of the rated value during summer and winter, and is particularly high during the high-yield summer months.

CHARACTERISTIC *I–V* CURVES FOR MODULES

Apart from differences in efficiency, the flatter current-voltage curves of thin-film modules are also noticeable, particularly for CdTe modules. One reason for this is that the less clearly defined MPP point makes higher demands on the control technology, and the flatter characteristic curves result in lower filling factors than with crystalline modules.

*Table 2.3
Comparison of typical filling
factors of PV modules
Source: manufacturers' specifications,
measured data LEEE-TISO,
TÜV and ECN*

Module type	Filling factor
Crystalline silicon modules	0.75 to 0.85
Amorphous silicon modules	0.56 to 0.61
CIS modules	0.64 to 0.70
CdTe modules	0.47 to 0.64

Furthermore, thin-film modules are generally more flexible in terms of their geometric dimensions. With crystalline modules, the module dimensions are determined by the geometry of the given silicon wafer. The nominal voltage is a multiple of the series-connected cells. As already described earlier, in thin-film technology the cells consist mostly of 0.5cm to 2cm wide cell strips. Cell and module manufacturers using thin-film technology have greater freedom in choosing the length and number of interconnected cell strips when designing the modules. The module design then determines the power and, thus, the current and voltage of the module. By enlarging the module area, the power can be increased almost infinitely (see Figure 2.73).

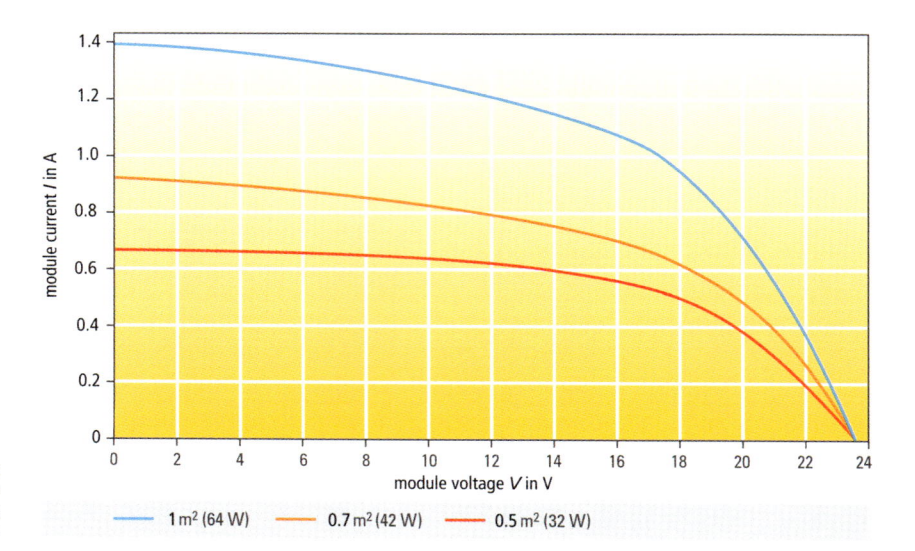

Figure 2.73
Typical characteristic I–V curves of amorphous thin-film modules
Source: Uni-Solar

As an alternative to continuous cell strips across a module, some manufacturers of thin-film modules use separate large-scale cell arrays with an area of, for example, 34cm × 12cm. In each cell array, several approximately 0.5cm wide cell strips are interconnected with one another. The cell arrays are then interconnected via bypass diodes to form a large module. The division into separate cell arrays and the interconnection with bypass diodes optimizes the shading tolerance of these modules.

SHADING

Compared with crystalline modules, thin-film modules have a much greater shading tolerance. With standard modules from individual silicon wafers, complete shading of a cell generally leads to failure of half the module (see section '2.1.9 Hot spots, bypass diodes and shading'). In contrast, the stripe-shaped individual cells of thin-film modules help to prevent cells from becoming completely shaded. The power therefore reduces only proportionately to the shaded area. The shading losses are often lower than with crystalline silicon modules.

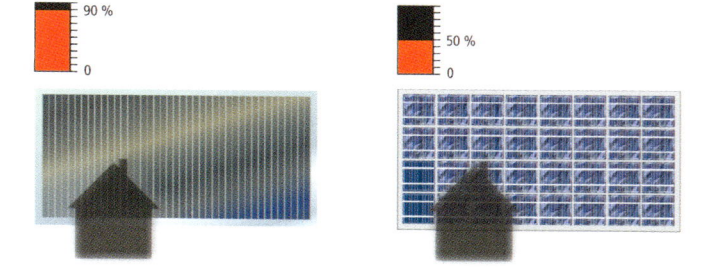

Figure 2.74
Comparison of the shading characteristics of thin-film modules and crystalline silicon modules
Source: Solarpraxis

When designing thin-film systems, attention should be paid to the different effect of shadows parallel and perpendicular to the cell strips. The shadow to the left in Figure 2.75 causes greater losses and should therefore be avoided.

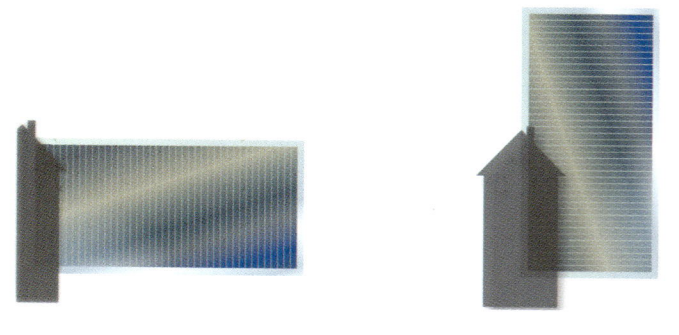

Figure 2.75
System planning taking shading into account with thin-film modules
Source: Solarpraxis

LOW LIGHT CONDITIONS

As described in section '1.5.3 Spectral sensitivity' in Chapter 1, thin-film cells are able to absorb visible light with short and medium wavelengths better than crystalline cells. This spectral sensitivity enables thin-film cells to use low solar irradiance more effectively. This means that in outdoor tests, thin-film modules can achieve greater efficiency in lower irradiance classes. Measurements conducted by TÜV in Germany on CdTe single cells have shown efficiencies up to 14 per cent higher than the STC value with irradiance below 500W/m^2. If stacked cells, as with amorphous cells, are deployed, this effect can be increased to more than 30 per cent above the STC efficiency. The additional yield is attained, above all, by optimizing the top cells to the energy-rich blue light of the solar spectrum. This portion of the solar spectrum is able to penetrate clouds so that when there is greater diffuse radiation, as is often the case with cloudiness and lower irradiance, the efficiency of the cells increases.

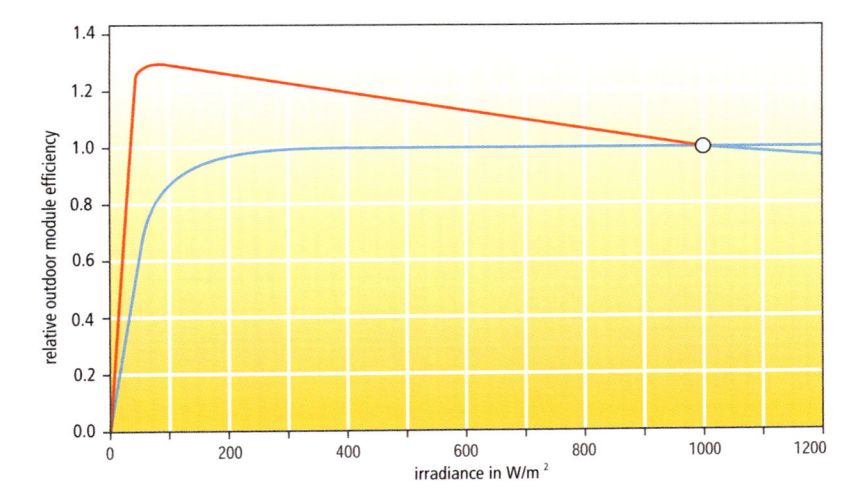

Figure 2.76
Module efficiency under open air conditions with amorphous triple cells compared to crystalline modules
Source: based on measurements carried out by Dutch research institute ECN

Figure 2.77 shows a slight increase in the MPP voltage compared to crystalline cells under low light conditions. With low incident angles for the sunlight and/or high air mass values, CIS modules have slightly increased efficiency (up to 2 per cent), while amorphous silicon loses up to 3 per cent efficiency. The efficiency of crystalline modules falls by around 1 per cent (Durisch et al, 2004).

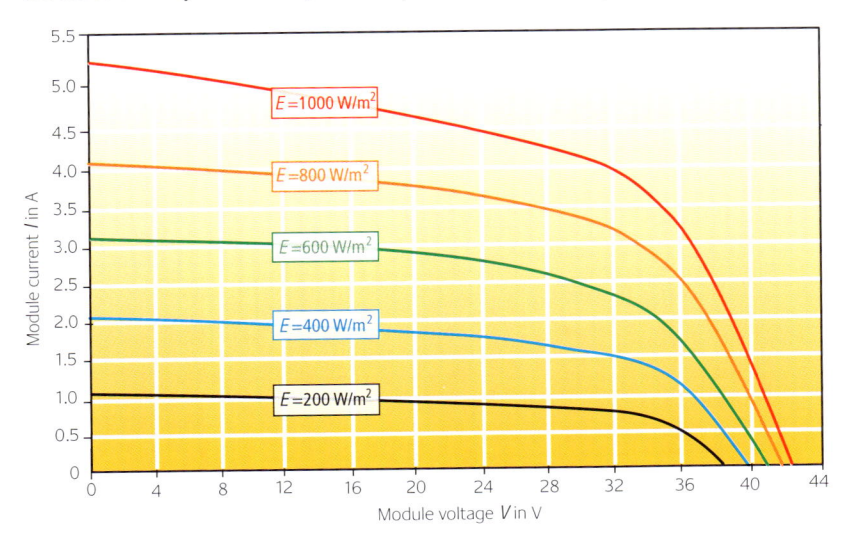

Figure 2.77
Current/voltage curve with different irradiance and constant temperature for amorphous triple cell modules
Source: Uni-Solar

TEMPERATURE BEHAVIOUR

Generally, thin-film modules do not react as significantly to increases in temperature. Because of their larger band gaps, thin-film cells lose less power at higher temperatures. CIS modules show similar temperature behaviour as crystalline silicon modules. The power coefficient is only somewhat smaller than with crystalline modules.

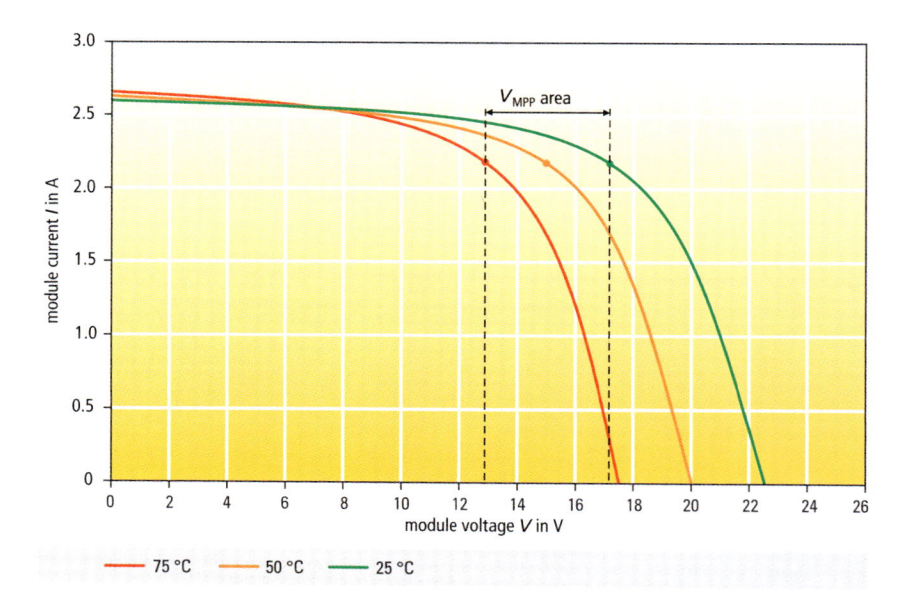

*Figure 2.78
Temperature dependence of
CIS modules
Source: Shell Solar*

The power reduction for each degree Celsius increase in temperature is, compared with crystalline silicon modules, up to 0.3 per cent less with CdTe modules and up to 0.4 per cent less with amorphous modules. The MPP points at different temperatures lie closer together. The strength of thin-film technology therefore lies in building integration where it is usually difficult to guarantee good module ventilation and a minimum of shading.

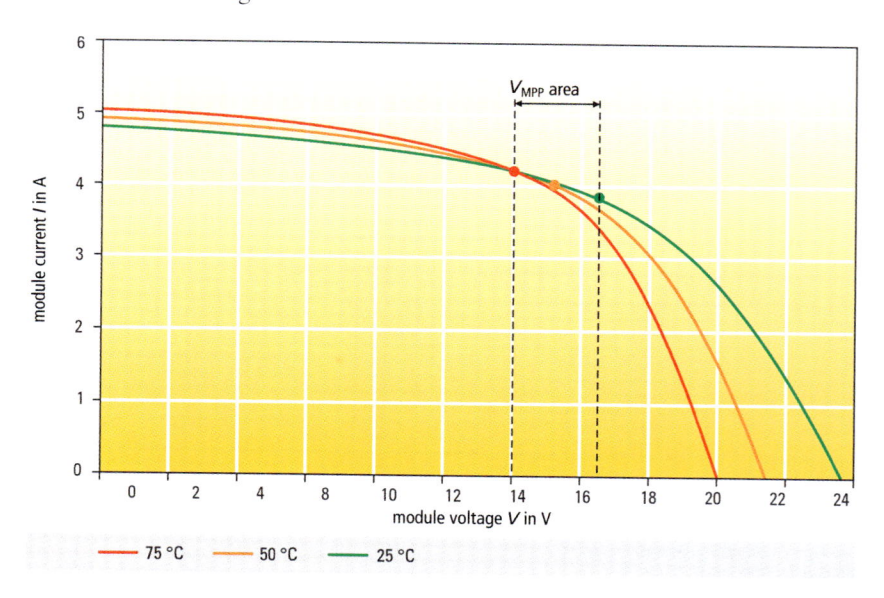

*Figure 2.79
Temperature dependence of
amorphous modules
Source: Uni-Solar*

With amorphous modules and low irradiances, the temperature coefficient for the power can even assume a positive value. In this case, there can be even more power at higher temperatures than at 25°C. In contrast, for crystalline modules the temperature coefficient for all irradiances is negative.

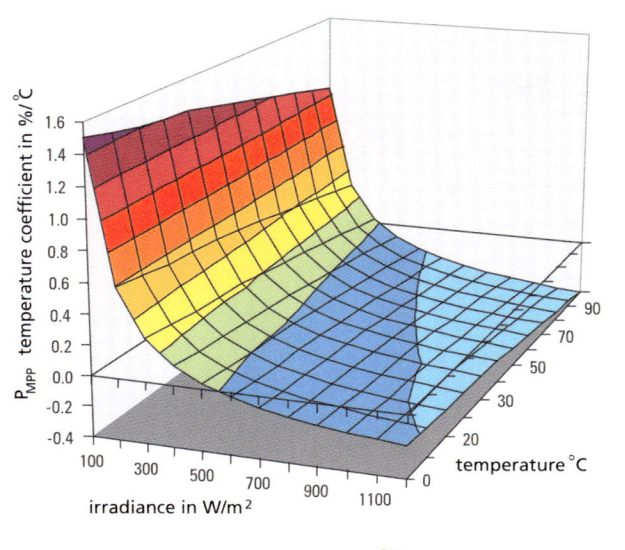

Figure 2.80
Power temperature coefficient for an amorphous tandem module
Source: King, 1997

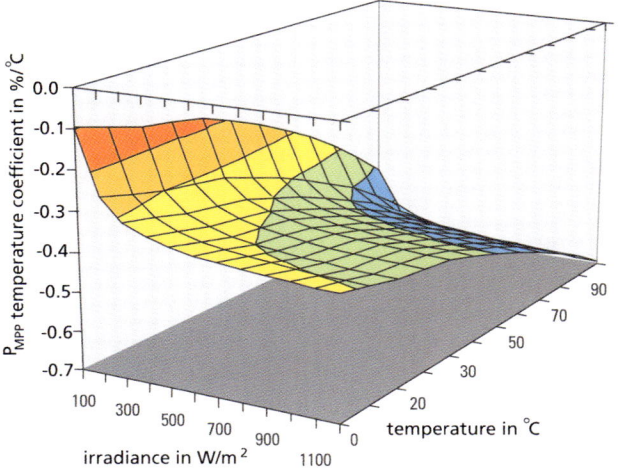

Figure 2.81
Power temperature coefficient for a polycrystalline module
Source: King, 1997

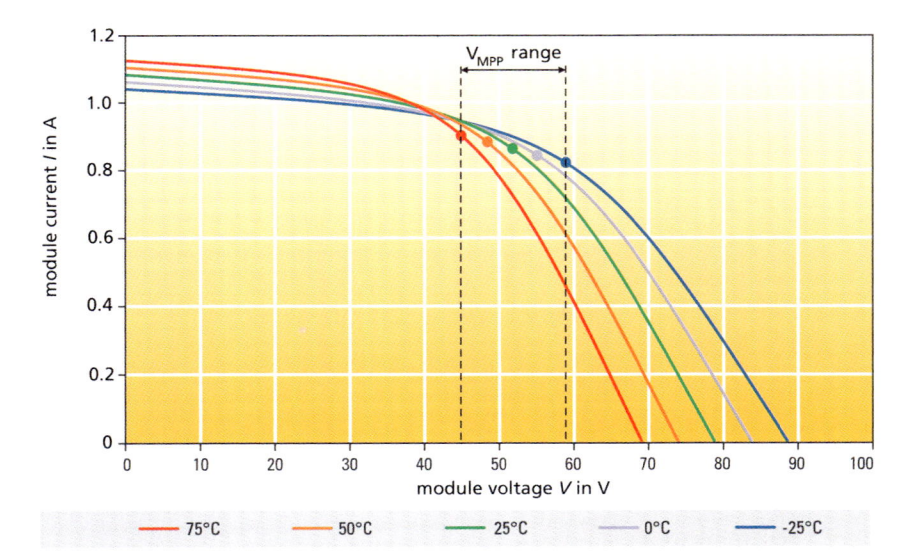

Figure 2.82
Temperature dependence of CdTe modules
Source: AntecSolar measured data

CdTe modules generally have a high nominal voltage. These lie between 30V and 70V. At irradiance levels of 100W/m², the voltage does not collapse as it does in crystalline modules. CdTe modules still achieve 10 per cent of their nominal power. With increasing temperature (E = constant), the MPP voltage does not fall as much. Measurements conducted by FirstSolar showed a slight MPP voltage increase with low irradiance and a temperature of 25°C (see Figure 2.83).

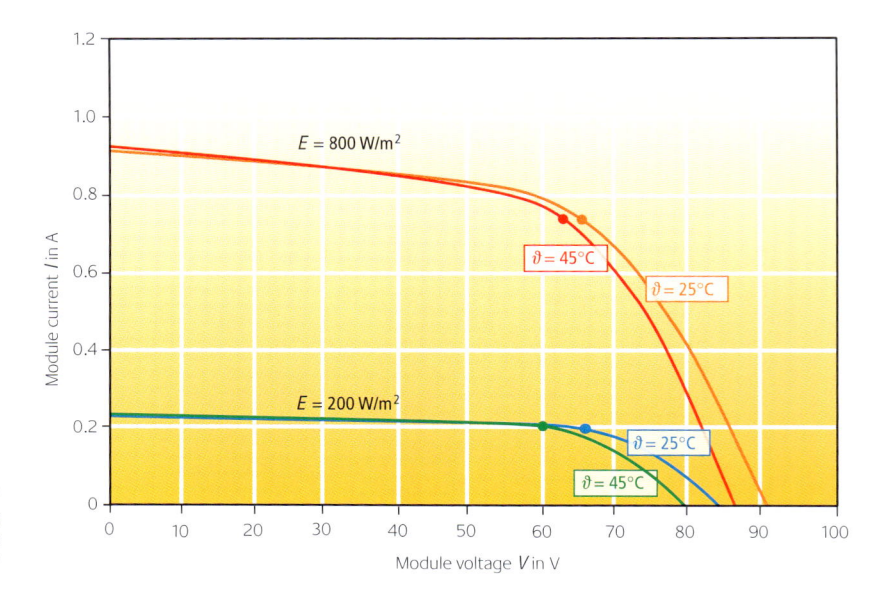

Figure 2.83
Combined temperature and irradiance
dependence of CdTe modules
Source: FirstSolar measured data

Typical temperature coefficients for thin-film modules at STC can be seen in Table 2.4.

Table 2.4
Comparison of typical
temperature coefficients
Source: manufacturers' specifications,
measured data LEEE-TISO,
TÜV and ECN

Temperature coefficient	Amorphous modules	CIS modules	CdTe modules
Open-circuit voltage	−0.19 to −0.5%/°C	−0.26 to −0.5%/°C	−0.22 to −0.43%/°C
Short-circuit current	+0.01 to +0.1%/°C	+0.01 to +0.1%/°C	+0.02 to +0.08%/°C
MPP power (STC)	−0.1 to −0.3%/°C	−0.33 to −0.6%/°C	−0.18 to −0.36%/°C

Differences in electrical qualities, such as behaviour with diffuse light, increased temperature and shading, ensure that with many implemented systems, it is thin-film modules (particularly those consisting of amorphous stacked cells) that achieve a high-energy yield in kWh per kWp and thus a high-performance ratio (see section '4.8 Yield forecast' in Chapter 4). The specific additional yield for amorphous triple modules relative to crystalline PV modules in Central European locations is 10 per cent to 20 per cent.

2.1.11 Quality certification for modules

For laymen it is difficult to judge the quality of a module. For this reason, certification marks provide a certain amount of security when making assessments. The certification of the testing institutes to IEC is generally recognized. A further important assessment criterion is the length of the manufacturer's warranty period.

CERTIFICATION AND APPROVAL TESTING

A special testing procedure has been developed for modules at the European Commission's Joint Research Centre in Ispra (Italy). The resulting Test Specification No. 503, 'Crystalline silicon terrestrial photovoltaic (PV) modules – Design qualification and type approval', was adopted in 1993 as the standard IEC 61215 by the International Electrotechnical Commission (IEC). For amorphous modules, the specification has been extended to take into consideration the degradation of the cell material. Thus in 1996 a further standard, IEC 61646, 'Thin-film terrestrial photovoltaic (PV) modules – Design qualification and type approval' was specified. Modules approved by this test are considered to be very reliable and durable.

For the module certification, eight modules are selected at random from the production line. One module is used as a control, whilst the other seven modules are subject to various testing procedures. These include the following tests:

- visual inspection;
- performance under different conditions;
- (STC, NOCT and at T = 25°C and E = 200W/m^2);
- insulation test;
- measurement of temperature coefficients;
- outdoor exposure test;
- hot spot endurance test;
- thermal cycling test and UV test;
- humidity–freeze test;
- damp–heat test;
- robustness of terminations test;
- mechanical stress and twist tests;
- hail resistance test.

The various tests and measurements are partly defined in other standards. The basis for the performance measurement are specified in the various sections of standard IEC 60904-1 'Measurement of photovoltaic current–voltage characteristics'. This standard specifies the requirements for test and measurement procedures and defines artificial light sources to be used in the test. For instance, it specifies the spectral distribution corresponding to a reference solar spectrum. They also take into account standard IEC 60891, which describes the procedures for temperature and irradiance corrections to the measured I–V characteristics of crystalline cells and modules to other temperatures and irradiances. The UV test is conducted according to standard IEC 61345. The mechanical load test in relation to impact damage is specified in the standard IEC 61724.

The approval certificate to IEC 61215 or IEC 61646 has become generally accepted in recent years as one of the quality marks for modules – particularly in Europe. It is now demanded by most authorizing authorities for national and international support programmes. Standard modules are usually certified to IEC 61215 or IEC 61646. It is rather unusual for special and custom-made modules to be certified because of the high costs for such a test and the low number of modules.

Unfortunately measurement under STC conditions says little about the modules' actual operating performance. It would make sense to specify the mean performance or yield under actual operating conditions (field conditions). This requires measurement and weighting of the efficiencies at various irradiances and temperatures. Meanwhile, various institutes throughout the world are working on standardized procedures for determining the energy yield under field conditions. The greatest progress so far has been made by the European Commission's Joint Research Centre in Ispra, which has led to the draft standard IEC 61853 'Performance testing and energy rating of terrestrial photovoltaic modules'. This draft defines six reference days with typical irradiance and temperatures courses in which the yield is determined.

1. Hot day in desert regions: high irradiance and high temperature (with maximum values of 1100W/m^2 and 45°C).
2. Spring day in mountainous regions: high irradiance and low temperature (with maximum values of 1000W/m^2 and 6°C).
3. Autumn day with a cloudy sky: average irradiance and average temperature (with maximum values of 350W/m^2 and 15°C).
4. Hot, humid summer day with slight cloudiness: average irradiance and high temperature (with maximum values of 600W/m^2 and 30°C).
5. Winter day in northern countries: low irradiance and low temperature (with maximum values of 200W/m^2 and 0°C).
6. Summer day in cool coastal regions: normal irradiance and low temperature (with maximum values of 1000W/m^2 and 18°C).

As all required data are recorded as part of the measurement of the modules to IEC 61215, this only needs to be combined with the weather data of the reference days using a standardized procedure to determine the yield. It is perfectly possible that a module with greater efficiency under standard test conditions can have a lower energy yield than a comparison module.

PROTECTION CLASS II TEST

This test is concerned with protecting people against electric shocks. This protection must be ensured for modules of Protection Class II by double or increased insulation that remains intact for the entire lifetime of the module. This test presupposes proof of suitability according to IEC 61215.

TÜV-PROOF CERTIFICATION MARK

The TÜV Rheinland has been awarding the TÜV-PROOF mark since 1999, which goes beyond testing to IEC 61215. The certification mark can be awarded for complete systems or individual components such as modules, inverters and DC main switch. The safety and quality of a finished product are tested. The TÜV-PROOF mark on a module signifies that, amongst others, the specified efficiency and conformity with Protection Class II and IEC 61215 were checked and approved.

MANUFACTURERS' WARRANTY PERIODS

The quality of a module is also testified by the length of warranty. Manufacturers' warranties normally last between 10 and 25 years. However, it should be noted to which power the warranty relates: the minimum power or the nominal power rating. A guarantee of 90 per cent of the minimum power with a performance tolerance of 10% is only equivalent to a guarantee of 80 per cent of the nominal power rating.

2.1.12 Interconnection of PV modules

PV modules are combined by series and parallel connection to form an electrically and mechanically larger unit, the PV generator. The series-connected modules are described as a string. To avoid power loss in the overall system, only the same type of modules should be used. A string of three PV modules and the resulting current–voltage curves are illustrated in Figure 2.84.

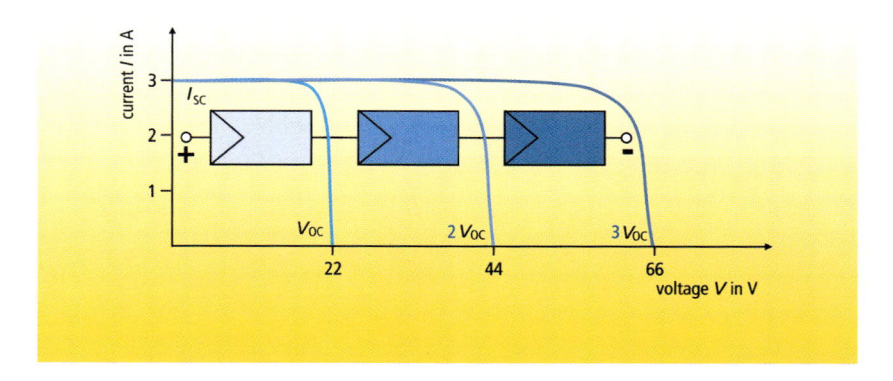

Figure 2.84
Series connection of three PV modules

The number of series-connected modules determines the system voltage of the grid-connected PV systems, which corresponds to the input voltage of the connected inverter. It should be noted that the modules' open-circuit voltage is always greater than their operating, rated and MPP voltages. This can lead to the permissible input voltages of downstream devices being exceeded.

Parallel connections with one PV module per string are often used for stand-alone systems (see Figure 2.85).

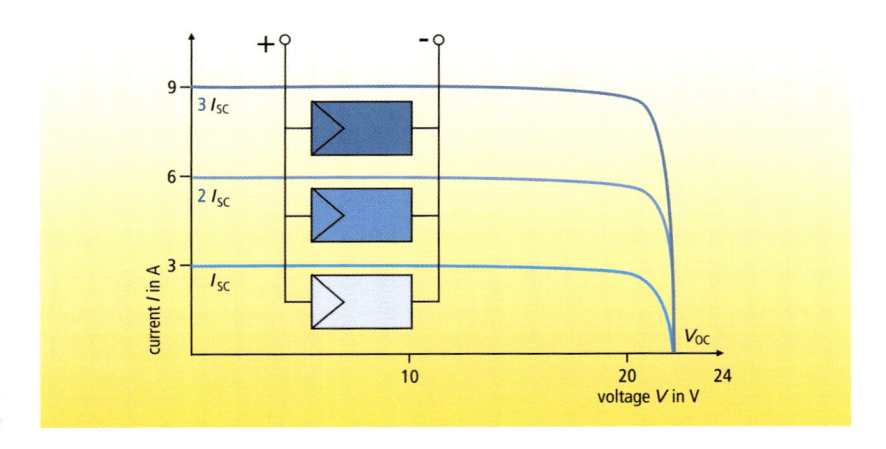

Figure 2.85
Parallel connection of three PV modules

With grid-connected systems several strings are connected in parallel, with the number of modules depending on the system voltage. This produces the characteristic curves shown in the graph in Figure 2.86.

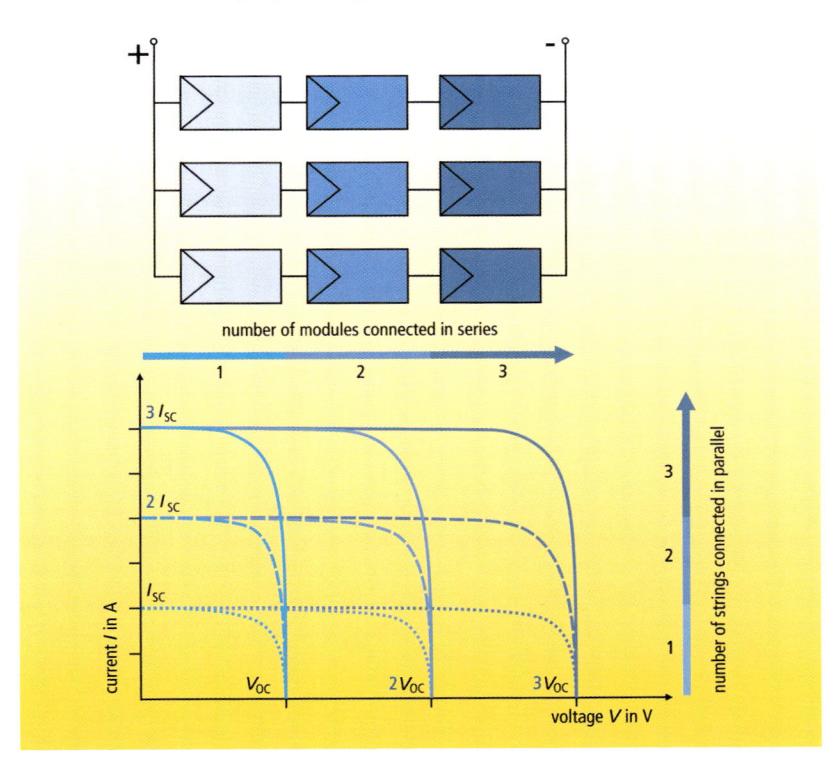

Figure 2.86
Interconnection of PV modules

2.2 PV array combiner/junction boxes, string diodes and fuses

The individual strings are connected together in the PV array combiner/junction box – as is the DC main cable and, if required, the equipotential bonding conductor(s).

The PV array combiner/junction box contains supply terminals and isolation points, and, if required, string fuses and string diodes. Surge arresters are often installed in PV combiner/junction boxes to divert excess voltage to earth/ground (see section '4.6 Selection and sizing of the PV array combiner/junction box and the DC main disconnect/isolator switch' in Chapter 4). This is why the equipotential bonding or earth/grounding conductor is run into the PV array combiner/junction box. The DC main disconnect/isolator switch (see section '2.5 Direct current load switch (DC main switch)') is also occasionally housed in it. Recently, string monitoring elements have begun to be used, mostly in larger systems. These elements notify the data monitoring system of any fault in the string, which allows troubleshooting to be initiated. The PV array combiner/junction box should be executed to Protection Class II and demonstrate a clear separation of the positive and negative sides within the box. If mounted externally, it should be protected to at least IP 54.

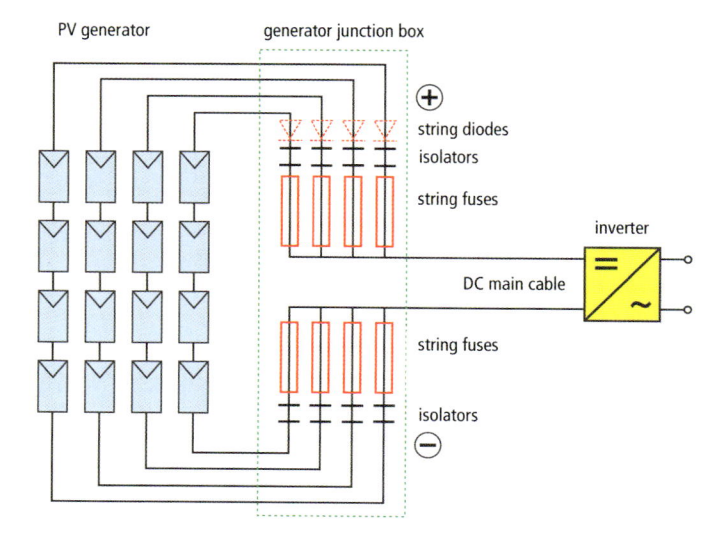

Figure 2.87
String diodes, string fuses and PV array combiner/junction box
Source: R. Haselhuhn

The string fuses protect the wiring against overloading. They must be designed for DC operation (also see section '4.6 Selection and sizing of the PV array combiner/junction box and the DC main disconnect/isolator switch' in Chapter 4). Device fuses (miniature fuses) are usually deployed as string fuses. Other fuse types are unsuitable owing to their high release currents. National codes and regulations need to be referred to when choosing all fuses and circuit breakers.

Figure 2.88
Various DC fuses

In order to decouple the individual module strings, string diodes can be connected in every string in series. Should a short circuit or shading occur in a string, the other strings can continue to work undisturbed. Without the string diodes, a reverse current would flow through the failed string. If string diodes are used, their blocking voltage must be designed to be twice the open-circuit voltage of the PV string at STC (in accordance with VDE 0100 Part 712 in Germany). String diodes are connected in the forward-biased direction. This enables the complete string current to flow through the string diodes (heat sinks are usually necessary). The current flow results in power losses (approximately 0.5 per cent to 2 per cent), caused by the drop in forward voltage at the string diodes of about 0.5V to 1V. Thus, even with shaded systems and using string diodes, the annual energy yield is not substantially greater than with systems without string diodes. The losses from the reverse currents are compensated for by the losses from the voltage drop. The failure of string diodes has proven to be problematic. Experience from the 1000 Roofs programme in Germany has shown that faulty string diodes – and, thus, completely failed PV strings – were often not discovered and repaired until very late.

For this reason, today, grid-connected PV systems are almost exclusively built without string diodes. Studies conducted by the Fraunhofer Institute for Solar Energy Systems ISE have shown that standard modules withstand reverse currents seven times the size of the module short-circuit current without being damaged (Laupcamp, 1998). String diodes are not necessary if only modules of the same type are used, if they adhere to the standards of Protection Class II, are certified to withstand 50 per cent of the nominal short-circuit current in the direction opposite to normal current flow, and if the open-circuit voltage does not deviate by more than 5 per cent between the PV array's individual strings.

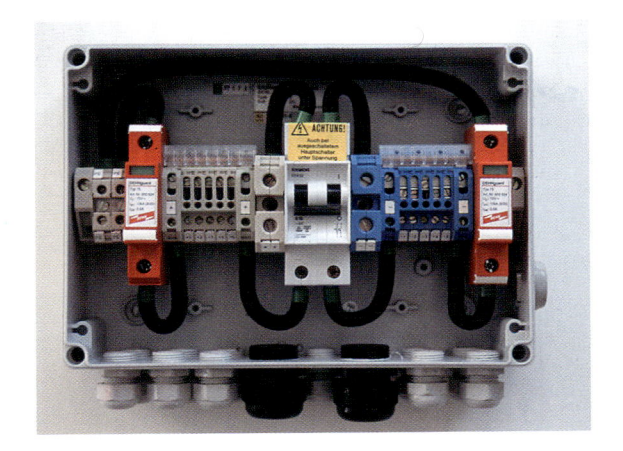

*Figure 2.89
PV array combiner/junction box*

To protect the module and string cables from overloading, string fuses are used in all unearthed/ungrounded cables (positive and negative cables). If string fuses are not used, the string conductors must be dimensioned to the maximum short-circuit current of the PV generator less the string current. To choose a suitable PV array combiner/junction box, see Section 4.6. In larger PV systems, besides the main PV array combiner/junction box, which includes switchgear and fuses, additional combiner/junction boxes are also sometimes used (see Figure 2.90).

*Figure 2.90
PV array combiner/junction box with supplementary combiner/junction box*

2.3 Grid-connected inverters

Grid-connected inverters are also known as grid-tied inverters. They are not to be confused with inverters in stand-alone systems (see Chapter 8). Occasionally, inverter-chargers (devices which invert DC power from batteries but are also able to charge batteries from a diesel generator or the grid) are called *grid connected*. However, inverter-chargers are only able to take power from the grid, not put power onto the grid, and are not to be confused with the grid-connected inverters discussed here.

2.3.1 Wiring symbol and method of operation

The solar inverter is the link between the PV array and the AC grid and AC loads. Its basic task is to convert the solar DC (direct current) electricity generated by the PV array into AC (alternating current) electricity and to adjust this to the frequency and voltage level of the building's electrical system.

The symbol on the left is used as the circuit diagram symbol.

It is also known as a DC–AC converter. Using modern power electronics, the conversion to grid-standard alternating current involves only small losses.

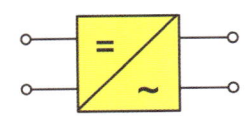

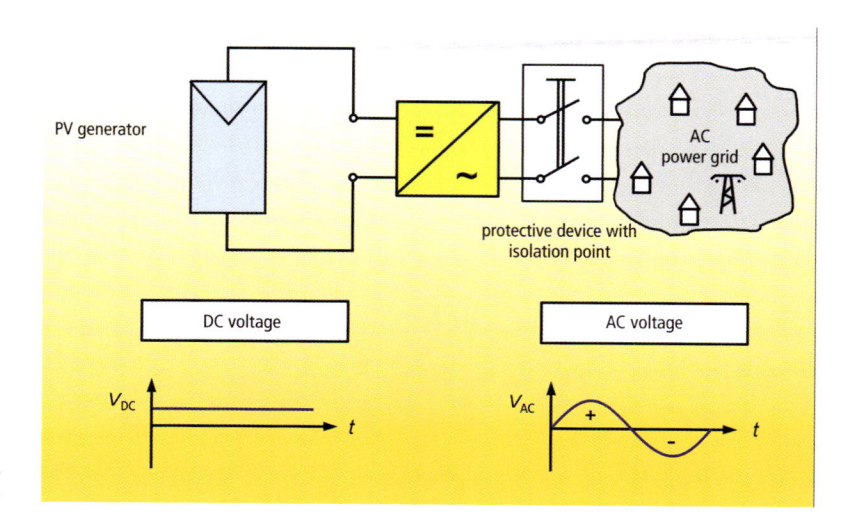

Figure 2.91
Principle of a grid-connected inverter
Source: R. Haselhuhn

In grid-connected PV systems, the inverter is linked to the mains electricity grid directly or via the building's grid. With a direct connection, the generated electricity is fed only into the mains grid. With a coupling to the building's grid, the generated solar power is first consumed in the building, then any surplus is fed to the mains electricity grid.

PV systems up to a power of 5kWp (or up to a size of approximately 50m^2) are generally built as single-phase systems. With larger systems, the feed is three phase: in other words, the feed is connected to the three-phase supply system. However, multiple single-phase inverters are also increasingly being used. These are shared equally between the three phases.

Figure 2.92 shows the principle of coupling PV systems with single- and three-phase inverters to the electricity grid.

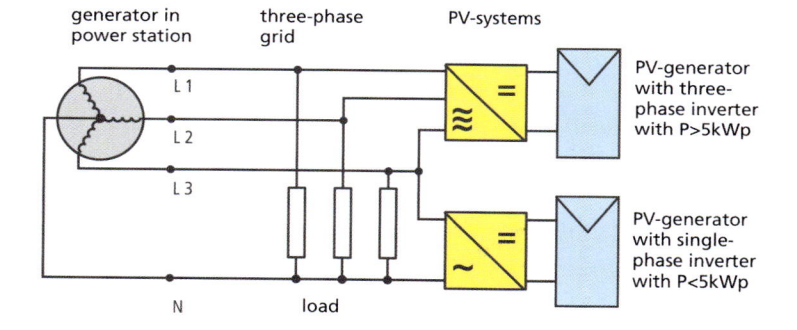

Figure 2.92
Principle of connecting PV systems to the grid with a single-phase and three-phase inverter
Source: R. Haselhuhn

In order to feed the maximum power into the electricity grid, the inverter must work in the MPP of the PV array. As already stated in section '2.1.8 Irradiance dependence and temperature characteristics', the MPP of the PV array changes according to climatic conditions. In the inverter, an MPP tracker ensures that the inverter is adjusted to the MPP point. Since the modules' voltage and current vary considerably depending upon the weather conditions, the inverter needs to move its working point in order to function optimally. To do this, an electronic circuit is used that adjusts the voltage so that the inverter runs at the point at which the PV array achieves its maximum power (MPP). This MPP tracker ensures that the greatest possible power is fed into the mains electricity grid. The MPP tracker essentially consists of an electronically controlled DC converter.

Modern grid-connected inverters are able to perform the following functions:

- conversion of the direct current generated by the PV modules into mains-standard alternating current;
- adjustment of the inverter's operating point to the MPP of the PV modules (MPP tracking);

- recording of the operating data and signalling (e.g. display, data storage and data transfer);
- establishment of DC and AC protective devices (e.g. incorrect polarity protection; overvoltage and overload protection; protection and monitoring equipment to keep within relevant national codes and regulations).

MANUFACTURERS OF GRID-CONNECTED INVERTERS

Up to 10kW: Aixcon, ASP, Conergy, Dorfmüller, Elettronica Santerno, Exendis, Fronius, G & H Elektronic, Ingeteam, Kaco, Karschny, Kyocera, Magnetek, Mastervolt, Pairan, Philips, Phoenixtec, RES, Siemens, SMA, Solar-Fabrik, Solar Konzept, Solarstocc, Solarworld, Solon, Solutronic, Sputnik, Sun Power, Sunset, Sunways, Total Energy, UfE, Victron, Würth Solergy, Wuseltronik, Xantrex.

From 10kW: ACE, Conergy, Elettronica Santerno, Energetica, Kaco, RES, SatCon, Siemens, SMA, Solar Konzept, Sputnik, Xantrex.

Depending upon their principle of operation, grid-connected inverters are divided into grid-controlled and self-commutated inverters.

2.3.2 Grid-controlled inverters

The basic assembly of a grid-controlled inverter is a bridge circuit with thyristors. The classical use of thyristor inverters in automation technology (drive technology, motor controllers, etc.) led to the first solar inverters being designed as thyristor devices. In larger PV systems, thyristor inverters are also used, as well as the predominant insulated gate bipolar transistor (IGBT) inverters.

For the single-phase inverters with lower powers (< 5kWp), there are now only a few manufacturers who still build inverters on this principle.

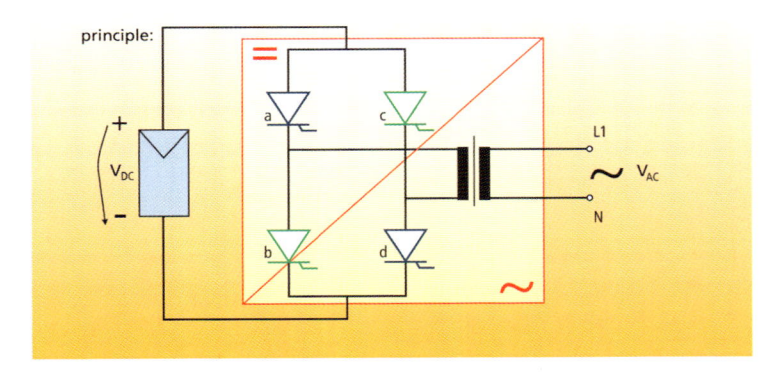

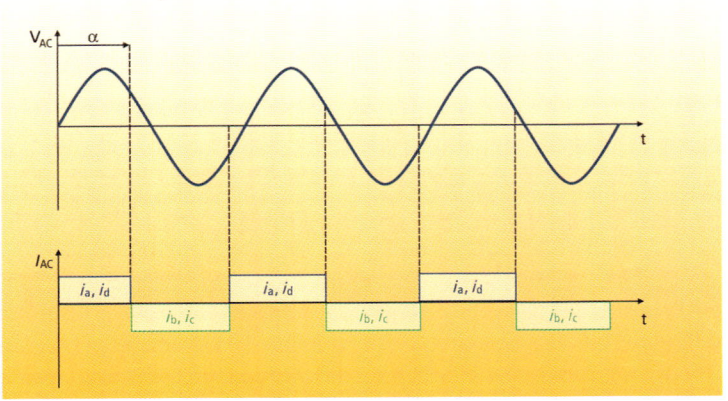

Figure 2.93
Principle of grid-controlled inverters
Source: R. Haselhuhn

The grid-controlled inverter uses the mains voltage to determine the switch-on and switch-off pulses for the electronic power switching devices. Each pair of thyristors in the bridge circuit switches the DC power first in one direction then in the other at a speed of 50Hz (50 times per second). At the moment of switching, the energy is stored in electrolytic condensers, which are connected parallel to the DC input. Since thyristors can only switch a current on, but cannot switch it off again, the mains voltage is required to switch off the thyristors (commutation). For this reason, these inverters are called 'grid controlled'. If grid losses are experienced, inverter shoot-through occurs, which means that stand-alone operation is not possible with grid-controlled inverters. As can be seen in Figure 2.93, square-wave currents are created – for this reason, these inverters are often called square-wave inverters. These deviations from the mains grid sine wave cause relatively high harmonic components (disturbance) with a simultaneous high take-up of reactive (blind) power from the grid. The limit values for harmonics are specified in IEC 61000-3-2, 3, 5 and 11. To limit the harmonic content, compensation equipment and output filters are required. A 50Hz mains transformer is used to electrically isolate the mains grid. In modern thyristor devices, the trigger pulses are formed by a microprocessor. By delaying the trigger pulse (delay-angle control), it is possible to implement MPP tracking.

2.3.3 Self-commutated inverters

In self-commutated inverters, semiconductor elements that can be turned on and off are used in the bridge circuit. Depending upon the system performance and voltage level, the following semiconductor elements are used:

- metal-oxide semiconductor power field effect transistors (MOSFETs);
- bipolar transistors;
- gate turn-off thyristors (GTOs) (up to 1kHz);
- insulated gate bipolar transistors (IGBTs).

These power-switching devices, using the principle of pulse width modulation, enable a good reproduction of the sinusoidal wave.

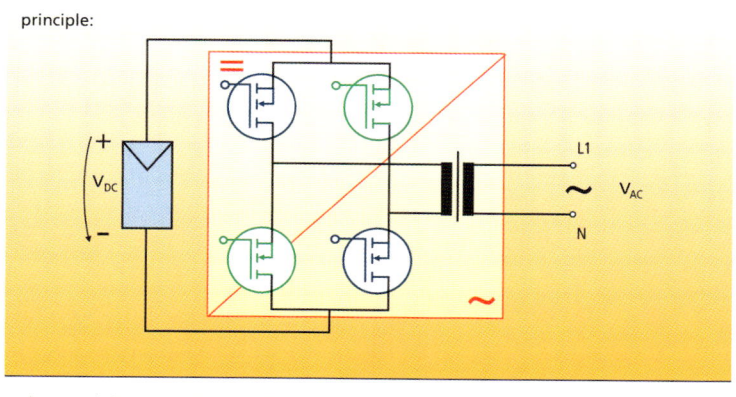

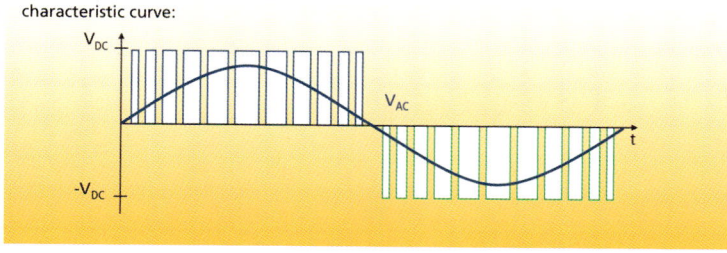

Figure 2.94
Principle of self-commutated inverters
Source: R. Haselhuhn

By rapidly switching the power-switching devices on and off at a frequency in the region of 10kHz to 100kHz, pulses are formed, the duration and spacing of which are shaped to correspond to the sine wave. Thus, after smoothing with a downstream low-pass filter, there is good agreement between the power being fed in and the sine wave of the grid. For this reason, the power that is fed in has only a small low-frequency harmonic component. The reactive power requirement of self-commutated inverters is low.

Because of the high frequency of switching to form the pulses, these devices create high-frequency interference. This means that the problems of electromagnetic compatibility (EMC) need to be taken into account in the equipment concept. This is achieved by using suitable protective circuits and screening the equipment. Self-commutated inverters bearing the CE mark and with European Community (EC) conformity declaration generally keep within the EMC limit values.

Self-commutated inverters are, in principle, suitable for stand-alone grids. If these inverters are connected to the mains grid, the frequency of the fed power must be synchronized with the mains frequency. The pulsing of the bridge circuit matches that of the mains frequency.

SELF-COMMUTATED INVERTERS WITH LOW-FREQUENCY (LF) TRANSFORMER

In self-commutated and in grid-controlled inverters, low-frequency (LF) transformers at 50Hz are often used to match the voltage to the grid. The magnetic field of the transformer separates (electrically isolates) the DC circuit from the AC circuit.

A typical self-commutated inverter with LF transformer comprises the following essential circuitry components:

- switching controller (step-down converter);
- full bridge;
- grid transformer;
- maximum power point tracker (MPPT);
- monitoring circuit with ENS (mains monitoring with allocated switching devices, or MDS) grid monitoring.

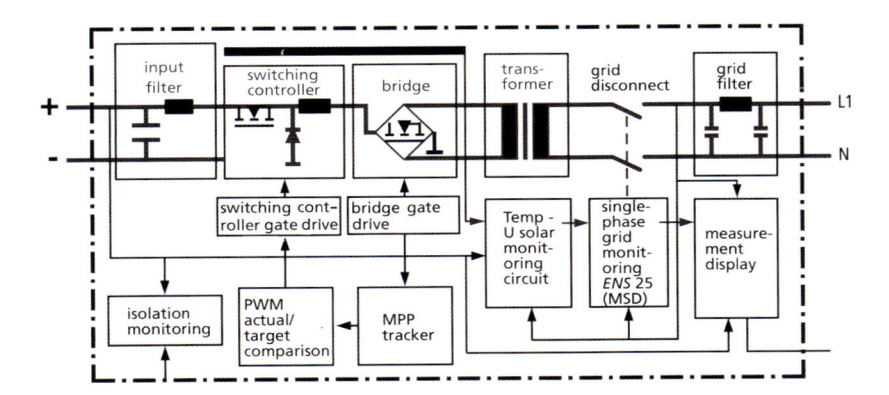

*Figure 2.95
Circuit design for an inverter
with LF transformer
Source: Skytron*

The electrical isolation by means of the transformer allows the PV array to be designed for safety extra-low voltage (SELV). In addition, no potential equalization for the PV array is then necessary (see the section on '4.7 Lightning protection, earthing/grounding and surge protection' in Chapter 4). The transformer also reduces electromagnetic interference.

Power losses can be caused by the transformer. In addition, the transformer results in increased size and weight, as well as increased costs for the equipment. For this reason, some manufacturers use a smaller transformer or omit it completely.

SELF-COMMUTATED INVERTERS WITH HIGH-FREQUENCY (HF) TRANSFORMER

High-frequency (HF) transformers now have a frequency of 10kHz to 50kHz. These transformers, compared to LF transformers, have lower losses, are smaller in size, weigh less and cost less. However, the circuitry in inverters with HF transformers is more complex so that the difference in price between these and inverters with LF transformers is not relevant.

Figure 2.96
Inverters with an HF transformer, not hidden away in the plant room but in the reception area at Solar-Fabrik in Freiburg, Germany

TRANSFORMERLESS INVERTER

For lower power ranges, inverters without a transformer are offered (see Figures 2.97 and 2.98).

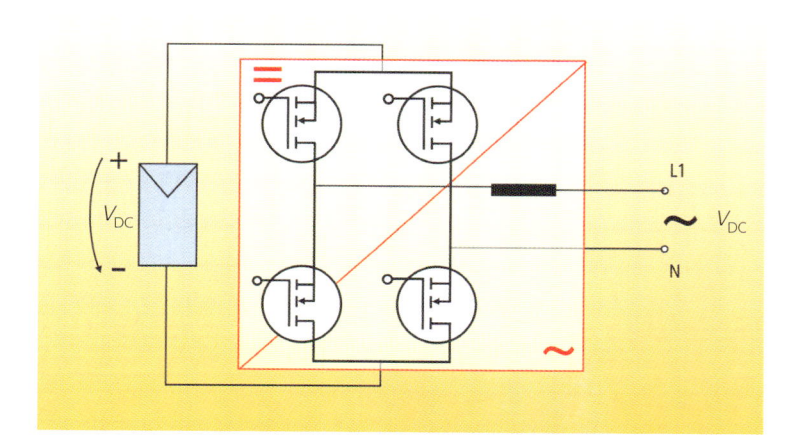

Figure 2.97
Principle of transformerless inverters
Source: R. Haselhuhn

The inverter's losses are reduced by the elimination of the transformer. In addition, the size, weight and costs are lowered. The PV generator voltage must either be significantly higher than the crest value of the grid voltage or be changed by using DC-to-DC step-up converters in the inverter. Where DC-to-DC converters are used, additional losses occur, partially cancelling out the transformer losses that are avoided. The lack of electrical isolation between DC and AC power circuits in transformerless inverters creates high requirements for the electrical safety concept. The Verband der Elektrizitätswirtschaft (VDEW) guidelines for the parallel operation of own-generation systems contains the following stipulation: 'in own-generation systems with inverter and without isolating transformer, DIN VDE 0126 requires the installation of a universal current (AC/DC) sensitive residual current device (RCD)'. 'Universal current sensitivity' refers to the protective device's reaction to faults on the DC and on the AC side. Today, universal current sensitive RCDs for transformerless inverters are offered that ensure electrical safety. It is possible to integrate the protective circuit within the ENS/MSD. In PV systems with transformerless inverters, capacitive discharge currents of over 30 milliamperes (mA) during normal operation can flow to earth via the PV modules. For this reason, conventional RCD protective devices, which trip at 30mA, cannot be used.

Figure 2.98
Transformerless inverter with the 'flying inductor circuit' process
Source: Siemens

Because of the absence of electrical isolation, it is easier to couple electromagnetic interference pulses (electro-smog) within the PV generator. This means that with transformerless inverters, there is a stronger electromagnetic impact on the environment. Measurements of the electromagnetic field showed that with a transformerless inverter, the limit value of the Bundes-Immisionsschutzverordnung (German Federal Emission Control Ordinance) was kept at a distance of 10cm from the PV generator, and the recommendations of engineering biologists were kept to a distance of 1m (Bopp, 1999). Using a special switching method for transformerless inverters (flying inductor circuit), it is possible to avoid electromagnetic grid alternating fields at the PV generator (Holz, 2000). This method was developed by the company Karschny and is used in devices in the Sitop series produced by Siemens.

The pros and cons of inverters with and without a transformer are shown in Table 2.5.

Table 2.5 Comparison of inverters with and without a transformer

	With transformer	Without transformer
Features	• Input and output voltage are electrically isolated • Widely used • Largely central inverters • Largely string and module inverters	• The PV generator voltage must either be significantly higher than the crest value of the grid voltage, or DC to DC step-up inverters must be used
Advantages	• Safety extra-low voltage possible ($V < 120V$; no dangerous body currents when touching the DC side) • Many years of operational use • Reduction of the electromagnetic interference • The potential equalization of the PV generator can be dispensed with	• Higher efficiency (in devices without DC–DC converter) • Reduced weight • Reduced size • In-string and module inverters: reduction of the DC installation
Disadvantages	• Transformer losses (magnetic and ohmic losses) • Increased weight • Increased size	• Use of additional protective equipment: DC-sensitive earth-leakage circuit breaker integrated within the inverter • Fluctuations of the operating point • Live parts must be protected • Entire installation to Protection Class II • Increased electromagnetic interference

2.3.4 Characteristics, characteristic curves and properties of grid-connected inverters

CONVERSION EFFICIENCY η_{CON}

The conversion efficiency describes the losses that arise when converting direct current into alternating current. In inverters, these comprise the losses caused by the transformer (in devices that have a transformer), the power switching devices and by own consumption for management, control, recording operating data, etc.

$$\eta_{CON} = \frac{P_{AC} \text{ Input real power (of fundamental component)}}{P_{DC} \text{ Input real pwer}}$$

The conversion efficiency is strongly dependent upon the input power. It also changes by a few percentage depending upon the inverter's input voltage. For a long time, this effect was neglected.

Two Swiss institutes are conducting intensive investigations into the voltage dependence of solar inverters: the laboratory for electronic measurement systems at the Neu-Technikum Buchs (NTB) and the photovoltaics laboratory at the Bern University of Applied Sciences (BFH) in Burgdorf. Franz Baumgartner at the NTB is conducting investigations into PV systems in operation and comparing the inverter efficiencies according to manufacturers' measurements with the PV generator energies under various weather conditions (temperature and irradiance). Meanwhile, Heinrich Häberlin at the photovoltaics laboratory is calculating the efficiencies in laboratory measurements using highly dynamic solar generator simulators. In laboratory experiments, tracking efficiency (as well as many other parameters), in particular, can be found with high accuracy.

TRACKING EFFICIENCY η_{TR}

The first grid-connected inverters had a fixed point controller (i.e. the inverter's operating point was set to a particular voltage). Any adjustment to the varying climatic conditions was possible only to a very limited degree.

A state-of-the-art grid-connected inverter in a grid-connected PV system has to ensure optimum adaptation to the characteristic curve of the PV array connected to it (I–V curve). During the day, the operating parameters in the PV array are constantly changing. The differing irradiance and temperature conditions change the PV array's maximum power point (MPP). In order to always transform the maximum solar power into alternating current, the inverter must automatically set and track the optimum operating point (MPP tracking). The quality of this inverter adjustment to the optimum operating point is described by the tracking efficiency:

$$\eta_{TR} = \frac{P_{DC} \text{ instantaneous input real power}}{P_{PV} \text{ maximum instantaneous PV array power}}$$

Here the fluctuation of the operating point caused by undesired coupling of the grid voltage frequency into the DC side should be as small as possible. This effect occurs particularly strongly in transformerless devices.

The adjustment speed of the MPP tracking system affects whether irradiance peaks of short duration can be utilized, such as occur on days with rapidly changing cloud conditions. However, high power peaks well above 1000W/m² are usually cut off by the inverter's power limit.

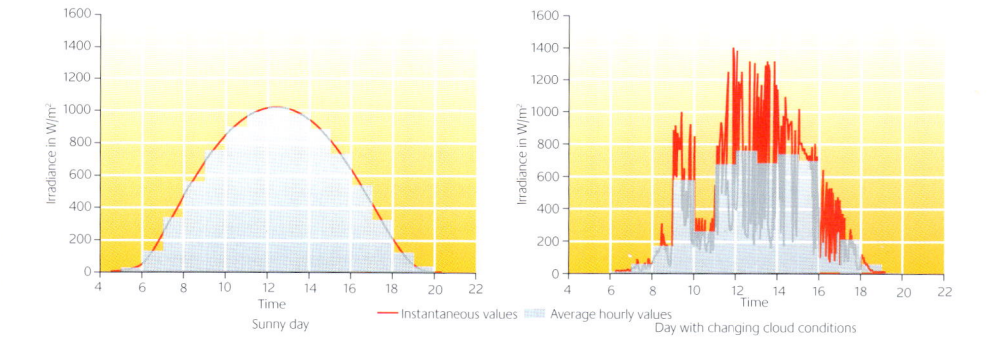

Figure 2.99
Instantaneous values (red line) of insolation compared to hourly values (blue line) on a cloudless day (left) and on a cloudy day
Source: Burger, 2005

STATIC EFFICIENCY η_{INV}

Static efficiency is formed as the product of conversion and tracking efficiency:

$$\eta_{INV} = \eta_{CON} \times \eta_{TR}$$

This static efficiency can be calculated for various load scenarios.

Generally, only the conversion efficiency that is achieved during operation in the inverter's nominal range (V_n and I_n) is stated as the nominal efficiency on the data sheets. In addition, the maximum efficiency is also often stated, which usually lies in the partial load range of 80 per cent to 50 per cent of the nominal power.

The nominal scenario and the maximum efficiency are reached only under certain irradiance and temperature conditions. The changing irradiance is responsible for the fact that the inverter often works in the partial load range and only very rarely in the nominal state. For the solar yield over the whole year, the dependency of the efficiency on different PV output and, hence, different loads on the inverter is a decisive factor. Therefore, efficiency characteristic curves (see Figure 2.100) provide a better picture than stating the nominal efficiency.

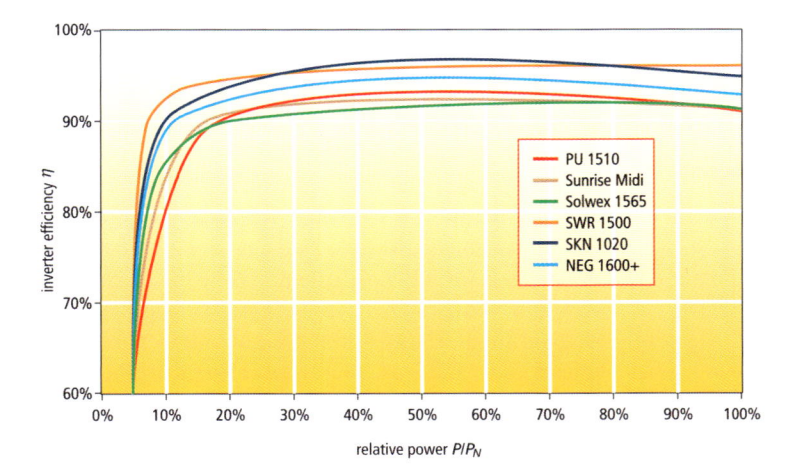

*Figure 2.100
Characteristic curves for various inverter types (according to manufacturers' specifications)
Source: R. Haselhuhn*

The efficiency characteristic curves are accurate at a certain ambient temperature for the inverter and depend upon the input voltage.

EURO EFFICIENCY η_{Euro}

In order to enable and facilitate a comparison of different inverters based on their efficiency, a European standard way of measuring efficiency, η_{Euro}, was introduced. This is a dynamic efficiency weighted for the European climate. Figure 2.101 shows the frequency and energy content of different irradiance classes of solar radiation based on one year in Germany.

Since the irradiance at the PV array fluctuates frequently during the day, the inverter is constantly under different power loads. The inverter's efficiency depends upon the instantaneous input power to the device. If the load is smaller, caused by low irradiance at the PV array, the efficiency is lower.

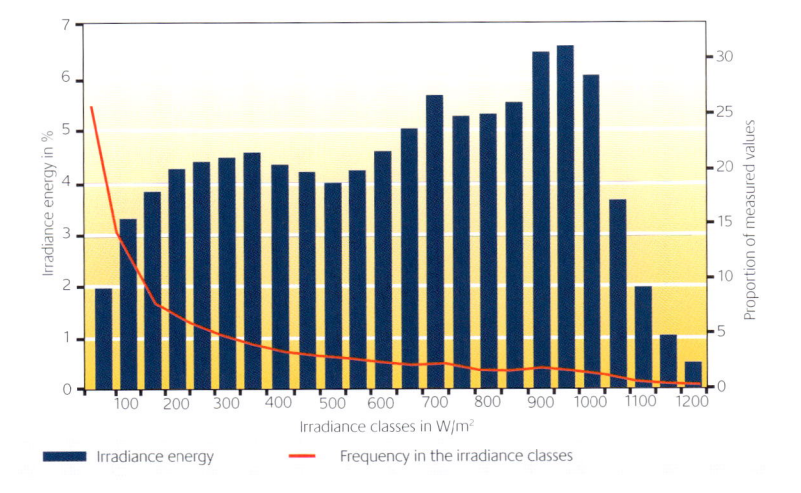

*Figure 2.101
Frequency and energy of different irradiance classes based on a south-aligned 30° tilted array in Munich, Germany
Source: five-minute values, solar laboratory Munich UAS*

In the Central European climate, the most energy is generated in the middle power range of a PV module's nominal power rating. For this reason, manufacturers optimize the inverter efficiency for partial load operation. Inverters are best assessed from their average operating efficiency. A good approximate description of this is provided by what is called the European efficiency (Euro efficiency) (Hotopp, 1991).

In order to take account of the different load conditions, particular parameters are used for the energetic weighting. To calculate the Euro efficiency, the following six efficiencies at different outputs are used:

$$\eta_{Euro} = 0.03 \times \eta_{5\%} + 0.06 \times \eta_{10\%} + 0.13 \times \eta_{20\%} + 0.1 \times \eta_{30\%} + 0.48 \times \eta_{50\%} + 0.2 \times \eta_{100\%}$$

The value $\eta_{100\%}$ gives the efficiency in the nominal case. The power of the PV array then corresponds to the inverter's nominal power ($P_{PV} = P_{nINV}$). On average, the 100 per cent inverter load is assumed for 20 per cent of the operating time over the year ($0.2 \times \eta_{100\%}$). The efficiency at half array power is given the strongest weighting in the value for the Euro efficiency. The 50 per cent inverter load is assumed for 48 per cent of the operating time over the year. The other four load cases are formed in a similar way. The Euro efficiency allows the efficiencies of different inverters to be compared; this is an efficiency weighted for the Central European climate. The latest inverters achieve a Euro efficiency of 92 per cent to 96 per cent. However, the Euro efficiency is usually only calculated by the manufacturers at nominal voltage, while the MPP operating range covers a wide voltage range. The efficiencies depend upon the ambient temperature and the input voltage. For this reason the efficiency should be measured by the manufacturers in line with the IEC 61683 standard 'Photovoltaic Systems–Power Conditioners–Procedure for Measuring Efficiency'. According to this, the efficiency is to be calculated at an ambient temperature of 25°C ±2°C and at the three following voltages:

1 minimum input voltage;
2 nominal voltage;
3 90 per cent of the maximum input voltage.

So far the definition of the Euro efficiency and the process for measuring tracking effectiveness have not been specified in a standard. A DKE 373 working group in Germany is working on standardized processes for calculating the efficiency of inverters at different voltages. In addition to measuring the partial output ranges of IEC 61683, this requires measurement of values at 5 per cent and 110 per cent of the nominal power rating. The 5 per cent value is required according to the definition of the Euro efficiency. The 110 per cent value is useful since a short-term output of 110 per cent of the rated power output may be fed into the electricity grid. The latter values would need to be incorporated within a new definition of the Euro efficiency. When determining the efficiency, the MPP tracking and, hence, the tracking effectiveness are taken into account. Measurement arrangements and voltage jumps are proposed here (Kremer and Fuhrmann, 2005).

A corresponding standard would significantly improve the estimates of the actual operating efficiency and would be helpful when comparing devices. Franz Baumgartner at the NTB and Heinrich Häberlin at the photovoltaics laboratory recommend calculating weighted averages for different voltages in each case, and from this they derive new definitions for overall efficiency (*Euro realo* and *total eta*, respectively) (Baumgartner, 2005; Häberlin et al, 2005). It remains to be seen whether these approaches will be adopted by the industry and incorporated within European standards. In principle, there would be no need to create a new definition of the Euro efficiency since the definition of this value has not yet been specified in a standard. The Euro efficiency should be measured at 25°C ambient temperature and the value should be specified with the associated input voltage.

Measurements on inverters in the Burgdorf photovoltaics laboratory showed deviations in the conversion efficiency of up to 2 per cent across wide power ranges with differences of 200V in the input power range. When measuring the quality of inverters' MPP tracking, the devices were found to be strongly voltage dependent by up to 10 per cent, especially in the lower partial power range (below 20 per cent). Happily, tracking effectiveness in the output range above 20 per cent was found to be 100 per cent. Heinrich Häberlin calculated the overall efficiency using the values for the conversion efficiency and tracking effectiveness for three or four different voltages. Referred to as 'total efficiencies', these approximate the actual operating efficiency of the devices by taking into account the voltage ranges.

These values allow planners to find the best system voltage for system configuration. This enables the optimum module wiring to be found for the inverter – as a result, this is desirable for planners if manufacturers start calculating the efficiencies at different voltages. The required inverter measurements do require more work than the efficiency measurement, which has been usual to date.

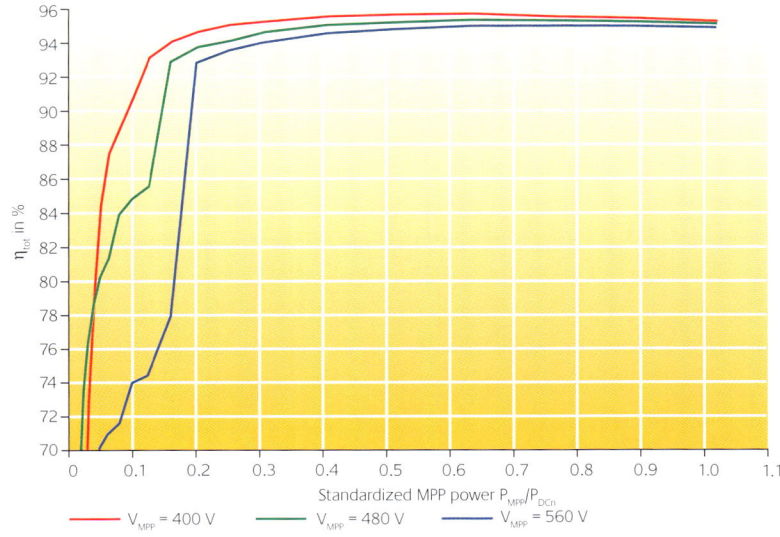

Figure 2.102
Measurements of the overall efficiencies ('total' efficiencies) for a type NT4000 transformerless inverter with different input voltages
Source: Häberlin et al, 2005

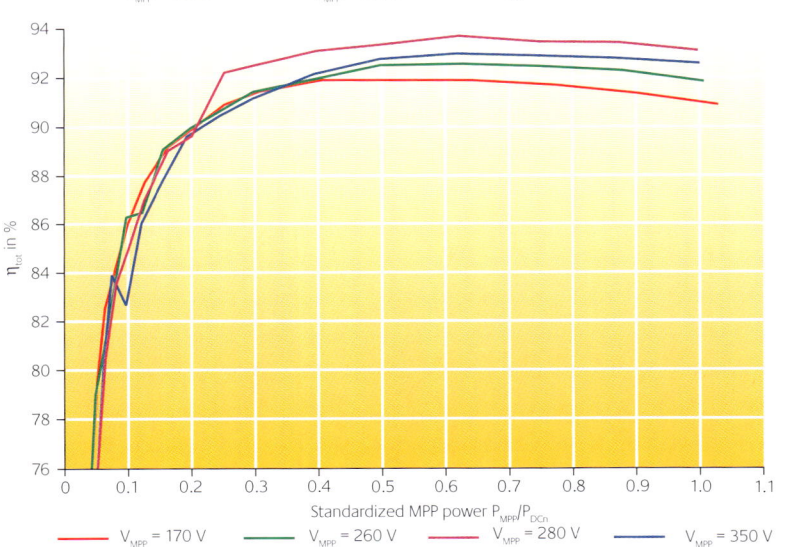

Figure 2.103
Measurements of the overall efficiencies ('total' efficiencies) for a type IG30 inverter with transformer for different input voltages
Source: Häberlin et al, 2005

OVERLOAD BEHAVIOUR

In systems that do not have the optimum orientation or systems with partial shading, it can make sense from technical and economic points of view to under-dimension the inverter somewhat (see the section on '4.4 Sizing the inverter' in Chapter 4). In order to have certainty in planning, it is important to consider the inverter's behaviour under overload conditions. If an inverter is subject to a load above its rated load, its electronic components are, in turn, subjected to a greater thermal load. A brief overload will not cause the components to heat up significantly and will generally be tolerated without damage to the device. The power reduction control is then thermally activated: as soon as a critical component temperature is reached, the power is reduced. As a result, the maximum power is also influenced by the inverter's ambient temperature. Many devices feature additional fans that push the power reduction threshold upwards. The power is often limited by shifting the working point – the inverter no longer operates at the MPP. It is also possible to restrict the power by switching the inverter alternately off and on. Inverters generally start reducing power at figures in excess of the rated power output. The power reduction behaviour of inverters and its effect on efficiency and yield were investigated at the Fraunhofer Institute ISE in Freiburg, Germany (Burger, 2005).

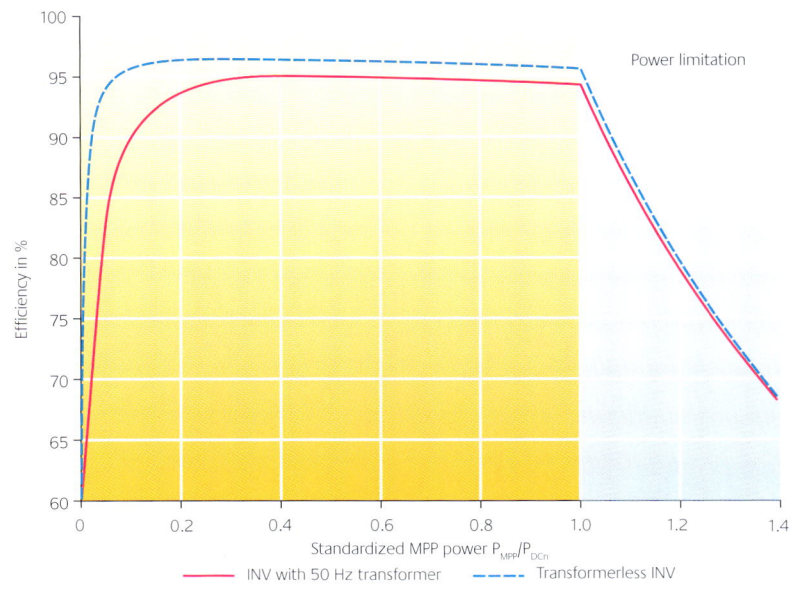

*Figure 2.104
Efficiency and power reduction for inverter with mains transformer (blue) and transformerless inverter (red)
Source: Fraunhofer Institute ISE in Freiburg, Germany; Burger, 2005*

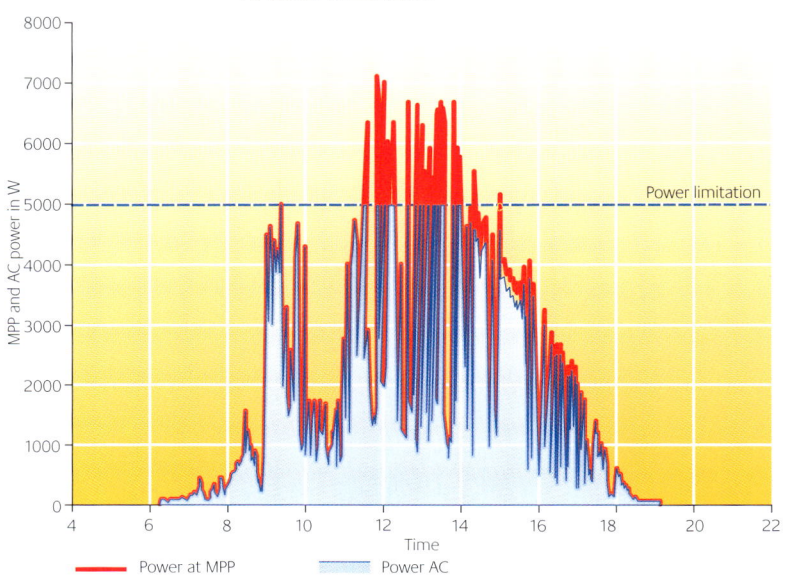

*Figure 2.105
Power restriction on a day with strong cloud/sun alternation
Source: Fraunhofer Institute ISE in Freiburg, Germany; Burger, 2005*

RECORDING OPERATING DATA

Almost all inverter manufacturers offer data capture functions directly integrated within the devices or as an optional add-on. The data can either be read on a display, represented by LEDs and/or sent to a PC. This enables the PV system to be monitored and evaluated.

Data recording generally covers the following values:

- input side: voltage V_{DC}, current I_{DC} and power P_{DC};
- output side: voltage V_{AC}, current I_{AC}, power P_{AC} and frequency f;
- inverter operating time;
- generated energy volume;
- device status and faults.

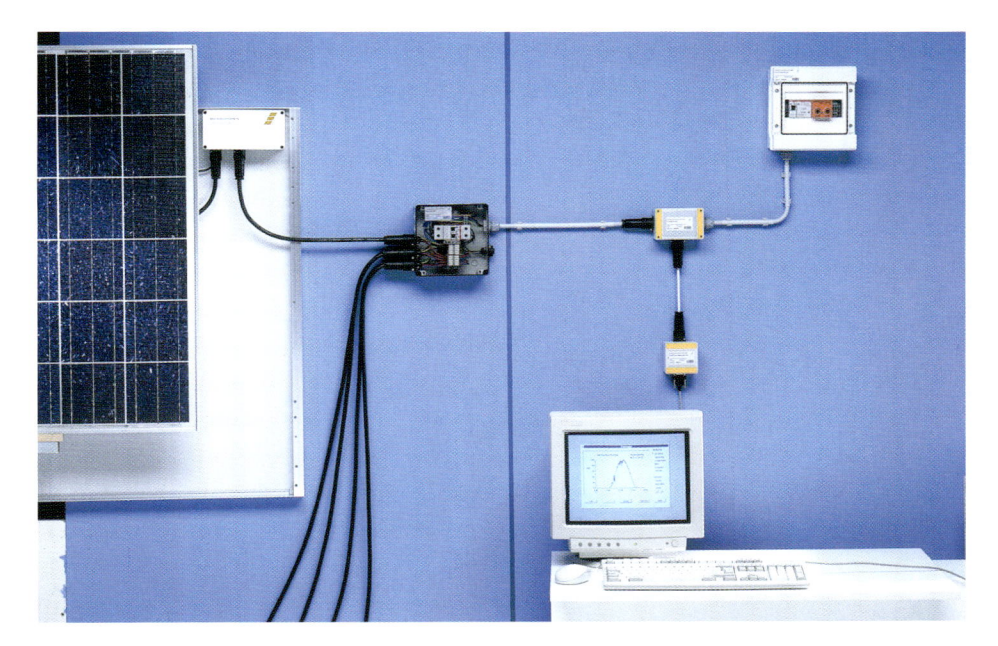

*Figure 2.106
Module inverter with operating data
recorder and PC interface
Source: Dorfmüller*

The number of values recorded is anywhere between five values per day and one value per minute. Operating time and energy volume are recorded as daily, weekly, monthly and/or yearly values. The data are either stored directly or sent to a data logger and output on a computer. Data is evaluated using suitable software that manufacturers provide for their devices. The storage capacity of commercially available devices is between 28 and 450 days. Some devices require an external data logger or a computer takes over this function. Most inverters have a serial or parallel PC interface: RS-232 or RS-485 interface.

The increasing modularity of inverters has led many manufacturers to offer external data loggers with multiple communications interfaces. These allow the data from multiple inverters to be recorded and evaluated centrally. Increasingly too, automated system monitoring concepts are being realized. Fault signalling contacts enable the output of acoustic alarm signals or visual signals on display units, as well as messages by fax, computer, email, text message or internet. Operating data evaluation systems are described in detail in '7.6 Monitoring operating data and presentation' in Chapter 7.

ADDITIONAL CHARACTERISTICS AND PROPERTIES

There are similar problems with inverters as there are with solar modules when it comes to specifying and calculating the technical data (see '2.1.10 Electrical characteristics of thin-film modules'). In addition, no standard has yet emerged for inverter data sheet specifications. With the specified Euro efficiency, it is worth bearing in mind that there are not yet any standards for finding and calculating this weighted value across different operating states. Often, it is only calculated at the optimum voltage. But since inverter efficiencies are critically dependent upon the input voltage, the efficiencies should be measured in accordance with IEC 61683 across the entire MPP range. The DGS/RAL solar quality auditing commission makes a corresponding requirement for inverters (see section 7.7.2 on 'Quality and reliability of inverters' in Chapter 7). As well as the complete datasheet information, meaningful installation and operating instructions, including extended guarantee and service provisions, are also important. The usual guarantee period is currently five years. Many manufacturers offer an optional extended guarantee at extra cost or as part of a service agreement for which an extra charge is made.

Table 2.6
Additional inverter parameters

	Parameter	Symbol	Unit	Description
Power data	Nominal power DC	$P_{n\,DC}$	W	PV power output for which the inverter is designed for continuous operation under the specified environmental conditions in order to output the nominal AC power
	Maximum PV power	P_{DCmax}	W	Maximum PV power that the inverter can accommodate
	Nominal power AC	$P_{n\,AC}$	W	AC power that the inverter can output continuously (40°C ambient temperature, nominal voltage conditions)
	Maximum AC power	P_{ACmax}	W	Maximum AC power that the inverter can output (40°C ambient temperature)
	Partial efficiency	$\eta_{5\%}$	%	Partial efficiency at 5% of the DC nominal power
	Partial efficiency	$\eta_{10\%}$	%	Partial efficiency at 10% of the DC nominal power
	Partial efficiency	$\eta_{20\%}$	%	Partial efficiency at 20% of the DC nominal power
	Partial efficiency	$\eta_{30\%}$	%	Partial efficiency at 30% of the DC nominal power
	Partial efficiency	$\eta_{50\%}$	%	Partial efficiency at 50% of the DC nominal power
	Partial efficiency	$\eta_{100\%}$	%	Partial efficiency at 100% of the DC nominal power
	Partial efficiency	$\eta_{110\%}$	%	Partial efficiency at 110% of the DC nominal power
	Euro efficiency	η_{Euro}	%	For definition, see section 2.3.4
	Reduced efficiency with increased ambient temperature	$\Delta\eta_T$	%/°C	Specifies the reduction in efficiency at ambient temperatures above 25°C
	Power factor	$\cos\varphi$		The power factor expresses the reactive power requirements and should be higher than 0.9 as soon as the delivered power is > 50% nominal power.
	Switch-on power	P_{on}	W	Specifies the PV power from which the inverter starts working
	Switch-off power	P_{off}	W	Specifies the PV power from which the inverter switches off
	Stand-by power	$P_{stand-by}$	W	Specifies how much power the inverter consumes when it is not feeding and has not yet switched into night shutdown mode
	Night power	P_{night}	W	Specifies how much power the inverter consumes when it is in night mode
Voltages	Nominal voltage DC	$V_{n\,DC}$	V	PV voltage for which the inverter is designed
	MPP voltage range	V_{MPP}	V	Specifies the input voltage range in which the inverter looks for the MPP point
	Maximum DC voltage	V_{DCmax}	V	Maximum PV voltage that may be present at the inverter
	Turn-off voltage	V_{DCoff}	V	Minimum PV voltage at which the inverter will operate
	Voltage range AC	V	V	AC voltage range in which the device feeds energy into the grid
Currents	Nominal current DC	$I_{n\,DC}$	A	PV current for which the inverter is designed
	Maximum DC current	I_{DCmax}	A	Maximum PV current that may flow into the inverter
	Nominal current AC	$I_{n\,AC}$	A	AC current that the inverter supplies to the grid at nominal power
	Maximum AC current	I_{ACmax}	A	Maximum AC current that the inverter can output
	Distortion factor	k	%	Quality factor for the feed-in current and voltage (calculated from the ratio of the effective value of the harmonics to the total effective value of the alternating voltage), should be less than 5%
Other	Dimensions	h, l, w	m	Height, length, width
	Weight	m	kg	Total weight of the inverter
	Noise level		dB(A)	Depending upon the type and the power class, various levels of operating noise may be generated; this should be taken into account when choosing the location of the inverter
	Temperature range	T	°C	Depending upon the type and the power class, different temperature range that should be taken into account when choosing the location of the inverter (e.g. loft or outdoors)

Table 2.7 Additional inverter properties

Properties	Description
IP protection category	Should be considered when choosing the location of the inverter: • IP 5 dust protection • IP 6 dust proof • IP 1 protected against dripping water • IP 3 protected against water spray • IP 4 splash proof • IP 5 hose proof • IP 7 waterproof An inverter outdoors must be protected at least to IP 54
Installation site	Note on the required properties of the installation site
Note on mounting	Note on the installation procedure and required properties of the installation site and the surface on which the unit is installed
Insulation monitoring	Specifies the method used to monitor the insulation resistance of the PV system via the inverter
Overload behaviour	Description of the overload behaviour (preferably efficiency diagrams at various voltages and with ambient temperature stated)
Short-circuit and open-circuit strength	Relates to how the inverter behaves in the event of a fault
Reverse polarity protection	Effective protection against damage to the inverter if polarity is accidentally reversed (e.g. during installation)
Conformity declaration in accordance with German electrical industry guidelines	Inverter manufacturer should provide proof of this
Mains monitoring with allocated switching devices (MSD) (German SFS and ENS)	MSD devices are automatically operating isolation points with two automatic parallel devices for mains monitoring, each with an allocated switching device in series; MSD devices are often integrated within inverters
Service; warranty	The warranty period should be at least two years
Recording operating data; data storage	Recording the relevant operating data and faults, fault-indicating and, if applicable, data storage
PC interface; software; power line modem	RS 232 or RS 485 PC interface; evaluation software; a power line modem can send operating data over electricity cables
Display; indicators; documentation	Sufficient instrumentation; ease of use; documentation for customer and installation engineer

2.3.5 Grid-connected inverter types and construction sizes in various power classes

Grid-connected inverters can be divided into three groups: central inverters, string inverters and module inverters. The associated system concepts that result are explained in detail in Chapter 4. Examples of different grid-connected inverters are shown below along with their dimensions.

CENTRAL INVERTER WITH HIGH POWER OUTPUT RANGE (THREE PHASE)

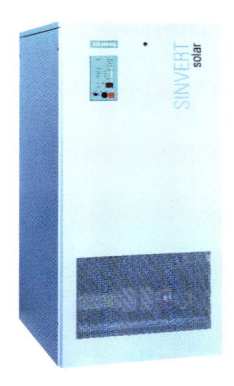

- *Type:* SINVERTsolar 100.
- Manufacturer: Siemens AG.
- Concept: self-commutated inverter with LF transformer.
- DC nominal power: 93kW.
- MPP voltage: 460V to 700V.
- Size: 13,725mm × 950mm × 850mm.
- Weight: 750kg.

CENTRAL INVERTER WITH LOW POWER OUTPUT RANGE (SINGLE PHASE)

- *Type:* Top Class III – TCG 2500/6.
- Manufacturer: ASP.
- Concept: self-commutated inverters with LF transformer.
- DC nominal power: 2.5kW.
- MPP voltage: 82V to 120V.
- Size: 456mm × 320mm × 211mm.
- Weight: 22kg.

STRING INVERTER

- *Type:* Sunny Boy 2100TL.
- Manufacturer: SMA Technologie AG.
- Concept: transformerless, self-commutated inverter.
- DC nominal power: 2kW.
- MPP voltage: 125V to 600V.
- Size: 295mm × 434mm × 214mm.
- Weight: 25kg.

MODULE INVERTER

- *Type:* DMI 150/35.
- Manufacturer: Dorfmüller Solaranlagen GmbH.
- Concept: self-commutated inverter with LF transformer.
- DC nominal power: 120W.
- MPP voltage: 28V to 50V.
- Size: 80mm × 200mm × 100mm.
- Weight: 2.8kg.

2.3.6 Further developments in grid-connected inverter technology

INVERTER WITH MULTIPLE MPP TRACKERS (MULTI-STRING CONCEPT)

Chapter 4 explains how the use of string inverters in shaded systems or systems with different module orientations results in avoidable energy losses. The use of multiple MPP trackers enables these losses to be reduced. The concept is also often called the 'multi-string concept'. Here, the system is designed so that modules having similar irradiance conditions are wired together in one 'string'. Each string has its own DC–DC converter with a separate MPP tracker for operating the respective string or string bundle at its maximum power. The strings' MPP points can therefore be different. Through a DC bus, a constant DC voltage is provided via the DC–DC converter of the actual inverter unit.

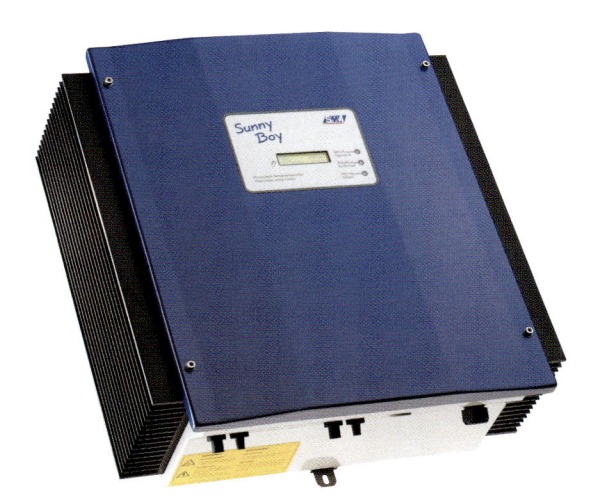

Figure 2.107
Multi-string inverter type SB 4200TL
Source: SMA

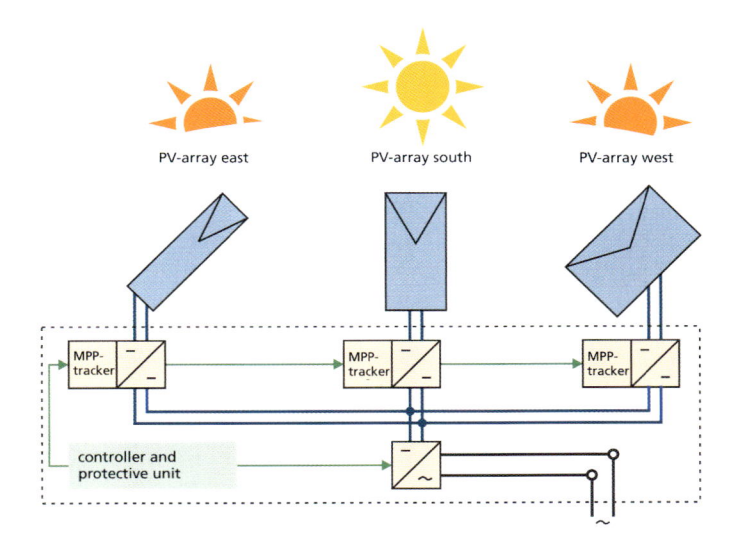

*Figure 2.108
Multi-string inverter
Source: SMA*

INVERTER WITH SEPARATE MPP TRACKERS (STRING CONVERTER CONCEPT)

In this concept the inverter is separated from the MPP trackers. A separate MPP tracker – also called a string converter – is used in each string. This consists of a DC–DC converter and supplies a constant high DC voltage (e.g. 850V), which is carried to the inverter via the DC cables. Thus, in devices produced by RES, it is possible to operate up to 30,000 string converters with one inverter unit, while the strings can have different string voltages (a maximum of 800V), cell technologies (polycrystalline, mono-crystalline, thin film, etc.), module temperatures, module arrangements and orientations, and shading conditions, without the associated mismatching losses occurring. Inverter manufacturer RES combines this concept with modular inverter units with power outputs of 1.5 kilovolt-amperes (kVA) or 5kVA. By connecting up to 256 units together in control cabinets, systems in the megawatt class can be achieved. RES has developed a new kind of automatic 'method of preventing unwanted stand-alone operation when operating private generating systems on the low voltage mains grid' in which there is no need for the 'disconnection device that is accessible at all times' even in systems with total power outputs greater than 30kVA. However, it remains to be seen if this will be acceptable to regulatory authorities. According to the manufacturer's information, the concept has already gained acceptance among several grid operators.

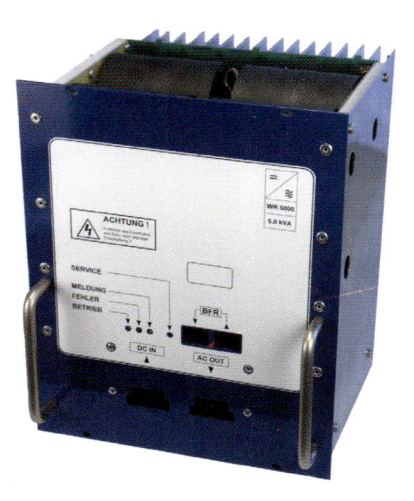

*Figure 2.109
MPP tracker/string converter
Source: RES*

*Figure 2.110
Inverter plug-in module
Source: RES*

Concepts are under development in which per PV module, a separate MPP tracker and DC–DC converter are connected via the DC bus line to the central inverter unit.

MASTER–SLAVE CONCEPT IN LOW POWER RANGES

Another step to optimize energy was developed out of the master–slave concept (see section '4.2.1 Central inverter concept' in Chapter 4). This concept, which is usually used for large inverters (20kW and more), was expanded for inverters in low power ranges (up to 5kW). Multiple small inverters work together as a master–slave unit. When irradiance is low, only the master device operates. As soon as the power limit of the master device is reached as irradiance increases, the first slave device switches in. As soon as its power limit is reached, the next slave device switches in. This concept allows an optimization of efficiency to be achieved. The characteristic curve of the master–slave unit comprises the characteristic curves of the individual inverters and, especially in the low power range, has greater efficiency than an individual device with the same overall power. It is possible to split the individual inverters of the master–slave unit to different sub-arrays or strings so that different MPP points of the individual inverters are possible as operating points. Other manufacturers also call this concept the team, mix or PSC concept.

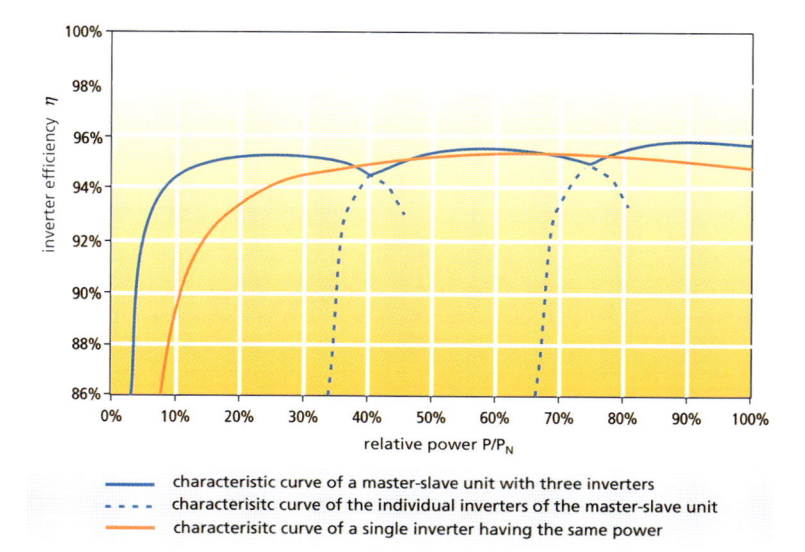

Figure 2.111
Efficiency curve of a master–slave unit with three inverters
Source: R. Haselhuhn

THREE-PHASE CONCEPT IN LOW POWER RANGES

A further possibility for optimizing devices in low power ranges is to use a three-phase feed. The advantages here, as well as the high efficiency, are an improvement in grid quality, the simple circuitry and the resilience and long life of these devices. Another advantage is that for these devices, instead of the ENS/MSD to protect the grid, the use of redundant protective equipment with three-phase voltage monitoring is also possible.

DEVELOPMENTS RELATING TO LARGE-SCALE GRID-CONNECTED INVERTERS

Conventional concepts first generate a 400V voltage via low voltage transformers and then step up this voltage via a subsequent medium voltage transformer. A trend among large inverters with power outputs of 500kVA and above is for a direct medium voltage feed. This eliminates the low voltage transformer, and after the inverter's power stage a medium voltage transformer directly converts the voltage up to medium voltage (e.g. 20kV). The transformer losses are reduced compared to a conventional low voltage feed with a subsequent medium voltage transformer. In addition, the elimination of a transformer reduces the costs.

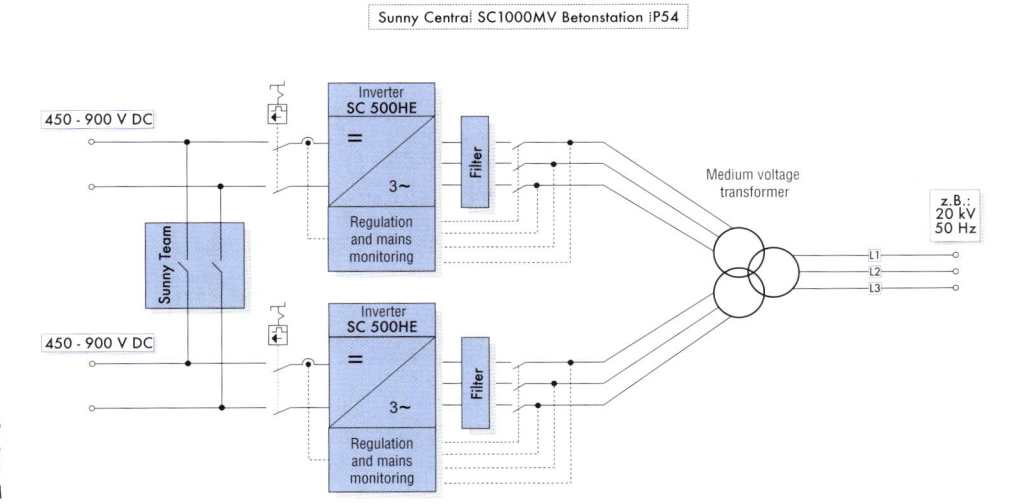

Figure 2.112
Large-scale grid-connected inverter with direct medium voltage feed
Source: SMA

As well as the classic master–slave mode, operation of the individual blocks is enabled for certain system operating scenarios. One operating scenario that can occur in damp weather is that the insulation resistance of the PV array, as a whole, falls below permitted levels. If the individual sub-arrays are operated separately, the insulation resistances of the sub-arrays have a higher value than the insulation resistance of the array as a whole. For system monitoring in large-scale systems, weather stations are used along with multiple measuring modules in the module tilt plane. Control and monitoring are increasingly implemented via visualization and control software solutions in the internet. Here, as well as the measured values, fault messages and characteristic parameters (e.g. the performance ratio) are also defined.

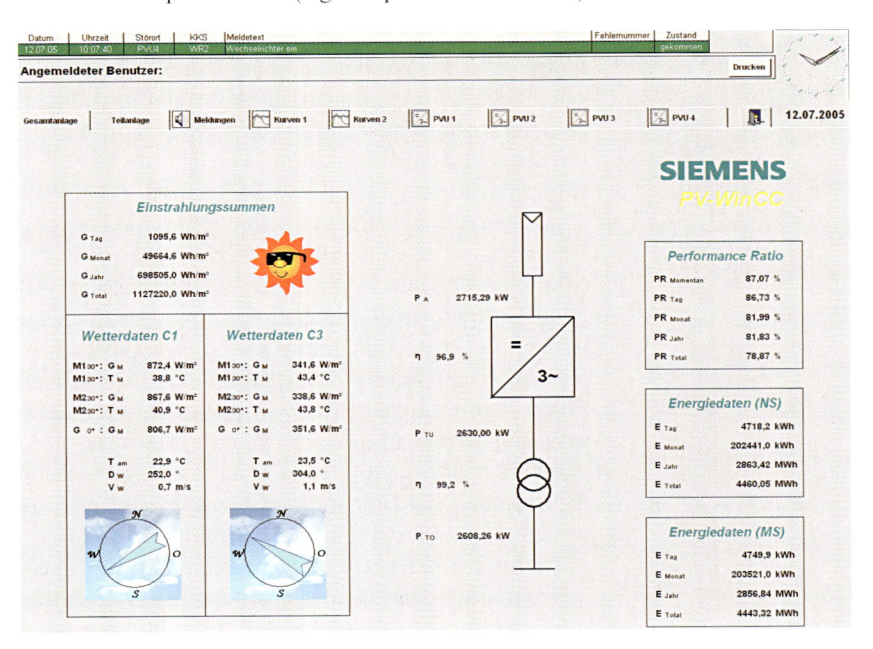

Figure 2.113
Internet-based user interface for system visualization, operating data monitoring and control of large-scale inverters
Source: PV-WinCC developed by Siemens

FURTHER DEVELOPMENTS

Inverter manufacturers are increasingly emphasizing the modularity of their devices. Multiple small inverters are linked together on the DC or AC side. This enables multiple smaller inverter units to be wired together in a modular system to give a desired higher output. Often, several power components can be connected to one management and control unit, in this way achieving higher overall power corresponding to the number of power components.

Figure 2.114
Power modules in the IG 500 inverter
Source: Fronius

Figure 2.115 and 2.116
Inverter with modular designed technology
Source: Fronius

As well as the power components, this modularity also applies to various functions that the devices can be equipped with optionally according to customer requirements. These include, in particular, displays, operating data recording, data loggers, communications interfaces, protective functions and so on. In this way, for example, the operating data of multiple inverters can be monitored centrally via their communications interfaces.

Aixcon has developed a grid-connected inverter that measures the complete I–V curve of the entire PV solar array in operation and can send this data to a computer. This curve enables the identification of installation faults such as reversed polarity of individual modules, cell breakage and shading, soiling or transfer resistance in the plug connectors. In this way, it is possible to carry out measurements to check correct commissioning, while subsequent measurement is also possible manually on site by connecting a PC with analysis software or automatically via a web adapter. The web adapter can automatically conduct measurements at defined time intervals and store the characteristic curve. Hence, it is not necessary for an electrical engineer to measure the characteristic curve on site using additional measurement equipment, which might otherwise have been required.

In order to increase the service life of grid-connected inverters in outdoor applications, in particular, some manufacturers in the low power range are once again using active cooling systems. With these, temperature-dependent speed control is used so that the fans are activated only when the thermal conditions make it necessary. Hermetic isolation of the cooling system and electronic components enables the IP protection class for exterior use to be achieved.

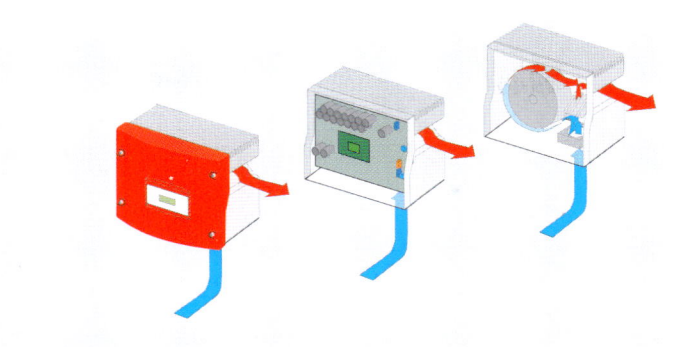

*Figure 2.117
Hermetic isolation of cooling system
and electronics for inverters in
outdoor applications
Source: OptiCool housing
concept by SMA*

Schematische Darstellung von OptiCod®

Most PV installation codes and regulations require a main DC circuit breaker disconnect/isolator switch to be installed between the DC side (the PV array and associated DC wiring) and the grid-connected inverter. Some manufacturers offer inverter-integrated electronic DC disconnect/isolator switches. This may mean that the installation engineer does not need to install a separate DC circuit breaker disconnect/isolator switch – national codes and regulations need to be referred to. The Electronic Solar Switch from SMA interrupts the current flow by breaking a short-circuit bridge inside the device. The power is reduced by an insulated gate bipolar transistor (IGBT) connected in parallel to such an extent that no arc is caused when the short-circuit bridge is broken.

*Figure 2.118
SMA inverter with integrated
electronic DC circuit breaker
Source: SMA*

The 'ebreak' circuit breaker by inverter manufacturer Aixcon also implements DC isolation (in accordance with VDE 0100 Part 712) in the device and also incorporates an emergency cut-out and surge protection.

Inverter manufacturers are reacting to the development of ever-larger solar cells (e.g. 6, 6+ or 8 inch cells) by producing devices that have a higher input power range.

2.4 Cabling, wiring and connection systems

2.4.1 Module and string cables

For the electrical installation of a photovoltaic system, only such wiring and cabling should be used as meet the requirements for this application. A distinction is made between module or string cables, the DC main cable and the AC connection cable. The electrical connecting cables between the individual modules of a solar generator and to the generator junction box are termed 'module cables' or 'string cables'. These cables are generally used outdoors. In order to ensure earth fault and short-circuit proof cable laying, the positive and the negative poles may not be laid together in the same cable. Single-wire cables with double insulation have proven to be a practicable solution and offer high reliability. Additional possibilities for earth fault and short-circuit proof cable laying can be taken from VDE 0100 Part 520.

Figure 2.119
Typical single-wire solar cables
Source: Multicontact

The frequently used double-insulated rubber-sheathed cable of type H07 RN-F is only permitted for operating temperatures of up to 60°C in its standard version and is therefore only suitable for PV systems to a limited extent. Roof tile manufacturers have measured temperatures of up to 70°C on roofs. For this reason, 'solar cables' are used in outdoor applications. Their main features are that they are UV and weather resistant and are suitable for a large temperature range (e.g. –55°C to 125°C). Some manufacturers offer cables covered with metal mesh, where the shielded cable not only provides protection against rodents but also improves protection against over-voltages. For in-roof installations, the standard version can be used.

Table 2.8
Cable requirements

1 Mechanical resistance	Compression, tension, bending and shear loads
2 Weather resistance	UV and ozone resistance when cables are laid outdoors without protection; heat and cold resistance (laying temperatures: 70°C on roofs, 55°C in lofts)
3 Earth fault-proof and short-circuit proof installation	Individual cable with double insulation

The modular connection boxes enable wires with a cross section of 1.5mm² up to 6mm² to be clamped. The cables are often supplied in the three colours of red, blue and black, enabling a more understandable generator setup.

Table 2.9 lists examples of some module line types from various manufacturers and their characteristics.

Table 2.9
Some module cable types and their characteristics
*Note: * V – maximum voltage between conductor and earth; and V0 – maximum voltage between outer conductors.*

Module cables for outdoor applications	Manufacturer	Nominal voltage V/V_0*	Temperature range	Cross sections in mm²
Lapptherm Solar Plus	Lapp cable	900V/1500V	–50°C–120°C	2.5; 4; 6; 10
Flex-Sol	Multi-Contact	600V/1000V	–40°C–90°C	2.5; 4; 6
Radox 125	Huber+Suhner	600V/1000V	–25°C–125°C	2.5; 4; 6
Siemens solar cable	Siemens AG	1800V/3000V	–40°C–120°C	2.5; 4; 6
Solar cable *C	Solar-Kabel GmbH	1800V/3000V	–25°C–90°C	2.5; 4; 6
Solarflex 101	Helukabel GmbH	600V/1500V	–30°C–125°C	2.5; 4; 6; 10; 16
TECSUN S1ZZ-F solar cable	Pirelli Kabel und Systeme GmbH	900V/1800V	–40°C–120°C	2.5; 4; 6; 10
Titanex 11 H07RN-F	ConCab cable	450V/750V	–35°C–85°C	2.5; 4

2.4.2 Connection systems

Connection of the module string cabling and other DC cabling should be made with extreme care. Poor contacts can lead to the occurrence of arcing and, hence, to an increased fire risk. Normally, four types of connection systems are used.

SCREW TERMINALS

For connecting to screw terminals, metal end sleeves (in accordance with DIN 46228 in Germany) are used on the end of flexible stranded wires.

POST TERMINALS

The connection to post terminals is made with cable lugs that are clamped between the nut and the bolt (the post).

SPRING CLAMP TERMINALS

In connection boxes that use spring clamp terminals, cables can be securely attached without metal end sleeves.

PLUG CONNECTORS

To simplify installation, PV modules with module-connecting leads and touch-proof plug connectors are increasingly being offered.

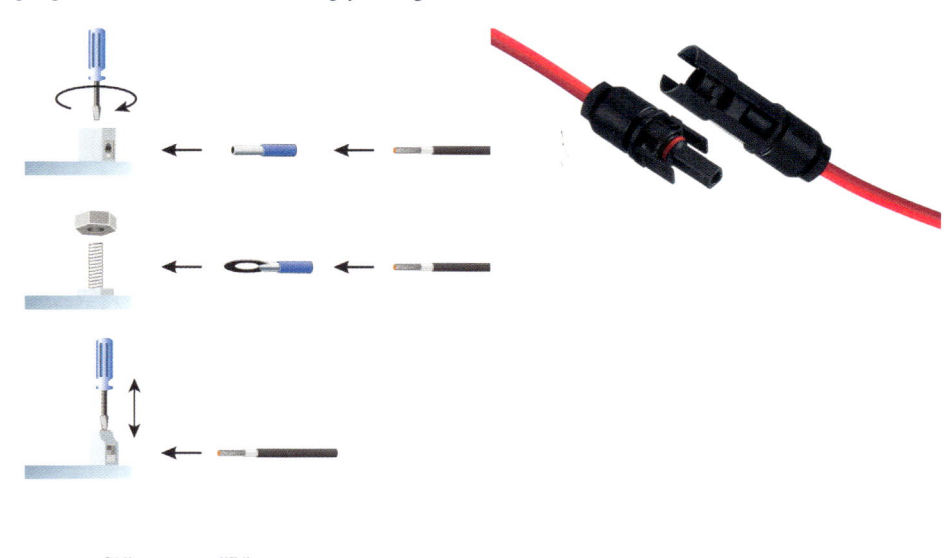

Figure 2.120
Cable connection systems
Source: Solarpraxis

Figure 2.121
Touch-proof plug connector enables the plug connectors to be fitted onto and removed from the cable using standard electrician's tools
Source: Tyco

To facilitate module installation, plug-and-play technology has become the standard in grid-connected systems. The vast majority of modules are factory fitted with touch-proof plug connectors. Plug connectors have a transfer resistance of less than 5 milliohms, so with a 5A current there is a very slight voltage drop of less than 0.025V.

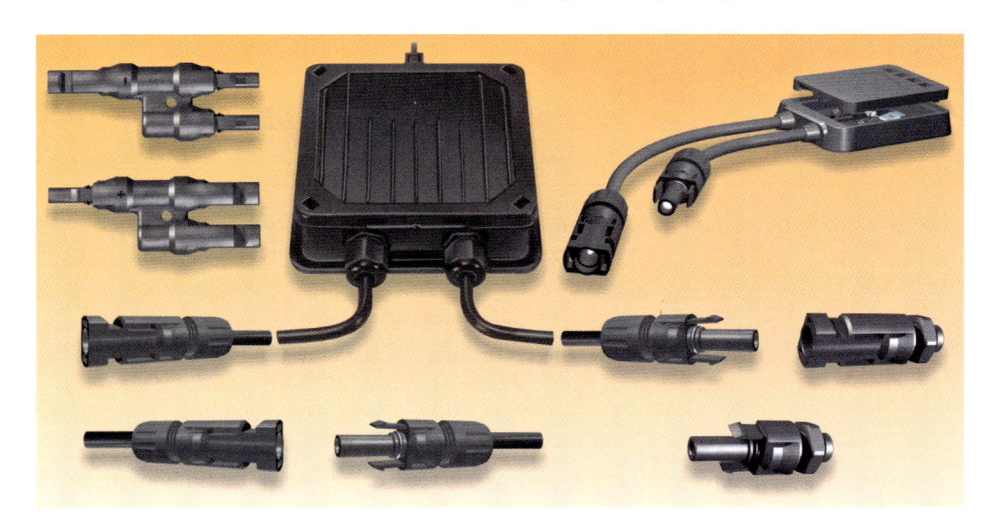

Figure 2.122
Various touch-proof plug connectors
Source: Multi-Contact

MANUFACTURERS OF MODULE PLUG CONNECTORS

Hirschmann, Huber+Suhner, Multi-Contact, Tyco.

2.4.3 DC main cable

The cable types named above can also be used for the DC main cable. The DC main cable connects the PV junction/combiner box with the inverter. In addition to the cable types named earlier, for cost reasons the common type PVC-sheathed cable with the code NYM or NYY is often used; but national codes and regulations need to be referred to and followed. If the PV combiner/junction box is located outdoors, these PVC-sheathed cables must be laid in a protective pipe as they are not UV resistant. If alternatives are available, PVC-sheathed cables should not be used outdoors. For the cables (as also for the installation material), halogenated plastic (PVC) is often used. For environmental considerations, halogen-free products should be selected (e.g. type NHMH-J).

To avoid short circuits and faults to earth/ground, single-wire sheathed cables are recommended individually for positive and negative cables. If multi-wire cables are used, the green/yellow earth/ground wire (European colour code) must not carry any voltage. For PV installations exposed to a lightning risk, screened cables should be used (see the section on '4.7 Lightning protection, earthing/grounding and surge protection' in Chapter 4). The cables should be routed so that mechanical damage (e.g. by rodents) is not possible. It should be possible to switch all poles of the DC main cable to zero potential. The DC main disconnect/isolator switch and the isolation points in the PV array combiner/junction box are used to do this.

2.4.4 AC connection cable

The AC connection cable links the inverter to the electricity grid via the protection equipment. In the case of three-phase inverters, the connection to the low voltage grid is made using a five-pole cable. For single-phase inverters, a three-pole cable is employed. The usual cables of type NYM, NYY or NYCWY can be used. National codes and regulations need to be followed.

2.5 Direct current load switch (DC main switch)

In the event of faults or in order to carry out maintenance and repair work, it must be possible to isolate the inverter from the PV generator. The DC load switch is used for this. According to the IEC 60364-7-712 standard 'Electrical Installations of Buildings – Solar Photovoltaic (PV) Power Supply Systems' (corresponds to VDE 0100 Part 712), an accessible load switch is required between the PV generator and inverter.

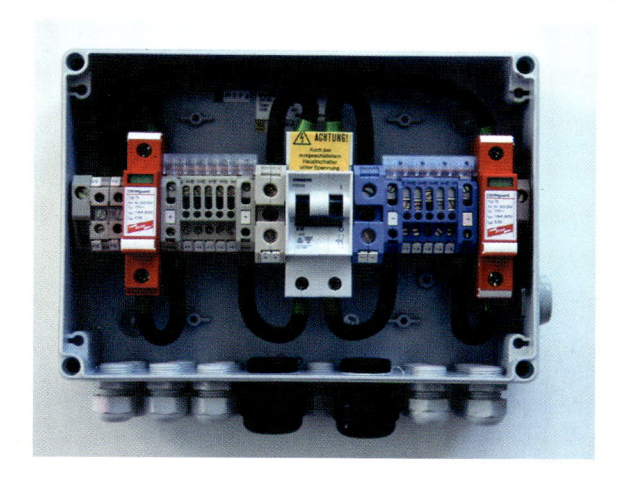

Figure 2.123
DC main disconnect/isolator switch in PV array combiner/junction box

The DC main switch should have load switching capability and be rated for the maximum open-circuit voltage of the solar generator (at −10°C) and for the maximum generator current (short-circuit current under STC). When selecting the switch, the switch must be able to move the relevant direct current. The DC main switch is often housed in the generator junction box. For safety reasons, it is better to install it directly before the inverter. Touch-proof plug connectors (e.g. on string inverters) may only function as isolators without load. As long as the irradiance is sufficient, the PV system supplies energy and is therefore under load. When separating a plug connector under load, the direct current can result in a long burning arc, which is a serious safety and fire risk. Some manufacturers offer DC load switches integrated within the inverter (see also '2.3.6 Further developments in grid-connected inverter technology').

MANUFACTURERS OF DC SWITCHES

ABB, AEG, Klöckner Moeller, Merlin, Santon, Siemens, Winkhaus.

2.6 AC switch disconnector

The AC switch disconnector must be double-pole, clearly labelled and lockable in the 'off' position only. Note that a second isolator is required adjacent to the inverter if the inverter is mounted in a separate room.

2.6.1 Miniature circuit breakers (MCBs)

Line circuit breakers are over-current protective devices that can be switched back on after they are triggered. They automatically isolate the PV system from the electricity grid if an overload or short circuit occurs. Automatic circuit breakers are often used as AC isolators.

2.6.2 Earth leakage circuit breakers

Earth leakage or residual current devices (RCDs) monitor the current flowing in the forward and return conductors in the electrical circuit. If the difference between the two currents exceeds 30mA, the RCD isolates the circuit within 0.2 seconds. The RCD will trip if there is an insulation fault, or if there is earth or body contact with one of the conductors. This is not always mandatory as most invertors switch off when there is an earth fault.

3 Site Surveys and Shading Analysis

3.1 On-site visit and site survey

In order to begin planning a grid-connected PV system and to be able to offer a quotation, a site visit is essential. This enables an assessment of the basic conditions for the PV system.

First, it is important to establish whether the building is suitable for installing a PV system. A thorough initial investigation avoids planning errors and miscalculations in the quotation that is submitted. The installation work for the PV array, the installation site (e.g. for the inverter), the wiring routes, the actual laying of cables and for expanding or modifying the meter cupboard can be better estimated and agreed in consultation with the customer.

Before planning gets under way, the customer should also be asked how much they expect to spend, and possible subsidies should also be taken into account since these will have a deciding influence in the size of the system (see Chapter 9). For work on roofs, a quote can be obtained from a roofer. The choice of roofing firm should be agreed with the building owner.

The following points should be borne in mind during the on-site visit and recording of data. They form the basis for good planning:

- customers' wishes with regard to module type, system concept and method of installation;
- desired PV power or the desired energy yield;
- the financial framework, taking the respective subsidy conditions into account;
- usable roof, façade and open space surfaces;
- orientation and angle of inclination;
- roof shape, roof structure, roof substructure and type of roofing;
- usable roof openings (vent tiles, free chimney flues, etc.);
- data on shading;
- installation sites for PV combiner/junction boxes, isolating facility and inverter;
- meter cupboard and space for extra meters;
- cable lengths, wiring routes and routing method;
- access, particularly when equipment is required for installing the PV array (crane, scaffolding, etc.).

Checklists of information for site surveys are provided at the end of this chapter. These checklists will help to record the data at the on-site meeting. They may be printed out, taken to the site visit and filled out there.

The following documents can assist in planning and some will be required when applying for subsidies and for registering with the distribution network operator:

- site plan of the building to ascertain its orientation;
- construction drawings of the building to ascertain roof slope, the usable area and the cable lengths;
- photographs of the roof and of the electric meter location.

It has proven useful to take the following items to the on-site meeting:

- checklists for recording the building survey (see '3.8 Checklists for building survey');
- information on photovoltaics:
  - general information on any government schemes that may be relevant;
  - company prospectuses and product descriptions;
  - photos of existing PV systems;
  - this volume;
- compass;
- protractor with plumb bob for finding the slope of the roof;
- folding ruler;
- tape measure;
- pocket torch (flashlight);
- shading analyser or sun path diagram on acetate (see '3.4 Shading analysis' and '3.5 Shade analysis tools using software');
- camera.

Figure 3.1
Shading analysis using a camera with special solar geometric equipment
Source: Sol-i-dar Architecture firm

3.2 Consulting with the customer

The planning and building of a photovoltaic system is generally initiated by a customer's enquiry. As well as the quotation, consultation with the customer is an important and essential step before commissioning the building of a photovoltaic system.

In a talk with the customer, the PV installer should inform himself of the customer's expectations and wishes. Consultation in this case, first and foremost, consists of helping the customer to make up their mind.

When it comes to solar technologies, expert advice to the customer is often crucial. As well as technical knowledge concerning structure, function, sizing and the installation of photovoltaic systems, the PV installer should also possess knowledge about costs/subsidies and the global significance of solar energy use.

The aim is to engage the customer as an active partner in the dialogue and to answer their questions in a way that is comprehensible to the non-expert. Diagrams used as explanatory aids can be helpful.

You should be prepared to answer the following questions from the customer:

- What is the difference between a PV module and a solar collector?
- How does a solar cell work?
- How much electricity will my system produce in a year?
- What happens to the power supply when the sun isn't shining?
- How much power does the system produce when the sky is cloudy?
- Where is the generated power used?
- What does 'kilowatts peak' power mean?
- Is my roof suitable for a system like this?

- What happens if the modules get dirty? Snow? Dust?
- Can hailstones damage the modules?
- Is it worth having a module-mount that tracks the sun?
- Are there any other module colours except black and blue?
- Do I need planning permission?
- What will my system cost, including installation?
- What subsidies are available?
- Can you help me to make the application for subsidies?
- How much money do I get for the electricity that I feed into the mains grid?
- Can you file the application with the network operator for electricity feed-in?
- How long is the approximate payback time for a PV system?
- What return can be achieved?
- What tax aspects need to be taken into account?
- Does the system require maintenance?
- Can lightning destroy the system?
- How long will the system last?
- How do you count the guarantee periods?

3.3 Shadow types

Ideally, PV arrays should be installed in a shade-free location. However, grid-connected systems are usually found in urban and suburban areas and the modules are usually installed on roofs, where some shading is sometimes inevitable. Shading can reduce the output of a PV array considerably and ideally should be avoided. As a result, this subject is discussed in depth here, and this section deals mainly with the issue in grid-connected systems (PV arrays for stand-alone systems are usually located in rural areas and have ground-mounted arrays – sufficient land around the building is usually available, so they can be situated where there is no shade).

A shadow cast on a PV system has a much greater effect on the solar yield than, for example, in the case of solar thermal systems. The operating results gained from the German 1000 Roofs programme showed that partial shading due to the circumstances of the location was observed in around half of all systems. In a large number of these systems, shading caused annual yield reductions of between 5 per cent and 10 per cent. Shading can be classified as temporary, resulting from the location, the building or caused by the system itself (self-shading).

Note: direct shadows can have a critical effect on PV-array output!

3.3.1 Temporary shading

Typical temporary shading includes factors such as snow, leaves, bird droppings and other types of soiling. Snow is a significant factor, especially in mountainous areas. Dust and soot soiling in industrial areas or fallen leaves in forested areas are also significant factors. Snow, soot and leaves collecting on the PV array cause shading. The effect of this will be less if the array self-cleans (i.e. if it is washed away by flowing rainwater). A tilt angle of 12° or more is usually sufficient to achieve this. Greater tilt angles increase the flow speed of the rain water and, hence, help to carry away dirt particles. This type of shading can be reduced by increasing the tilt of the PV array. Good self-cleaning takes places on the modules. Snow on a PV array melts faster than the surrounding snow, so that, generally, shading only occurs on a few days.

In snowy areas, arranging the standard modules horizontally (A) enables losses caused by snow to be reduced by half. In this way, shading caused by the snow generally affects only two and not four rows of cells on each module, as is the case in the vertical arrangement (B) (see Figure 3.3).

*Figure 3.2
Snow on a PV system*

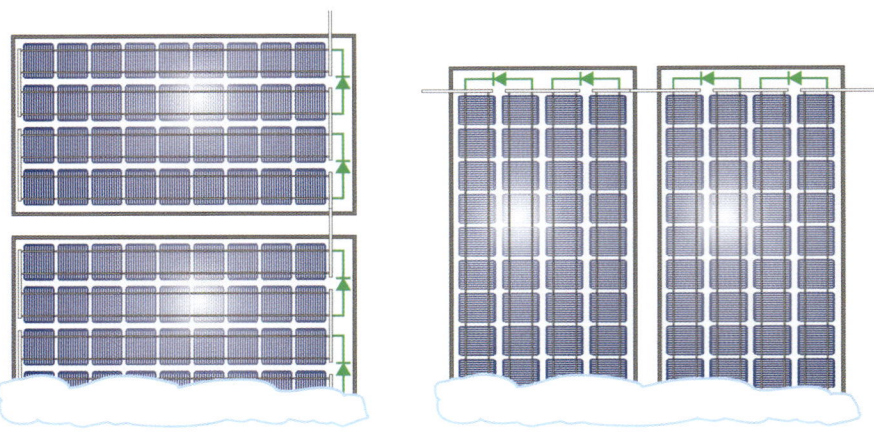

*Figure 3.3
Arrangement of tilted PV modules
in case of snow
Source: R. Haselhuhn, Solarpraxis*

A) horizontal arrangement

B) vertical arrangement

Shading caused by leaves, bird droppings, air pollution and other dirt has a stronger and longer-lasting impact. If a system is heavily affected by this, regular cleaning of the PV modules will noticeably increase the solar yield. In a normal location and with sufficient tilt, it can be assumed that the loss due to soiling amounts to 2 per cent to 5 per cent. Generally, this loss is acceptable (Quaschning, 1996).

*Figure 3.4
Bird droppings, which soil modules,
should be removed
Source: BSR*

If heavy soiling is present, the modules may be cleaned using water (hosed) and a gentle cleaning implement (a sponge) without using detergents. To avoid scratching the surfaces, the modules should not be brushed or wiped with a dry cleaning implement.

3.3.2 Shading resulting from the location

Shading resulting from the location covers all shading produced from the building's surroundings. Neighbouring buildings, trees and even distant tall buildings can shade the system, or at least lead to horizon darkening. It must be taken into account that, due to the growth of trees and shrubs, vegetation may shade the system only after a couple of years. Overhead cables running over the building also have a negative effect, casting a small but effective moving shadow.

Figure 3.5
In the case of façade systems, shading resulting from the location is relevant if there is only a small distance to neighbouring buildings

Figures 3.6 and 3.7
Due to the growth of a tree, the shade cast by it may change considerably over the course of 20 years

3.3.3 Shading resulting from the building

Shading resulting from the building involves direct shadows, which should therefore be viewed as particularly critical. Special attention should be paid to chimneys, antennae, lightning conductors, satellite dishes, roof and façade protrusions, offset building structures, roof superstructures, etc. Some shading can be avoided by moving the PV modules or the object causing the shading (e.g. antennae). If this is not possible, the impact of the shading can be minimized by taking it into account when selecting how the cells and modules are wired up and in the system concept (see Chapters 2 and 4, and the section on '3.6 Shading, PV-array configuration and system concept').

Figures 3.8 and 3.9
Shading due to vertical struts of the escape balcony at the Amica building in Aachen, Germany
Source: Saint Gobain

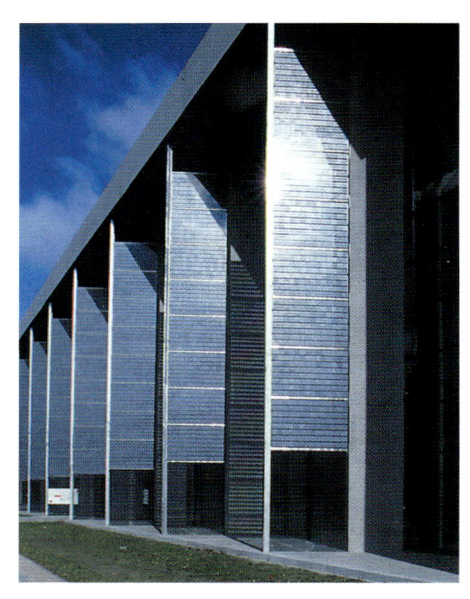

Figure 3.10
Shading due to eaves at the UTZ (Environmental Technology Centre) in Berlin, Germany
Source: Solon

3.3.4 Self-shading

With rack-mounting systems, self-shading of the modules may be caused by the row of modules in front (see also '3.7 Shading with free-standing/rack-mounted PV arrays'). Space requirements and shading losses can be minimized through optimization of the tilt angles and distances between the module rows. A poorly designed or installed mounting system may also cause micro-shading in sloping roof installations.

Figure 3.11
Distance between module rows well chosen: On 21 December, the shadow cast by the front row of modules in the early afternoon does not yet hit the lower modules

Figure 3.12
Unfavourable distribution of module tables causes relevant direct shading

*Figure 3.13 and 3.14
Avoidable micro-shading caused by module clamps (left) and projecting screws (right)*

3.3.5 Direct shading

Direct shading, in particular, can cause high losses of energy. The closer the shadow-casting object, the darker the shadow will be since the module is hit by the core shadow and the less diffuse light reaches the PV module. Thus, the core shadow cast by a near object reduces the energy incident on the cell by approximately 60 per cent to 80 per cent, while a partial shade leads to a reduction only half as high. The larger the distance to the shadow-casting object, the brighter the shadow is and the more shading losses are reduced. Figure 3.15 illustrates the creation of core and partial shadows.

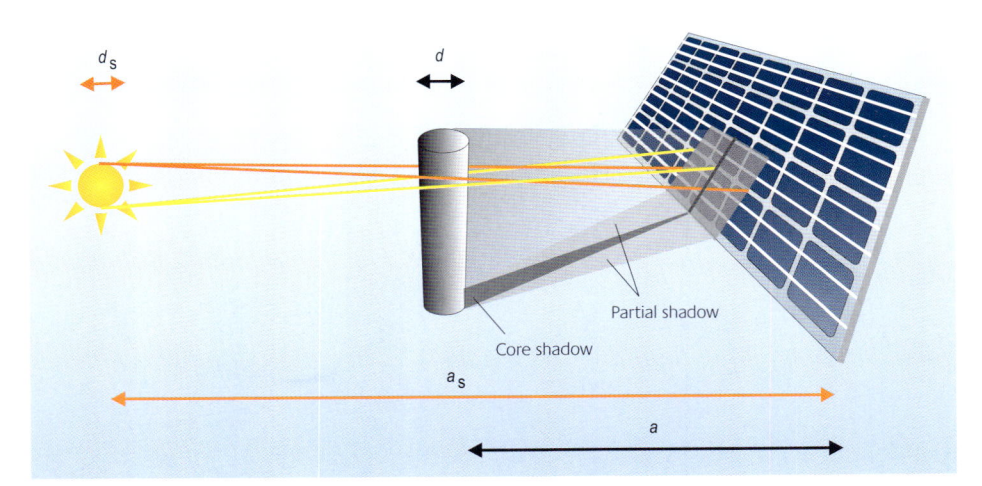

*Figure 3.15
Shading through core and partial shadow
Source: R. Haselhuhn*

Based on the thickness of the shadow-casting object d, the optimum distance aopti from the modules can be calculated. Since the sun is a plane source of light, the distance can be calculated on the basis of the similar triangle relations of the sun tangents that touch the object. The optimum distance a_{opti} from the modules is determined by:

$$a_{opti} = \frac{(a_s + a_{opti}) \times d}{d_s} \approx \frac{a_s + d}{d_s}$$

With $a_{opti} \ll a_s$ and with:

- a_s (distance Earth to sun): a_s = 150 million kilometres;
- d_s (diameter of sun): d_s = 1.39 million kilometres;

it follows that:

$$a_{opti} = \frac{a_s \times d}{d_s} = 108 \times a$$

Thus, for example, an overhead cable with a diameter d = 5cm must be at least 5.4m away so that no core shadow is cast onto the modules.

For wider objects, a distance at which the core shadow does not exceed the width of one cell may be tolerated. At a cell width of 10cm, the distance may be reduced by 1m; at a cell width of 20cm, it may be reduced by 2m.

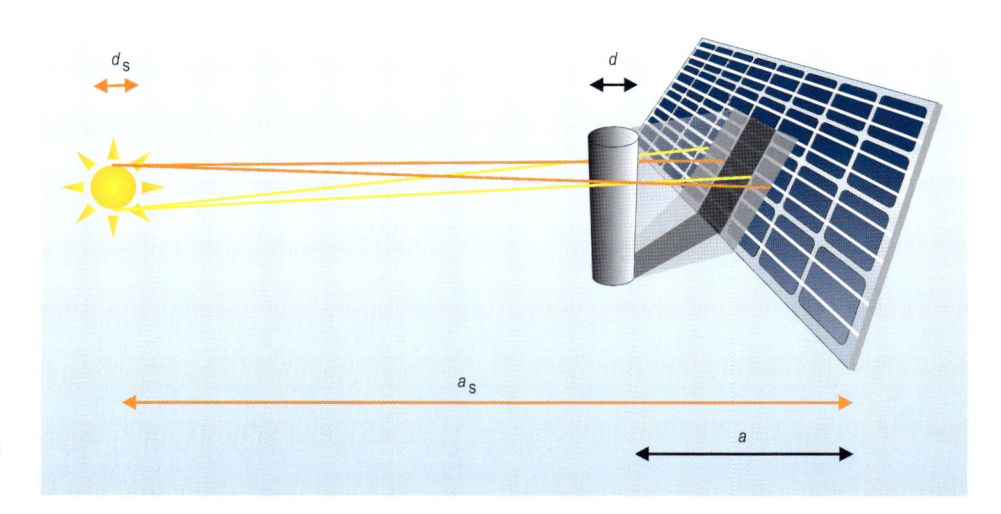

Figure 3.16
Width of core shadow equals width of cell
Source: R. Haselhuhn

Direct shading should be reduced in any case. The fluctuation of the shading depending upon the season and time of day, as well as the resulting losses, may be calculated using corresponding simulation programmes.

Figure 3.17
Critical direct shading and PV array orientation on the building of Deutsche Bank in Berlin, Germany
Source: Udo Siegfriedt

Figure 3.18 and 3.19
Avoidable direct shading caused by lighting rods (left) and safety ropes on the roof of the Energieforum in Berlin, Germany
Source: R. Haselhuhn

3.4 Shading analysis

In order to assess the shading resulting from the location, a shading analysis is performed. For this, the shadow outline of the surroundings is recorded for one point in the system: usually the centre point of the PV array.

In larger systems or if greater accuracy is desired, the shading analysis should be carried out for several different points.

The shadow outline for the surrounding area can be found using:

- site plan and sun path diagram;
- sun path diagram on acetate;
- shading analyser (a digital camera and software, or Solar Pathfinder).

3.4.1 Using a site plan and sun path diagram

When using a site plan and sun path diagram, the distance and the dimensions of shadow-casting objects are calculated. From this information, the azimuth angle and elevation angle are calculated.

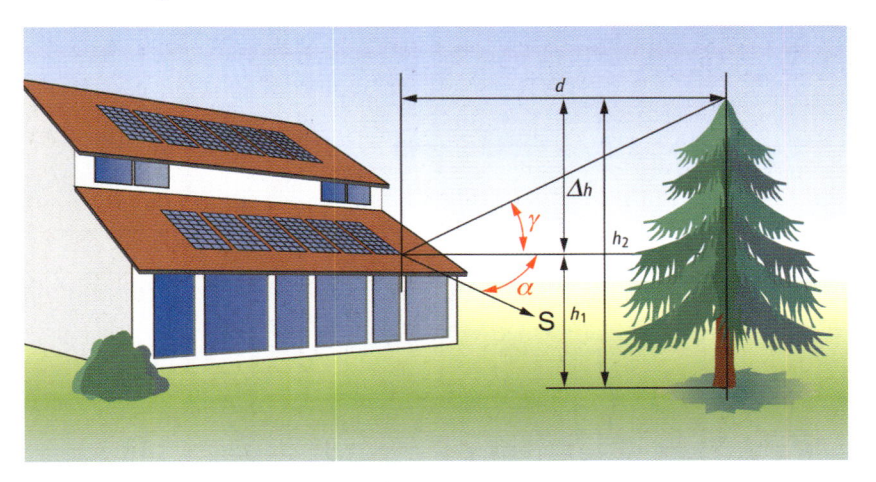

Figure 3.20
Calculation of an object's elevation angle and azimuth angle
Source: V. Quaschning

The elevation angle γ is calculated from the difference between the height of the PV system h_1 and the height of the shading object h_2 and its distance:

$$\tan\gamma = \frac{h_2 - h_1}{d} \rightarrow \gamma = \arctan\left(\frac{h_2 - h_1}{d}\right) = \arctan\left(\frac{\Delta h}{d}\right)$$

The elevation angle is worked out for all obstacles in the area surrounding the solar system, requiring the height and distance of the objects from the observer to be known. The azimuth of the obstacles can be calculated directly from the site plan or sketch.

3.4.2 Using a sun path diagram on acetate

A sun path diagram with a height axis in trigonometric divisions can also be used. This is printed onto acetate and arranged in a semi-circle. The observer, from the perspective of the system, now looks at the objects through this diagram and can directly read off and note down the elevation and azimuth angles. In order to be able to record a greater viewing angle, a wide-angle lens can be useful, such as is used in a door peephole. Figures 3.21 to 3.23 illustrate the use of this simple shading analyser (Quaschning, 2000).

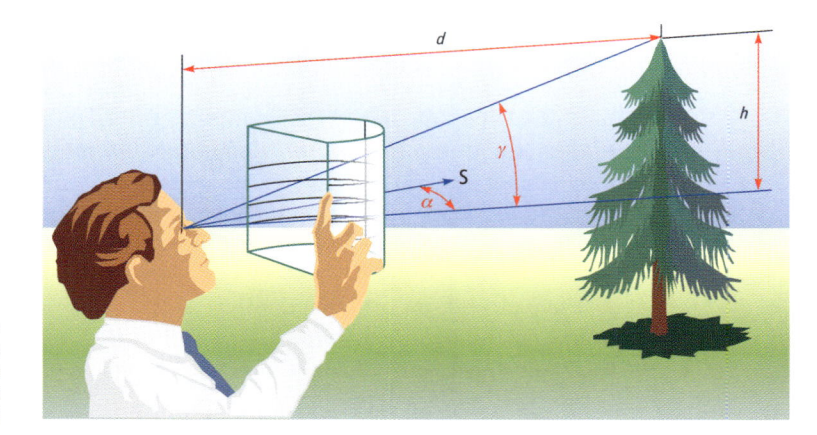

*Figure 3.21
Calculation of an object's elevation angle and azimuth angle using a sun path diagram on acetate
Source: V. Quaschning*

For shading caused by trees, a transmission factor is given:

- for coniferous trees: τ = 0.30;
- for deciduous trees in winter: τ = 0.64;
- for deciduous trees in summer: τ = 0.23.

The transmission factor specifies how much solar radiation passes through the tree. It is taken into account in some simulation programmes (e.g. PV-Sol).

*Figure 3.22
Surrounding areas with angular grid
Source: V. Quaschning*

The outcome of the shading analysis is the silhouette of shading caused by the surroundings in the sun path diagram.

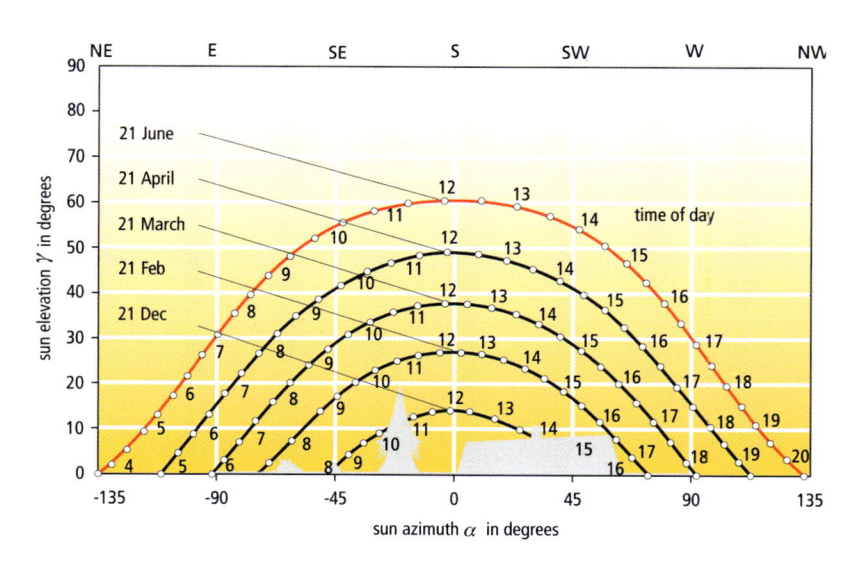

*Figure 3.23
Sun path diagram of Berlin with shadow outline
Source: V. Quaschning*

From Figure 3.23 it is possible to read the level of shading that occurs in a particular month. In the example shown, the location is 50 per cent shaded on 21 December. In the morning and early afternoon, the sun penetrates through for an hour or so on each occasion. From 21 February onwards, no more shading occurs. There will be no shading in the period from March to October.

Further evaluation can be performed graphically, by calculation or (the easiest way) using simulation programmes (e.g. SUNDI, PV-Sol, PVS and SolEm; see Chapter 5). The geometry of the PV array and the way in which the modules are wired is only taken into account in more complex simulation programmes (e.g. PV-Cad and PV-SYST; see Chapter 5).

Without software, irradiance totals are required for the individual months at the installation site. This allows the irradiance loss for the respective month to be estimated from the percentage share of shading, which is calculated from the sun path diagram.

3.5 Shade analysis tools using software

Several shade analysis tools that use software are available. These enable accurate shade analysis and are less prone to error than manual methods.

The *Solar Pathfinder* uses a highly polished, transparent, convex plastic dome to give a panoramic view of the entire site. All the trees, buildings or other obstacles to the sun are plainly visible as reflections on the surface of the dome. Because it works on a reflective principle rather than actually showing shadows, it can be used any time of the day, any time of the year, in either cloudy or clear conditions. The actual position of the sun at the time of the solar site analysis is irrelevant. Once the Solar Pathfinder has been properly set up, it can be used to provide shading data manually or digitally. When used manually, the outline of the horizon's reflection seen on the dome is traced onto the underlying diagram by inserting a white marking pen (included) through the slots on the side of the unit. The traced line shows exactly at what hours of the day and months of the year an obstacle will shade that particular location. A digital camera can be used to take photos of the dome and the photos are analysed by software. In this way, several sites can be easily and quickly assessed.

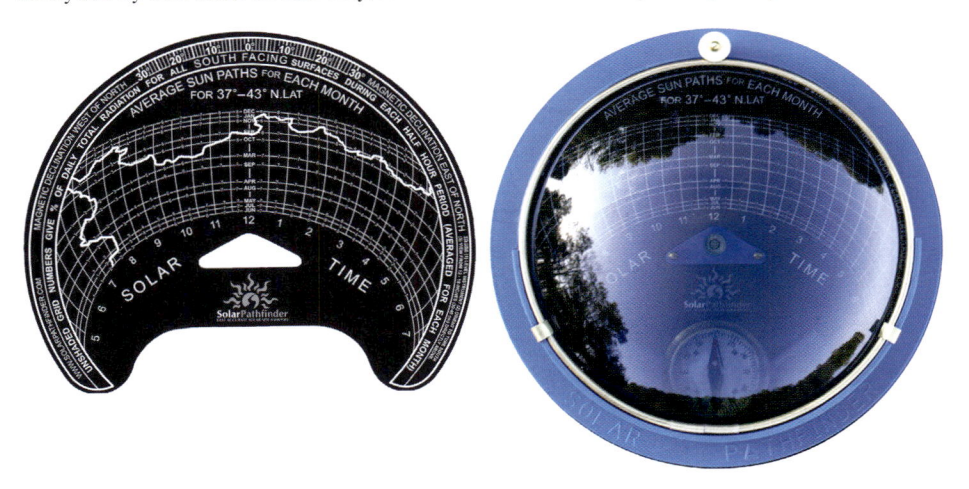

Figure 3.24
Solar Pathfinder
Source: Solar Pathfinder

The *Panorama Master* and the *HORIcatcher* provide systems for fixing the digital camera and corresponding software, which allows the horizon line to be created automatically. The elevation angle and azimuth angle of the objects are determined using a shading analyser (digital camera with software). However, the pictures should be taken horizontally, at an angle of at least 180° from the installation site or with a fisheye lens.

Figure 3.25
ShellSolar employee taking photographs to assess the horizon line for a ground mounting system using the Panorama Master by Swiss company Energiebüro
Source: R. Haselhuhn

Figure 3.26
Photographing the horizon with a spherical mirror using the HORIcatcher
Source: Meteotest

The Panorama Master is a spirit-level system for holding and aligning a digital camera. It allows the surrounding area to be photographed in several shots at defined angles through a 360° circle. The specially designed *horizON* software (see also section '5.5.4 Supplementary programmes and data sources' in Chapter 5) is used to stitch together the individual photographs into a complete panoramic image. It also automatically finds and draws in the horizon line. The horizon line can be exported in an appropriate file format to be processed in simulation programmes (e.g. PV*SOL and PVS). This means that designers do not need to go through the laborious process of entering the data by hand. The HORIcatcher (see also section '5.5.4 Supplementary programmes and data sources' in Chapter 5) makes it possible to digitize the horizon line from a single photograph taken using the spherical mirror. The supplied software allows the data files to be imported within other simulation programmes.

The outcome of the shading analysis is the shading silhouette caused by the surroundings of the building in the sun path diagram. It is possible to read off from this diagram the level of shading that occurs in a particular month.

In the example given in horizON, the site is shaded by the high-rise building in the middle for just under two hours a day in winter (from 9.30 am to 11.20 am). In addition, shading occurs for approximately half an hour during sunrise and sunset. From March to September, the shade cast by the building on the right moves across the site for one to two hours after 5.00 pm (Haselhuhn, 2005).

Figure 3.27
Creating the horizon line using the horizON programme

Further evaluation of the diagram is performed via suitable simulation programmes. Most simulation programmes calculate the irradiance losses and from these compute rough values for the yield losses (e.g. PV-Sol, PVS and SolEm). Here, the shadow outline is determined at one point on the PV array generator (often the centre point) and is entered. This accuracy is sufficient in many cases. For the horizon line in Figure 3.27 and a typical PV system, an irradiance reduction of 9 per cent and yield losses of 10 per cent were determined by the PV*SOL simulation programme (Haselhuhn, 2005).

This is, nevertheless, based on the assumption that the shading covers the entire PV array equally. The yield reductions generally turn out to be greater than one would suspect based on the shaded surface area. The more complex simulation programmes PVSYST, PVcad and 3DSolarwelt enable a three-dimensional shading analysis that also takes inhomogeneous shadow outlines into account.

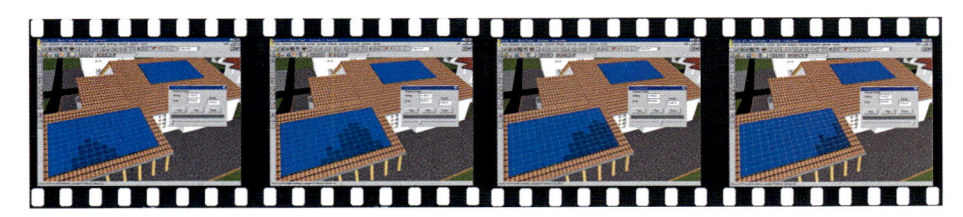

Figure 3.28
Course of shading illustrated by the 3Dsolarwelt software
Source: Solarschmiede GbR

3.6 Shading, PV-array configuration and system concept

The effect of shading on PV systems depends upon the following factors:

- number of shaded modules;
- cell and bypass diode interconnection (see section '2.1 PV modules' in Chapter 2);
- degree of shading;
- spatial distribution and the course of shading over time;
- interconnection of the modules (see section '4.2 System concepts' in Chapter 4);
- inverter design.

As already described in the previous section, when predicting yield, the irradiance reduction is normally determined for the PV array area (i.e. the PV array). What this fails to take into account, however, is that the PV array characteristic I–V curve changes as a result of shading (see Figure 2.68 in Chapter 2). This causes the maximum power point (MPP) to shift. The operating point of the inverter tries to track the MPP point. The change of the MPP determines the power reduction relative to the unshaded array. The input voltage range of the inverter determines the interconnection concept of the solar modules. With string inverters with high input voltages, often all of the modules are connected in series. If the inverter has a low input voltage, then this leads to a PV array being used with several parallel strings.

A scientific study on the problem of shading has been conducted at the Technical University Berlin using different system designs (Siegfriedt, 1999). The PSpice electronic simulation programme was used to determine the PV array characteristic I–V curves and the expected power losses with different shading situations. A PV array with a total of 20 modules connected in series was compared with an array with four strings connected in parallel. With an irradiance of 1000W/m^2, two, four, six and eight modules were shaded by reducing the irradiance on these modules to 500W/m^2. Whereas with series connection, the characteristic curves do not depend upon the position of the shaded modules, the parallel-connected PV array produces different characteristic curves for different shading situations.

Figures 3.29 to 3.33 demonstrate that the power curves with shading show a maximum with small voltages and a second maximum with higher voltages. The factors named at the beginning determine how pronounced the maxima are; whether they even exist or not, or are just a slight bump on the characteristic curves; where the MPP lies; and whether this falls at all in the inverter's tracking range. The amount of power loss with shaded arrays depends upon the input voltage range of the inverter and, thus, upon the sizing (see section '4.4.2 Voltage selection' in Chapter 4). Furthermore, the MPP tracking concept of the inverter is also decisive. Depending upon the tracking concept and the course of shading over time, the system is operated in one of these maxima. There are fundamental differences between series and parallel connection.

3.6.1 Connection in series (string concept)

With connection in series, both power maxima are possible operating points for the inverter if they lie within the operating range of the MPP tracker. Which of these two points is reached depends upon the course of shading over time and the behaviour of the tracker.

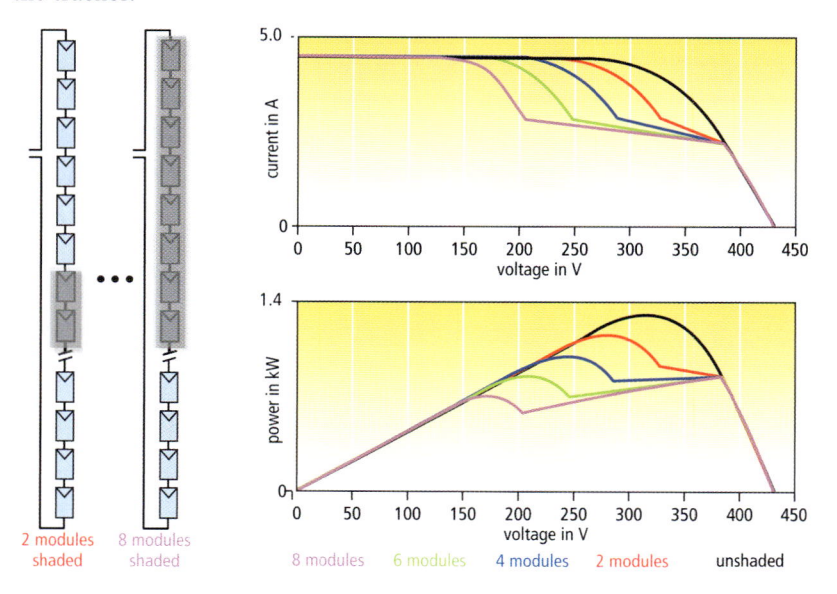

Figure 3.29
Shading situations and characteristic I-V curves in the case of connection in series
Source: U. Siegfriedt

With the initially unshaded PV array, the inverter is in the only maximum of the characteristic curve. The gradual extension of shadows across the array surface causes increasingly more modules to be shaded. The left maximum, which initially still represents the MPP, shifts towards smaller voltages. Since this maximum is well pronounced, the inverter also remains in the left maximum even if the MPP lies in the right maximum when there are a larger number of shaded modules.

If the modules are already shaded during the morning, and there is already shading at the moment of switching on, the inverter tracks from the open-circuit voltage to the right maximum. If this is sufficiently pronounced and the tracking movement is not too substantial, the inverter remains at this point regardless of whether the MPP is situated here. The voltage is somewhat greater in this maximum than with an unshaded array.

3.6.2 Connection in parallel

The shading situations observed with connection in parallel produce completely different characteristic curves. The most favourable behaviour occurs when the shaded modules are either situated in the same string or are distributed across only a few strings. The left power maximum here lies at half or less than half the open-circuit voltage of the PV array and is therefore almost always outside the inverter's tracking range. It is hardly pronounced, so that the inverter almost certainly tracks to the right maximum. This almost always represents the MPP.

With an increasing number of shaded strings the left power maximum becomes even more pronounced. With severe shading it is possible for the MPP to be situated here. The left maximum lies at half the open-circuit voltage of the PV modules and is therefore outside the inverter's operating range. The right maximum lies somewhat above the MPP voltage of the unshaded array.

Figure 3.30
Shading situations and characteristic I–V curves in the case of connection in parallel and shading in two strings
Source: U. Siegfriedt

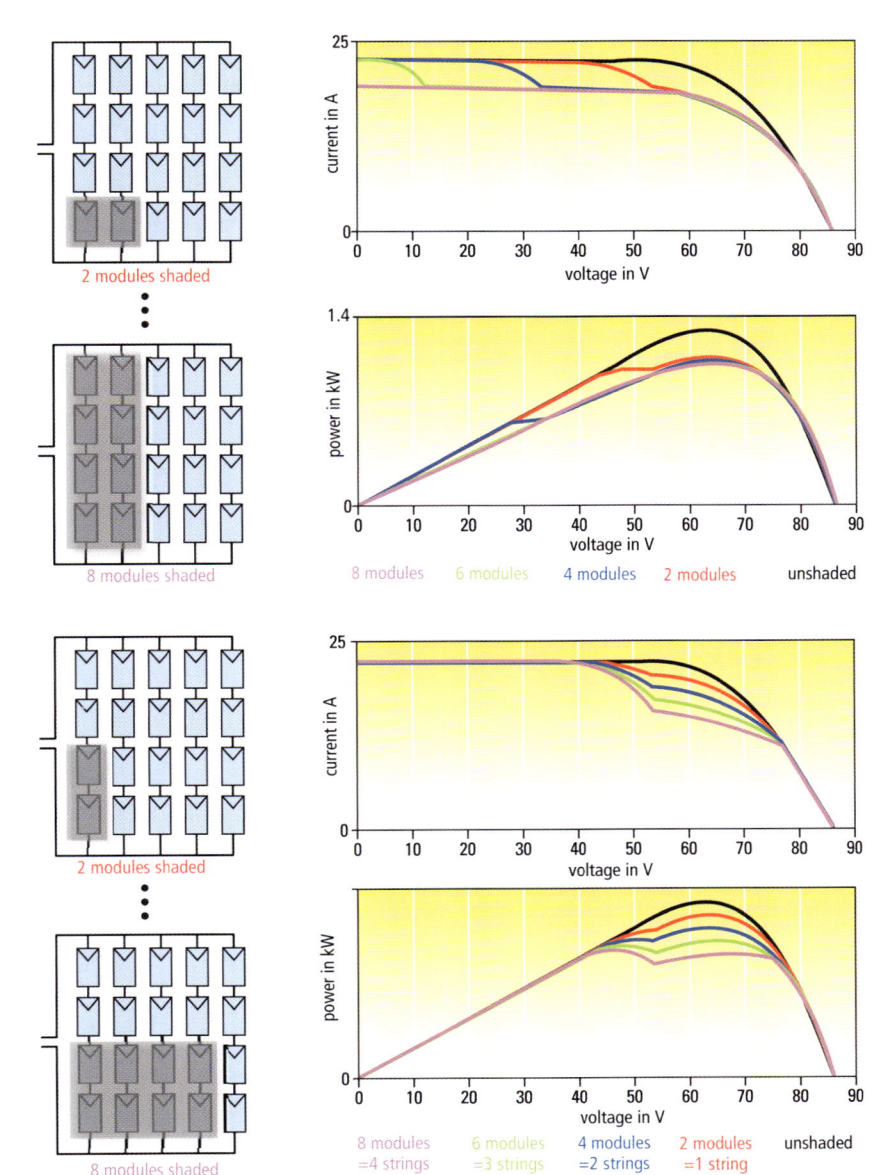

Figure 3.31
Shading situations and characteristic I–V curves in the case of connection in parallel and shading in one to four strings
Source: U. Siegfriedt

3.6.3 Comparison of connection concepts

With the series connection, both power maxima for the shaded characteristic curve are clearly pronounced; when fewer modules are shaded, the voltages are within the tracking range of the inverter. For this reason, both operating points must be taken into account in the following comparison. With parallel connection, the inverter can only track to the right power maximum effectively since the left is weakly pronounced and the voltage is too low. It would only be possible to track to the left maximum if there is severe shading in many strings. In this case, there would be a somewhat lower power loss than in the right maximum.

With parallel connection it can be clearly seen that the power loss only depends upon the number of shaded strings. With shading in two strings, despite the increase from two shaded modules to eight, power loss remains almost constant. The series connection shows considerably greater power losses. In the left maximum, the loss increases with every additionally shaded module. In the right maximum, there is a higher constant power loss with extensive shading (Siegfriedt, 2000).

The reduction in the energy yield depends upon the duration of the shading throughout the entire year. The shading effect with a series and parallel configured array was studied and compared over a long-term basis using the PV façade at

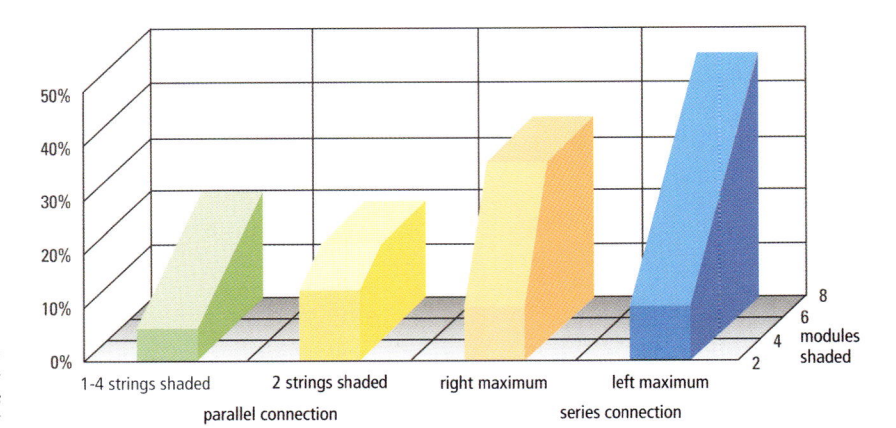

*Figure 3.32
Power losses with different
connection concepts
Source: U. Siegfriedt*

Saarland University of Science and Technology. For comparison purposes, both designs were implemented on the same PV system by switching between the two. In order to limit the currents with the parallel connection, DC–DC converters were connected to every module and linked with a central inverter via a DC energy bus. The DC–DC converters, which are also responsible for tracking the MPP, are prototypes. It can be expected that manufacturers will pick up on this approach. The long-term measurements revealed that there was an up to 30 per cent higher energy yield for the parallel connection compared with the series connection for this façade (Freitag and Weber, 2000). For MPP tracking of individual strings via DC–DC converters, series devices are now available (see the section on '2.3.6 Further developments in grid-connected inverter technology' in Chapter 2).

With no or low shading, the possible yields are independent of the PV array configuration. Here, owing to simpler and more cost-effective mounting, string inverters can provide the better solution economically. If it is not possible to have an unshaded system, parallel-connected modules allow for lower power and, therefore, yield losses, especially if careful planning leads to shading occurring in just a few strings. The disadvantages of such array configurations, such as cable losses resulting from higher currents or increased installation expenditure, are more than compensated for by the increased yields, especially as other yield-reducing effects, such as mismatching, have a greater effect with connection in series than with connection in parallel. Standard simulation programmes are unable to take sufficient account of these complex situations so that, particularly when there is direct shading, the simulation results should be judged critically.

3.7 Shading with free-standing/rack-mounted PV arrays

Frequently PV array are mounted on flat areas (e.g. flat roofs or open spaces). They can be mounted horizontally; but the highest energy yield is attained when there is optimum inclination. In Berlin, a 30° angle would lead to an increase in yield of 12.5 per cent compared with horizontal installation. In addition, horizontal systems must be cleaned more frequently – otherwise, significant losses may occur as a result of soiling (see section '3.3.1 Temporary shading').

To specify the utilization of a specific area, the area exploitation factor f is used. This is defined by the ratio of the module width to the module row distance:

$$f = \frac{b}{d}$$

This usually results in an area exploitation factor of between 0 and 1, or 0 per cent and 100 per cent. A factor of 100 per cent exploitation of the area would, however, cause considerable mutual shading of the individual module rows.

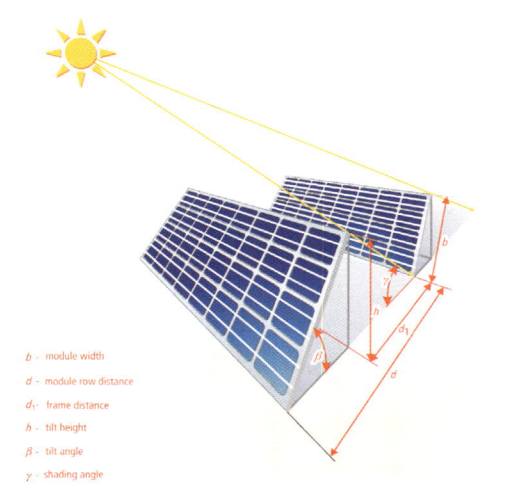

*Figure 3.33
Shading with rack-mounted PV arrays
Source: Solarpraxis*

b – module width
d – module row distance
d_1 – frame distance
h – tilt height
β – tilt angle
γ – shading angle

With a low tilt angle β, there is less shading and the area can be better exploited. However, solar yield across the year then drops. For this reason, a tilt angle of 30° is usually chosen (at Central European latitudes), as well as an area exploitation factor of between 30 per cent and 40 per cent. The distance between the module rows depends upon the module widths, as well as the tilt and shading angles:

$$d = b \times \frac{\sin(180° - \beta - \gamma)}{\sin\gamma}$$

As a good compromise, the zenith angle of the sun in winter (Berlin 14°) is often selected as the shading angle.

3.7.1 Reducing the mutual shading losses of rack-mounted PV modules

In general, two rules of thumb have proven themselves in Germany with a tilt angle β of 30°:

1 To reduce losses: distance $d_1 = 6 \times$ height h.
2 Taking into consideration optimum area exploitation: distance $d = 3 \times$ module width b.

Figure 3.34 can be used to determine shading losses. The process is as follows: first, a specific tilt angle (e.g. 30°) and an area exploitation factor (e.g. 50 per cent) are selected.

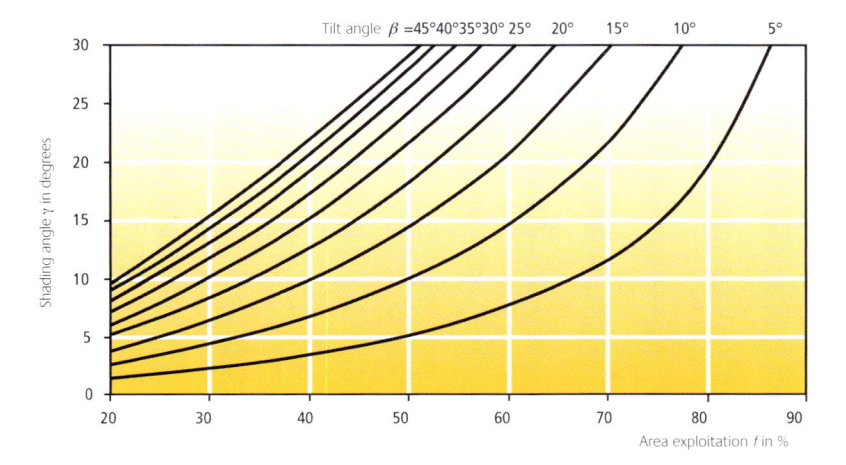

*Figure 3.34
Shading angle as a function of the degree of land use f and incline β*

A shading angle of 24° is obtained from the point of intersection of the β = 30° line with the 50 per cent line. With this value, a point of intersection can be found on the β = 30° line in Figure 3.34, from which a shading loss of 12 per cent can be derived. The resulting yield losses are also affected by the cell and module shading and the system design.

In general, if the area exploitation factor is reduced below 33 per cent, then it is almost impossible to increase energy yield. In addition, it is also recommended that the tilt angle should only be reduced below 30°, with area exploitation factors above 50 per cent (in Germany) (Quaschning, 2000).

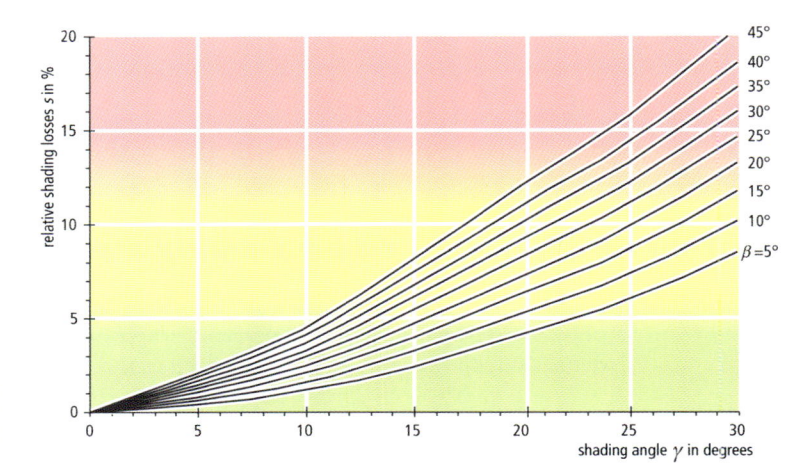

Figure 3.35
Shading losses in relation to shading angle γ and tilt angle β
Source: V. Quaschning, R. Haselhuhn

3.8 Checklists for building survey

The following checklists have been devised for a grid-connected PV system on the roof of a single- or two-family house. Copies of the construction documents (floor plan, section, roof plan and site plan) should be obtained from the customer if available. Larger PV systems require a more detailed survey. If necessary, develop your own checklists based on this model. If the system will be partly shaded, the additional checklist may also be used.

In order to assess the effect of shading on the PV system, it is sensible to make a rough sketch, as demonstrated in Figure 3.35. This can be sketched in on the shading checklist in the field with the points of the compass or on the copy of the site plan. With a new building, additional factors to consider include whether there will be later planting and other new buildings in the immediate vicinity. In addition, the extent of tree growth should also be taken into account.

The following checklist should be marked on the sketch or a copy of the site plan (with additional photographs, if necessary):

- roof area, taking orientation into account;
- usable area for the PV system;
- chimneys, antennae and satellite dishes;
- buildings situated close by (approximate distance and height);
- trees (approximate distance and height) – labels: deciduous tree (D) and coniferous tree (F);
- overhead cables (electricity and telephone) if these could shade the PV system;
- other shading: building projections, dormer windows, etc.

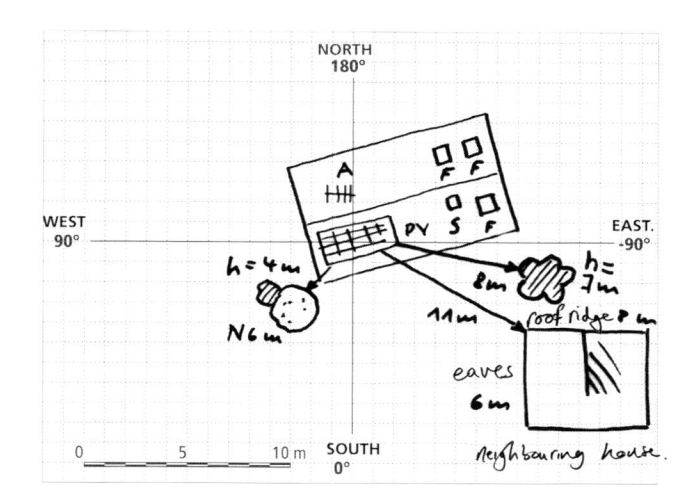

Figure 3.36
Sample drawing
Notes: Usable area for PV = PV;
roof light = L; antenna = A; chimney = C;
coniferous tree = F; deciduous tree = D.
Source: Solarpraxis

3.8.1 PV system checklist

Name of customer: _____

Street, number: _____

Postcode, city: _____

Phone (private): _____

Phone (work): _____ from: _____ to: _____

Fax: _____

Address of construction site (if different): _____

_____

If required: Architect: _____

            Electrician: _____

            Roofer: _____

Figure 3.37
PV system checklist – relevant measurements

Available roof area :   Length =_____ m × width =_____ m = _____ m²

with the following roof elements: ☐ Chimney   ☐ Antenna

                                     ☐ Skylight   ☐ Lightning conductor

                                     ☐ Dormer window   ☐ Other: _____

USEFUL DOCUMENTS

Construction plans ☐ Site plan ☐ Floor plan ☐ Roof plan
☐ Elevations ☐ Sections ☐ Building description
☐ Photographs ☐ Roof ☐ View of house with chosen roof area
☐ Meter location ☐ With shading: shading situation

CUSTOMER WISHES AND EXPECTATIONS

☐ Roof-mounted ☐ Roof-integrated ☐ Other
☐ PV module type ☐ Mono-crystalline ☐ Polycrystalline
☐ Amorphous ☐ Thin-film

PV power (approximately) _____ kWp
Maximum investment €_____
Desired energy yield _____ kWh/a
Maximum area _____ m²
Other _____

ROOF

Roof shape ☐ Gable roof ☐ Flat roof ☐ Monopitch roof
☐ Hipped roof ☐ Broach roof ☐ Mansard roof
☐ Saw-tooth roof ☐ Gable roof with partial hip
☐ Other: _____

Roof covering ☐ Slate ☐ Corrugated Eternite ☐ Roofing felt
☐ Tiles ☐ Gravel ☐ Bitumen
☐ Pantile ☐ Plain tile ☐ Roman tile
☐ Other: _____

Roof construction Heat insulation? ☐ yes ☐ no
Roof substructure _____
Distance between rafters = _____ m
Accessibility of the roof ☐ Crane necessary ☐ Scaffolding necessary
Vehicle access ☐ yes ☐ no
Roof openings that can be used for laying cables? ☐ yes ☐ no
☐ Ventilation tiles ☐ Other roof inlets

3.8.2 PV generator, inverter and meter

Orientation of the PV generator
  from –90° (east) via 0° (south) to +90° (west): _____°

Tilt angle of the PV generator
  from 0° (horizontal) to 90° (vertical): _____°

Is there lightning protection? ☐ yes ☐ no

Where can the PV generator be earthed? _____

Place for generator junction box? _____

Where is the electricity meter? ☐ Cellar ☐ Corridor ☐ Living room
☐ Storage room ☐ Outside the building: distance = ___ m
☐ Other location: _____

Meter connection? _____

Is there still room for a meter? ☐ yes ☐ no

Is there space for the inverter too? ☐ yes ☐ no

Place for the inverter? _____

Place for the DC main switch? _____

3.8.3 Lines and installation

Approximate cable length distance between PV generator and junction box: _____ m

Distance between PV generator and equipotential busbar: _____ m

Distance between generator junction box and inverter: _____ m

Distance between inverter and mains supply: _____ m

Location and type of installation for the DC main cable: _____

Location and type of installation for the AC connection cable: _____

Must the roof be broken through? ☐ yes ☐ no ☐ Number: _____

3.8.4 Other

Annual electricity consumption? _____ kWh/year

With new building   Enquire about future planting and new buildings in the immediate vicinity.

With shading   Use shading checklist!

Is the house a listed building or in a conservation area? ☐ yes ☐ no

3.8.5 Shading checklist

Sketch in (take photos if required)

- roof area (note orientation);
- usable area for the PV system (place the centre of the PV system at the origin of the co-ordinates);
- chimneys, antennae, satellite dishes;
- buildings situated close by (approximate distance and height);
- trees (approximate distance and height). Labels: deciduous tree (D) and conifer (F)
- overhead lines (electricity/telephone);
- other shading: building projections …

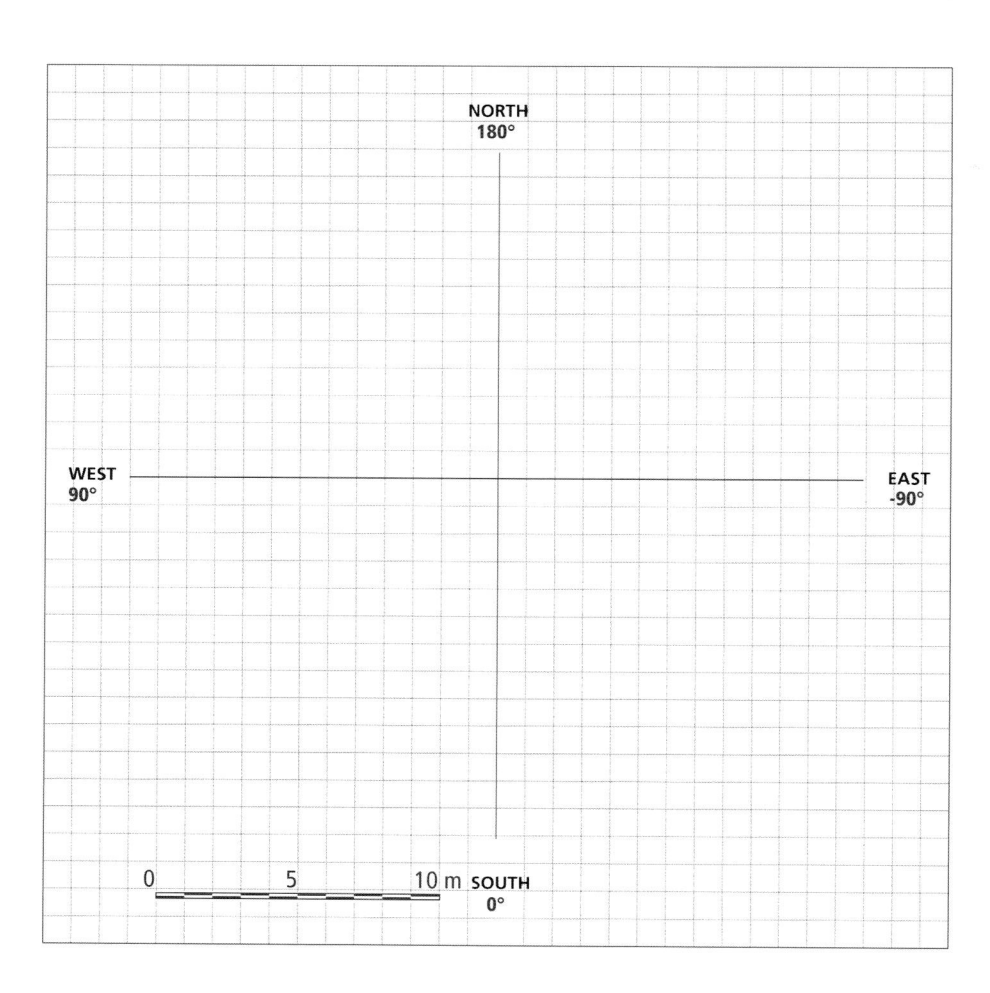

Figure 3.38
Basis for sketch of shading considerations
Labels: PV = Usable area for PV
C = Chimney
W = Roof window
C = Conifer
A = Antenna
D = Deciduous tree

4 Planning and Sizing Grid-Connected Photovoltaic Systems

4.1 System size and module choice

A suitable area for the PV system is agreed upon with the customer during a site visit. The checklist for the building survey includes the system specifications: orientation, tilt, available area, type of mounting, shading, cable lengths, inverter location, etc.

The modules are then chosen according to the:

- cell material: mono-crystalline, polycrystalline, amorphous, CdTe or CIS, or thin-film technology; and
- module type: standard module with/without frames, glass–glass module, PV tile, etc. (see also Chapters 2 and 6).

A specific module type is selected based on these specifications. The module's technical specifications determine the rest of the system sizing. First, the rough number of modules that can be accommodated in the area is determined. This number enables the rough overall power of the PV system to be determined:

- Rule of thumb: 1kWp = approximately $10m^2$ PV area.

Table 4.1 can be used for a more precise estimate of the area requirement depending upon the cell material.

Table 4.1
Area requirement of PV systems
Source: Solarpraxis/DGS Berlin BRB

Cell material	Required PV area for 1kW$_p$
Mono-crystalline	$7m^2$–$9m^2$
High performance cells	$6m^2$–$7m^2$
Polycrystalline	$7.5m^2$–$10m^2$
Copper indium diselenide (CIS)	$9m^2$–$11m^2$
Cadmium telluride (CdTe)	$12m^2$–$17m^2$
Amorphous silicon	$14m^2$–$20m^2$

The area requirement when using semi-transparent modules will increase roughly in proportion to the area of the modules that is transparent. The following points need to be taken into consideration for the actual roof planning:

- number of modules corresponding to multiples of module width and height in relation to the available roof width and height;
- the distance between the modules and the edge of the roof, which should be three times the vertical distance between the module surface and the roof;
- an expansion gap between the modules, generally of between 6mm and 10mm;
- roof superstructures (e.g. chimneys, roof vents and antennae) and the shadows they cast;
- local surroundings with regard to shading

For the design, the mounting and architectural integration of the PV array in the building also play an important role. These subjects are dealt with in detail in Chapter 6.

4.2 System concepts

The system concepts are determined by the inverter system components. This results in central and decentralized system concepts. The connection of modules to form strings and their parallel connection should be optimally coordinated with the inverter. Depending upon the module tolerance, greater or lesser mismatch losses result when the modules are connected together in strings. Werner Hermann at the TÜV Rheinland PV Certification Laboratory in Germany calculated the dependency of mismatch losses on module tolerance and on pre-sorting the modules. The results are shown in Figure 4.1.

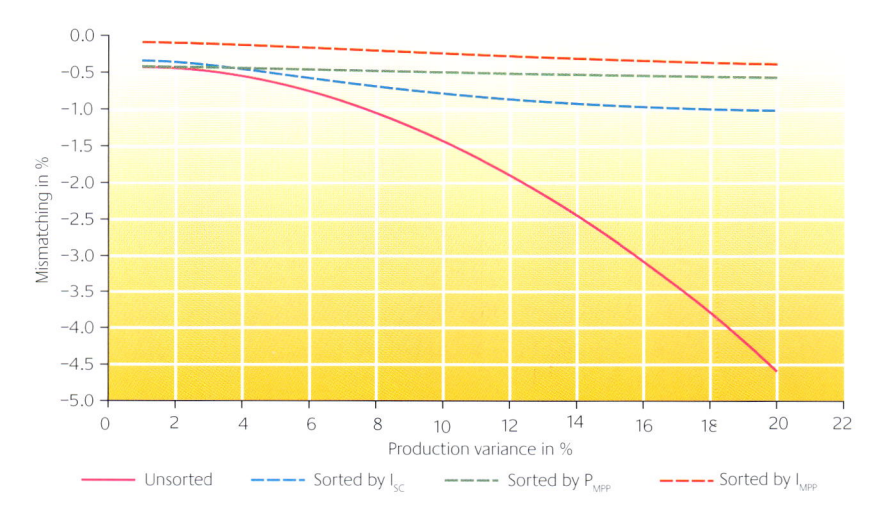

Figure 4.1
Mismatch in a PV array with 8 strings and 14 modules of 150W connected in series, depending upon production variance (total of 112 PV modules)
Source: TÜV Rheinland

According Figure 4.1, if modules with a production tolerance of ±5 per cent are connected in series unsorted, the losses are less than 1 per cent. If modules are sorted by current, the mismatch losses are reduced to approximately 0.2 per cent. With production variance greater than 8 per cent, sorting by MPP currents can be recommended as a standard practice (Herrmann, 2005).

Inverters are available as central inverters for an entire system, as string inverters for a string, and as module inverters for an individual module. Each of the three concepts has advantages and disadvantages. Which concept is chosen depends upon the type of application. De-central inverter concepts should be considered for systems consisting of sub-array areas with different orientations and tilts, or for systems that are partially shaded.

4.2.1 Central inverter concept

LOW-VOLTAGE CONCEPT

In the low voltage range (UDC ≤120V), only a few modules (three to five standard modules) are connected in series in a string. One advantage of these short strings compared with longer strings is that shadows have less effect since the module with the largest shading in a string determines the entire string current. In addition, the loss depends upon the number of shaded strings, where the number of shaded modules is less significant. If only a few strings are shaded, the losses remain low.

Furthermore, with a voltage of less than 120V, it is possible to execute the design to Protection Class III (see Table 4.2). The disadvantage of this concept are the resulting high currents. Relatively high cable sections have to be deployed to reduce the ohmic losses. For this reason, the concept is rarely implemented. Typical applications are building-integrated systems with custom-made modules.

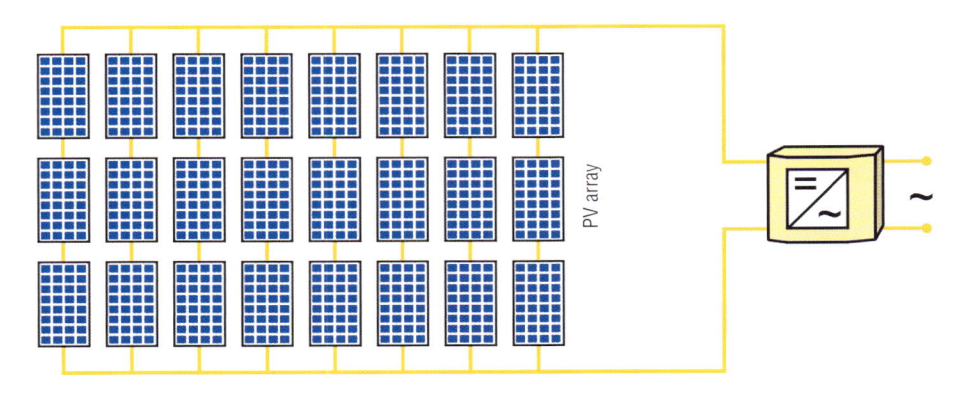

Figure 4.2
Low voltage concept with central inverter
Source: Solarpraxis

Table 4.2
Protection classes

Electrical protection classes		Symbol
	Device is earthed/grounded	⏚
Class II	Protective insulation (double/reinforced insulation)	▢
Class III	Safety extra-low voltage (maximum AC: 50V; maximum DC: 120V)	⬦

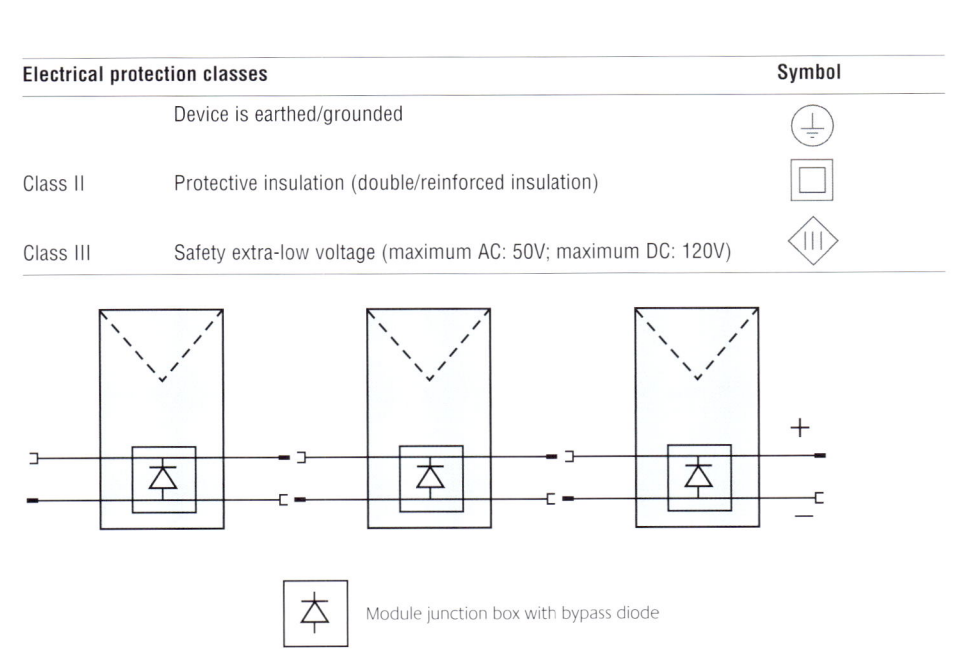

Figure 4.3
Parallel connection of modules with four connection cables

Module junction box with bypass diode

The Dutch company OKE-Service has developed a concept in which frameless crystalline modules are wired in parallel without bypass diodes. The key feature in this concept is that the metal module mounting frame is used to conduct electricity and for direct parallel connection of the modules. Called 'PV-wirefree', this concept dispenses with the DC wiring and other DC components (such as fuses and PV combiner/junction box). Other advantages are the minimization of shading losses and that it is possible to dispense with Protection Class II. This makes the concept well suited to integration within buildings and locations with direct shading. However, to date, it has hardly been used. The reason for this is that the electrical module-to-mounting-frame connecting element exists only as a prototype and that hardly any suitable modules and inverters with the low voltage range are currently in production. The company Multi-Contact has been named a development partner for the connecting elements. It remains to be seen when the first systems using the PV-wirefree concept will come onto the market.

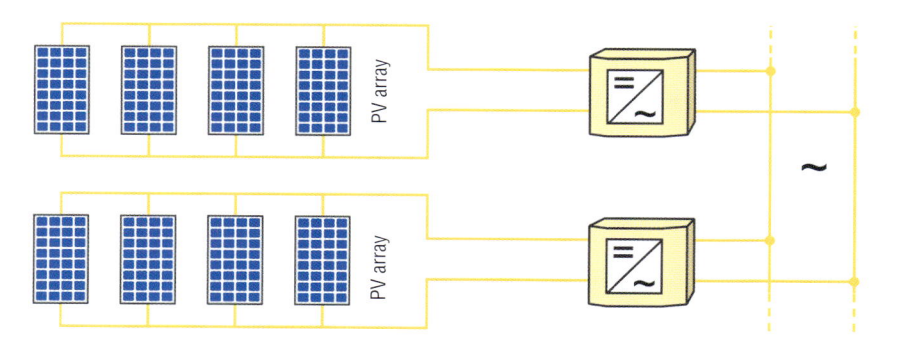

*Figure 4.4
Parallel connection concept
Source: Solarpraxis*

CONCEPT WITH HIGHER VOLTAGES

Protection Class II is required for concepts with longer strings and their associated higher voltages ($U_{DC} > 120V$). The advantage of this concept is that smaller cable cross sections can be used as a result of the lower currents. A disadvantage is the greater shading losses due to the long strings.

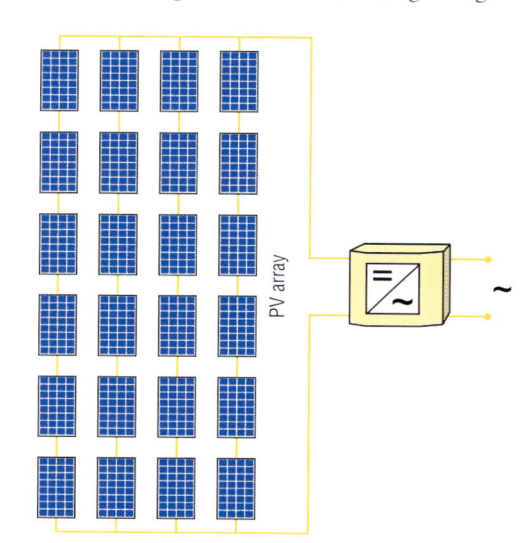

*Figure 4.5
Concept with central inverter
and higher voltages
Source: Solarpraxis*

MASTER–SLAVE CONCEPT

Larger PV systems often use a central inverter concept based on the master–slave principle. This concept uses several central inverters (mostly two to three). For the sizing, the total power is divided by the number of inverters. An inverter is the master device and operates in lower irradiance ranges. With increasing irradiance, the power limit of the master device is reached and the next inverter (slave) is connected. In order to load the inverters equally, the master and slave are swapped over (rotating master) in a specific cycle.

The advantage of this concept is that with lower irradiance, only one inverter operates (master); thus, the efficiency – particularly in lower power ranges – is greater than if only one central inverter is used. However, the investment costs increase in comparison with central inverters.

An example of the master–slave concept is the 1MW PV system on the roof of the Neue Messe exhibition centre in Munich, Germany.

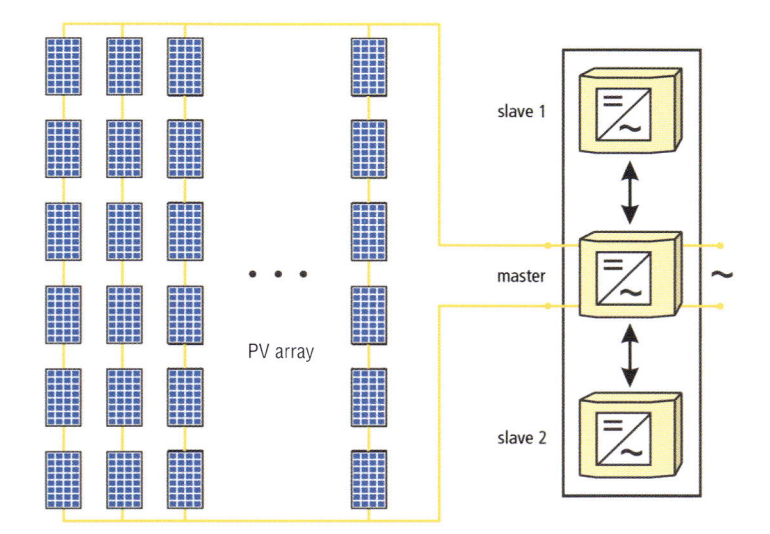

*Figure 4.6
Master–slave concept with a central inverter unit
Source: Solarpraxis*

*Figure 4.7
1MW PV system of the Neue Messe exhibition centre in Munich, Germany
Source: Shell Solar*

4.2.2 Sub-array and string inverter concept

Systems with outputs up to 3kW are generally built with string inverters. In most cases, the entire PV array forms just one string. Medium-sized systems often have two or three strings connected to the inverter, resulting in a sub-array concept. With a system with variously oriented sub-arrays or with shading, a sub-array and string inverter concept enables better power-related adaptation to the irradiance conditions. An inverter is used per sub-array or per string. Care should be taken to ensure that only modules with the same orientation, angle and freedom from shade are connected together in a string. With strings that are too long, shading can cause greater power losses since the module with the least irradiance determines the entire string current.

Using string inverters facilitates easier installation and can considerably reduce installation costs. The inverters are often mounted in the immediate vicinity of the PV array and are connected string wise; they are currently available for powers of approximately 500W to 3000W.

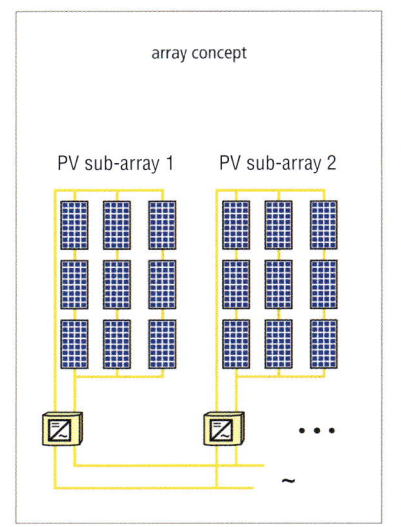

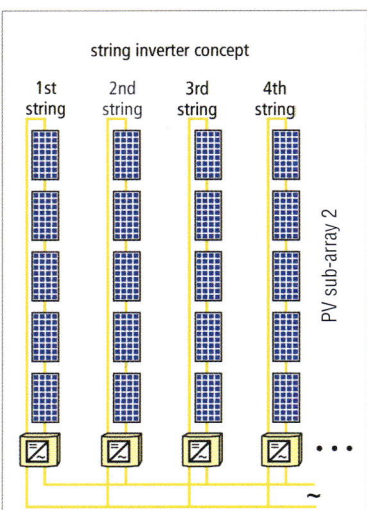

Figure 4.8
Sub-array and string inverter concept
Source: Solarpraxis

Figure 4.9
String inverters are used in the 1MW photovoltaic system on the Mont Cenis training academy in Herne, Germany
Source: Scheuten Solar

Since the inverters are connected directly to the module strings, this provides the following advantages and cost savings compared with central inverter concepts:

- omission of the PV combiner/junction box;
- reduction of the module cabling to series interconnection and omission of the DC main cable.

The 1MW PV system on the Mont Ceris Training Academy in Herne, in North Rhine-Westphalia, Germany, is an example of a hybrid concept using sub-arrays and string inverters. Here, 569 string inverters are used. The architectural concept envisaged the use of six different PV modules. A total of 16 different string configurations were formed so that the voltage levels are sometimes dissimilar. The string inverters enable adjustment to the different MPP voltages of the strings.

4.2.3 Module inverter concept

A prerequisite for high system efficiency is that the inverters are optimally adjusted to the PV modules. It would be most advantageous if every module was operated permanently at maximum power point. The MPP matching is more successful if PV modules and inverters are conceived as a unit. These module inverter units are also called AC modules. Some devices are so small that they can be stored in the module junction boxes.

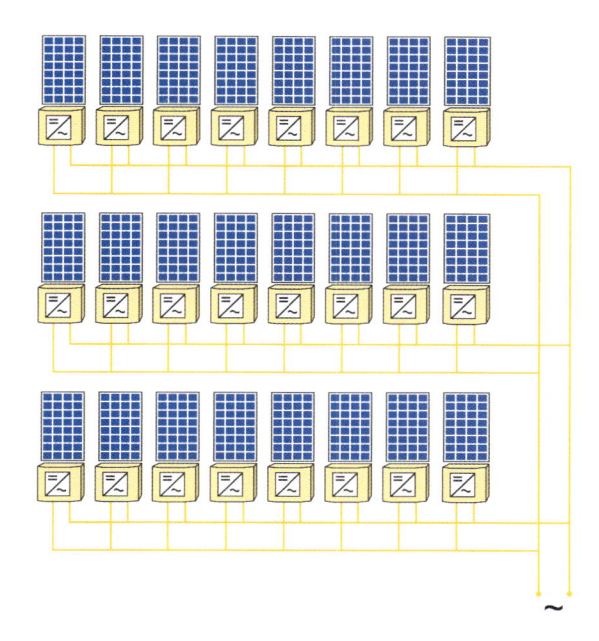

Figure 4.10
Module inverter concept
Source: Solarpraxis

Another advantage is the ease with which the PV systems can be extended. Other concepts cannot be so simply broadened. Module inverters enable PV systems to be extended as desired, even with only one inverter module.

It is often claimed that the disadvantage of module inverters is the lower efficiency. There is, in fact, not such a great difference to central inverters, as is shown in Figure 4.11. In addition, the lower efficiency is compensated for by the greater yield that results from improved matching with the respective module's MPP point.

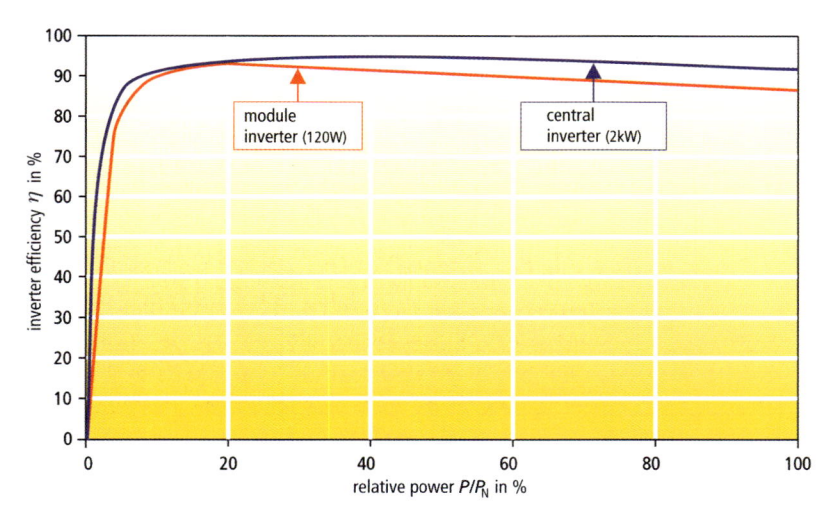

Figure 4.11
Efficiency curve of a central inverter and a module inverter
Source: V. Quaschning

Module inverters are currently still relatively expensive. There will only be cost advantages once AC modules and module inverters are widely available on the market.

When mounting AC modules, it should be ensured that faulty inverters can be easily replaced. Just as important with this concept is monitoring the individual inverters by recording the relevant operating data, faults and fault signals, and storing the data. Manufacturers offer systems that can be monitored with a PC; the software displays the recorded data.

The module inverter concept is advantageous for façade-integrated systems, particularly if there is considerable partial shading of the façade by the surroundings, or projections and recesses in the façade. Façade-integrated isolation glass modules with module inverters, which are fed into the sub-distribution units of the apartments, are used in an office and apartment building at Moritzplatz in Berlin, Germany.

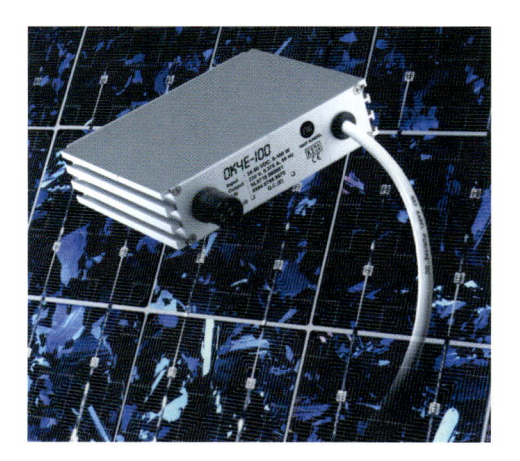

Figure 4.12
Module inverter
Source: NKF

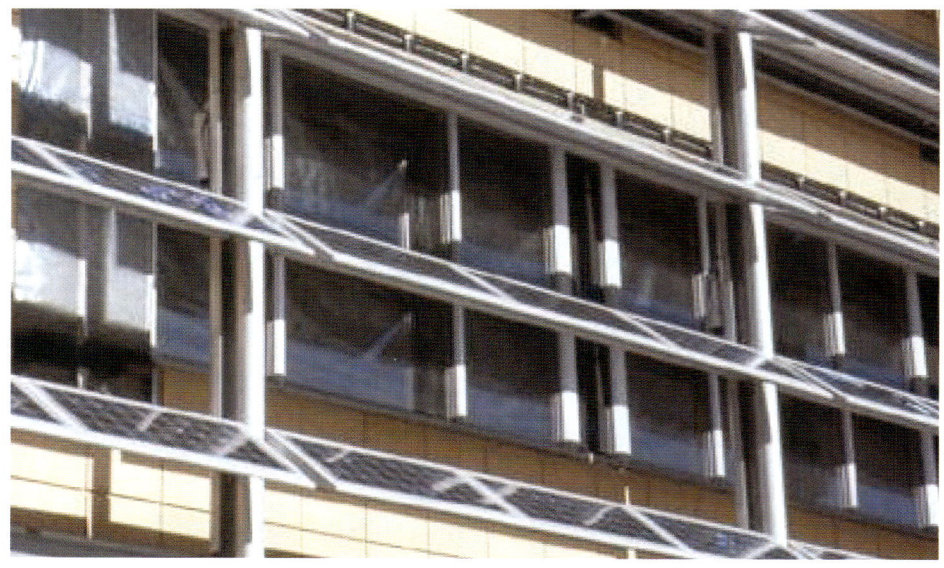

Figure 4.13
PV façade of the ECN building in
The Netherlands with string inverters
Source: OJA-Services:
Copyright: Tjerk Reijenga

4.3 Inverter installation site

When choosing the installation site, it is crucial that the environmental conditions specified by the manufacturer are maintained (essentially humidity and temperature). The ideal installation site for inverters is cool, dry, dust free and indoors. It makes sense to install inverters next to the meter cupboard or close by. If the environmental conditions permit, the inverter can be installed close to the PV array combiner/junction box. This reduces the length of the DC main cable and lowers the installation costs.

The ventilation grilles and heat dissipaters need to be kept uncovered to ensure optimum cooling. For the same reason, the devices should not be installed right on top of each other if this can be avoided. The noise produced by the inverter should also be taken into account when choosing the installation site. The units should be protected from aggressive vapours, water vapour and fine particles. For example, in barns or stables, ammonia vapours can arise that may damage the inverter. Larger central inverters are often installed in a separate inverter cabinet – together with protective equipment, meters and switchgear.

String inverters are increasingly being used in exterior applications on roofs or elsewhere outdoors. These devices have IP 54 protection and can withstand outdoor weather conditions. However, it is still recommended that inverters be protected from direct sunlight and rain in order to exert a positive influence on their service life. In addition, accessibility must be guaranteed in the event of a fault to enable repairs to the inverter. If environmental conditions permit, it makes sense to install the PV inverter close to the PV array combiner/junction box. This reduces the power losses through the DC main cable and lowers the installation costs.

4.4 Sizing the inverter

Inverters' technical specifications provide important information on sizing and installation. It is essential that they are observed when sizing. The system and connection concept determines the number, voltage level and power class of the inverters.

4.4.1 Choosing the number and power rating of inverters

The number and power rating of inverters is determined by the overall power of the PV system and the chosen system concept.

A single-phase parallel feed according to the VDEW guideline in Germany is allowed up to an apparent feed power of $S_{AC} = 4.6\text{kVA}$; according to this, the nominal output power of the inverter P_{nAC} must correspond to this value. Above 4.6kVA, the feed must be multi-phase. This can be accomplished using multiple single-phase inverters spread equally between the three phases, if possible (maximum unbalanced load of 4.6kVA). However, according to VDN bulletin 03/2004 relating to the VDEW guideline, it is permissible to feed a maximum power of 10 per cent above the inverter's nominal output power into the electricity grid for a period of ten minutes (VDN, 2004). The inverter manufacturers guarantee these values and certify them in the conformity declaration:

Maximum inverter output power (AC) = $S_{\text{max 10 min}} \leq 1.1 \times S_N$

The solar array and inverter(s) have to be optimally matched to each other's output values. The nominal power of inverters can be ±20 per cent of the PV array output power (under STC), depending upon the inverter and module technology and the local conditions, such as regional insolation and orientation of the modules.

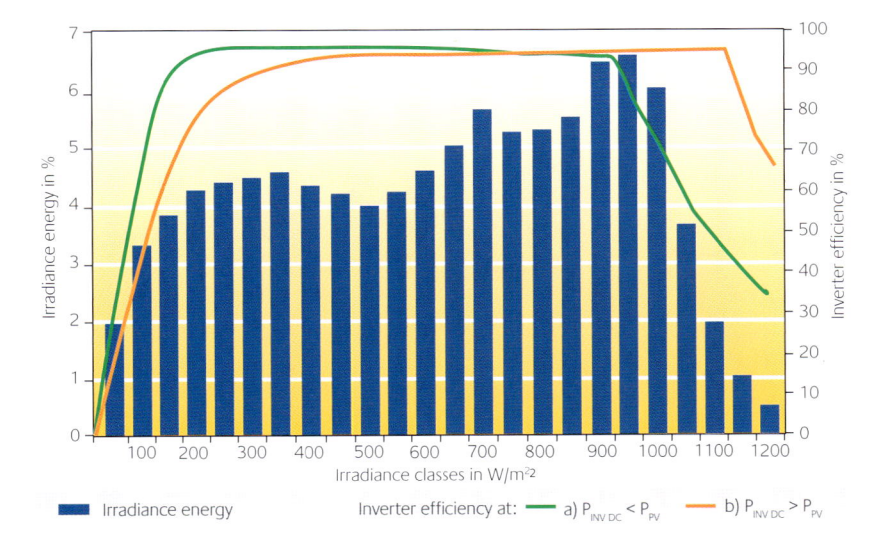

Figure 4.14
Distribution of annual solar irradiance at the module plane of a system in Munich, Germany (30° south), and efficiency curves for a smaller (−10 per cent) and a larger (+10 per cent) inverter; irradiance classes are based on five-minute measured values
Source: R. Haselhuhn

The under sizing of inverters by up to 40 per cent, which was common in the past, was recommended based on the use of hourly irradiance data. Hourly data indicated that there were only small amounts of irradiance energy at irradiance levels above 850W/m². However, data since gathered on a minute/second basis indicates that that significant quantities of energy are also available at an irradiance level of 1000W/m².

As a guide, a ratio between the PV array power and inverter power of 1:1 is used for sizing. Since inverters are available at specific power levels, and the number of modules and, thus, the power of the solar array are determined by the usable area, a deviation from the 1:1 sizing is generally the rule. However, the inverter manufacturer's stated maximum connectable PV power should not necessarily be trusted. Experience shows that in some cases very high values are stated, with the result that the devices often work in the overload range. The consequences are avoidable energy losses owing to the power limit control and possible premature ageing of the devices. A more reliable method is to calculate the DC power via the inverter's nominal efficiency from the AC nominal power. The inverter manufacturer has to state the AC nominal power in the conformity declaration. The nominal AC power is the power that the inverter can continuously feed into the grid without cutting out at an ambient temperature of 25°C (±2°C). On average, the DC power rating is around 5 per cent higher than the inverter's nominal AC power. The following power range can be specified for the sizing range:

$$0.8 \times P_{PV} < P_{INV\ DC} < 1.2 \times P_{PV}$$

The ratio of PV array power rating (Wp) to the inverter's nominal AC power is known as the inverter sizing factor c_{INV}:

$$c_{INV} = \frac{P_{PV}}{P_{INV\ AC}}$$

The sizing factor describes the inverter's level of utilization. A typical sizing factor is 1 within a range of:

$$0.83 < c_{INV} < 1.25$$

When inverters are being installed in lofts or outdoors, they may need to be undersized because they may be operating in conditions of high ambient temperature. Some locations may not be suitable because of extremes of temperature and other environmental factors. Manuals should be referred to or suppliers asked when there is any doubt as to the suitability of a location.

Investigations by Bruno Berger at the Fraunhofer ISE in Freiburg, Germany, showed additional losses of 0.5 per cent to 1 per cent, with a sizing factor c_{INV} of 1.1 to 1.2 as a result of the inverter cutting out during brief irradiance peaks. With sizing factors of 1.2 to 1.3, additional losses of between 1 per cent and 3 per cent occurred. These investigations were based on a 30° tilted south-aligned ventilated PV array. The annual efficiencies of inverters with and without transformer were calculated in relationship to the sizing factor (see Figure 4.15). Curves are shown for instantaneous values and average minute and hourly values. Hence, the difference in yield from the instantaneous value tracking can be read off from the graph when sizing with a simulation programme on an hourly interval basis. However, the quality of the inverter's tracking system still needs to be taken into account (i.e. the tracking efficiency).

Figure 4.15
Simulated annual efficiencies of inverters with and without transformers depending upon the sizing factor
Source: Fraunhofer ISE

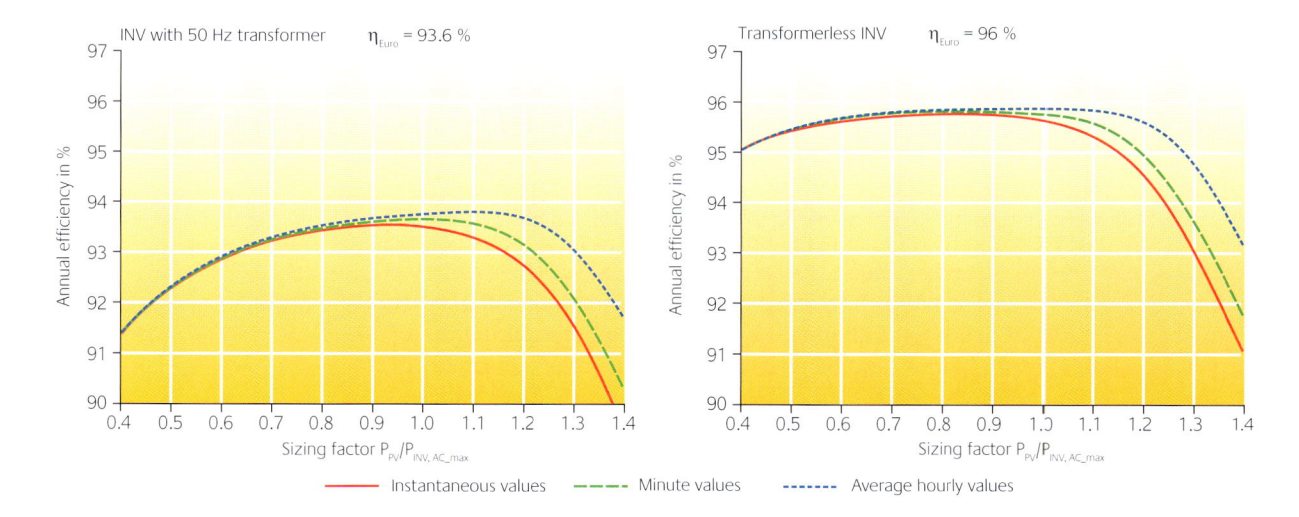

As well as the inverters' tracking efficiency, the cable losses and a possible negative deviation from the module nominal power have also not yet been taken into account (Burger, 2005).

In systems without optimum alignment or systems with partial shading, it can make sense from a technical and economical point of view to size the inverter somewhat smaller. Here the different overload characteristics of inverters should be borne in mind (see the section on '2.3.4 Characteristics, characteristic curves and properties of grid-connected inverters' in Chapter 2). With frequent continuous overloading the service life of the device diminishes rapidly. On no account, however, may the maximum input voltage of the inverter be exceeded. That is why the following is crucial for sizing.

When using amorphous modules, the degradation of the modules must be taken into account during sizing. Amorphous modules can have a power that is around 15 per cent higher in the first months of use before reducing to a constant nominal value as a result of the initial light degradation (see the section on '2.1.10 Electrical characteristics of thin-film modules' in Chapter 2). This effect must also be taken into account with the subsequent voltage and current sizing of the inverter. During this period, the operational voltage can be around 11 per cent higher than the nominal value and the operating current can be around 4 per cent higher.

4.4.2 Voltage selection

The magnitude of the inverter's voltage is the sum of the voltages of the series-connected modules in a string. Since the module voltage (see the section on '2.1.8 Irradiance dependence and temperature characteristics' in Chapter 2) and the voltage of the entire PV array depends upon the temperature, the extreme cases of winter and summer operation are used when sizing.

In order to enable inverters to be optimally matched to the solar array, it is important to take the modules' temperature and irradiance operating parameters into account. The PV array voltage is strongly dependent upon the temperature. The operating range of the inverter must be matched with the I–V curve of the PV array. The MPP range of the inverter should, as can be seen in Figure 4.16, incorporate the MPP points of the array I–V curve at different temperatures. In addition, the turn-off voltage and the voltage resistance of the inverter must be taken into account.

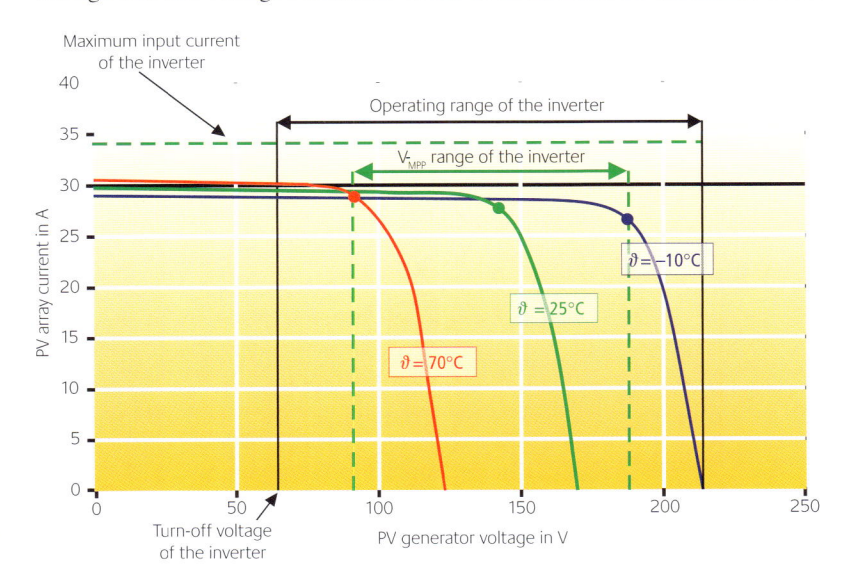

Figure 4.16
PV array I–V curves and an inverter's operating range
Source: R. Haselhuhn

MAXIMUM NUMBER OF MODULES IN A STRING

The first limit is defined by a winter temperature of –10°C (Germany). At low temperatures, the module voltage increases. The highest voltage that can occur in an operating condition is the open-circuit voltage at low temperatures. If the inverter is switched off on a sunny winter day (e.g. because of grid failure), this can lead to the open-circuit voltage being too high when it is switched back on again. This voltage must be lower than the maximum DC input voltage at the inverter; otherwise the inverter could be damaged. Thus, the maximum number of series-connected modules is derived from the quotient of the maximum input voltage of the inverter and the open-circuit voltage of the module at –10°C:

$$\eta_{max} = \frac{V_{max\,(INV)}}{V_{OC\,(module\,-10°C)}}$$

The open-circuit voltage of modules at –10°C is not always specified on module manufacturers' data sheets. Instead, information is often provided on the voltage change ΔV as a percentage or in mV per °C. The change of voltage is prefixed with a negative sign. This enables the open-circuit voltage at –10°C to be calculated from the open-circuit voltage under STC conditions $V_{OC(STC)}$ as follows.

With ΔV in percentage per °C:

$$V_{OC\,(module\,-10°C)} = (1 - 35°C \times \Delta V/100) \times V_{OC\,(STC)}$$

With ΔV in mV per °C:

$$V_{OC\,(module\,-10°C)} = V_{OC\,(STC)} - 35°C \times \Delta V$$

Here it should be ensured that ΔV has a negative symbol.

If neither figure is given, it is possible to use Figure 2.61 in Chapter 2 to determine the value. This shows that the open-circuit voltage of a mono-crystalline or polycrystalline module at –10 °C increases by around 14 per cent compared with STC conditions:

$$V_{OC\,(module\,-10°C)} = 1.14 \times V_{OC\,(STC)}$$

MINIMUM NUMBER OF MODULES IN A STRING

During summer, modules on a roof can easily heat up to around 70°C (in Germany, it can be higher; see below). This temperature is generally used as a basis when determining the minimum number of modules in a string. With well-ventilated systems, a maximum temperature of 60°C can be assumed in Germany.

With full irradiance in summer, a PV system has a lower voltage than at STC conditions (nominal voltage on the module data sheet) owing to the increased temperatures. If the operating voltage of the system drops below the minimum MPP voltage of the inverter, this would no longer feed the maximum possible power and, in the worst case, would even switch itself off. For this reason, the system should be sized such that the minimum number of series-connected modules in a string is derived from the quotient of the minimum input voltage of the inverter at the MPP and the voltage of the module at the MPP at 70°C. The following formula provides the lower limit value for determining the number of modules in a series:

$$\eta_{min} = \frac{V_{MPP\,(INV\,min)}}{V_{MPP\,(module\,70°C)}}$$

If the voltage of the module at MPP at 70°C is not specified on the module manufacturer's data sheets, this can be calculated as follows from the MPP voltage under STC conditions $V_{MPP(STC)}$ using the figures for the voltage change ΔV in percentage or in mV per °C.

$$V_{MPP\ (module\ 70°C)} = (1 + 45°C \times \Delta V/100) \times V_{MPP\ (STC)}$$

With ΔV in mV per °C:

$$V_{MPP\ (module\ 70°C)} = V_{MPP\ (STC)} + 45°C \times \Delta V$$

In general, it can be assumed that the MPP voltage of a mono-crystalline or polycrystalline module at 70°C will drop by around 18 per cent relative to STC conditions:

$$V_{MPP\ (module\ 70°C)} = 0.82 \times V_{OC\ (STC)}$$

The maximum occurring temperature is determined by the location of the system. This should be taken into account when determining the voltage change. With roof and façade-integrated PV systems without ventilation, temperatures can increase to 100°C. In this case, the voltage V_{MPP} at 100°C is used for determining the minimum number of modules in a string. Again, Figure 2.61 in Chapter 2 can be used. With system concepts with long strings, severe shading can cause a considerable drop in the MPP voltage. This should be taken into account when sizing. By checking the voltage limits and determining the frequency of the occurring voltages, simulation programs can provide information for optimizing the sizing.

VOLTAGE OPTIMIZATION

When optimizing the sizing, it should be borne in mind that the inverter's efficiency depends upon the voltage (see the section on '2.3.4 Characteristics, characteristic curves and properties of grid-connected inverters' in Chapter 2). Specifications and/or graphs of the voltage dependency are required here. However, currently, only very few manufacturers provide data for the efficiency at different inverter voltages. A number of devices (e.g. by inverter manufacturers Fronius, Siemens, SMA and Sunways) were measured in detail at the photovoltaics laboratory of the Bern University of Applied

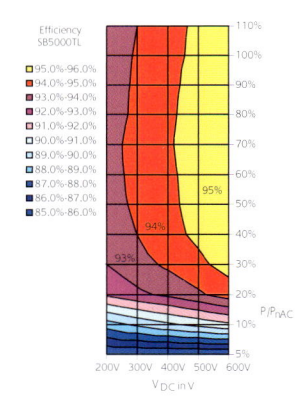

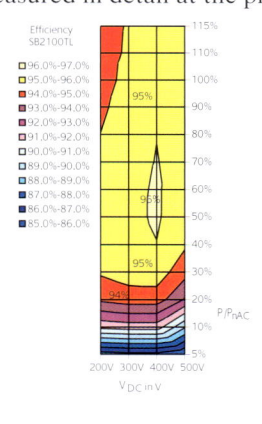

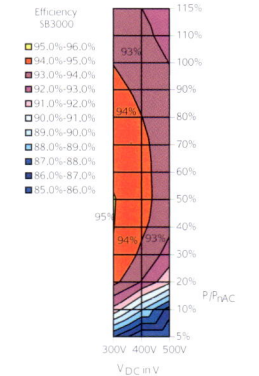

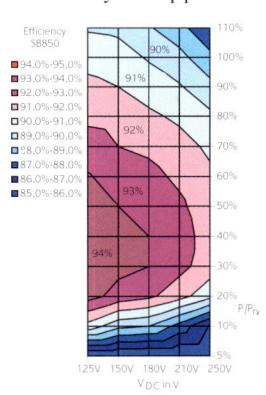

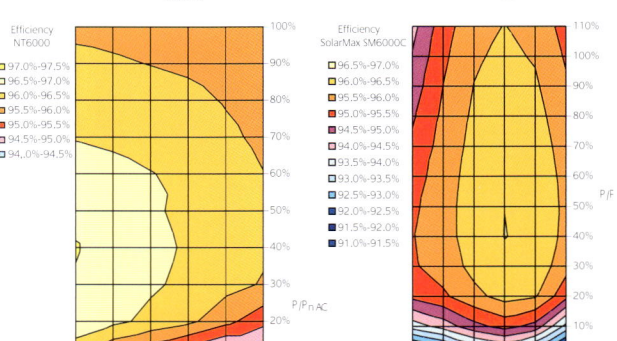

*Figure 4.17
Efficiency in relation to the input voltage for various types of inverter manufactured by SMA, Sputnik and Sunways
Source: Baumgartner, 2005*

Sciences in Burgdorf, Switzerland (see section 2.3.4 in Chapter 2). A comprehensive description of the test and technical data is available on the internet at www.pvtest.ch.

For some types of inverters produced by SMA, Sputnik and Sunways, the following figures can be used directly to optimize voltage sizing. These are based on manufacturers' measurements and investigations conducted at the laboratory for electronic measurement systems at Neu-Technikum Buchs (Buchs College of Technology, NTB) in Switzerland.

By closely matching the PV array to the inverter, the yield can be increased by several per cent. An inverter with better efficiency in the PV array's normal voltage operating range can therefore be a good investment. The inverter usually accounts for only 10 per cent of the costs of a grid-tied PV system, so any increased inverter costs can be rapidly recovered.

4.4.3 Determining the number of strings

On completing the sizing, one should ensure that the maximum PV array current does not exceed the maximum inverter input current. The maximum number of strings is as large as the quotient from the maximum permitted DC input current of the inverter and the maximum string current:

$$\eta_{string} = \frac{I_{max\ INV}}{I_{n\ string}}$$

If the inverter is undersized, one should check how frequently the inverter is situated in the excess current range. Estimates can thus be made on whether there is slight or high overloading. This can be achieved using suitable simulation programs. Figure 4.18 depicts an overloaded inverter: this can lead to premature ageing of the inverter or to destruction of electronic components.

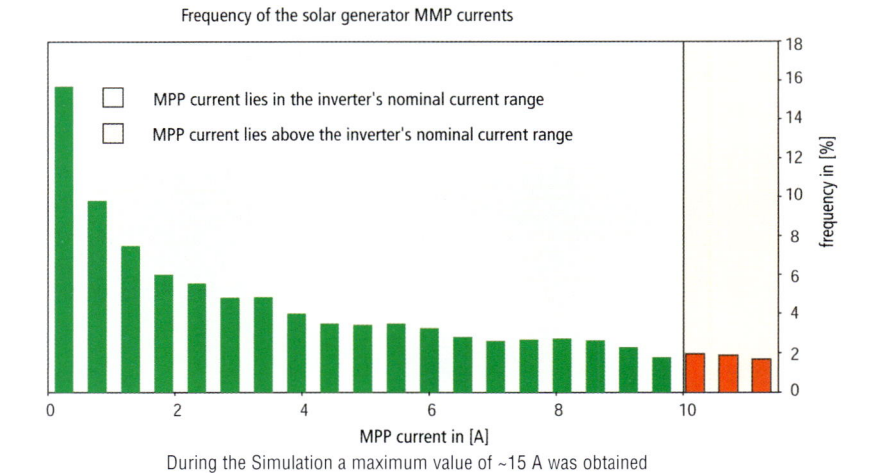

Figure 4.18
Calculating the current load using the SolEm simulation programme

During the Simulation a maximum value of ~15 A was obtained

4.4.4 Sizing using simulation programs

As already mentioned, inverters can be sized using suitable simulation programs. Simulation programs and sizing programs provide warnings of incorrect sizing if the limit values are exceeded. By way of example, Figure 4.19 shows the error message displayed by the PV*SOL program when the PV system is incorrectly sized. However, such programs should not be trusted blindly. Programs can contain errors, and even if the dimensioning limits are adhered to, the sizing may still not be optimal. In addition, many simulation programs only check the sizing by using hourly irradiance values. Simulation programs are described in detail in Chapter 5.

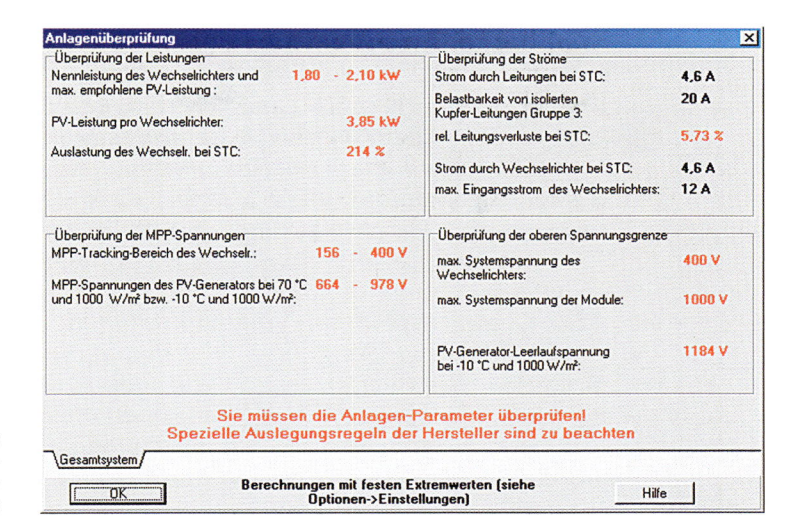

*Figure 4.19
Error message displayed by
PV*SOL program
Source: Siemens*

4.5 Selecting and sizing cables for grid-tied PV systems

National codes and regulations specifying the types and sizes of DC cables permitted in grid-tied PV systems need to be referred to when selecting and sizing DC cables for these systems. The following example is based on practice in Germany. Practices in other countries may and do differ. The same applies to fuses and other wiring accessories.

Based on a roof plan showing the module layout and the building survey (see Chapter 3), the approximate wiring lengths are worked out. For the module wiring, adept arrangement of the modules can minimize wiring lengths and, hence, cable losses and surge coupling (see section '4.7.2 Indirect lightning effects and internal lightning protection').

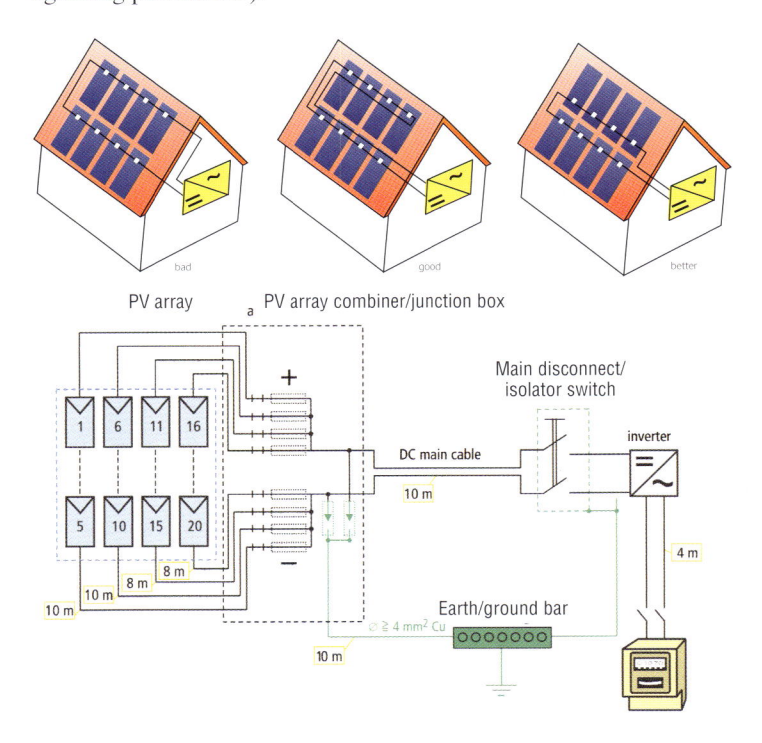

*Figure 4.20
(a) Module wiring variants: the variant on the right optimises wiring lengths and surge coupling;
(b) wiring diagram for a 2kW system with central inverter (example)
Source: R. Haselhuhn*

A system wiring diagram should be drawn up. This can be used as a guide during installation and kept as part of the system's documentation.

When sizing the cables, three essential criteria should be observed: the cable voltage ratings, the current carrying capacity of the cable and the minimizing of cable losses.

4.5.1 Cable voltage ratings

PV systems do not generally exceed the voltage ratings of standard cables (nominal voltages of 450V to 1000V) used in Germany and other European countries. With large PV systems and long module strings, the voltage rating of the cable should be checked, taking into account the maximum open-circuit voltage (at –10°C) of the PV string or array to which it is to be connected.

4.5.2 Cable current carrying capacity

The cable cross section is then sized in accordance with the maximum current. Here, the values for the current-carrying capacity of the cable, which are listed in IEC 60512 Part 3, must be maintained. The maximum current that may flow through the module or string cable is the generator short-circuit current minus the short-circuit current of one string:

$$I_{max} = I_{SC\ PV} - I_{SC\ String}$$

The cable is either designed for this current or string fuses are used to protect the cables from overloading. The maximum current I_{max} must be lower or equal to the current-carrying capacity of the cable or the protective device I_z:

$$I_{max} \leq I_z$$

Table 4.3 Current-carrying capacity of standard PV cables according to manufacturer's specifications
Source: VDE

Cable	Cable cross-section in mm²	Maximum current in A Individually laid			Accumulation: Six cores in a bunch	
		30°C	55°C	70°C	55°C	70°C
AEG Solar module	2.5	42	32	24	17	13
Cable (maximum 90°C)	4	56	42	32	22	17
Radox 125	2.5	49	38	34	20	18
(maximum 125°C)	4	66	51	45	27	24
Titanex 11	2.5	33	24	17	13	9
(maximum 85°C)	4	45	33	23	17	12

When using the string concept, it should be considered that the generator short-circuit current approximates the nominal current of the string. Since the fuses are triggered only by multiple surge currents, it is not possible to protect the cable using string fuses. In accordance with IEC 60364-7-712, the string cable must be able to carry 1.25 times the generator short-circuit current and be laid such that it is earth-fault proof and short-circuit proof.

With the cable sizing, the routing requirements according to IEC 60512 Part 3 must be observed. The current-carrying capacity of the cables is influenced by the ambient temperature, the bunching with other cables, and the method of installation (in the cable conduit, in timber partitions, behind plaster, etc.). Roof tiles can have an ambient temperature of 70°C, so this temperature must be applied when sizing the module or string cables for roof-mounted installations.

String fuses are used mainly with PV systems with several strings, and must be employed for systems with four or more strings where fault conditions could lead to significant module reverse currents. Fuses or miniature circuit breakers (MCBs) are deployed. The cable cross section of the module or string cables can be dimensioned with the trigger current of the string fuses. Here, the permitted current-carrying capacity of the cable $I_{z\ Cable}$ must be the same as or greater than the trigger current of the string fuse:

$$I_{z\ Cable} \geq I_{a\ String\ fuse}$$

$$I_{max\ String} \geq I_{z\ Cable}$$

The fuse should be triggered at twice the string short-circuit current at STC:

$$2 \times I_{SC\ String} > I_{a\ String\ fuse} > I_{SC\ String}$$

In order to avoid false trips, the nominal current of the fuse must be at least 1.25 times larger than the nominal string current:

$$I_{n\ String\ fuse} \geq 1.25 \times I_{n\ String}$$

Since an error can occur both on the positive and the negative voltage sides, the fuses must be installed along all unearthed cables. The fuse or MCB must be rated for DC operation.

4.5.3 Minimizing the cable losses/voltage drops

The sizing of the cable cross sections takes into consideration the need for as little cable loss/voltage drop as possible. The 1998 German draft standard VDE 0100 Part 712 recommends: 'The voltage drop in the direct voltage circuit should be no greater than 1 per cent of the nominal voltage of the PV system at standard test conditions (STC).' This limits the loss power through all DC cables to 1 per cent at STC. Practice has shown that this 1 per cent recommendation for PV systems with inverters operating with a higher DC input voltage ($V_{MPP} > 120V$) can be maintained with standard cable cross-sections without any problems. System designs with string inverters are a good example of this.

For PV systems with inverters operating with lower V_{MPP} values (e.g. low voltage concept), it is possible that the voltage drop with the string or module cable exceeds the 1 per cent limit, even when using a 6mm² cable, particularly when there are greater distances between the inverter and the PV generator. With such system designs, a 1 per cent voltage drop in the string cables and an additional 1 per cent drop with the DC main cable is acceptable.

The current is, depending upon the irradiance conditions, almost always lower than the nominal current of the system, which is only attained under STC conditions. With half the current, the cable loss resulting from $P = I^2 \times R$ is only one quarter of the loss under nominal conditions. For this reason, when using a 2 per cent voltage drop dimensioning limit under STC conditions, it can be expected that the annual yield reduction on the DC side will be approximately 1 per cent. The advantages of this system concept more than compensate for these losses, particularly with shading situations (see section '4.2.1 Central inverter concept'). If there is difficulty working with a cable, then under certain circumstances it is possible to use the next smallest cross section. This accepts that there will be greater power losses.

A nominal current load of 2A to 3A per square millimetre cross section can be used as a guide value for DC cables. This should only be used for rough estimations and checked when sizing the cable.

4.5.4 Sizing the module and string cabling

Table 4.4
Electrical parameters for sizing the DC cable

Electrical parameters	Symbol	Unit
Simple wiring length for module and string cabling	L_M	m
Line loss of the module and string cabling under STC	P_M	W
Cable cross section of the module and string cabling	A_M	mm²
Electrical conductivity (copper $\kappa_{CU} = 56$; aluminium $\kappa_{AL} = 34$)	k	m/Ω 3 mm²
Power of string under STC	P_{St}	W
String voltage	V_{MPP}	V
String current	I_{St}	A
Number of strings of the PV generator	n	

After sizing the cross section, taking into consideration the current-carrying capacity, the cross section with the 1 per cent recommendation can be selected. The following three formulae enable the cross section of the module or string cables to be calculated with almost the same string cable lengths. Here, it is assumed that there will be a cable loss of 1 per cent in relation to the string power at STC:

$$A_M = \frac{2 \times L_M \times P_{St}}{1\% V_{MPP^2} \times \kappa}$$

$$A_M = \frac{2 \times L_M \times I_{St^2}}{1\% P_{St} \times \kappa}$$

$$A_M = \frac{2 \times L_M \times I_{St}}{1\% V_{MPP} \times \kappa}$$

The calculated value for the cross section of the module or string cable AM is rounded up to the next highest value for standard cable cross sections (2.5mm², 4mm² and 6mm²). The following formulas are used to calculate the overall losses in all module and string cables for the selected cable cross-section:

$$P_M = \frac{2 \times n \times L_M \times P_{St^2}}{A_M \times V_{MPP^2} \times \kappa}$$

$$P_M = \frac{2 \times n \times L_M \times I_{St^2}}{A_M \times \kappa}$$

The layout of the PV system normally leads to different string cable lengths and, thus, different cross sections for the module and string cables. In these cases, the following general formula can be used:

$$P_M = \frac{2 \times I_{St^2}}{\kappa} \times \left(\frac{L_1}{A_1} + \frac{L_2}{A_2} + \frac{L_3}{A_3} + \frac{L_4}{A_4} + \ldots \right)$$

The lengths of the module and string cables (copper) can be derived from Figures 4.21 to 4.26 for the usual cross sections. Here the permitted 1 per cent loss has been taken into account.

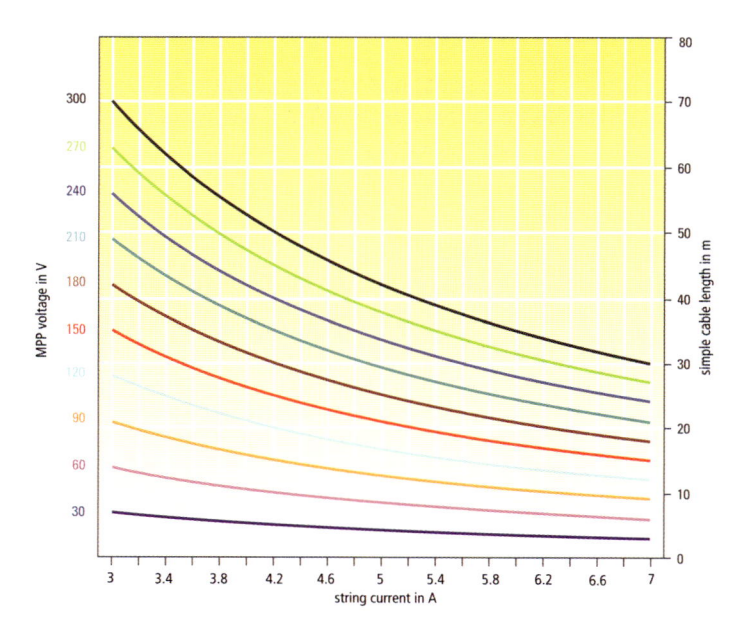

Figure 4.21
Recommended wiring lengths of string cabling with cable cross-section $A = 2.5mm^2$ up to an MPP voltage of 300V
Source: R. Haselhuhn

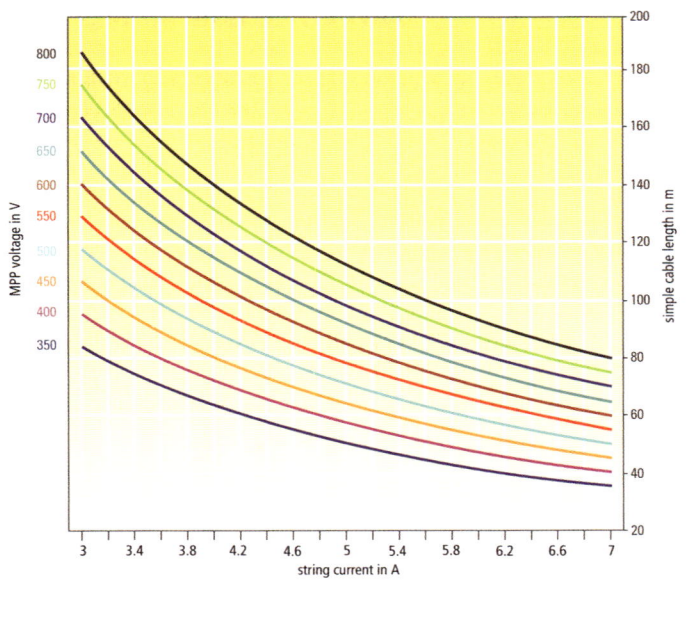

PLANNING AND SIZING GRID-CONNECTED PHOTOVOLTAIC SYSTEMS 169

Figure 4.22
Recommended wiring lengths of string cabling with cable cross section $A = 2.5mm^2$ for MPP voltages above 300V
Source: R. Haselhuhn

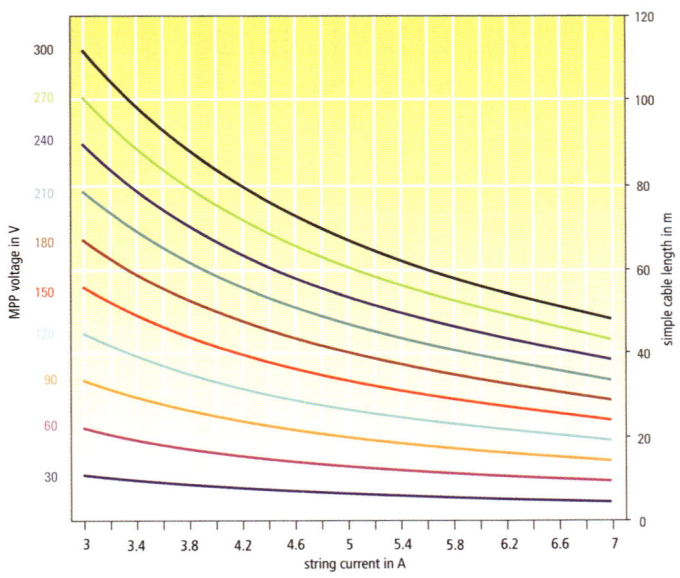

Figure 4.23
Recommended wiring lengths of string cabling with cable cross section $A = 4mm^2$ up to an MPP voltage of 300V
Source: R. Haselhuhn

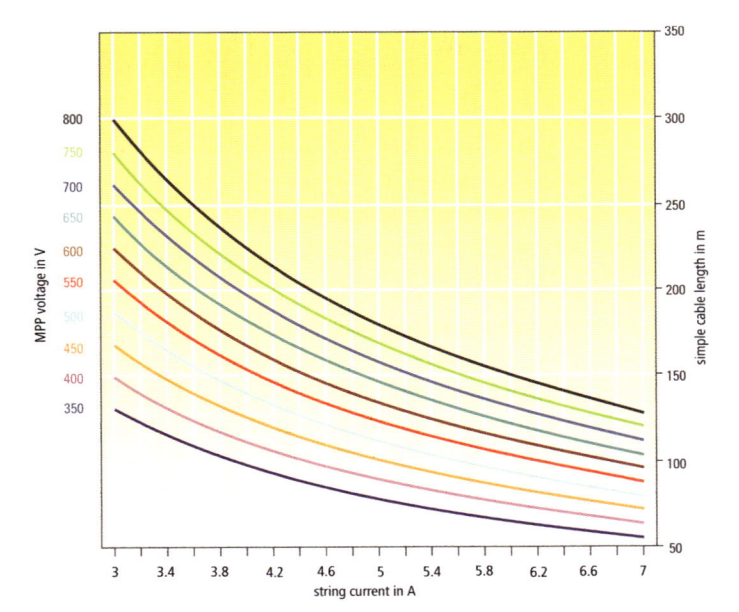

Figure 4.24
Recommended wiring lengths of string cabling with cable cross section $A = 4mm^2$ for MPP voltages above 300V
Source: R. Haselhuhn

*Figure 4.25
Recommended wiring lengths of string cabling with cable cross section $A = 6\,mm^2$ up to an MPP voltage of 300V
Source: R. Haselhuhn*

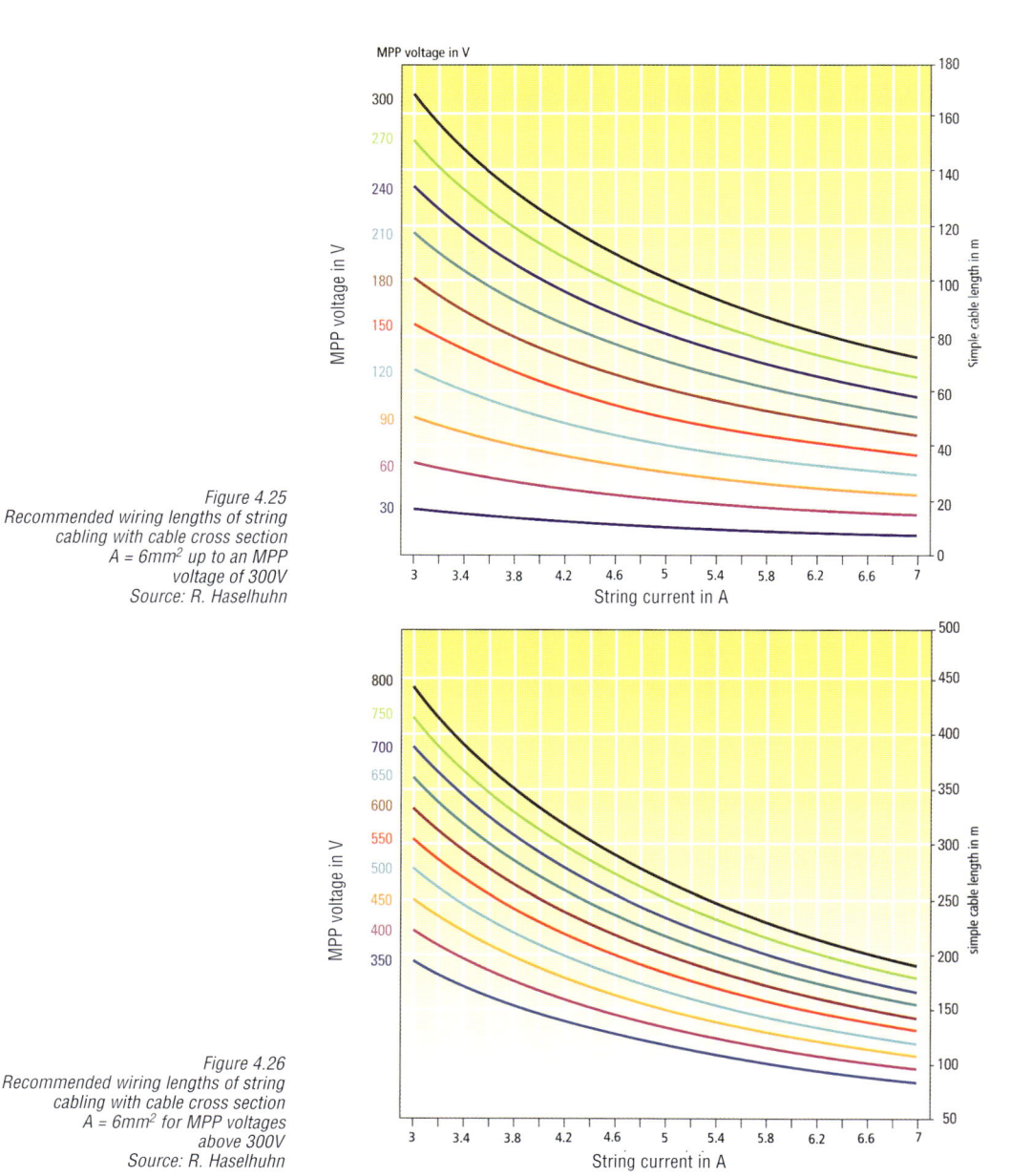

*Figure 4.26
Recommended wiring lengths of string cabling with cable cross section $A = 6\,mm^2$ for MPP voltages above 300V
Source: R. Haselhuhn*

4.5.5 Sizing the DC main cable

*Table 4.5
Electrical parameters for sizing the DC main cable*

Electrical parameters	Symbol	Unit
Simple line length of the DC main cable	$L_{DC\,cable}$	m
Line loss of the DC main cable	$P_{DC\,cable}$	W
Cable cross section of the DC main cable	$A_{DC\,cable}$	mm^2
Electrical conductivity (copper $\kappa_{CU} = 56$; aluminium $\kappa_{AL} = 34$)	κ	$m/\Omega \times mm^2$
Nominal power of the PV module/array	P_{PV}	W_P
Nominal voltage of the PV module/array	V_{MPP}	V
Nominal current of the PV module/array	I_n	A

The DC main cable and the DC bus cables from PV sub-arrays must be able to carry the maximum occurring current produced by the PV array. Since the short-circuit current of the PV array is only slightly higher than the nominal current, this will not trigger fuses. As protection against isolation and earth/ground faults, it is possible to use a DC-sensitive earth/ground fault current leakage circuit breaker.

In general, the DC main cable is sized to 1.25 times the PV array short-circuit current at STC in accordance with IEC 60364-7-712 (although national codes and regulations need to be referred to):

$$I_{max} = 1.25 \times I_{SC\ PV}$$

The cross section of the cable must be selected according to the permitted current carrying capacity of the cable. Here again, the temperature reduction factors and, with cable bunching, the accumulation factors need to be taken into consideration.

The cross section of the cable can then be optimized in terms of energy by using the following formulas. Here it is again assumed that there will be a cable loss of 1 per cent in relation to the nominal power of the PV array.

The cross section $A_{DC\ cable}$ is derived from:

$$A_{DC\ cable} = \frac{2 \times L_{DC\ cable} \times I_{n^2}}{(v \times P_{PV\_PM}) \times \kappa}$$

with the loss factor $v = 1$ per cent, or $v = 2$ per cent with the low voltage concept.

The calculated value for the cable cross section for the DC main cable $A_{DC\ cable}$ is rounded up to the next highest value for standard cable cross sections (2.5mm², 4mm², 6mm², 10mm², 16mm², 25mm², 35mm², etc.).

The actual cable loss from the DC main cable $P_{DC\ cable}$ is calculated for the selected cable cross section as follows:

$$P_{DC\ cable} = \frac{2 \times L_{DC\ cable} \times I_{n^2}}{A_{DC\ cable} \times \kappa}$$

$$P_{DC\ cable} = \frac{2 \times L_{DC\ cable} \times P_{PV^2}}{A_{DC\ cable} \times V_{MPP^2} \times \kappa}$$

For reasons of earth/ground fault-proof and short circuit-proof installation, individual single-core sheathed cables are recommended for positive and negative cables. If multi-core cables are used, the green/yellow wire (earth/ground – European colour code) must not carry any voltage. For PV installations exposed to a lightning risk, screened cables should be used (see section '4.7 Lightning protection, earthing/grounding and surge protection'). If DC and AC cables are laid together, the laying requirements specified in national codes and regulations must be observed and the cables labelled accordingly.

4.5.6 Sizing the AC connection cable

Table 4.6
Electrical parameters for sizing the AC supply cable

Electrical parameters	Symbol	Unit
Simple line length of the AC connection cable	$L_{AC\ cable}$	m
Line loss of the AC connection cable	$P_{AC\ cable}$	W
Line cross section of the AC connection cable	$A_{AC\ cable}$	mm²
Electrical conductivity (copper $\kappa_{CU} = 56$; aluminium $\kappa_{AL} = 34$)	k	m/Ω 3 mm²
AC nominal current of the inverter	I_{nAC}	A
Nominal grid voltage (single phase: 230V; three phase: 400V)	V_n	V
Power factor (between 0.8 and 1)	cos φ	–

The calculation of the cross section of the AC connection cable assumes a voltage drop of 3 per cent relative to the nominal grid voltage (German regulations).

For the cross section $A_{AC\ cable}$ with a single-phase feed:

$$A_{AC\ cable} = \frac{2 \times L_{DC\ cable} \times I_{n\ AC} \times \cos\varphi}{3\% V_n \times \kappa}$$

and with a symmetrical three-phase feed for the three-phase AC cable:

$$A_{AC\,cable} = \frac{\sqrt{2} \times L_{AC\,cable} \times I_{n\,AC} \times \cos\varphi}{3\% \, V_n \times \kappa}$$

With PV systems up to 5kW, this produces cable cross sections $A_{AC\,cable}$ up to 6mm². With single-phase inverters and feeding, the standard cable type NYM-J 3×1.5mm² to 6mm², for example, is used. With three-phase feed, the cable type NYM-J 5mm² $\times$ 1.5mm² – 4mm² is used.

To calculate the cable loss $P_{AC\,cable}$ for the selected cable cross section, the following formulas are used.

With single-phase feed:

$$P_{AC\,cable} = \frac{2 \times L_{AC\,cable} \times I_{n\,AC^2} \times \cos\varphi}{A_{AC\,cable} \times \kappa}$$

and with three-phase feed:

$$P_{AC\,cable} = \frac{\sqrt{3} \times L_{AC\,cable} \times I_{n\,AC}}{V_n \times \kappa \times \cos\varphi} \times P_n = \frac{3 \times L_{AC\,cable} \times I_{n\,AC^2}}{A_{AC\,cable} \times \kappa}$$

With asymmetrical three-phase feed from multiple single-phase units, the cable losses are calculated from the total of the individual cable losses of the different phases and the neutral conductor:

$$P_{AC\,cable} = P_{L1} + P_{L2} + P_{L3} + P_N$$

For the calculation, the currents in the individual phase cables L1, L2, L3 and N need to be found. An asymmetrical load is only permitted up to a power difference between the phases of 4.6kVA (German regulations).

In addition, the grid impedance, also known as the loop resistance, should be no higher than 1.25Ω at the inverter's input. This results in a specific resistance for the inverter supply cable. This resistance is determined by the length (distance from the feed-in point) and the cross section of the AC connection cable.

4.6 Selection and sizing of the PV array combiner/junction box and the DC main disconnect/isolator switch

For many system configurations, PV array combiner/junction boxes can be purchased from solar wholesalers as ready-made components. Module and inverter manufacturers offer various types that are suitable for standard systems. Externally mounted PV array combiner/junction boxes should be protected to IP 54 and be UV resistant. In addition, it is recommended that an installation site should be selected that protects the junction box from rain and direct solar irradiance.

When choosing, ensure that there are a sufficient number of terminals for the strings. The PV combiner/junction box should adhere to Protection Class II regulations. The combiner/junction box should be easily accessible for any possible later maintenance work. With combiner/junction boxes with screw terminals, ensure that the connections are correctly executed since faulty execution can cause an entire string to fail. Connection boxes with spring clamp terminals or with other suitable terminal systems do not require metal end sleeves and are easy to use.

With standard modules with short-circuit currents of 3A, it is usual to deploy 4A fuses. However, many modules on the market these days have a higher short-circuit current of 4A to 18A. The suitable string fuses are selected correspondingly. The fuses must be designed for DC operation. The string diodes for decoupling the individual strings are only used for very severely shaded systems with central inverters or when employing PV modules that do not adhere to Protection Class II regulations. They are integrated in the combiner/junction box while ensuring that their operational heat can be dissipated. For overvoltage protection, surge arresters in the combiner/junction box are connected to positive and negative poles in order to protect against earth/ground potential (see section '4.7 Lightning protection, earthing/grounding and surge protection'). For this reason, the earthing/grounding conductor is connected to the main earth/ground terminal (earthing/grounding regulations differ from country to country and national codes and regulation relating to the earthing/grounding of grid-connected PV systems need to be referred to). Likewise, the DC main disconnect/isolator is often also integrated within the PV combiner/junction box.

It is also sensible to have a separate DC main disconnect/isolator switch directly before the inverter. This generally prevents any accidental activation by another person – for instance, when maintenance work is being conducted on the inverter. In addition, this also enables the DC main cable to be isolated.

According to IEC standard 60364-7-712, an accessible disconnect/isolator switch is required between the PV array and inverter. The dual pole DC main disconnect/isolator switch is designed for the maximum open-circuit voltage for the solar array at $-10°C$ $V_{OC\,(PV\,-10°C)}$ and for 125 per cent of the maximum array current (short-circuit current with full irradiance $I_{SC\,PV}$):

$$I_{DC\,MS}\ 1.25 \times I_{SC\,PV}$$

When selecting the main DC disconnect/isolator switch, ensure that it is rated for the relevant direct current. Touch-proof plug connectors (e.g. on string inverters) may only function as isolators without load and are not permitted as replacements for DC main switches (see also the section on '2.5 Direct current load switch (DC main switch)' in Chapter 2).

The PV array combiner/junction box can also be built from standard electrical components in an accordingly shockproof housing. When constructing the switching, it is possible to use terminal blocks mounted on top-hat rails. The positive and negative sides must be rigorously separated to prevent any possibility of earth/ground faults and short circuits.

With large systems, several PV combiner/junction boxes are often required. With a system with string inverters, the PV combiner/junction box can be dispensed with as the strings are connected directly to the inverter. The surge arresters (varistors) are integrated within the string inverters.

4.7 Lightning protection, earthing/grounding and surge protection

National codes and regulations dealing with lightning protection, in general, and with regard to grid-connected PV systems differ considerably from country to country and even from region to region within a country. In this section, the installation and practices discussed are those followed in Germany.

Generally, the following applies: photovoltaic systems do not increase the risk of buildings being struck by lightning. Therefore, there is no need for an additional lightning protection system. However, in its VDS 2010 directive, the German property insurance association Verband der Schadensversicherer requires lightning and overvoltage protection systems, which should be installed in accordance with Lightning Protection Class III and VDE standards. (According to VDE 0185, the degree of protection for a building is classified using four Lightning Protection Classes I to IV. The requirements for lightning and overvoltage protection decrease from Lightning Protection Class I to IV. The legal requirements for lightning protection in Germany are stipulated in the building regulations provided by the various federal states. The highest protection classes apply, for example, to airports, hospitals, plants at risk of fire and explosion, as well as industrial plants. With public buildings and venues, the regulations stipulate Class III lightning protection.)

However, if a specialist can prove, in accordance with VDE V 0185 Part 2, that the risk of lightning strikes and damage is minimal, the insurers accept solutions that deviate from the VDS 2010. This is already practised, for example, where there are high buildings close by that have lightning protection.

Lightning protection systems are particularly aimed at providing personal protection when there are direct lightning strikes. If the PV system is in an exposed location, then a suitable lightning conductor must be used. For example, rack-mounted photovoltaic systems on flat roofs of buildings that are exposed to lightning strikes must be equipped with their own lightning protection system since the PV generator, as a projecting roof structure, presents a preferential point of impact. Lightning protection systems are constructed according to VDE V 0185 Parts 1 to 5. If a lightning protection system already exists on the building, the photovoltaic generator must be connected to this.

Table 4.7 provides an overview of the necessary measures to be taken for lightning and overvoltage protection for PV systems on buildings in which no valuable data technology is housed.

Table 4.7 Selection of measures to be taken for lightning and overvoltage protection for PV systems on buildings without any valuable data technology
Source: R. Haselhuhn

PV systems on buildings … without a lightning protection system
Inverter with transformer
Transformerless inverter
PV modules to protection Class II or low voltage concept?
Yes
No
1 No surge arresters in the PV array combiner/junction box and
2 No potential equalization of the PV rack required
1 No surge arresters in the generator junction box required
2 Potential equalization of the PV rack: cross section of the equipotential bonding conductor = cross section of the DC main cable; however, at least 4mm^2
3 With a transformerless inverter: earthing and potential equalization according to section 4.7, if required

With a lightning protection system
PV array in protection area
PV array outside protection area
Is the safety distance to the lightning protection system kept?
1 Connect PV array to lightning protection system using the shortest cable route possible with a diameter of at least 16mm^2
2 Surge arresters Class I and II with equipotential bonding conductor (minimum 16mm^2)
PV modules to protection Class II or low voltage concept?
1 Surge arrester Class II
2 Potential equalization of the PV rack: cross section of the equipotential bonding conductor = cross section of the DC main cable; however, at least 4mm^2
1 Surge arrester Class II
2 No potential equalization of the PV rack required

The following comments generally apply for lightning and overvoltage protection for PV systems:

- PV systems do not generally increase the risk of buildings being struck by lightning.
- If a lightning protection system exists on the building, the PV generator must be connected to this. The internal lightning protection should be carefully executed.
- If the PV system is in an exposed location, then a suitable lightning conductor must be used.
- If no lightning protection system exists, the PV generator must be earthed and connected to the potential equalization, unless:
  - with smaller systems (< 5kW); and
  - when using PV modules to protection class II; or
  - with electrical insulation and a safety extra-low voltage (SELV) design.
- Suitable surge arresters on the DC side in the generator junction box are recommended.
- Overvoltage protection on the AC side is also generally recommended.

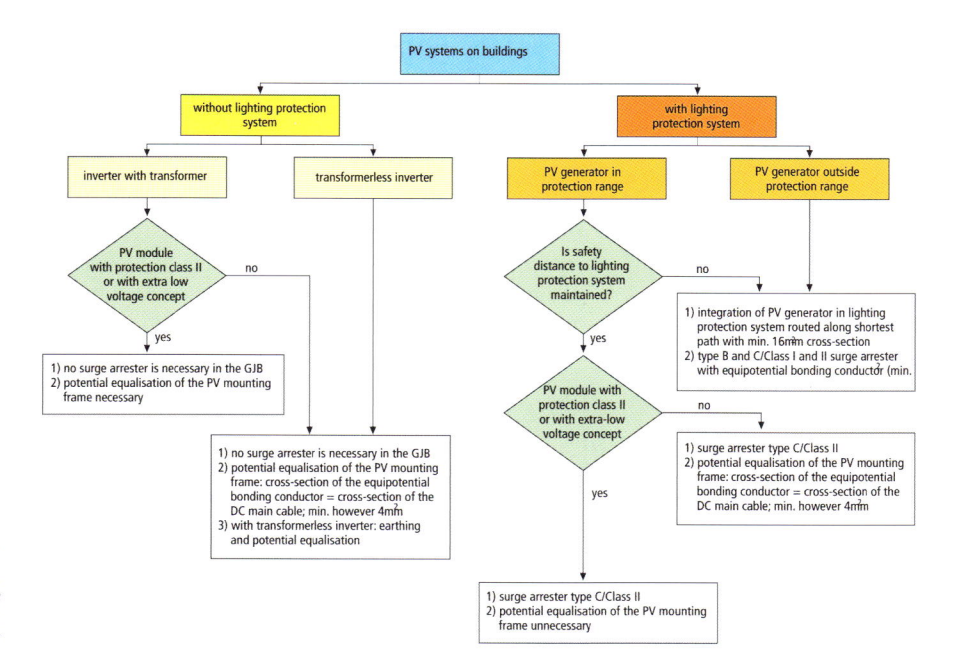

*Figure 4.27
Selection of measures for lightning and overvoltage protection for PV systems on buildings without valuable data technology equipment*

4.7.1 Lightning protection – direct strikes

The probability of a direct lightning strike can be calculated using the building's dimensions, information on the environment and the average number of lightning strikes for the respective region. Standard VDE 0185 Part 2 and its respective risk management software are used to determine the risk of lightning strikes and the technically and economically optimal protection measures. Table 4.8 can be used to judge whether a PV system is being installed in an area where there is a high probability of lightning strikes (i.e. where there is a high lightning density). For an average house in an urban area, the probability of a lightning strike is around once every 1000 years. For example, the probability of a strike on an isolated farmhouse on a mountain ridge in an area of increased storm activity increases to one strike around every 30 years. If this farmhouse stands in an non-exposed location in a rural area with the 'normal' number of thunderstorms, the probability drops again to one strike around every 500 years (Becker, 1997).

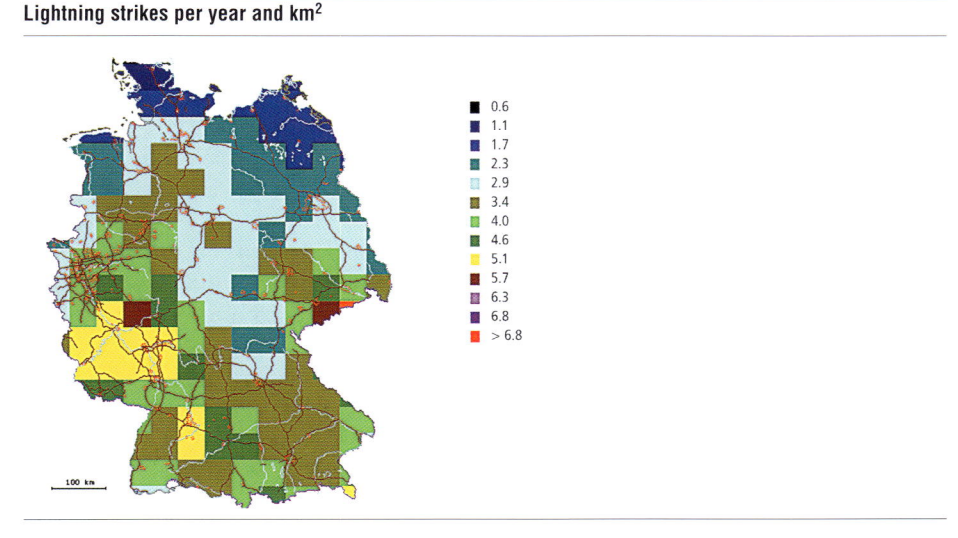

*Table 4.8
Average number of lightning strikes from 1999 to 2004
Notes: The external lightning protection comprises all equipment and measures for arresting and conducting lightning. A lightning protection system consists of an arrester device, the lightning conductor (at least 16mm² copper conductor) and the earthing system. It should be constructed to VDE 0185 Part 3 and VDE 0185 Part 100.
Source: Siemens Lightning Information Service, www.blids.de*

Lightning strikes per year and km²

The following criteria for external lightning protection must be observed during the installation of a PV system:

- public use (risk of panic, public venues, etc.);
- conductive roof superstructures that protrude from the building;
- surface of the solar generator >15m^2;
- the building contains valuable data technology with high availability requirements and backup systems;
- protection of important electrical safety technology (fire alarms, burglary alarms and security technology).

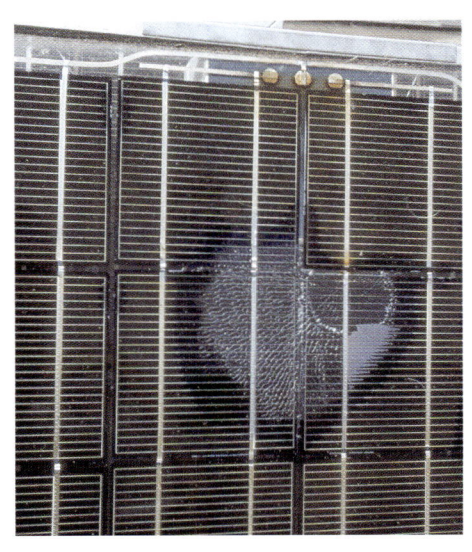

Figure 4.28
Lightning damage
Source: Phönix Contact

4.7.2 Indirect lightning effects and internal lightning protection

Every lightning strike creates indirect effects in its surrounding area within a perimeter of around 1km. The probability of a building being affected by indirect lightning is therefore much greater than the likelihood of the building being struck by direct lightning. It can be assumed that during a PV system's service life, it will be affected numerous times by lightning strikes in the surrounding area.

The indirect lightning effects are essentially inductive, capacitive and galvanic coupling. These couplings generate surge voltages from which the building's electrical systems must be protected. The internal lightning protection encompasses all measures and equipment in the building concerned not just with protecting, for example, electronic devices from indirect lightning effects, but also from the effects of switching in the public mains supply. The greater the risk of lightning for the house and the more valuable the data technology, the more extensive are the measures that must be taken for the internal lightning protection. A prerequisite for the functioning of the internal lightning protection is seamless potential equalization to IEC 364-5-54. All potentially conductive systems (e.g. water, heating, gas pipes, etc.) must be connected with the earth/ground system.

Lightning can induce voltage surges in the PV modules, the module cables and the DC main cable. Voltage levels induced in PV modules with metallic frames are about half the value of those induced in frameless PV modules. In order to reduce surges in the module cables, each string's positive and negative cables (+ and –) should be routed as close to each other as possible.

Here it should be ensured that the cable laying is short-circuit proof. The smaller the yellow labelled areas for the open loops in the PV array circuit, the lower the induction voltage generated by lightning current in the module cables (see Figure 4.30).

Induced voltage surges in the DC main cable should be minimized by laying the positive and minus cables as close together as possible. Shielded individual cables are

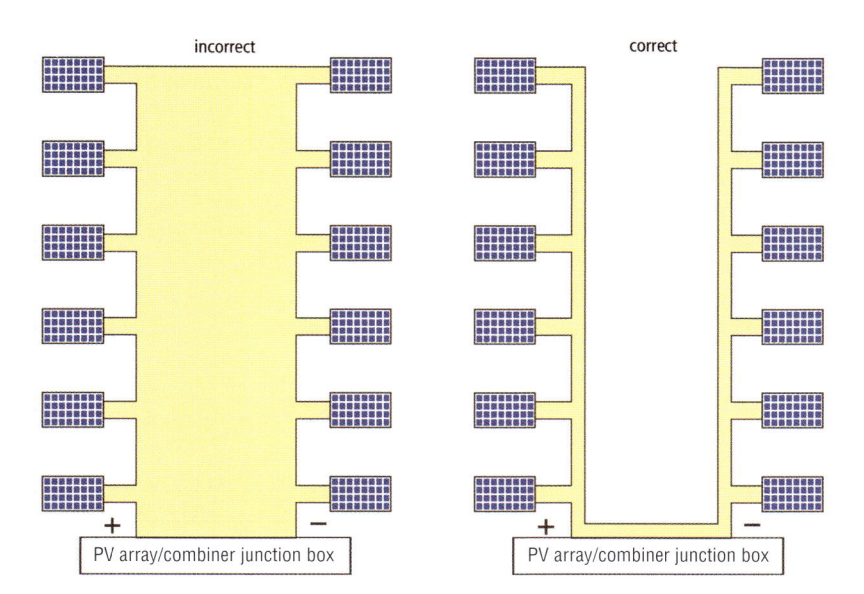

Figure 4.29
Loop formation in module wiring
Source: Solarpraxis

recommended in systems exposed to the risk of lightning. The cross section of the shield should be at least 16mm² copper (Cu). The upper shield end should be well connected to the metallic substructure and the PV module frames along the shortest path possible. It is also possible to use a metallic protective pipe. If no shielded cables are used, surge arresters with a nominal leakage current of around 10kA must be connected to the active conductors. With shielded cables, it is sufficient to use surge arresters with a nominal leakage current of around 1kA.

Surge arresters are used to protect PV systems and downstream electronic devices from capacitive and inductive coupling and from grid overvoltage. Usually, the surge arresters are installed in the PV array combiner/junction box. With systems at risk of lightning, other surge arresters are installed before and after the inverter.

DIN VDE 0675 Part 6 (Germany) differentiates between two classes of arresters: I and II. Class I arresters can discharge direct lightning currents and are used when there is an increased risk of lightning. Class II surge arresters are usually used on the DC and AC side with surge current capabilities of 1kA (standard surge 8/20) per 1kWp. The operating voltage UC (DC) of the surge arresters must correspond to at least the open-circuit voltage of the PV array. Table 4.9 shows the types and the rated voltages for AC and DC surge arresters.

Table 4.9
Types and rated voltages for AC and DC surge arresters

Type	V_c (AC)	V_c (DC)
75	75V	100V
150	150V	200V
275	275V	350V
320	320V	420V
440	440V	585V
600	600V	600V

A surge arrester must be connected between each pole and the earth/ground. With Class II surge arresters, the inception and build-up voltages should be 1.4 times the maximum PV voltage (E VDE 0126 Part 31, Germany). With systems at risk of lightning, it is important to install only types with thermal isolating devices and fault indicators. The system operator should make a visual inspection of the surge arresters after every thunderstorm, at least, or once every six months. If the installation site of the surge arresters is not easily accessible, the arresters should be installed with a remote control for the fault indicator. The visual fault indicator is mounted at a point that can be easily seen by the system operator (e.g. in the immediate vicinity of the meter cupboard). With inverters with isolation monitoring, the triggering of the surge arresters can also be registered so that it is not necessary to have separate remote monitoring.

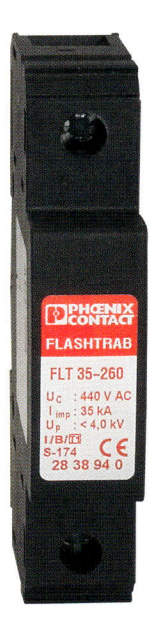

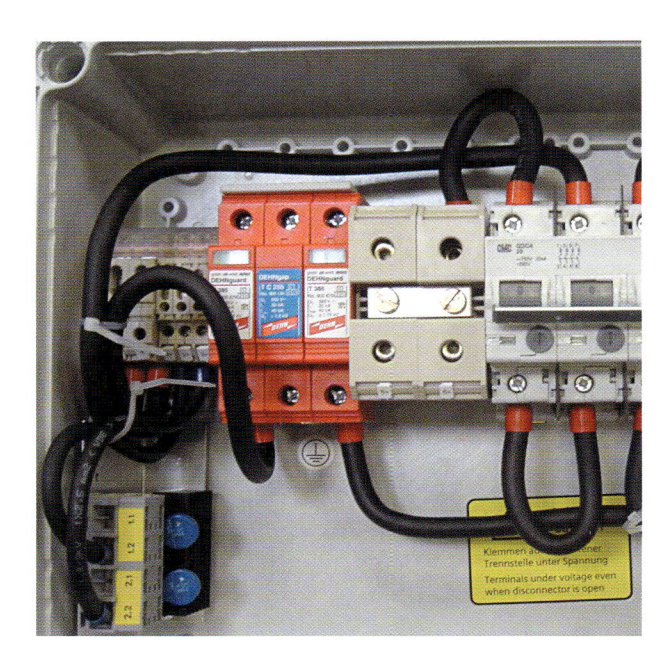

*Figure 4.30
Surge arrester*

*Figure 4.31
Surge arrester in
PV combiner/junction box*

With PV components for which the manufacturer has provided overvoltage protection (mostly varistors), it is also possible to dispense with external protection from atmospheric overvoltage. The isolation monitoring of the inverters recognizes if any varistors are triggered.

*Figure 4.33
Effect of indirect lightning strike on inverters: input electrolytic condenser has exploded due to high surge voltage
Source: Phönix Contact*

4.8 Yield forecast

In order to create a yield forecast (an estimate of what will be the annual electrical output of the system in kWh per year), the location and the overall efficiency of the PV system need to be assessed. To do this, the individual loss factors of the PV system are deducted from the theoretically expected energy yield of the PV array, E_{ideal}. Figure 4.34 shows the individual loss factors and the average percentage that they account for in relation to the energy E_{ideal} produced by the PV array.

As a measure for the installation quality, a location-independent parameter is used – known as the performance ratio. The performance ratio (PR) is defined as the ratio between the actual energy output of a system and nominal energy generation potential of a system (system yield in ideal/STC conditions) – a product of the overall system efficiency. The PR is an indicator of the system's actual output compared to an ideal system that operates without losses:

$$PR = \frac{E_{real}}{E_{ideal}}$$

The PV array's theoretically expected energy yield E_{ideal} is also called the ideal energy output. The PV array's ideal energy yield E_{ideal} is the result of the product of the solar irradiance on the PV modules surface A_{PV}, and of the efficiency of the PV modules:

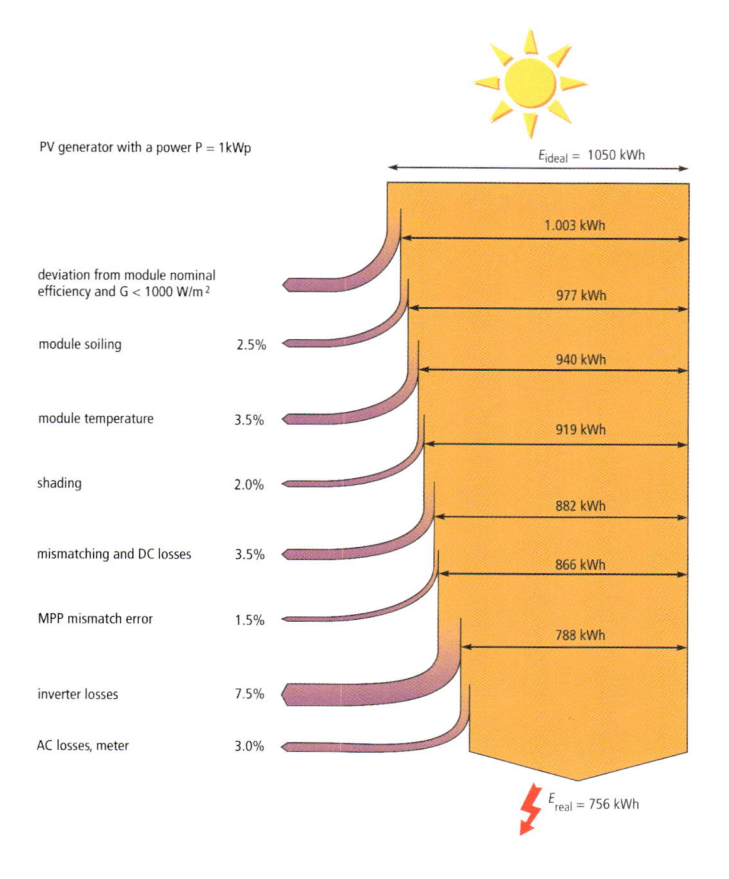

PLANNING AND SIZING GRID-CONNECTED PHOTOVOLTAIC SYSTEMS 179

Figure 4.33
Energy flow diagram of a grid-connected PV system
Source: R. Haselhuhn

$$E_{\text{ideal}} = g_{\text{PV}} \times \eta \times A_{\text{PV}}$$

Because:

$$\eta = \frac{P_{\text{PV}}}{1000 W / m^2 \times A_{\text{PV}}}$$

the PR can be calculated more simply:

$$PR = \frac{E_{\text{real}}}{g_{\text{PV}}} \left[\frac{kWp}{m^2} \right]$$

where e_{real} is the specific annual solar yield measured at the system output meter in kWh/kWp.

It should be borne in mind that g_{PV} is the specific irradiance at the tilted array surface and not the horizontal global irradiance.

Irradiance on the tilted plane is calculated from a yield forecast based on the long-term average values for horizontal global radiation for the location, which are provided by meteorological services and then converted for the PV array plane. As a result, simulation programmes often use 20-year monthly values that are converted into hourly values using suitable methods. The average values calculated by meteorological services are based on pyranometer measurements from weather stations and measurements by weather satellites.

In some cases, for system monitoring the irradiance on the tilted plane is measured and the PR calculated from this. Temperature-compensated PV irradiance sensors are generally used here (see the section on '1.2.7 How solar radiation is measured' in Chapter 1). Owing to the spectral deviation of a PV sensor, lower irradiances are measured than when a pyranometer is employed. Hence, this PR calculation based on the local irradiance calculation is more accurate than using meteorological services measured values, but tends to show a higher performance ratio. Exact figures can only be provided by using expensive pyranometers to find the irradiance – which then also have to be regularly cleaned and recalibrated.

The performance ratio enables a comparison between PV systems at different locations. To compare the technical quality of systems, the shading loss must be found and eliminated from the calculation.

Customers will expect a forecast of yield. The use of simulation programs for yield forecasts has become standard practice (see Chapter 5).

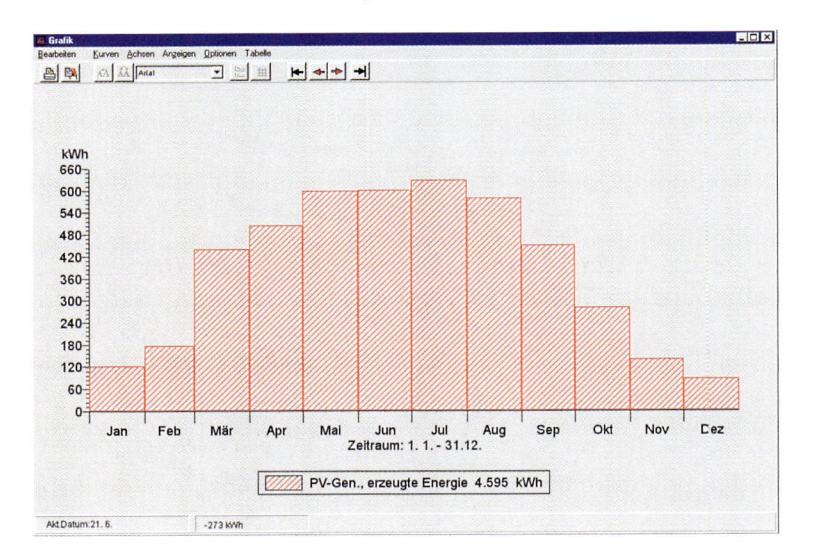

*Figure 4.34
Yield forecast with simulation program PV*SOL
Source: Valentin*

If no simulation program is used, an irradiance diagram of the location can provide an estimate. Using this irradiance diagram, first the annual irradiance g_{PV} onto the PV array is calculated from the azimuth and tilt. If there is shading, the shading losses need to be deducted from the irradiance (see the section on '3.3.2 Shading resulting from the location' in Chapter 3). Multiplied by the performance ratio (PR), this gives the PV system's annual energy yield in kWh per kW_p:

$$e_{real} = g_{PV} \times PR \left[\frac{kWp}{m^2} \right]$$

Depending upon the installation quality, the performance ratio can be assumed at between 70 per cent and 85 per cent. A very good PV system will achieve even higher values.

In order to compare operating results from different systems, the specific yield in kWh per kWp and year is calculated, as well as the performance ratio. Typical specific yields in Germany are shown in Chapter 8. An additional variable for assessing the system is the full load hours, which in the English-speaking world correspond to the final yield factor FY. The full load hours are the result of the quotients of the yield over a particular time period and the nominal power of the PV array. The reference timeframe can be a day, week, a month or a year:

$$FY = \frac{E_{real}}{P_{N\,PV}}$$

5 System Sizing, Design and Simulation Software

5.1 Use of sizing, design and simulation programs

There are numerous uses for software in photovoltaics. For example, in planning, it is a case of designing and optimizing the PV system. Sizing programs and simulators enable threshold values and operating states to be checked and finally simulate operation itself in many different variations. For exact yield forecasts and yield reports, the use of simulators is required. Simulators have traditionally also been employed in research and development, or again by component manufacturers. If the aim is to improve, optimize or develop new components and system concepts, simulation software can be used. This helps to reduce undesirable developments and can also lessen the scope of experiments. As well as these applications, programs can also be put to good use for education and training purposes.

Many installation engineers or planners who have been working for longer periods with particular PV modules and inverter types may, indeed, have past values for the sizing and yield, but will soon find themselves up against their limits when it comes, for example, to shading affecting the system. Generally, the sizing and simulation programs make complex circumstances transparent quickly and conveniently. Designing grid-connected PV systems, for instance, is by no means as easy as it first appears. Every inverter has a corresponding MPP region on the DC side. At the same time, each inverter has its own current and voltage limit values, which clearly define the permissible operating range for the equipment. The PV modules now have to be wired up as a generator in such a way that the operating ranges of the system components (PV modules and inverter) match one another. Every inverter, together with a specific PV module, can be wired up in a large number of different ways. For each possible wiring configuration, simulation programs can forecast and evaluate the system's behaviour based on the weather profile and the orientation and tilt of the PV array. When looking for the configuration that will deliver the highest yield or the most economically viable solution, and for the detailed design of more complex PV systems – not to mention predictions of the operating behaviour – rules of thumb and experience with previous systems are often unhelpful. System sizing and a yield calculation can be carried out swiftly and precisely if you are acquainted with a simulation program. A choice of different variants can be simulated to find out the best solution from energetic, economic and ecological points of view.

The operating behaviour of stand-alone systems is more complex compared with grid-connected PV systems. When planning these systems, it is important to find a balance for a given location (weather profile) in the relationships between the variables for the PV modules/energy store/loads (load profile) and to optimize their interaction in line with the system specification. For system design and optimization, it should be remembered that supply reliability, the service life of system components (batteries – cycle depth of discharge/number of cycles) and the way in which the system is used determine the economic efficiency of the PV system. In stand-alone systems, incorrect sizing may result in systems that are incapable of functioning, with total failure or faster battery ageing. Simulation programs make the operating characteristics of individual components and the system configuration transparent, analysable and, hence, also able to be optimized.

As well as the planning support provided by simulation programs, the results can also be used for marketing purposes and included in quotations to customers. Where good feed-in tariffs are offered for PV-generated electricity fed onto the grid (as in the case of Germany, Spain, France and Korea), customers will want to know what the expected yield will be and what the situation is regarding the economic efficiency of different system variations. Potential investors and operators of PV systems will ask about the optimum system solution, the level of energy savings and emission reductions. Working out the yield, calculating the economic efficiency and stating the emission reduction all highlight the advantages of PV systems and provide sales arguments. Some programs compile the layout diagram, the characteristics and the calculation results into a representative report ready for printing.

5.2 Checking the simulation results

For simulation calculation, it is a good idea to make a preliminary design and yield estimate. This is because as well as a powerful simulator, the planner's technical knowledge is called for in order to obtain realistic results through the simulation and optimization of larger or more complex systems. The results of the simulation are only ever as good as the realistic inputting of parameters and the simulation method. If incorrect data are input on a computer, the outcome will merely be a very impressive set of nonsense data in return. Simulation results should be thought through critically and not blindly trusted. It is important to choose the right program for the particular task since even the time spent learning to use the program will be of little use if that program's limitations mean that it is not suitable for the task in mind.

By using PV programs, it is possible to accelerate the planning process and avoid planning errors. At the same time, however, the software creates additional ways for the user to make errors. A description of the planned PV system that is as correct and accurate as possible is required as a base. Input errors and associated erroneous calculations are not infrequent with input forms that are complex, as found in some programs, especially if users are inexperienced. Some programs do carry out a plausibility check on important input parameters; but even that does not enable all errors to be ruled out. If new components are added into a program's databases, the data from the manufacturers' data sheets should be checked for plausibility prior to use. A check of the details on datasheets from module and inverter manufacturers conducted by the solar laboratory of the Munich University of Applied Sciences, Germany, discovered numerous errors (Zehner et al, 2005). On some datasheets, basic and essential information for system design, such as temperature coefficients, had been completely omitted. Other datasheets were inconsistent with themselves, or in the specification of parameters and the relationships between them; individual values were also occasionally physically implausible. For example, the specified nominal power did not correspond to the product of current and voltage, the stated I_{SC} value was the same as, or too similar to, I_{MPP}, or current temperature coefficients were suspect.

Insufficient calculation algorithms in the programs can also cause erroneous or unrealistic results. It is easy to understand why computer programs do not necessarily work completely without errors – one simply has to think of widespread operating systems or office packages that have hundreds of programmers involved in their development. Simulation programs in solar energy, on the other hand, are developed by individual persons or small teams. The number of units sold is relatively low, feedback on errors is sporadic and cost pressure only rarely permits constant maintenance and development once the software is released. Plausibility checks are always important. When using simulation programs the result should always be checked against past values. For grid-connected systems, the performance ratio (PR) or the power-related (specific) annual yield in kWh/kWp provide very good reference points (see Chapter 4). These assessment parameters are shown in most programs. For PV systems in Germany, the performance ratio should be greater than 0.7 and the annual yield should be at least 700kWh/kW$_p$. However, with unfavourable roof angles or if there is shading, the yield may work out lower. If the results are significantly higher than 1000kWh/kW$_p$ or significantly below 500kWh/kWp for façade systems, or less than 700kWh/kW$_p$ for rack-mounted standard systems on flat roofs, it may generally be assumed that input parameters are incorrect or there has been a fault in the calculation.

The calculation results for stand-alone PV systems are more difficult to assess. The results here can be checked based on rules of thumb for the relationship between PV modules, energy store and loads (see Chapter 8), or based on past values from previously implemented systems.

5.3 Simulation of shading

Shading situations present a particular challenge in the planning process. Shading occurs at many sites for PV systems. For example, in flat roof or ground-mounting systems, the module rows shade each other. The environment, edge shading or elevated horizons also have an influence. Shading has an effect on the system yield and on optimized system technology (bypass and string diodes, module wiring and inverter behaviour). Simulation programs are essential for calculating shading losses and optimizing the electrical and geometric system design.

In the German 1000 Roofs program, shading came on top for causing reduced yields and was responsible for considerable reductions of as much as 30 per cent and more in PV systems. The yield reductions generally turn out to be greater than one would suspect based on the shaded surface area (see Chapters 3 and 4). The shading analysis is therefore a sensitive point in simulations of PV systems. At the same time, the quality in the individual programs varies greatly. While in SOLDIM and GOMBIS the shading losses are estimated by the user, many programs (such as PVS, Greenius and SolEm) have a feature for entering the horizontal shading graphically. It is, nevertheless, assumed that the shading covers the entire PV array equally. Two software applications (horizOn and HORIcatcher) are available that support the specification of shading information and create a horizon line simulation from digital photographs.

The programs PVSYST, Pvcad, 3DSolarWelt and Solar Pro allow three-dimensional shading analyses. However, this greater accuracy requires a more complex description of the surrounding area. For more detailed analyses, the programs PVSYST and Solar Pro are recommended.

5.4 Market overview and classification

In order to create an overview of the market, the simulation programs may be classified according to the programming and calculation methods.

The programming method determines accuracy, user input requirements, flexibility, scope of application, computation time (no longer an issue today with ever-more powerful computers) and, finally, the price for the program. The more flexibly a program can be used, the higher the demands made of the user.

Software and simulation programs for photovoltaics

Calculation programs
- NSol
- PV F-Chart
- PVProfit
- RETScreen
- SolINVEST

Time-step analysis programs
- DASTPVPS
- Greenius
- Homer
- PVcad
- PV Design Pro
- PV*Sol
- PV*Express
- PVSYST
- RAPSIM
- Solar Pro
- SOLDIM
- SolEm
- 3DSolarwelt

Simulation systems
- INSEL
- Smile
- TRNSYS

Auxillary programs and data

Irradiance and climate
- HORIcatcher
- horizON
- METEONORM
- SUNDI

Data libraries
- CD-ROM with PV component overview from *Photon Magazine*
- Weather data on the internet (see Table 5.1)

Interpretation programs

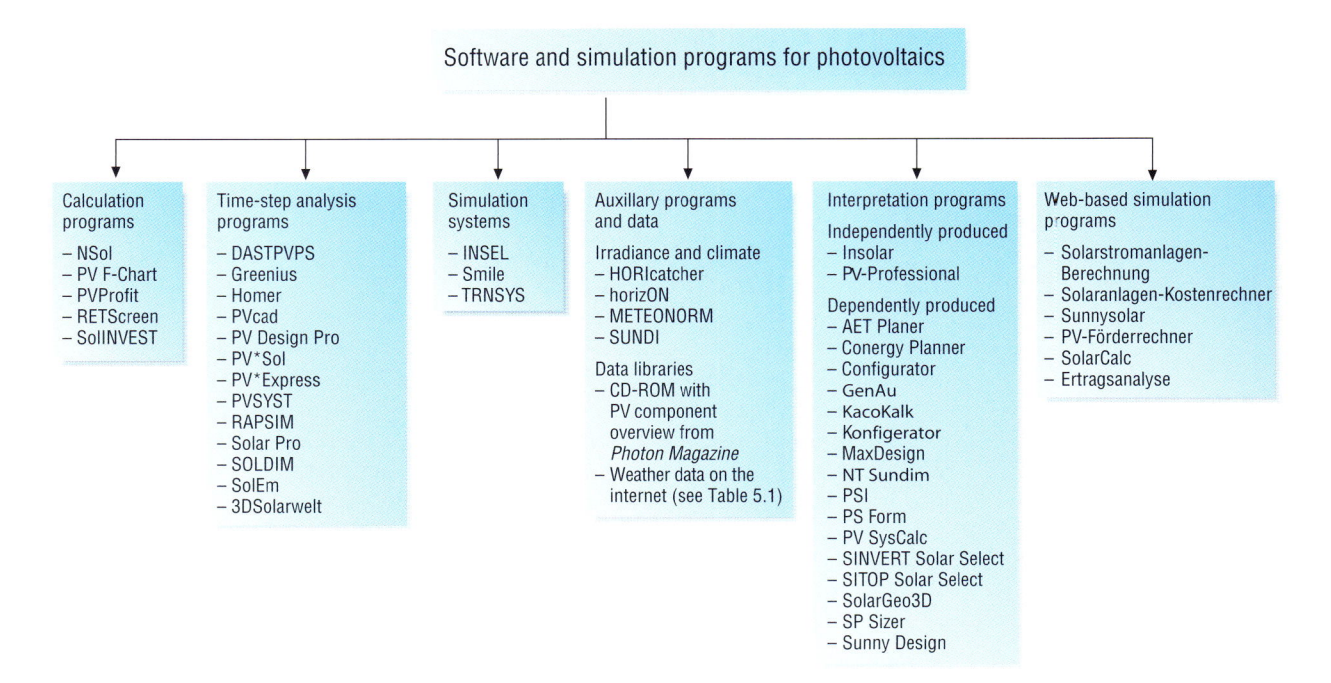

Independently produced
- Insolar
- PV-Professional

Dependently produced
- AET Planer
- Conergy Planner
- Configurator
- GenAu
- KacoKalk
- Konfigerator
- MaxDesign
- NT Sundim
- PSI
- PS Form
- PV SysCalc
- SINVERT Solar Select
- SITOP Solar Select
- SolarGeo3D
- SP Sizer
- Sunny Design

Web-based simulation programs
- Solarstromanlagen-Berechnung
- Solaranlagen-Kostenrechner
- Sunnysolar
- PV-Förderrechner
- SolarCalc
- Ertragsanalyse

Figure 5.1 Classification of sizing and design software for photovoltaic systems

5.5 Program descriptions

This section presents a selection of available, practicable and the most important simulation programs. As well as general information on the program, the performance scope, ease of use and special program features are all discussed. Example screen shots (a view of the software's user interface) are provided and are intended to give a visual impression to help get a better idea of the program. A demo version of almost all the programs is available at no charge and can, in some cases, be downloaded from the internet. In selecting a simulation program, as well as the simulation process, the particular solar application, the performance scope and the program's possible uses will, of course, all be important factors. Here, it is important which type of system or which system configuration one wants to simulate. This depends upon whether the aim is to simulate an on-roof system, a roof-integrated system, a stand-alone system, a PV hybrid system, a PV pump system or a grid-connected system. The program data, the parameters, weather and component libraries, and details on the system environment and the functional scope should be taken from the short descriptions and tables. The widely used programs PV*SOL and PVS are discussed in greatest depth, although none of these have a computer-aided design (CAD) interface. With the increased integration of photovoltaics within buildings, programs that offer CAD interfaces will be frequently used. For this reason, module manufacturers are being called upon to provide CAD files for their modules as a service via their websites. It should be borne in mind that programs are being constantly updated. It has also not been possible here to describe all available software, and it is likely that many of those included will have been upgraded before going to press.

5.5.1 Calculation programs

The first category comprises the program group of calculation or rough analysis programs. These programs are based mainly on statistical methods in combination with simple calculations. For the most part, the results are based on monthly values. Calculation programs are application oriented and deliver results very quickly. As a rule, however, they are less flexible and can only be used for standard systems. The program PV F-chart is briefly discussed below. Because of their simple structure, the computation processes in the calculation programs are suitable for web-based simulation programs.

PV F-CHART

PV F-chart is a system analysis and design program for photovoltaic systems. It provides monthly average performance estimates for each hour of the day. There is a Windows version available, and the program features weather data for over 300 locations (with the option to add additional weather data), hourly load profiles for each month, statistical load variation, buy–sell cost differences, and life-cycle economics with cash flow; calculations are possible in both imperial and SI units. The calculations are based on methods developed at the University of Wisconsin.

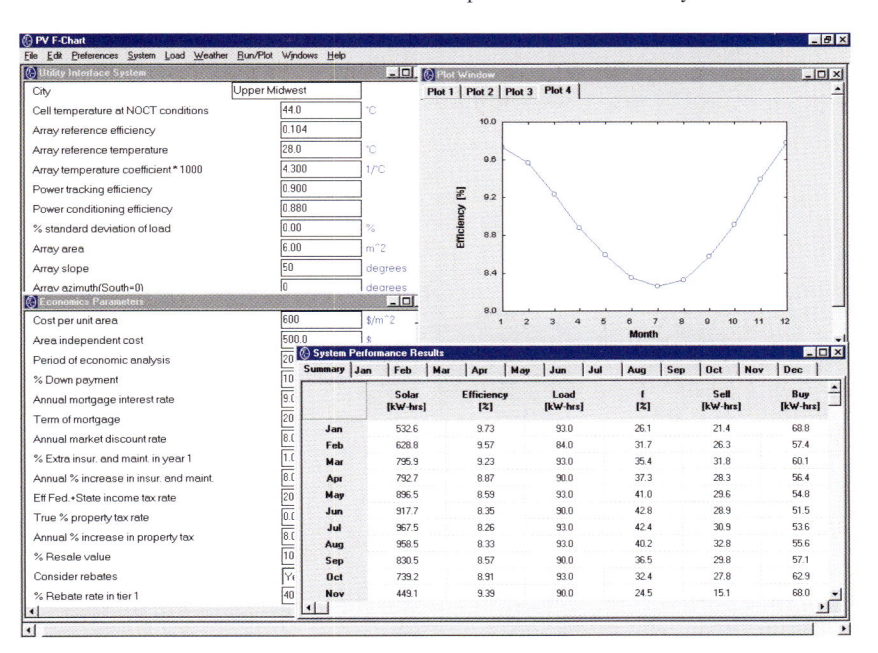

Figure 5.2
Input economic parameters for PV F-chart program

5.5.2 Time-step simulation programs

Time-step simulation programs are widely used as a result of their broad range of applications. These programs use models that are intended to reproduce the real system as exactly as possible. The systems' behaviour is calculated based on time series of meteorological input data, which usually have a resolution of one-hour intervals. Models have been implemented for various components, such as PV modules, inverters, batteries and loads for numerous specified system variants. The respective system is simulated in hourly or shorter time intervals using solar irradiance data, temperatures and, if applicable, consumption values for a typical simulation period (usually one year). These programs require a longer computing time than calculation programs. However, the increasing processing power in the latest computer systems means that this previous disadvantage has ceased to be an issue. Time-step simulation programs are considerably more flexible than calculation programs. Nevertheless, time-step simulation also has its limits because of the implemented methods. For simulating new types of system variants or investigating highly specific parameters, there is often no choice but to describe the individual system in a simulation system.

DASTPVPS

DASTPVPS is a software package for sizing, simulating and troubleshooting photovoltaic pump systems. The program was developed at the Universität der Bundeswehr (University of the Federal Armed Forces) in Munich, Germany. The optimum sizing for a photovoltaic pump system is a complex affair. As a result, the PC software tool DASTPVPS (Design and Simulation Tool for PV Pumping Systems) was developed. The DASTPVPS package comprises five modules: training; sizing; AC simulation; DC simulation; and diagnostics. The PV generator, inverter (where applicable), AC or DC motor, centrifugal, eccentric screw or piston pump and well, together with piping, can be sized using the program, and the complete PV pump system can be simulated. The training module is used to explain the basic way in which a PV system works and supports the engineer in sizing a suitable system. In the AC simulation module, the operating behaviour of an alternating-current PV pump system is calculated for various configurations and irradiance data. The DC simulation module enables the simulation of DC pump systems with direct connection to the solar generator (without MPP tracker). The diagnostics module offers system analysis and validation capabilities. The results are displayed in the form of graphs and tables. DASTPVPS includes an extensive library of irradiance data, modules and motor/pump units. But the users can also enter their own data into the component database in a straightforward way. DASTPVPS is still a DOS-based application, but also runs under Windows.

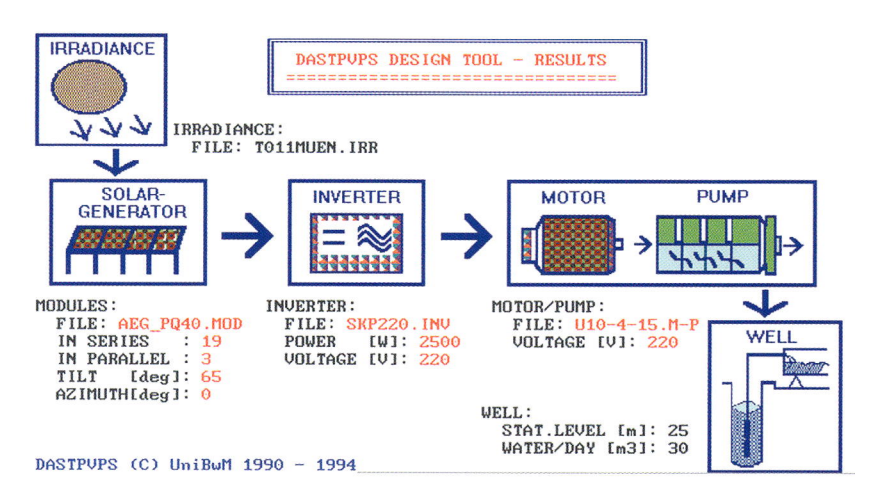

Figure 5.3
Results of dimensioning in DASTPVPS

GREENIUS

The Greenius simulation program has been on the market since July 2002. The program was developed by the Deutsches Zentrum für Luft- und Raumfahrt e.V. (German Aerospace Centre, DLR) at their outpost Plataforma Solar de Almería in Spain, and was subsidized under the European Union's (EU's) Altener program. The application area for this simulation program is mainly large commercial renewable power station projects. As well as photovoltaic systems, the program can also simulate wind farms and various types of solar thermal power stations. In Greenius, the power station is defined using location data, technology parameters and economic parameters. Data for different sites can be accessed from the Greenius weather database. Alternatively, users can feed in their own meteorological data. The technical simulations take place over one year at hourly intervals and show, for example, the hourly power output of the power stations. As well as the technical simulations, extensive economic calculations can be performed. This makes Greenius an important tool for designing and planning renewable power station projects. A sizing tool for grid-connected PV systems is included in the program. The program is aimed at project developers who, in addition to detailed technical results, also require economic analyses with key parameters through to extensive cash-flow analyses. In comparison with the other programs, the calculation of economic efficiency in this program is one of the most versatile. Numerous interfaces are available for exporting results and graphs to other Windows applications. Greenius enables comparisons of the technology of different renewable energy sources. This makes it particularly suitable for firms of designers who concentrate on the international market. A low-cost version of the program is available for the training market.

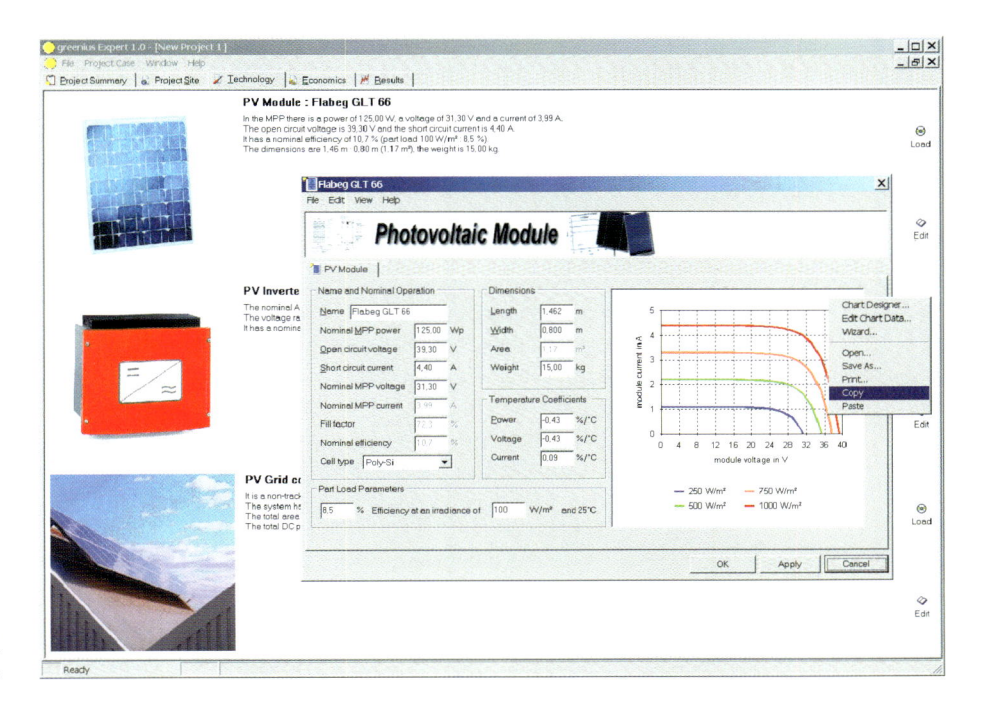

Figure 5.4
Specifications of a PV power plant in Greenius

PV-DESIGNPRO (SOLAR STUDIO SUITE)

The PV-DesignPro simulation program comprises three variants for simulating stand-alone systems, grid-connected systems and photovoltaic pump systems. For stand-alone systems, a reserve generator and a wind generator can be integrated within the photovoltaic system and a shading analysis can be carried out. The system can be optimized by varying individual parameters, and detailed calculations are performed for operating data and characteristic curves. The module and climate databases are very comprehensive. The PV-DesignPro simulation program is included in the Solar Studio Suite, along with programs for calculating solar thermal systems, tools for calculating the position of the sun and climate data. Although the Solar Studio Suite is the most comprehensive program package in this overview, its price undercuts most competitor products. This program package was developed in Hawaii.

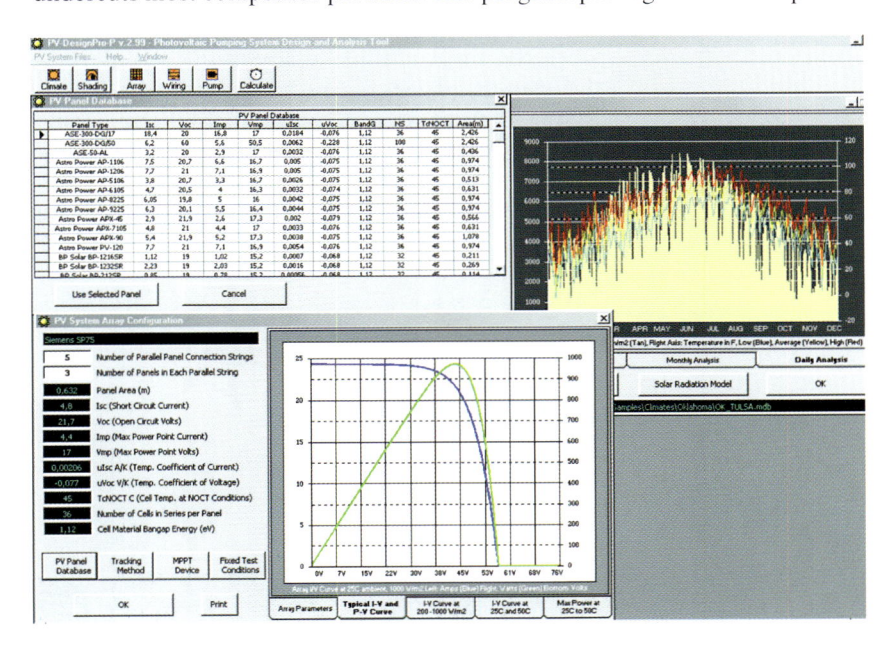

Figure 5.5
Graph of characteristic curves for a generator wiring design in PV-DesignPro

PVS

PVS for Windows is one of the longest-established time-step simulation programs on the PV software market. It was developed at the Fraunhofer Institute for Solar Energy Systems in Freiburg, Germany, and is marketed from Freiburg by a company called Econzept. PVS is a professional menu-driven program for simulating and dimensioning grid-connected PV systems and stand-alone PV systems. The influence of input variables on the operating behaviour of a system, such as irradiance, module temperature and consumption, and that of significant interdependencies such as the operating method of the solar generator and the role of the control system, can be observed in PVS with the aid of the simulation results. The behaviour of the system components is described with efficiency models, which are characterized by a few parameters needing to be specified by the user. Typical system configurations such as DC systems, AC systems with and without additional generator, and grid-connected systems can be simulated. For stand-alone systems, the load-shedding can be defined, for example, and the frequency distribution of the battery charge levels can be displayed.

In the databases for the PV modules and inverters, new products can be added with their characteristics. Via access to an internet database, the latest component data (currently only a module database) can be constantly updated. It is possible to enter two differently orientated arrays; however, only one inverter type can be used per system. Sizing is supported in that the input parameters are checked. If the inverter and PV generator are incorrectly sized, an error message appears.

The input data required for the irradiance simulation is provided by the radiation processor that is integrated within PVS. For the input data, only the monthly average values for the daily irradiance and temperature are required. These are provided from the accompanying database. A tool for comparing variations with different tilt angles is also integrated. To take shading into account, a shading editor is available. This allows the shadow outline to be input using a cursor. A menu for calculating the economic efficiency rounds out the software. The results are output by printer in three formats: as a short report, as a presentation and as a representation of the system behaviour. In addition, there is a system comparison feature. The economic efficiency is shown in a report. Going beyond this, it is possible to save the hourly simulation values in an ASCII file, enabling further evaluation using other software applications. The new version is currently under development. A newly structured program core and comprehensive redevelopment work on the sub-program that simulates PV stand-alone systems is anticipated. The main focus of this will be the battery model (taking the operating modes, ageing behaviour, losses and control systems into account), as well as the integration of a costs analysis and automated cost optimization, with the capability of custom configuring complex stand-alone systems.

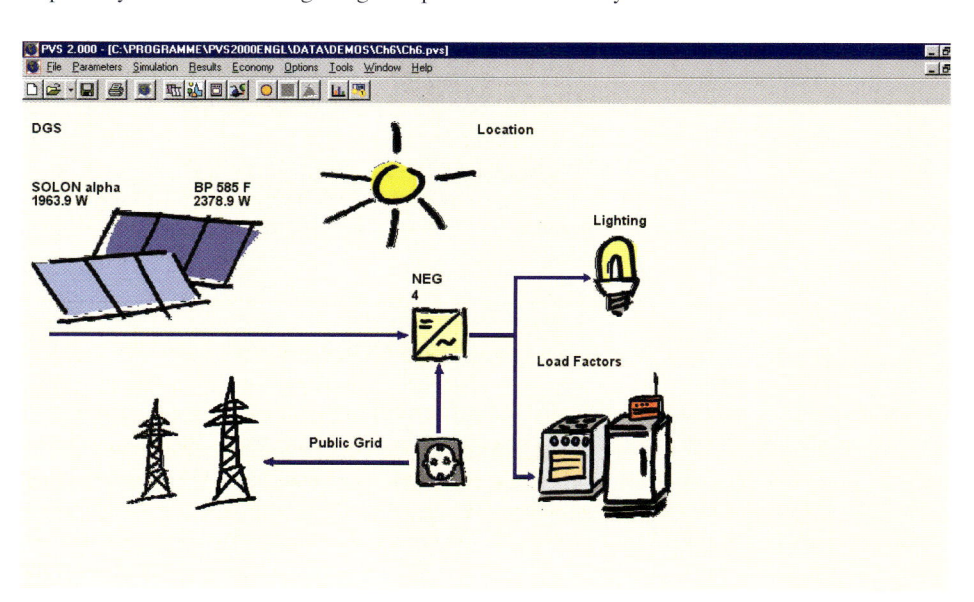

Figure 5.6
Main menu for the grid-connected system simulation in PVS

SYSTEM SIZING, DESIGN AND SIMULATION SOFTWARE   189

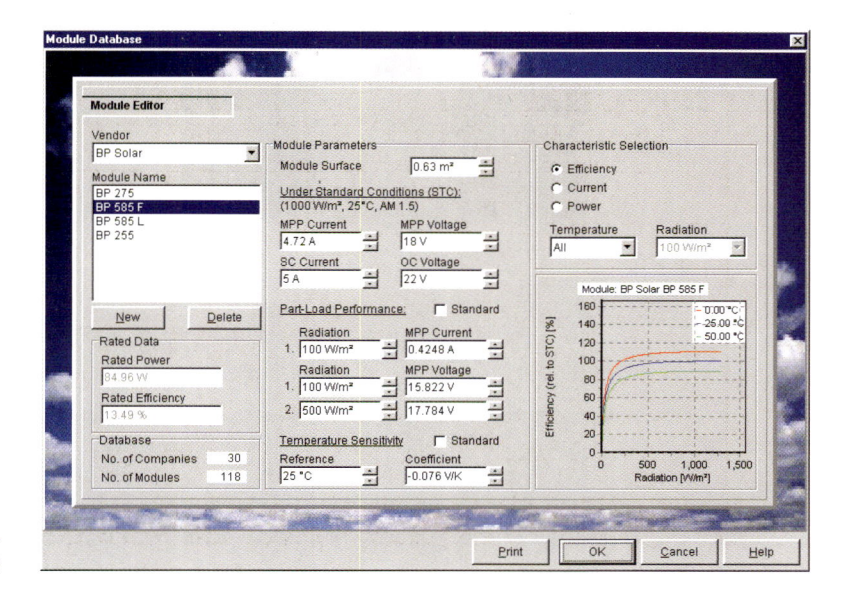

Figure 5.7
The module database menu in PVS

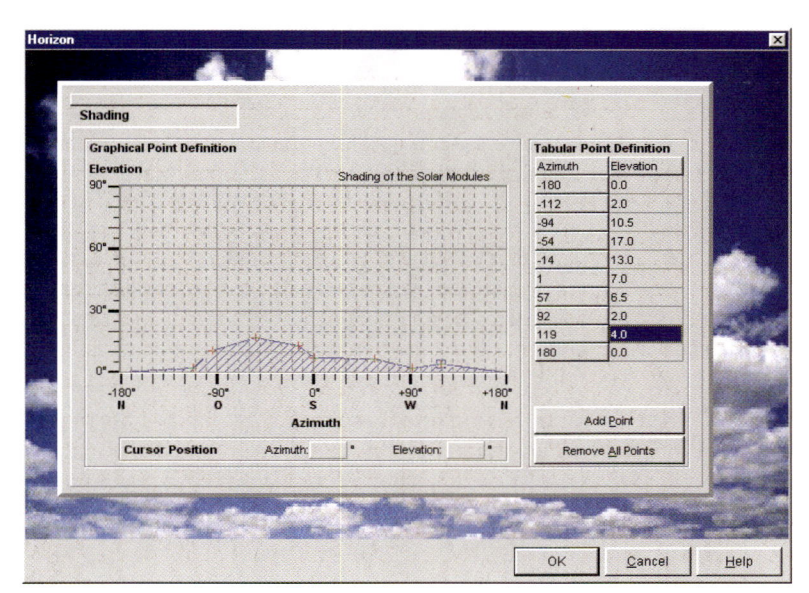

Figure 5.8
Shading editor in PVS

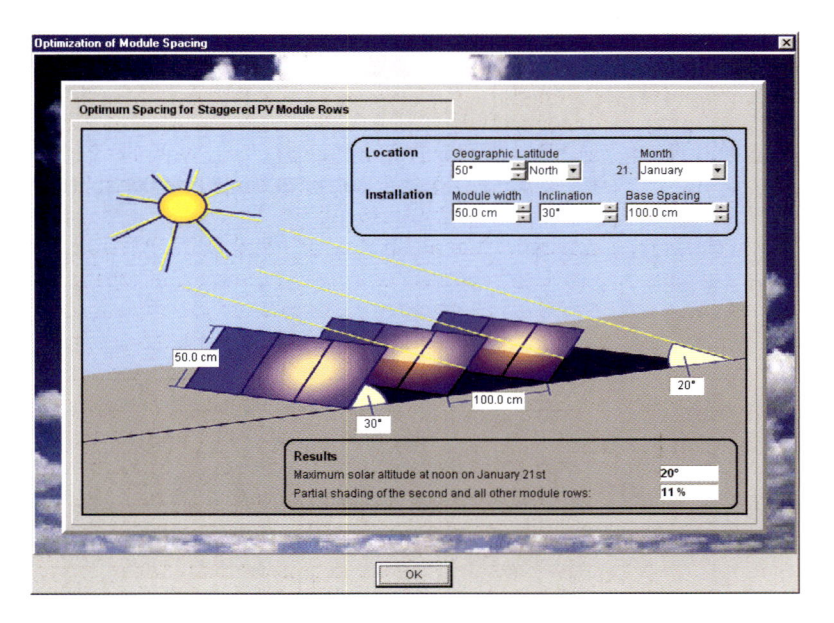

Figure 5.9
Spacing optimization for tilted systems

PV*SOL

One time-step simulation program that has now become widely used is PV*SOL by Valentin Energie Software GmbH from Berlin. This company also developed the well-known T*SOL program for solar thermal systems. PV*SOL enables the design and simulation of grid-connected PV systems and stand-alone systems. Throughout recent years, the program has undergone constant advancement, reaching a functional scope that makes PV*SOL a highly useful aid for professional PV work. With its quick design tool, the program enables easier system sizing and delivers the most important results in simulation runs at an impressive speed. Well-versed users benefit from the many program features for system optimization. Temperature and mismatch effects and the scattering of characteristic data can all be taken into account.

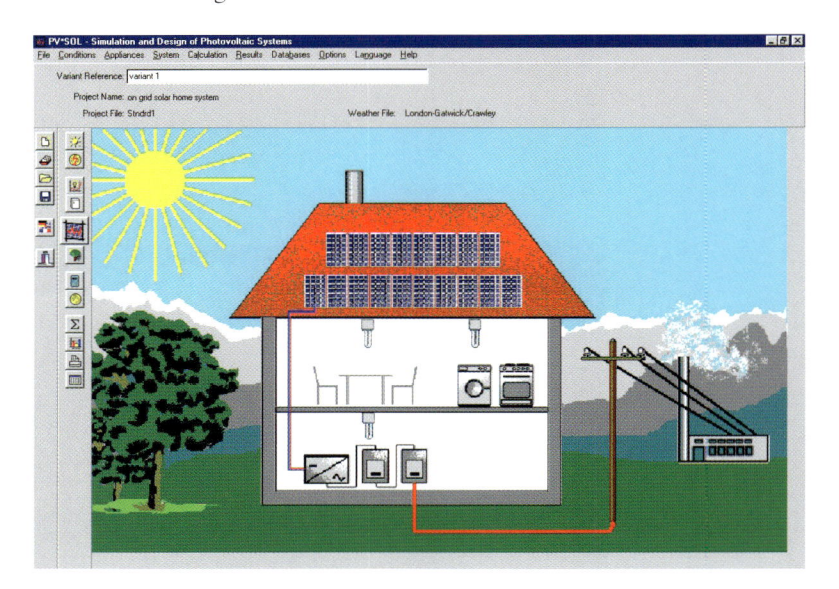

Figure 5.10
*Main menu for a grid-connected PV system simulation in PV*SOL*

The photovoltaic system being simulated can be subdivided into arrays with different orientations, modules and inverters. The program enables the simulation of all kinds of system concepts (central, string and module inverters). Just as in PVS, shading analyses can be carried out, although in more complex shading situations their accuracy is not a match for that of PVSYST. The predefined shadow-casting objects in the PV*SOL shading editor are helpful. After entering the parameters for these objects, the shading silhouette is drawn automatically. In the unshaded scenario, realistic results can also be expected since the partial-load behaviour of PV modules and inverters is taken into account. Temperature factors can be incorporated by using a dynamic temperature model. After the simulation, it is possible, for example, to display the temperature graph for a non-ventilated module on any particular day.

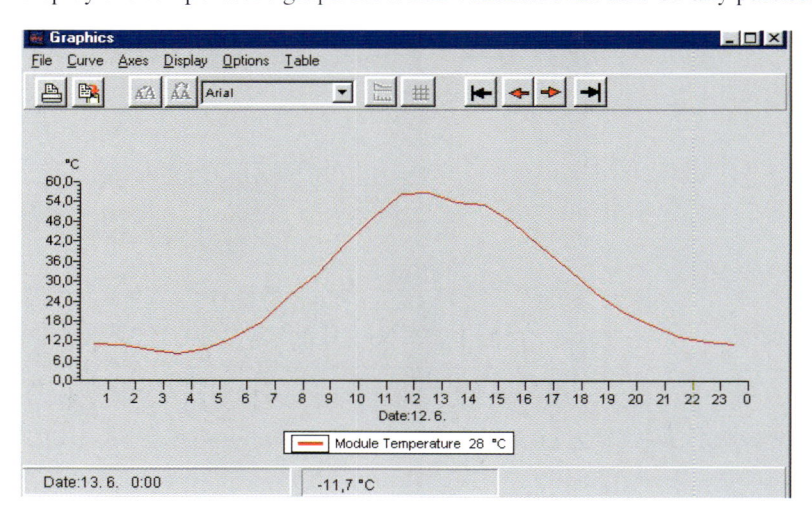

Figure 5.11
Temperature graph for a module on a summer's day

In the simulation, the various losses in PV systems are taken into account, such as mismatching, temperature, voltage and diode losses, and albedo. PV*SOL carries out a plausibility check on input data, which can avoid input errors at the earliest stage. Incorrect sizing is detected and the user alerted.

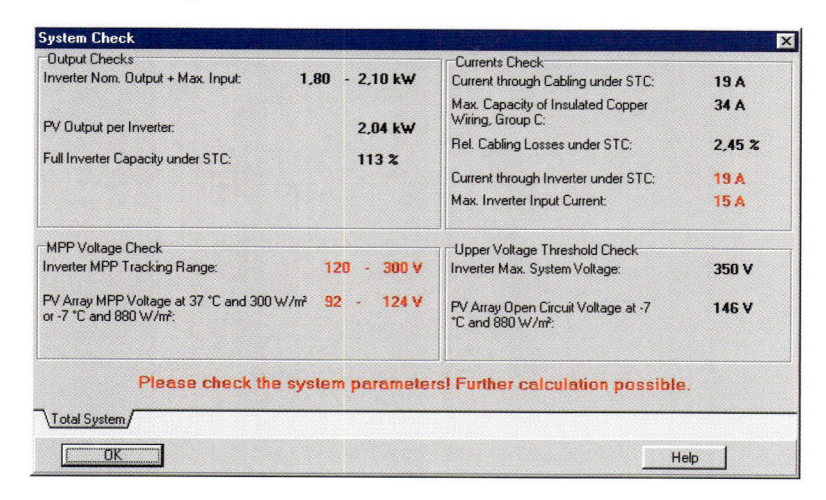

Figure 5.12
Error message for a design error

When selecting an inverter, if desired, only those models that are suitable for the selected module and the corresponding wiring set-up are displayed. Comprehensive up-to-date libraries are available for selecting components: approximately 500 module types, around 200 inverters, various accumulator types and predefined loads are currently available. As well as the component database, there are databases with numerous predefined consumption profiles, electricity tariffs and feed-in models. It is extremely easy to specify individual load profiles for consumers using the well-designed menu.

Around 250 sets of weather data for European sites are contained within the program, as well as data for nearly every state in the US. An unlimited number of additional weather datasets can be loaded. All databases can be edited for future updating. As results, PV*SOL supplies the calculation of the usual assessment variables for PV systems, as well as extensive presentations of results in the form of reports and graphs. The results can be shown as tables or graphs in up to one-hour resolution.

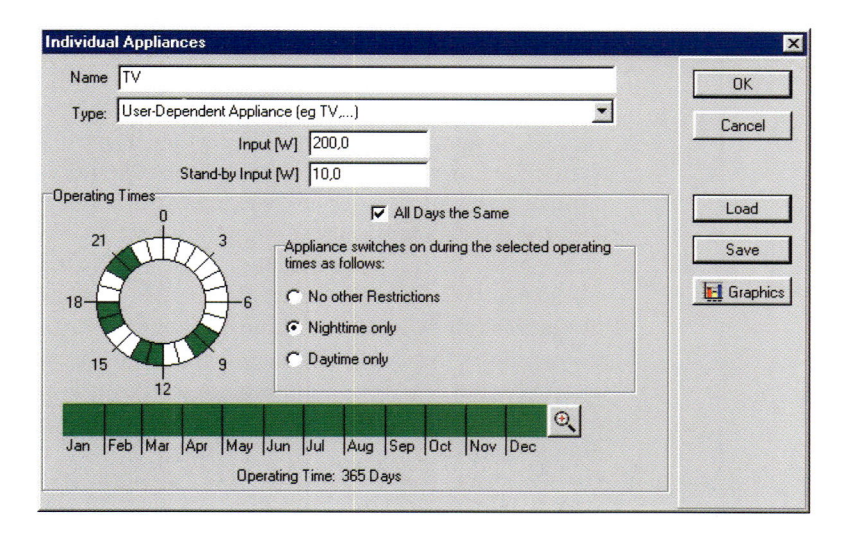

Figure 5.13
Load profile definition menu

The simulation results can be output as an extensive project report or processed in other applications. As well as exhaustive economic efficiency calculations taking all possible different subsidy, tariff and feed-in models into account, PV*SOL also enables a calculation of pollutant emissions. The program has direct interfaces to the

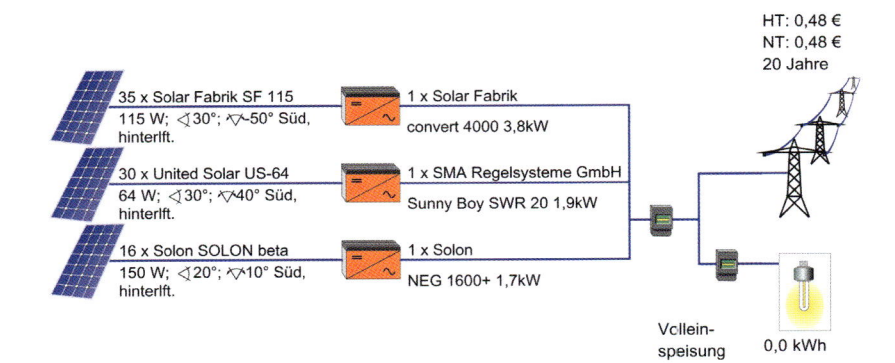

Figure 5.14
*Presentation of results in PV*SOL*

METEONORM program for weather data synthesis. PV*SOL is available as a downsized 'version N' purely for simulating grid-connected PV systems. In addition, the 'professional' version contains the models and libraries for simulating stand-alone PV systems. A multi-language version of PV*SOL is also available. Here, the user can switch languages at will while the application is running.

PVSYST

The extensive functionality of PVSYST, which is continually undergoing further development by the University of Geneva in Switzerland, makes it one of the most powerful and comprehensive programs in this overview. PVSYST, however, is also relatively complicated to use. In the current version, the user friendliness and operation of the program have been considerably improved compared with previous versions. PVSYST now works with a 'multi-level approach'. There are three different application levels with different functionalities corresponding to the various user groups such as architects, PV specialists, engineers and scientists, with their different expectations and PV knowledge. The program has a whole range of features, such as a 3-D tool to calculate shading, the ability to import system measurement data for directly comparing measured and simulated values, and a toolbox for solar geometry, meteorology and PV operational behaviour. It is intended that the next release of PVSYST, version 3.2, will also simulate amorphous solar modules. The program is only available in English and French. The online user support for PVSYST is also a useful feature. It is possible to contact the program author quickly and directly via email and through an online user forum. For testing purposes, a full version can be downloaded from the internet that works for ten days. Besides simulations of stand-alone systems with reserve generators and grid-connected systems, PVSYST is also able to make special analyses. For instance, it can be used for calculating module characteristic curves when there is partial shading, enabling, for example, the thermal loading on solar modules to be determined. Moreover, for the simulation, it is possible to determine and display numerous parameters, such as meteorological data, electrical voltages, currents, energies and performances. PVSYST also allows three-dimensional shading analysis.

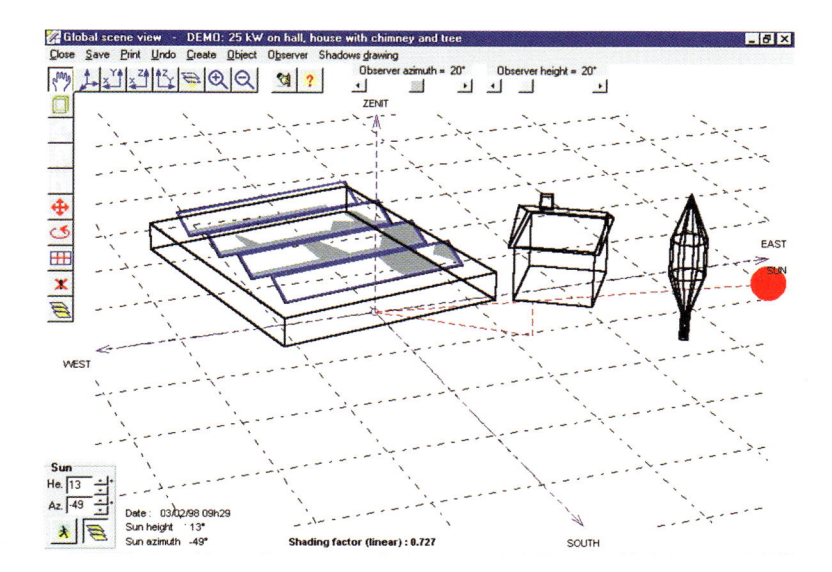

*Figure 5.15
Graphical user interface for three-dimensional shading analysis with PVSYST*

SOLDIM

The SOLDIM program can be used for designing PV stand-alone and grid-feeding systems or for customer acquisition. SOLDIM consists of the STASYS and IN-GRID modules, as well as databases and tools for sales support. IN-GRID was developed for planning and economically analysing grid-connected PV systems. The STASYS module can be used for designing PV stand-alone systems. Strictly speaking, SOLDIM represents a mixture between a calculation program and a time-step analysis program. Monthly mean values and some selected daily variations to the nearest hour are used for the simulation. The rapid calculation times, individual settings in the input windows, databases and priced item lists make SOLDIM a useful support when in discussions with a customer and for consulting. The program enables trained users to make rapid, reliable calculations and compare system configurations. Users can either purchase SOLDIM as a complete software package or simply buy the separate modules. SOLDIM and its modules are available in German and English. A new SOLDIM version, Visual PV Studio, is under development at the time of writing. In this version, the modular structure of SOLDIM is being developed further to provide a package that includes modules for designing stand-alone systems, hybrid systems, PV pumping systems, grid-feeding systems and a toolbox.

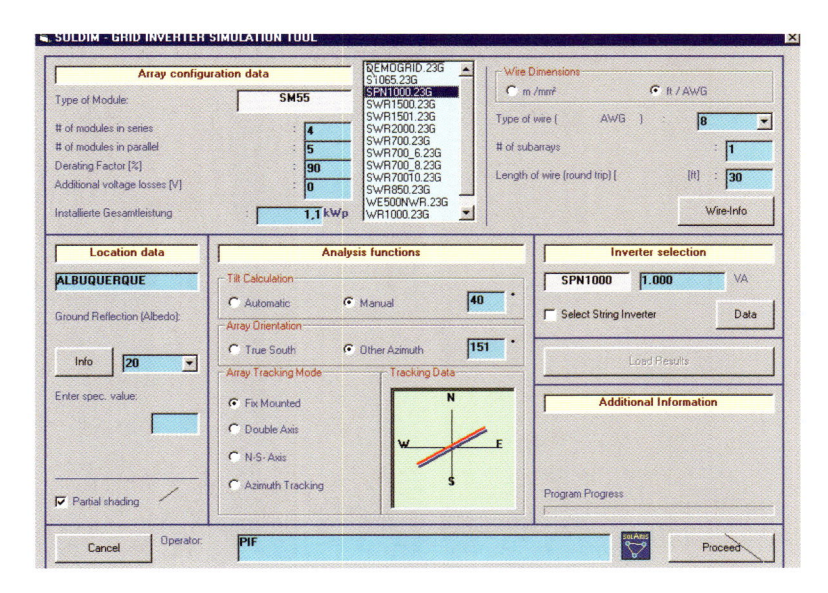

*Figure 5.16
Main menu of SOLDIM for calculating grid-connected PV systems*

5.5.3 Simulation systems

Simulation system programs are needed to simulate systems that go beyond the limits of time-step simulation programs or if it is intended to calculate completely new components and systems variants. These programs enable individual simulation modules to be written and implemented in calculations. Here, the user defines a simulation task using a formula or diagram-oriented simulation language. The most well-known simulation system for PV systems is INSEL. SMILE, which was developed at the Technical University of Berlin, also falls into this program category. PV systems can also be simulated with TRNSYS, a simulation system used in the building simulation and solar thermal sectors. Simulation systems used in the electronic sector, such as PSpice, can also achieve good simulation results if the equivalent circuits for solar cells are entered. It requires considerable training, however, to be able to make the most of the advantages provided by simulation systems such as high flexibility. With professional time-step simulation programs with user-friendly interfaces, even inexperienced computer users can produce a system simulation within a few hours. With simulation programs, on the other hand, the training time can last for several days or weeks. These are therefore more suitable for research and development purposes.

INSEL

The INSEL simulation environment, which was developed at the University of Oldenburg, has been on the market since the early 1990s. A block-oriented simulation language was designed for INSEL, which is specially tailored to simulate renewable electrical systems. Numerous models are implemented within the various blocks, including solar radiation calculations, solar cells, inverters, batteries, wind generators, pumping systems and solar thermal power stations. The various blocks can be visually combined with the HP VEE interface. INSEL includes a radiation database that contains the monthly mean values from around 2000 sites worldwide. INSEL is particularly suited for research purposes, simulating special applications, detailed simulation analyses, or for professionals who need considerable flexibility.

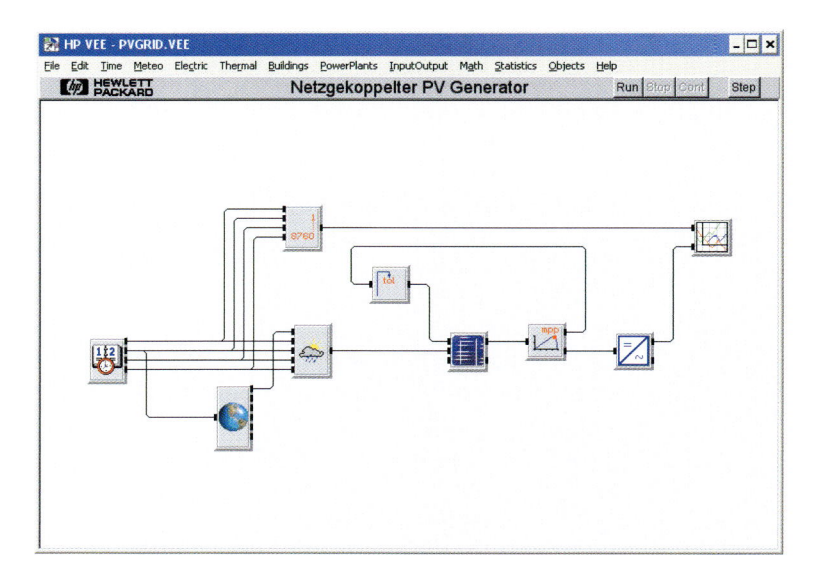

Figure 5.17
The INSEL simulation environment

SMILE

The Technical University of Berlin and the company GMD First have been developing the SMILE simulation environment since 1990. It is used mainly for simulating and optimizing complex energy-converting systems. The SMILE simulation environment includes an object- and formula-oriented simulation language with translator, run-time system, various numeric solvers, an optimizing framework and a component library. This component library provides a basic set of models for describing almost any energy converter and their combinations and allows the user to assemble, simulate and optimize an energy technology system using the 'building block principle'. The formula and building orientation enables various modules to be easily integrated and existing modules to be easily expanded. The application areas of SMILE extend to thermal solar energy use, heating and air-conditioning technology, building simulation, hydraulic networks and power station technology. In the field of photovoltaics, various models for PV modules and inverters are implemented. For SMILE, however, there are no graphical interfaces and resulting depictions. Currently, SMILE is only available for UNIX platforms. For non-commercial use, SMILE can currently be downloaded from the internet for free.

TRNSYS

TRNSYS was developed in 1974 at the Solar Energy Laboratory of the University of Wisconsin in the US and has been continually improved upon since then. Today, TRNSYS is the market leader among thermal simulation systems and evaluates concepts for rational energy consumption and systems for active and passive solar energy consumption. Although the simulation focuses on thermal systems, the TRNSYS simulation environment also includes photovoltaic models. However, the graphical user interface hardly supports the simulation of PV systems. Since it takes considerable time to become trained on TRNSYS, it is only recommended for experienced users.

5.5.4 Supplementary programs and data sources

These categories include programs for irradiance calculations and shading analyses, as well as component libraries and weather data. Programs for supplementing location data for the simulation include, for example, METEONORM, SHELL SOLAR PATH and SUNDI, which enable the generation of radiation data and depictions of sun path diagrams or shading analyses. Additional weather data are available online.

METEONORM

Meteotest in Bern was commissioned by the Swiss Federal Office of Energy (BEW) to develop the METEONORM program for the purpose of conducting calculations with meteorological data. Although most programs have an extensive site library, locations often have to be simulated for which there are no data available. Using METEONORM, it is possible to calculate the necessary global radiation and temperature data for any place in the world. Apart from these parameters, it is also possible to determine the relative humidity and the wind velocity and direction. In METEONORM, various quality-tested databases have been combined to create a worldwide database for simulating energy systems. Using spatial interpolation based on this comprehensive database, which now comprises weather data from 2400 weather stations throughout the world, required data can be calculated for desired locations at hourly intervals. The data output for the subsequent simulations is given at hourly intervals in 16 selectable data formats, as well as in user-defined formats. The site data generated can also be presented in graphical form and printed. For individually recorded irradiance and temperature data, the program enables statistical calculations to be used to create time series on a monthly or hourly basis for any period of time for the individual location. It is also possible to take account of sloped areas and shading.

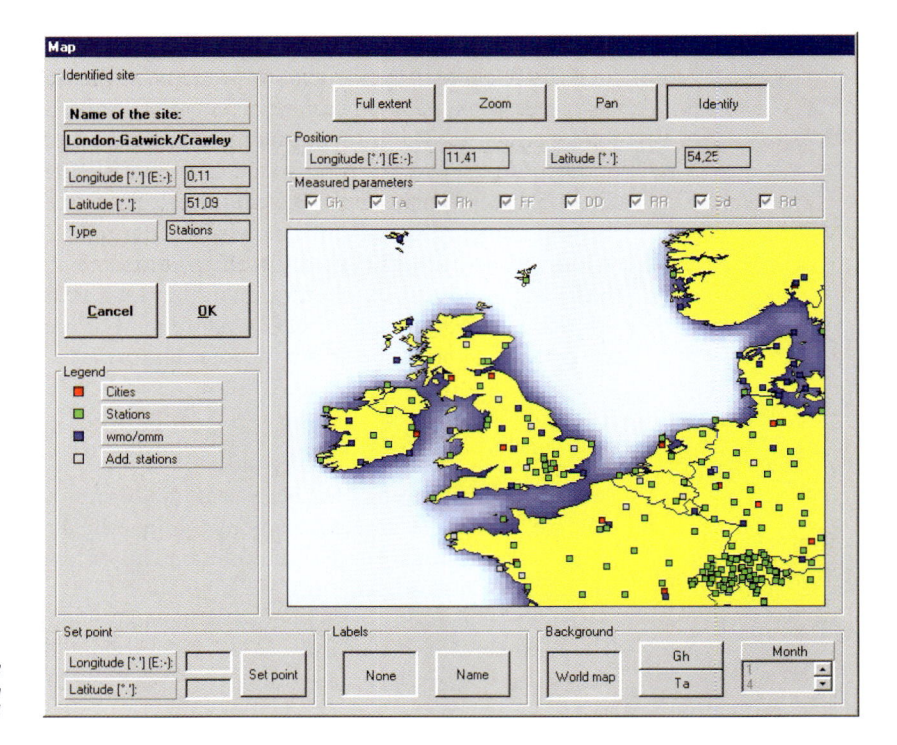

*Figure 5.18
Visual selection of a site in
METEONORM*

SUNDI

SUNDI was developed at the Institute for Electrical Energy Technology at the Technical University of Berlin. The program calculates the course of the sun's path and enables shading analyses. This can be done for any locations on Earth, which are either selected from the database or are input individually. In addition, using its own measurement values for global radiation, the software can determine direct and diffuse radiation and irradiance on an area with any orientation. All calculations can be conducted for a specific time, day or year. The results are provided in tabular or graphical form. Sun path diagrams can be displayed on the screen with the input shading. Horizontal, unshaded and shaded irradiance can be exported in half-hour intervals to other programs for further use. The program is already used by universities, colleges and training facilities. SUNDI can be downloaded free of charge from the internet.

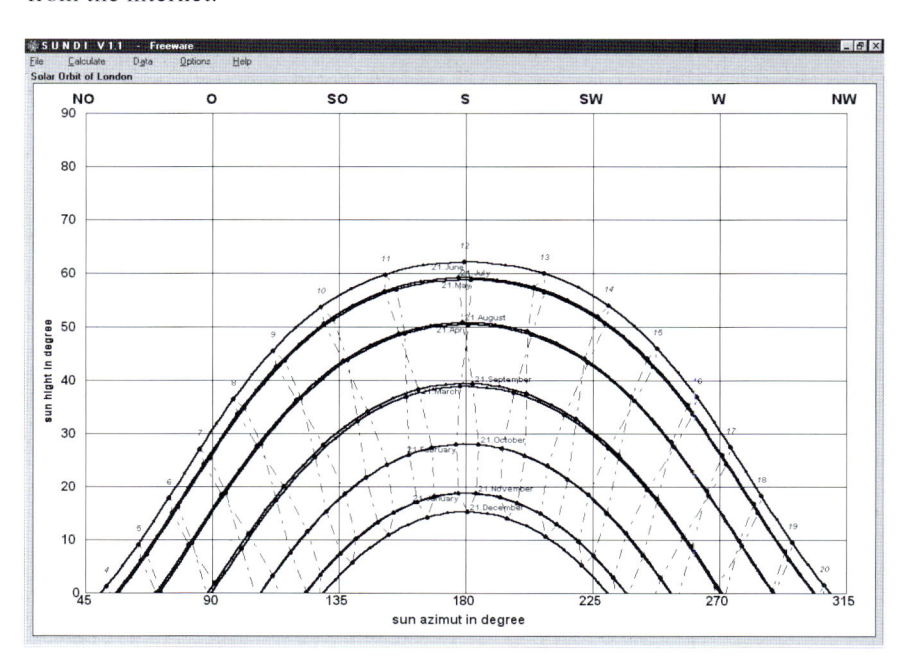

*Figure 5.19
Visual depiction of a solar path diagram
with shading elements for a location
in London with SUNDI*

SHELL SOLAR PATH

The Shell Solar Path program was developed at the University of Bochum in Germany, and the current release, version 2.0, is sold by Shell Solar. It enables the depiction of solar-altitude diagrams for any location in the world, taking shading into consideration. It can also determine the direct irradiance duration for sloped surfaces. As an additional feature, it is possible to depict sunrises and sunsets and day lengths. The time series generated for the solar altitude can be exported as a text file with a minimum time interval of 1 minute. However, although the duration of the shading can be determined, it is not possible to calculate the irradiance energies and shading losses. Shell Solar Path therefore does not compare with the SUNDI program in terms of functionality.

WEATHER DATA ON THE INTERNET

The following websites are available on the internet for researching weather data:

Table 5.1 Internet addresses for researching weather data

URL (Uniform Resource Locator)	Region	Description
http://www.satellight.com	Europe	Global radiation data based on satellite images, free
http://eosweb.larc.nasa.gov.sse	Worldwide	Global radiation data from the NASA database, free
http://wrdc-mgo.nrel.gov/	Worldwide	Global radiation data from the NREL (US), free

5.5.5 Design and service programs

Besides simulation programs that replicate and analyse the overall system, there are also design and service programs that provide information on grid-connected PV systems and support the design process.

Various inverter manufacturers offer users other design programs free of charge via the internet. The aim of inverter manufacturers is to provide clear information beyond that supplied in the product handbook on the individual device, its operating behaviour and any possible wiring configurations. These programs are mostly implemented as Microsoft Excel tables and have various functionalities. Perhaps the best-known program is offered by the inverter manufacturer SMA and is provided as the Excel table GenAu (see www.sma.de). GenAu has a database of PV modules available on the market and SMA inverters. The program enables the modules to be combined with the inverters using various possible generator interconnections. Here, the different generator inverter combinations are tested to ensure that they conform with the most important limiting values. In the program it is possible to choose between German, English, Italian and Spanish. The inverter manufacturer Fronius (see www.fronius.at) offers the Configurator program for sizing PV systems with its own devices. This program functions in a similar way to GenAu, but is somewhat more developed in terms of handling, has a customer administration feature, and allows the results to be printed out.

A comprehensive service program is offered by the inverter manufacturer Siemens. It is called SITOP solar select (see www.siemens.de/sitop/solar). This program has a whole series of functions that make it considerably easier to choose the best possible wiring configurations for a PV system using Siemens inverters. The program calculates all sensible interconnection possibilities, assesses them and then allows the selected systems to be analysed in detail. As is the case with online simulation software, service programs are in no way a replacement for getting consultation from a professional PV specialist. Another design tool is the SolarSizer program. A demo version is available at www.solenergy.org/html/about/SolarSizer.html. It is a useful tool for designing and sizing photovoltaic systems. It has a graphical interface to select components, and provides corresponding cost and energy calculations. The program was developed by the Centre for Renewable Energy and Sustainable Technology with assistance from Solar Energy International. Other available programs are Insolar (available at www.elektropraktiker.de/software) and PV-Professional.

5.5.6 Web-based simulation programs

The internet continues to develop as a simulation platform. There are now a whole series of web-based programs for online PV simulations. These free services are usually very restricted in what they offer. The programs are often used to try and enhance the service character of specific websites or portals. These are usually graphically well-made, simple programs that provide approximate results. In general, however, online simulation software is very limited in terms of functionality and accuracy. The programs are useful, however, for providing initial information for standard PV systems and a rough estimate of the yield. Although they relieve solar firms of a lot of the general work in obtaining information, it should be pointed out in the respective website that the calculations are only rough and limited in scope, and that they in no way replace consultation with a professional PV specialist.

A comprehensive online tool is PVWATTS developed by the National Renewable Energy Laboratory in the US. The website calculates electrical energy produced by a grid-connected photovoltaic system, although it is only usable for systems based within the US. Non-experts are able to quickly obtain performance estimates for grid-connected PV systems.

6 Mounting Systems and Building Integration

6.1 Introduction

Most building surfaces are suitable for the installation of photovoltaic arrays: sloping and flat roofs and façades. A distinction can be made between additive and integrative solutions.

Figure 6.1
Potential locations for installing a PV system

Types of installation

for sloping roofs		for flat roofs		for facades		as sun shading		for open spaces
on the roof	roof integration	on the roof	roof integration	in front of the facade	cold/warm facade	glass roof	canopies/ louvres	

In an *additive* solution, photovoltaic modules are secured to the roof or onto the façade using a metal structure. As a result, the photovoltaic system is an additional technical structural element on the building with the sole function of generating power.

In an *integrative* solution, building components of the roof or façade are replaced with photovoltaic components – this is also known as building-integrated photovoltaics (BIPV). The photovoltaic system becomes part of the building shell and, in addition to the function of generating power, performs functions such as weather protection, heat insulation, noise insulation, sun shading and safety. The diverse types of PV modules are described in section '2.1.4 Design options for PV modules' in Chapter 2 in terms of their design and function. This enables the exploitation of synergy effects and the implementation of optically upscale installation solutions.

This chapter is devoted to the basics of roof and façade construction and gives an overview of additive and integrative mounting systems for sloping roofs, flat roofs and façades, as well as glazed roofs and sunshade devices. The chapter concludes with a description of mounting systems for free-standing installations.

Figure 6.2
In the New Sloten housing estate in Amsterdam, The Netherlands, photovoltaic elements have been integrated into the buildings in different ways

6.2 Roof basics

6.2.1 The roof's tasks

So far we have seen the following tasks that a roof performs:

- upper delimitation of the building;
- take the loads resulting from roof coverings, wind, rain and snow;
- keep the effects of weather away from the inside of the building;
- heat insulation;
- noise insulation;
- fire protection;
- design (shape, colour, material, surface structure).

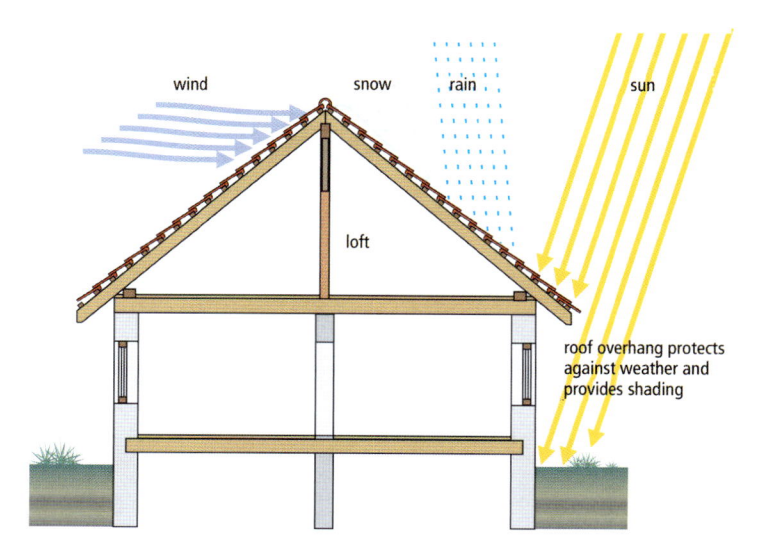

Figure 6.3
Tasks of the roof

In the future, the roof will increasingly incorporate energy-conversion elements to convert sunlight into power and/or heat. This means that in the future, the roof skin (and the façade) – as far as the material and appearance are concerned – will be subject to big changes.

6.2.2 Roof shapes

Roofs can be roughly classified according to their slope:

Flat roofs:	slope less than 5°
Slightly sloping roofs:	slope 5–22°
Normally sloping roofs:	slope 22–45°
Steep roofs:	slope greater than 45°

As well as the roof shapes shown in Figure 7.4, there are a number of arched roof shapes (e.g. barrel roof) and special forms of sloped roofs.

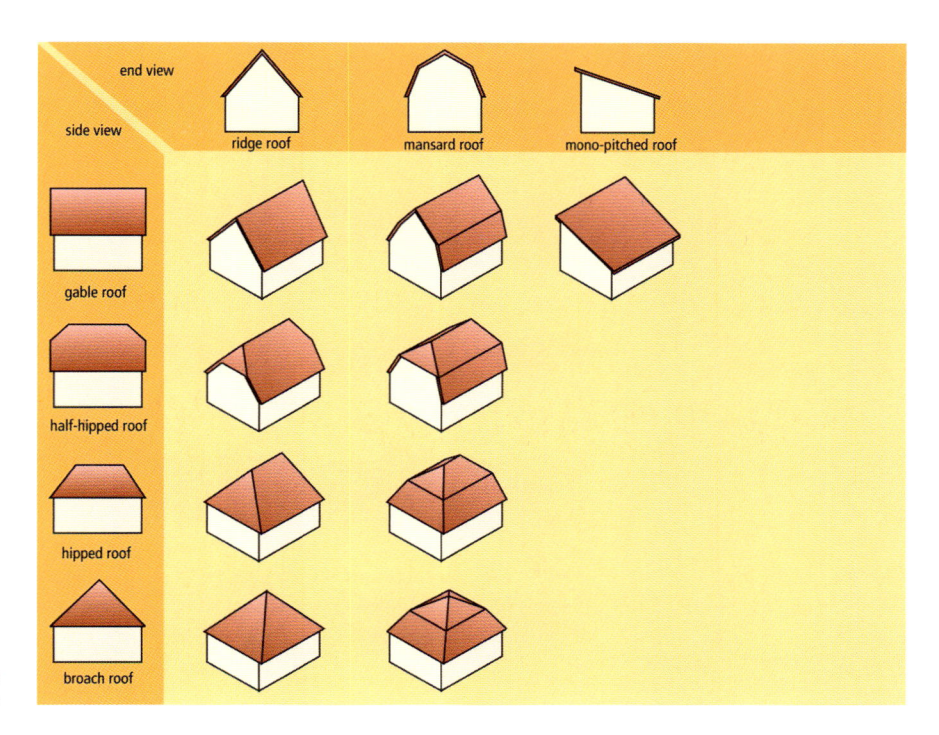

Figure 6.4
Basic shapes of pitched roofs

6.2.3 Roof constructions

TIED RAFTERS

In this roof the rafters span from the ridge to the eaves and they are tied at eaves level by the ceiling joists forming a triangle. This form of roof often has purlins as intermediate supports for the rafters.

PURLIN AND RAFTERS

In purlin roofs, the rafters rest as sloping bars on horizontal purlins that are independent of one another. The purlins run along the slope of the roof and are supported by vertical members and/or walls. The purlins take the vertical loads and transfer them to the supporting structure. Unlike a rafter roof, replacing rafters for roof fixtures and structures poses no strength problems.

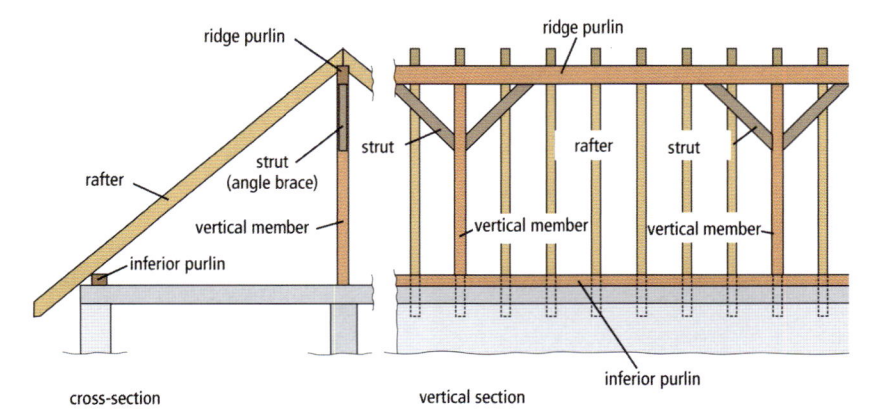

Figure 6.5
Purlin truss

RAFTER AND RAISED TIE OR COLLAR BEAM ROOF

In the rafter roof, rafters and the ceiling lying below them form a rigid triangle but the ceiling joist is raised above eaves level and there is a section of sloping ceiling. The rafters have to take higher loads over the sloping section and there is a tendency for the thrust at the eaves to create excessive deflections or eaves spread. The further the tie is raised above eaves level the greater the effect.

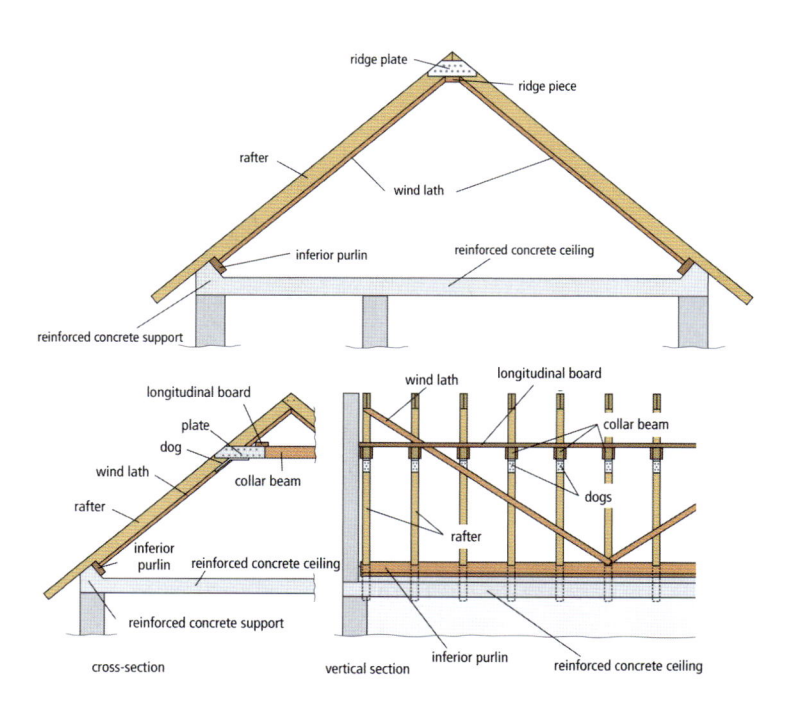

*Figure 6.6
Rafter truss (top) and
collar beam truss (bottom)*

TRUSS CONSTRUCTIONS

The roof constructions listed so far are not common in modern construction except purlin and rafters for larger buildings. Typically, roof trusses are used. These are self-supporting constructions (glued truss, nail plate truss and similar). Sloping ceilings can be accommodated and 'room in the roof' construction can use attic trusses. Here no changes may be made without the approval of a structural engineer.

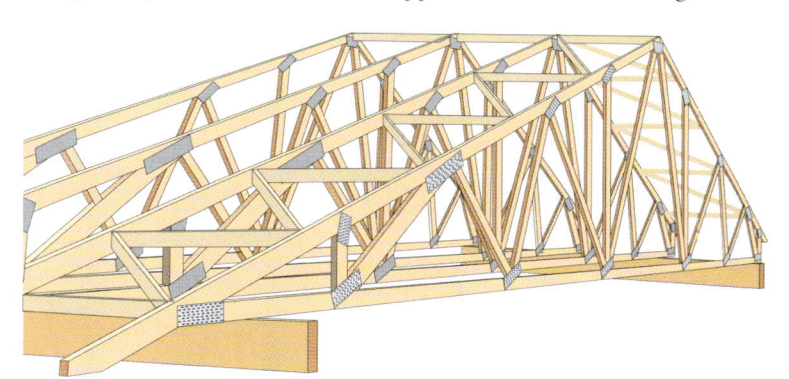

*Figure 6.7
Nail plate truss*

FLAT ROOF CONSTRUCTIONS

A wooden joist (wooden truss) ceiling can be used for the roof construction, but in larger buildings solid reinforced concrete slabs, profiled sheets or concrete girders are generally used, which are simultaneously the ceiling closing the room below and a component of the roof supporting structure.

6.2.4 Roof skin

A distinction is made here between roof covering and roof sealing.

ROOF COVERING – DRAINAGE COVERING USED IN SLOPING ROOF APPLICATIONS

This consists of individual elements such as clay and cement tiles, stone slabs, fibre-cement and natural slates, shingles and (profiled or corrugated) sheets. They are laid counter to the rainwater flow direction and require a minimum roof slope specified according to the covering type.

*Figure 6.8
Plain tile
Source: Arbeitsgemeinschaft
Ziegeldach e.V.*

*Figure 6.9
Wood shingles
Source: Arbeitsgemeinschaft
Ziegeldach e.V.*

*Figure 6.10
Natural slate
Source: Arbeitsgemeinschaft
Ziegeldach e.V.*

*Figure 6.11
Sheet metal roofs
Source: Holzapfel, 1999*

ROOF SEALING – SEALING COVERING USED IN FLAT ROOF APPLICATIONS

This is a fully waterproof layer over the entire surface of the roof, and can be made of, for example, bitumen roofing felt, plastic roof sheeting, or plastics that are applied as a fluid and then harden. It is absolutely essential for roofs with less than 5° slope. Connections and terminations, openings and the joint formations are also part of the roof sealing.

*Figure 6.12
Fitting bituminous sheeting
Source: A. W. Andernach*

*Figure 6.13
Plastic roof sheeting
Source: Saar-Gummiwerk*

Table 6.1
Overview of various types of roof skin
Note: a R = here the roofer is responsible.
b Asbestos problem, hazardous waste if manufactured before 31 December 1990.

Type	Examples	Material	Method of fixing, working	Points to be aware of
Tiles	Double-trough grooved tile, S-tile Plain tile	Clay	Dry laying, sometimes clamped, laid in a paper insulating foundation, in mortar, foam-filled	Good ventilation, otherwise frost damage; medium (plain tile) to high breaking strength; Blooming, green formation
Concrete tiles	'Frankfurter Pfanne' type	Concrete	Dry laying, sometimes clamped, laid in a paper insulating foundation, in mortar, foam-filled	Surface weathering, high breaking strength, blooming
Shingles	Slate I	Natural slate	Nailed	Discoloration, weathering, low breaking strength, R[a]
	Slate II	Fibre cement slate [b]	Nailed	Shrinkage (dryness), swelling (moisture) → drilled holes need to be made bigger, bleeding (streaking), R[a]
	Wood shingles	Wood – e.g. larch, cedar	Nailed	'Dishing' with changing humidity, for roof slope < 18° chemical wood preservative required
	Round cut scale shingles	Bitumen	Nailed, clamped	Draw moisture
Sheets	Trapezoidal sheets	Steel, galvanized	Bolted, riveted	Galvanized trapezoidal sheet, in some cases with additional plastic coating, bolt always in upper member
	Standing seam covering	Zinc, titanium, copper, aluminium	Secured with hold-down clips, crimped	Lay movably, non-clamping grooved joint, corrosion problem
Corrugated panels	Ridge covering, chimney seal	Lead, zinc	Soldered, crimped	Good ventilation, high thermal expansion, brittleness
	'Berliner Welle' type	Fibre-cement [b]	Bolted	Not seamed, water penetration through wind pressure
	Translucent corrugated sheets	Polyester	Bolted	Not seamed, water penetration through wind pressure
	Bitumen corrugated sheets	Bitumen	Nailed	Not seamed, water penetration through wind pressure
Flat panels	Toughened safety glass, insulating glass, laminated safety glass, solar protection glass PV modules	Glass, noble gases, plastic films, metal coatings, solar cells	Point-fixed, linearly or whole surface: glued, pressed, clamped, bolted, padded, depending on glazing system	Combines very well with PV modules as the same substructures and fixing can be used; Observe engineering standards for construction types with glass
Liquid plastics	Roof impregnation, roof terrace sealing	Polyurethanes, acrylates, resins	Poured, spread	Application subject to weather conditions, carefully prepared substrate
Roof sheeting	Plastomer welded sheeting	Bitumen	Glued, welded	Lifespan maximum 20 years, may be shorter than the solar installation. Bitumen corrosion in conjunction with zinc sheets
	Plastic sheeting	EPDM	Joined with hot air	Contains plasticizers, embrittlement, pay attention to compatibility with bitumen
Vegetable material	Thatched roof	Reeds	Joined/tied with wire to battens/sewed	No photovoltaic installation known to date, a conceivable possibility would be e.g. in-roof installation on dormer windows
	Green roof	Foil, substrate, special plants	Laid in place	Only mounted/tilted frames possible
Edge and special elements	Vent tiles	Concrete, clay	Laid dry	Can be moved
	Verge tiles	Concrete, clay	Bolted	Prevent the installation of PV modules
	Edge tiles, ridge tiles	Concrete, clay	Clamped (new buildings), mortar (old buildings)	Prevent the installation of PV modules
	Roof pantiles with vent pipe	Plastic, clay	Laid dry	Prevent the installation of PV modules
	Roof pantiles with antenna mount	Plastic, metal, clay	Bolted	Prevent the installation of PV modules
	Roof pantiles with tread step holder	Plastic, metal	Bolted	Prevent the installation of PV modules
	Roof drain	Plastic, metal, brick	Bolted, seamed, crimped, soldered, hold-down clips	Prevent the installation of PV modules
	Vent/extractor fittings, hoods and caps	Plastic, bitumen, fibre cement	Bolted	Prevent the installation of PV modules
	Connecting plates	Metal	Bolted, seamed, glued, soldered	

a R = here the roofer is responsible
b Asbestos problem, hazardous waste if manufactured before 31 December 1990

MOUNTING SYSTEMS AND BUILDING INTEGRATION 205

EXAMPLES OF SHAPED TILES, ROOF HOOKS, SNOW TRAPS AND STEP GRATES

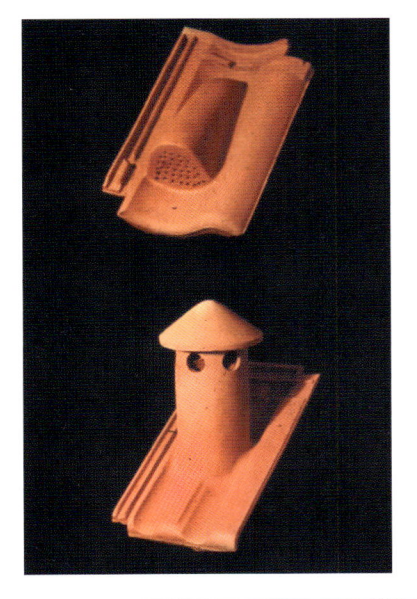

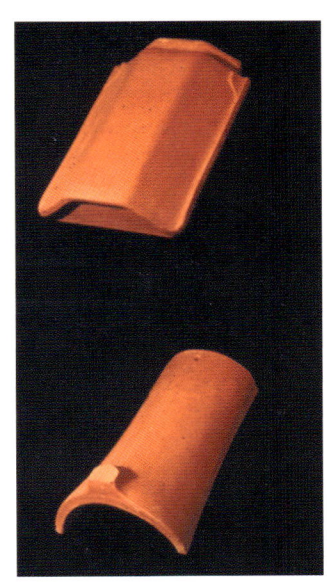

Figure 6.14
Vent tiles, vent pipe and various ridge tiles
Source: Arbeitsgemeinschaft Ziegeldach e.V.

Figure 6.15
Roof hook

Figure 6.16
Tread step
Source: Holzapfel, 1999

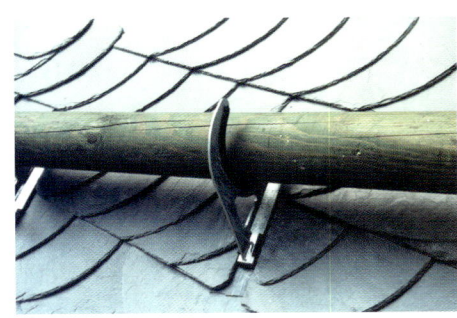

Figure 6.17
Round timber holder
Source: Holzapfel, 1999

Figure 6.18 Snow trap grid
Source: Pfleiderer

Figure 6.19
Solar opening tile for routing solar wiring
Source: Fleck GmbH

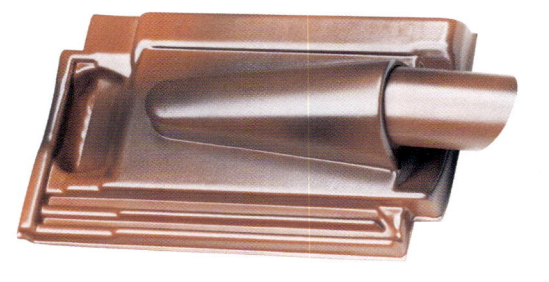

6.2.5 Sloping roof

As a rule, sloping roofs are built as a ventilated cold roof and structured as follows from the exterior to the interior:

- roof covering = 1st drainage level: drain off water;
- rain water and moisture are collected at the lowest point and led away via a rainwater drainpipe;
- roof battens;
- counter-battens (not usual for tiled roofs, but required if sarking boards are used);
- underfelt (sarking membrane or felt) or sub-roof = 2nd waterproof layer: stops water completely;
- normally one layer of sarking felt is used, but for special cases or local conditions, such as at less than the minimum roof slope or use of the loft area as living space, sarking boards are required. This is a somewhat unique case for roofs in Scotland;
- rafters;
- ceiling joists with thermal insulation.

If the ceiling of the top floor is not thermally insulated or if the loft is going to be used as living space, the roof needs to be insulated. Thermal insulation can be laid in three basic ways that can be combined:

- below the rafters;
- between the rafters;
- on the rafters (not shown in the diagrams). Here problems arise with securing the modules, as in this case the standard components cannot be used.

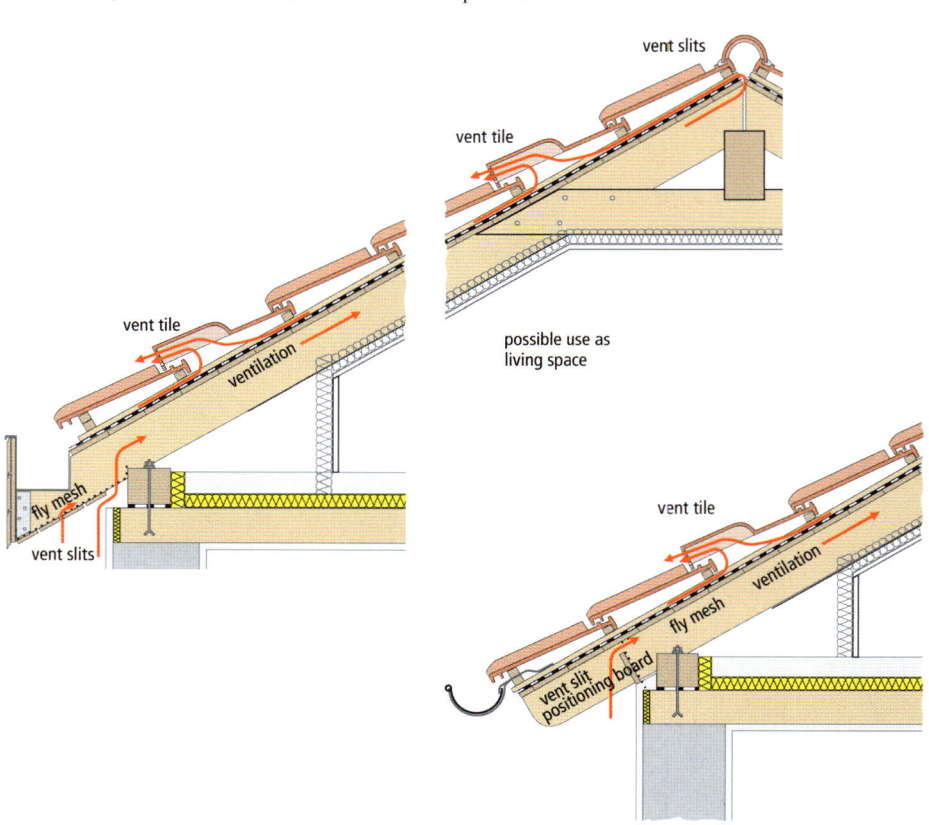

*Figure 6.20
Eaves and ridge of a roof with substructure.
(Left) Lined roof overhang
(Right) Roof overhang with visible rafter heads*

ROOF INSTALLATIONS AND SUPERSTRUCTURES

Skylights, dome lights, dormer windows and roof terraces serve to illuminate and aerate converted loft spaces and bring extra living space. When it comes to installing solar arrays, the available roof area can be reduced and shaded by these fixtures and structures.

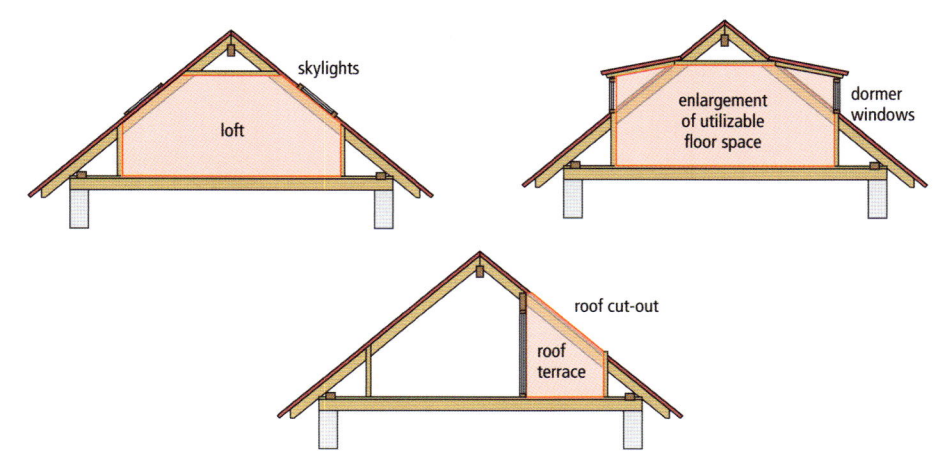

Figure 6.21
Use of the loft space as a living area

6.2.6 Flat roof

The definition of what roof slope constitutes a flat roof is somewhat flexible. It is usually defined at between 5° and 11°. In structural terms, pitched roofs have an overlapping roof covering, whereas a flat roof has roof sealing in place of the covering. This is the only drainage level. Roof drainage is generally solved via a roof drain, which discharges via a downpipe inside the building or via weep holes at the edge of the roof.

Depending on the roof structure, there is a distinction between the double-skin ventilated roof, also known as a 'cold roof', and the single-skin unventilated roof, known as a 'warm roof'. The warm roof is the most common roofing system for flat roofs.

THE VENTILATED ROOF: COLD ROOF

There is constant ventilation between the heat insulation and roof skin (roof sealing). Firstly, it is to avoid dew formation in the roof structure and moisture damage that would occur as a result. Secondly, the idea is to avoid transferring thermal deformations of the roof skin as the sun shines on it to the layers below. The ventilation zone must be at least 150mm high (pay attention to swelling of insulating material resulting from increased humidity) as the ventilation, unlike with pitched roofs, is not supported by thermal gradients and has to rely solely on the wind.

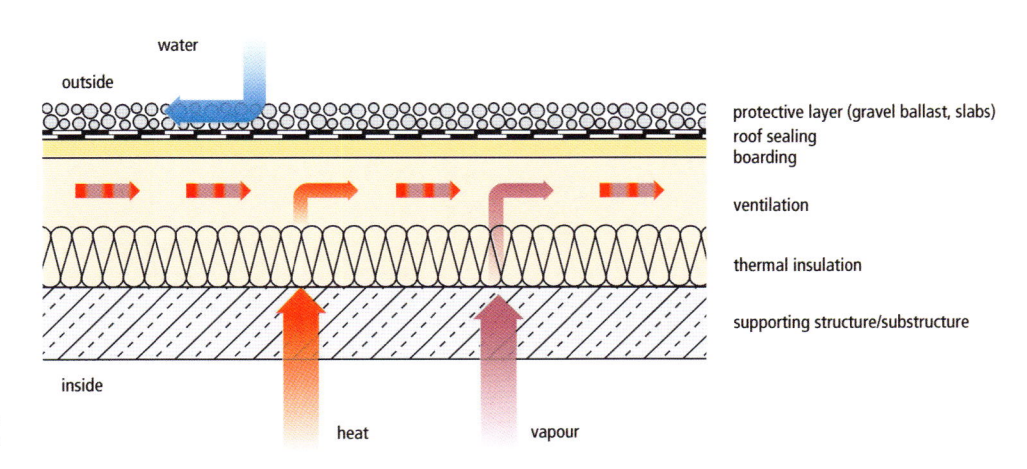

Figure 6.22
Cold roof

THE UNVENTILATED ROOF: WARM ROOF

Here the ventilation is omitted in favour of a simpler roof structure and a lower overall construction height. The heat insulation is located directly between the supporting structure and the roof skin. This requires the installation of a vapour barrier above the supporting layer in order to avoid wetting of the roof as the result of condensation in the area of the insulating layer.

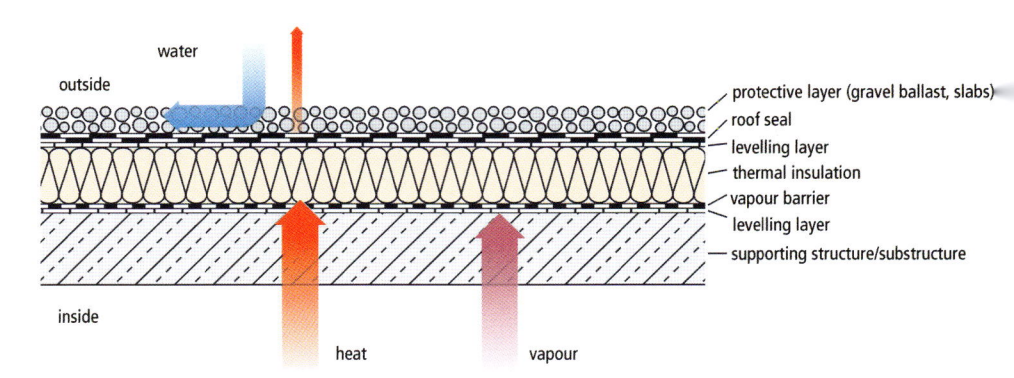

Figure 6.23 Warm roof

THE REVERSE ROOF

With a warm roof, the roof skin is subjected to heavy stresses resulting from rapid changes in temperature, frost–dew alternation and intensive solar radiation. These lead to the ageing phenomena and eventually to damage. In the face of these problems, a special form of warm roof was developed as the reverse roof, which is used increasingly often. In a reverse roof, the arrangement of the thermal insulation and sealing layer is reversed. The roof skin now lies protected from large temperature fluctuations and environmental influences below the thermal insulation. A waterproof insulating layer over the thermal insulation improves the heat insulation of the reverse roof. It prevents rainwater flowing on the roof skin and conducting heat away.

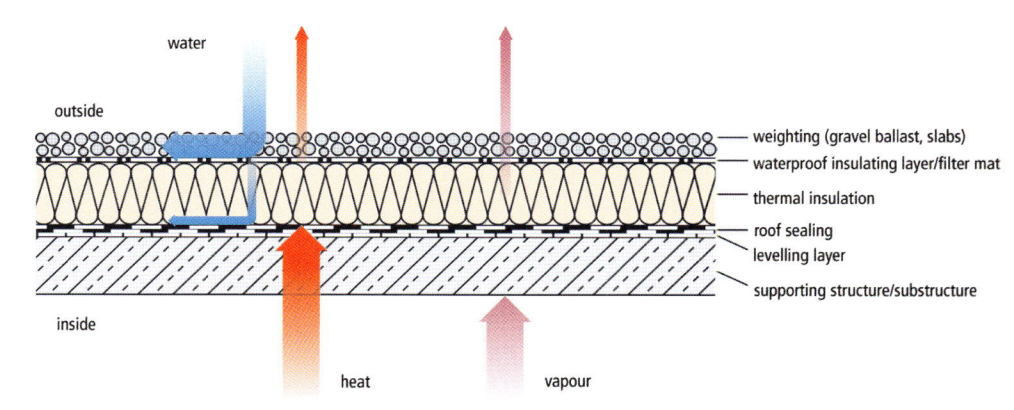

Figure 6.24 Reverse roof

However, it must be ensured that the thermal insulation is waterproof, frost-proof and dimensionally stable, and that it is resistant to rotting, can be walked on, and is accurate to dimension. The insulating effect must be maintained even with higher moisture content. Only rigid foam slabs made from extruded polystyrene meet these requirements, and are usually laid loose. They need to be weighed down with an extra load layer, such as gravel ballast, with a slab covering, or, indeed, photovoltaic modules.

6.3 Sloping roofs

While flat roofs leave the planner of a PV system a certain amount of freedom, sloping roofs dictate the orientation and tilt for the modules. For this reason the roof's suitability must be checked before planning begins (see Chapter 3).

6.3.1 On-roof systems

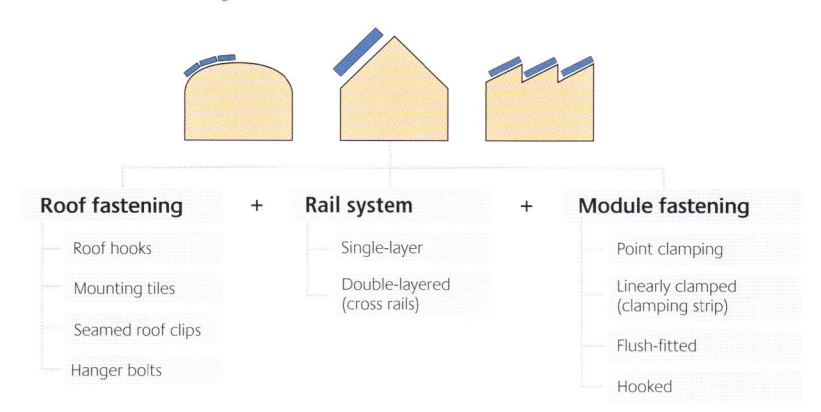

Figure 6.25
On-roof system mounting
Source: B. Batran/Solarpraxis

Roof fastening	+	Rail system	+	Module fastening
Roof hooks		Single-layer		Point clamping
Mounting tiles		Double-layered (cross rails)		Linearly clamped (clamping strip)
Seamed roof clips				Flush-fitted
Hanger bolts				Hooked

In on-roof mounting, the modules are fitted above the existing roof covering using a metal substructure. The roof covering is retained and continues to perform its waterproofing function. For fitting new arrays to existing roofs, on-roof mounting is usually the most cost-effective option since the mounting and material expenses are low. However, the disadvantage, apart from the aesthetic aspect, is that all components, including the fittings, electrical connections and cables, are exposed to the elements.

So that the system should integrate as harmoniously as possible within the existing roof environment, the modules should be arranged so that a self-contained PV array surface is created. Arrangements scattered across the roof or with a step-like appearance can appear clumsy. For complicated roof shapes or where there is shading that needs to be avoided, the most contiguous or the same types of roof surfaces should be chosen for the PV array.

The metal structure to support the modules consists of three main components: the roof mounts, the mounting rails and the module fixings. Using the roof mounts, a rail system is anchored to the roof structure beneath the roof covering or is fixed directly to the roof cover itself (but only if the roof covering is structurally strong enough). The modules are fixed to the rails with system-specific fixing elements. A detailed description is provided according to the information on structural requirements.

The substructure must be able to withstand the forces that occur on the PV array and transfer them to the roof structure without, at the same time, supporting itself on the roof covering. Apart from the thermal load at the height of summer, the modules are mainly exposed to mechanical stresses.

Figure 6.26
Rack-mounted photovoltaic array on the federal office of Bündnis 90/Die Grünen (Green Party) in Berlin, Germany

STABILITY AND STRUCTURAL REQUIREMENTS

PV arrays are considered to be building structures, and national building codes and regulations must be referred to and followed. The mounting system must be designed so that the expected loads at the specific location do not cause the PV array to lift up, overturn, slip down, etc. That means, for example, that the number of roof mounts and the depth of the support rails and fixing elements must be determined in accordance with the location, the roof and building geometry, as well as the module arrangement. When fixing the mounting frame with edge clamps, profiled metal decking clamps, etc., it is important to ascertain whether the roof structure and roof covering can support the additional loads. See also European Standard EN 1991, which is also described as Eurocode 1 (EC1).

LOADS

Push factors and pull factors are exerted on the modules. The loads result from the snow load, the impact pressure of the wind and the individual weight of the modules and the substructure. Pull factors result from the pulling effect of the wind. On-roof structures can work like the aerofoils of airplanes.

In order to minimize the forces exerted on the PV array, the following should be taken into account during planning:

- The gap between the module surface and the roof covering should not be too large. On the other hand, it must be big enough to ensure sufficient ventilation without trapping leaves, which would prevent the drainage of rain.
- Modules should not extend beyond the vertical and horizontal lines of the building (roof ridge, eaves and gable). The distance of the array from the edge of the roof should be at least five times as great as the height of the array above the roof surface.
- Module surfaces should have the same slope as the roof.
- If the modules are not mounted flush, but with a small gap at the sides, the pressure compensation is easier to achieve. This also avoids whistling sounds caused by the wind.

STRUCTURAL SPECIFICATIONS

Structural specifications verify that all components of a mounting system are suited for specific module weights, sizes and layouts, with various snow and wind loads in accordance with applicable standards. Many assembly instructions include tables based on structural specifications from which it is possible to derive the minimum number of roof fixings per square metre, the maximum span (= spacing between supports) and the maximum length of cantilever of the sections, etc. It should be noted that the structural tables often only apply for the central areas and do not provide values for the edge and corner areas. The structural specifications do not encompass the roof structure or the module frames. These must be taken into account on the building site and when using the assembly instructions provided by the module manufacturers.

CORROSION

Since photovoltaic systems are designed to last for a period of 20 years or more (and because, in the case of rack-mounted systems, all mechanical fittings are exposed to the elements) it should be ensured that only high-quality metal is used at the fixing points. For any type of metal (e.g. aluminium or V2A stainless steel), there are substantial differences in terms of quality depending upon chemical composition. Therefore, particular attention should be paid to the respective alloy or material number. Combinations of metals may only be used if there is no risk of electrochemical reactions taking place. If necessary, fixing points where there are different types of metals must be protected from moisture. Under some circumstances, insulation between metals with a high potential difference should be considered.

Other existing components (e.g. gutters, masonry covering, etc.) must also be taken into account in order to rule out any electrolytic corrosion. If the supporting structure is made at the installation site, good corrosion protection should be ensured. Hot-dipped galvanized material should not be drilled or shortened as the cold galvanizing subsequently applied to un-galvanized places is not as durable as the hot-dipped galvanizing.

In addition to technical corrosion protection, attention should also be paid to structural corrosion protection. The mounting system should be constructed so that no corners and nooks are created where dirt, leaves, needles or other deposits can accumulate and, thus, create starting points for corrosion. Avoid standing water.

ROOF FASTENING

Fixing points must be created in the roof surface to receive the mounting frames. The choice of mounting depends upon the existing roof covering. There are rafter-dependent and rafter-independent solutions. Rafter-independent solutions are fixed to the roof battens and offer more scope for positioning on the roof; but structurally they cannot take such high loads as rafter-dependent solutions.

ROOF HOOKS FOR TILES, CONCRETE ROOF TILES, PLAIN TILES OR SLATES

This is a type of hook design that is passed through the roof covering and is screwed tightly to the rafter (see Figure 6.29). The roof hooks are positioned so that their flanges sit within the hollow part of the respective roof tile below and are spaced 5mm away from the surface and face. They may not be laid or pressed against the roof tiles. If necessary, the roof hook must be shimmed with wood. A cut-out must be made in the tile covering the roof hook at the point where the roof hook comes through, using an angle grinder or hammer. With plain tile roofs, either a cut-out must be made in the tile covering each roof hook or, alternatively, the tile must be replaced with a half-sized or one third-sized roof tile. In each cut-out area, a titanium zinc sheet is placed beneath the roof hook, this sheet being fitted so that it has at least a 2cm overlap on each side. Alternatively, a special metal sheet can be inserted below and above the roof hook (see Figure 6.30).

Figure 6.27
Roof hooks for various roof coverings (from top to bottom): Standard pantiles, plain tiles, slates
Source: Solarzentrum Allgäu

Figure 6.28
Curved roof hook that does not press against the tiles even with high snow loads since the bow can only be deformed in an upward direction when exposed to loads
Source: Solarzentrum Allgäu

Figure 6.29
Adjustable roof hooks to compensate for considerable irregularities in the roof surface and variability in the thickness of battens
Source: Schüco

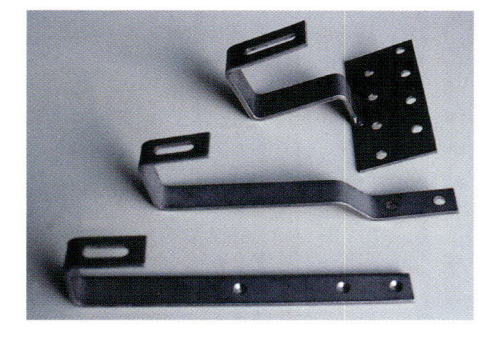

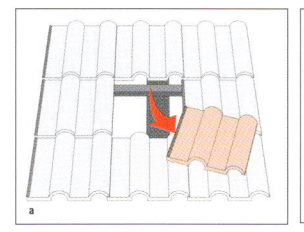

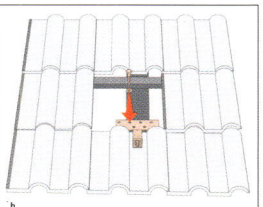

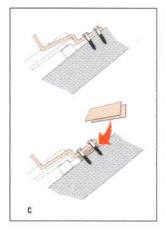

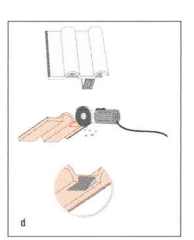

Figure 6.30
Installing a roof hook
Source: Solarpraxis

Figure 6.31
Roof hook for plain tiles with metal plate
Source: MHH Solartechnik GmbH

Figure 6.32
Rafter-independent roof hook for installation in additional, wider mounting battens
Source: Schüco

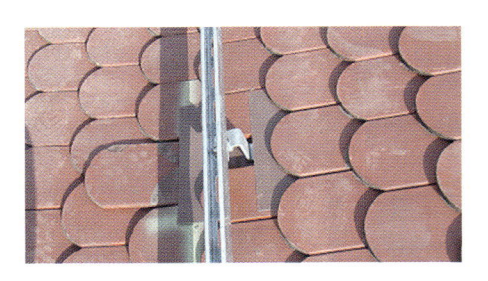

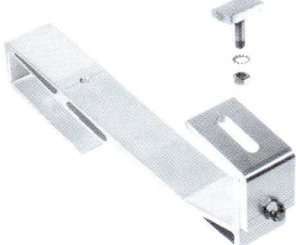

Figure 6.33
Roof hook add-on kit for sloping roofs with insulation above rafters
Source: Schletter

MOUNTING TILES FOR TILES, CONCRETE ROOF TILES OR PLAIN TILES

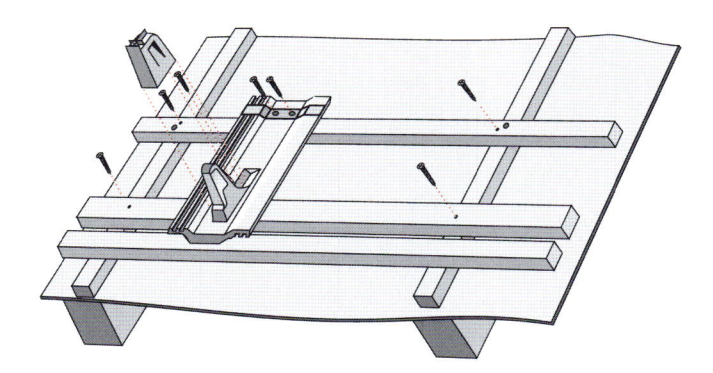

Figure 6.34
Installing a mounting tile

Various manufacturers offer special tiles to which photovoltaic arrays can be fixed. These are modified tiles made from plastic or sheet metal, which were originally intended to take snow trap grids or roof step grates. These are installed independently of the rafters in the roof covering and are additionally screwed to, or wedged between, the battens. Mounting tiles are only suitable for use in standard roof coverings that have the corresponding design and dimensions. These replace conventional roof tiles. This means that it is not necessary to trim the roof tiles to allow for the roof hooks. The tiles remain rain proof and stable, and there is no risk of damage due to roof hooks lying on the tiles. In structural terms, however, mounting tiles are unable to support high loads. For plain tiled roofs, metal roof panels with roof hooks are available, which are screwed to the timber formwork.

Figure 6.35
Mounting tile made of plastic
Source: Klöber

Figure 6.36
Mounting tile made of aluminium
Source: Phönix Sonnenstrom AG

Figure 6.37
Roof hook kit for plain tiles with stainless steel roof plate
Source: Phönix Sonnenstrom AG

Figure 6.38
Quickstocc profiled metal roof tiles are anchored between two roof battens with a wedging device
Source: Solarstocc AG

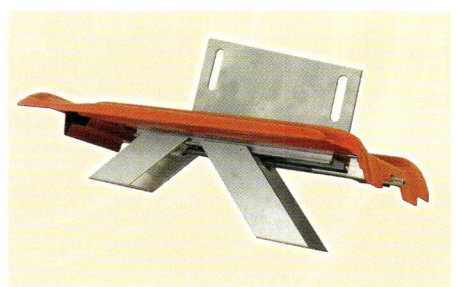

EDGE CLAMPS AND CLIPS FOR METAL ROOFS

If self-supporting standing seam, Kalzip or trapezoidal sheet roofs are sufficiently load bearing and can resist the effect of wind suction forces, mounting frames can be fixed directly to the metal sections. Suitable edge clamps for standing seam or round seam sections are placed on the sheet seams and clamped tight. Generally, a clamp should be placed on every second standing seam. For trapezoidal sheet roofs, there are special clips available that are fixed with self-drilling screws. The frames can be screwed tight using the holes provided in the clips and clamps. This type of roof mounting cannot take such high structural loads.

Figure 6.39
Standing seam clamp

Figure 6.40
Standing seam roof at the Bavarian Parliament Building in Munich, Germany
Source: Schletter

Figure 6.41
Round-headed clip for Kalzip roofs and façades
Source: HaWi Energietechnik

Figure 6.42
Profiled metal decking clamp with self-drilling screws and seal
Source: Schletter

HANGER BOLTS FOR ETERNITE CORRUGATED ROOFING AND TRAPEZOIDAL SHEET ROOFS

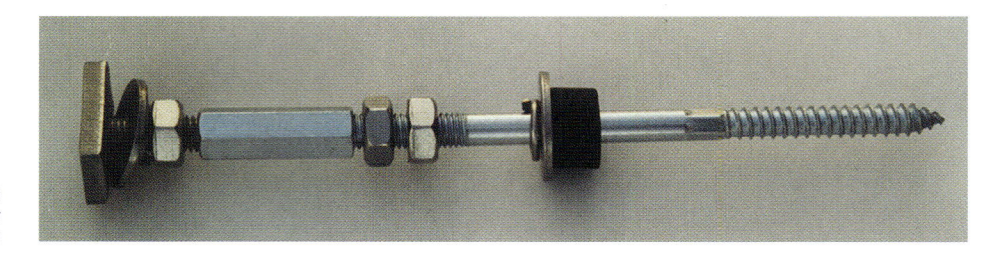

Figure 6.43
Corrugated cement-asbestos roof bolt
Source: Phönix Sonnenstrom

This roof fixture, specially developed for wave or trapezium corrugated sheets and made from stainless steel, is suitable for numerous sectioned roof types. The roof covering is drilled through at the roof fixing points and the hanger bolt is screwed into the rafter. The lower nut presses a sealing washer against the roof covering and seals the hole. The distance of the support rails to the roof is set using two counter nuts, which hold a mounting plate or mounting angle for fixing the frame. In structural terms, this kind of roof fixing can take substantially higher loads than edge clamps or clips. When working on asbestos, national regulations need to be followed. For example, in Germany, asbestos-based fibre cement panels may only be drilled through or processed by trained specialists upon being granted individual permits.

Figure 6.44
Mounted hanger bolt with connecting plate for fixing the rails
Source: IBC Solar AG

Figure 6.45
Fixing solution for steel purlins: The pressure plate is screwed into the purlin using a spacer sleeve and adhesive sealing tape
Source: Schletter

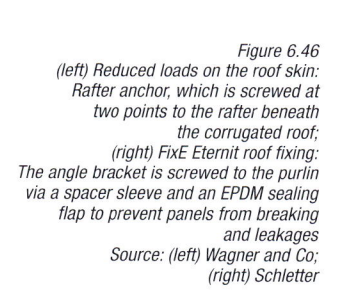

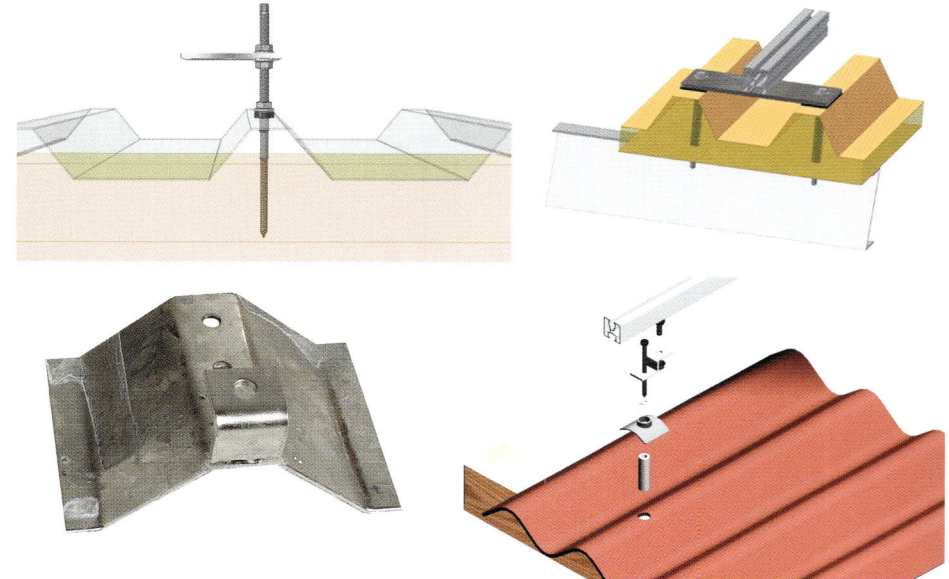

Figure 6.46
(left) Reduced loads on the roof skin: Rafter anchor, which is screwed at two points to the rafter beneath the corrugated roof;
(right) FixE Eternit roof fixing: The angle bracket is screwed to the purlin via a spacer sleeve and an EPDM sealing flap to prevent panels from breaking and leakages
Source: (left) Wagner and Co; (right) Schletter

PENETRATING METAL ROOFS

Figure 6.47
Mounting support that penetrates the metal roof: A soldered-on shoe ensures that the roof remains sealed

If, for structural reasons, it is necessary to fix the PV array to the roof supporting structure in the case of metal roofs, the penetration point must be sealed in accordance with the engineering directives for sheet roof coverings.

RAIL SYSTEM

Rails, mounted on the roof fixing points (roof hooks, mounting tiles, etc.), support the solar modules. Depending upon the system, the module support rails lie directly on the roof hooks or crosswise on a second layer of rails. In order to obtain a flat array surface, existing unevenness in the roof must be smoothed by the sloping roof base frame. It should, therefore, be ascertained before fitting whether it is necessary to adjust the height (depending upon the mounting system, this can be achieved using either adjustable roof hooks or shimming with washers or spacers) or whether cross-rail mounting would be the sensible solution. The sloping roof base frame should enable easy removal of individual modules since repair to the roof below the module may be required or a defective module may need to be replaced.

In the simplest and most commonly used mounting solution, each row of modules is laid vertically on two parallel horizontal support rails and the modules are mostly clamped at four points (e.g. at the mounting holes for the module frames). The distance between the rails depends upon the possible roof fixing points (e.g. upon the spacing between the rows of roof tiles and the specifications of the module manufacturer regarding areas where the modules may be fixed).

Figure 6.48
Mounting modules vertically on horizontal support rails
Source: Solarpraxis

If the modules are mounted horizontally, or if the roof substructure runs horizontally, it is more convenient to mount the module support rails vertically on the roof fixings. In many shading situations, it is advantageous to mount the modules horizontally (see Chapter 3).

Figure 6.49
Mounting modules horizontally on vertical support rails
Source: Solarpraxis

If the substructure does not provide any suitable fixing points at the required rail distances, or if the roof surface is very uneven, using a second rail system mounted at right angles is recommended. This so-called cross-rail mounting makes it easier to create a plane array surface area. Roof fixing points can sometimes be dispensed with since their spacing is independent of the modules sizes, which means that the structurally maximum-permitted distances can be exploited. Cross-rail mounting generally involves greater material expenditure.

Figure 6.50
Module assembly on cross rails: For horizontal modules, or for better alignment with the roof structure, the rail and module layout can be turned by 90°
Source: Solarpraxis

Group assembly is particularly sensible with large-scale arrays. Here, several solar modules are pre-mounted on rails and cabled. This work can be carried out on the ground. Using a crane or inclined hoist, module groups prepared in this way can then be hoisted on the roof and mounted on the lower rail system. Group mounting requires a cross-rail system.

Figure 6.51
Pre-mounted module group for assembly by crane: The rail and module layout can be turned by 90°
Source: Solarpraxis

For mounting systems where the modules are inserted in supporting sections or are clamped linearly, it is also necessary to use a double-layered rail system since the distance between the supporting sections must precisely correspond to the module length. The additional material expenditure reduces in proportion to the number of rows.

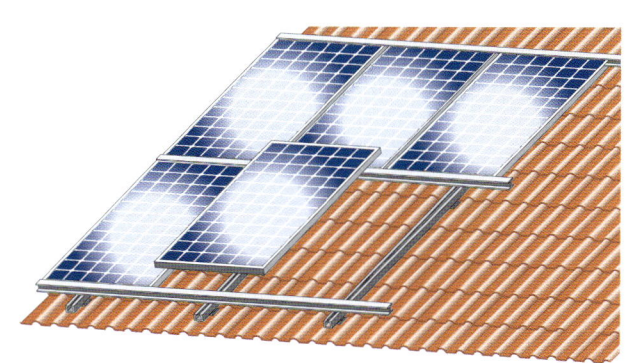

Figure 6.52
Linear module fixing on a cross-rail system
Source: Solarpraxis

MODULE FIXING

Point clamping

The double-sided centre clamps between two modules and the one-sided end clamps on the external modules on the ends of a row are usually fixed with sliding blocks, T-head bolts or threaded plates, which grip the grooves in the support rails. The screw length, or ideally the clamp height, is selected in accordance with the module frame height.

Figure 6.53
End and centre clamps
Source: (above) Solarzentrum Allgäu;
(below) MHH

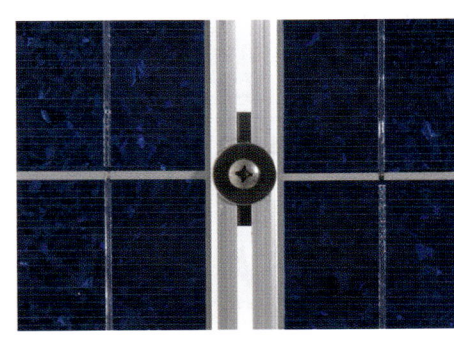

Figure 6.54
Centre clamps
Source: Solar-Fabrik, Solarworld

Figure 6.55
Laminate clamps with EPDM inlays
for frameless modules

Often, an anti-slip device is mounted (e.g. a stop bracket or simple screw in the mounting holes of the module frame) so that the modules do not slip from the rails when being mounted or replaced. This makes it easier to align the modules when mounting since they can be loosely laid on the rails and then shifted into position.

Linear clamping

Instead of point clamps, here modules are clamped with continuous clamping strips. Using the manufacturer's assembly instructions, it is important to check whether modules may be fixed outside the normally envisaged mounting areas. Advantages include attractive appearance and simple mounting, as the modules can be easily inserted in the sections without slipping and do not have to be individually aligned and fixed. However, the substructure is more elaborate as two rail systems are required.

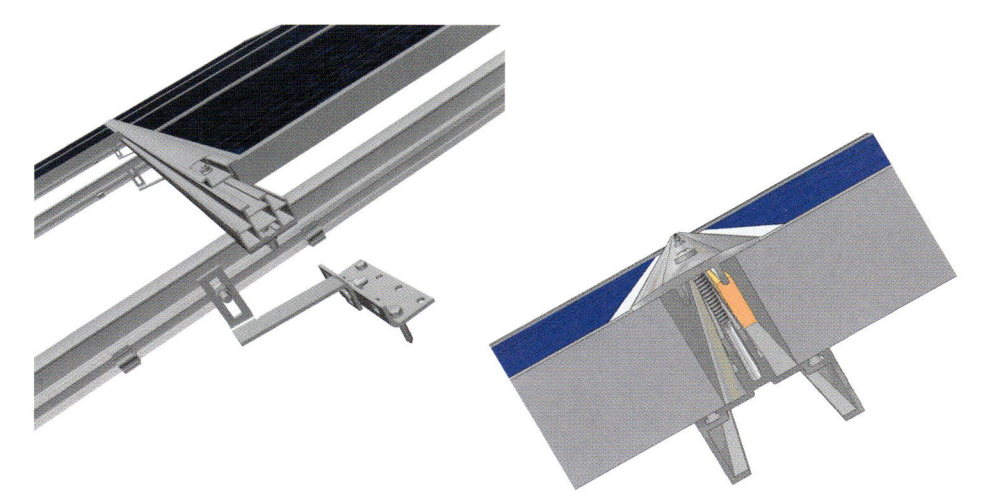

*Figure 6.56
Clamping strip mounting with clipped-on covering profile
Source: Energiebau Solarstromsysteme*

*Figure 6.57
Module mounting with clamping strip: The plastic resin acts as a spacer
Source: IBC SOLAR AG*

FLUSH-FITTED SYSTEMS

Flush-fitted systems have similar advantages and disadvantages as clamping strip mounting. Here, the modules are inserted in the support sections without using clamps and screws. Their own weight and friction prevents them from coming under tension. This means that no tools are needed to install the modules and they are easy to replace. Since there are no cross rails beneath the modules to prevent convection (they lie at the same level), the modules are well ventilated. However, there is risk of dirt collecting or damage due to frost if water is unable to drain off (e.g. via water drain holes or other structural measures). The edges of modules must be certified for linearly supported installation.

*Figure 6.58
Using a template to lay the horizontal sections at a distance equal to the module heights
Source: SolarMarkt*

*Figure 6.59
Insertion of the modules without tools
Source: Solarstocc*

HANGER SYSTEMS

With hanger systems, the modules are slotted to the substructure with pre-mounted retaining clamps. Even with severe temperature fluctuations, the modules remain tension free and are not subjected to any mechanical loads. However, because of the greater expenditure, suspension systems are seldom used.

THEFT PROTECTION

For screwed module fixings, various safety components are used to prevent PV modules from being stolen. For example, safety screws are utilized that can only be screwed tight and unscrewed with a special tool, or standard socket head screws with a safety device that prevents them from being loosened.

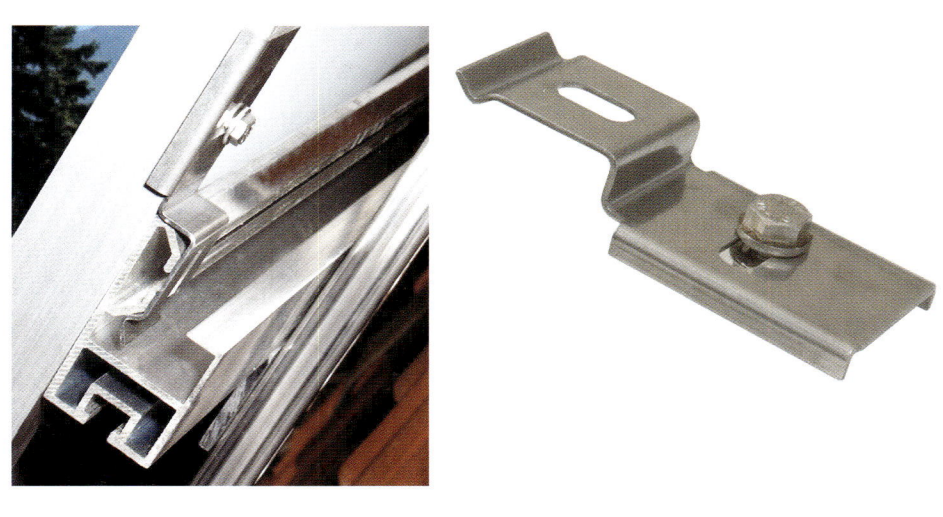

Figure 6.60
The retaining clamp screwed to the module frame engages in the supporting section
Source: HaWi Energietechnik

Figure 6.61
Theft protection by using safety screws with special hexagon slot, unscrewing the nuts and using special keys and bits
Source: Altec Solartechnik

Figure 6.62
'SecuFix' theft prevention: Stainless steel ball driven in to secure the module
Source: Schletter

Figure 6.63
Cutting and unscrewing the screw to release the module (e.g. with defective modules)
Source: Schletter

ON-ROOF MOUNTING SYSTEMS FOR SLOPING ROOFS

There are numerous sloping roof frames available on the market. If work on roofs used to involve a lot of trouble and time, new systems are showing a development towards material savings and simplified mounting. Supporting rails used to include

Figure 6.64
PV system mounted on roof
Source: Schüco

simple square-, U-, C- and L-shaped sections (e.g. Halfen rails or more elaborate rail systems with special sections). The appropriate bolts, clip holders, clamps, clamping strips, hooks and clips for securing the modules are ready supplied. Variable systems can be used universally and can take various different modules (type and dimensions). Standard materials are aluminium and stainless steel, and occasionally galvanized steel. Because of the risk of contact corrosion in direct contact with the aluminium frame, the use of steel in aggressive air (industrial air or close to the sea), even if it is galvanized, is not recommended.

Figure 6.65
Barrel roof with roof-mounted PV system
Source: SunTechnics

For on-roof mounting, frameless and framed glass–film modules or glass–glass modules can be used. Generally they are laid end to end; but there are systems in which frameless modules overlap from the top to the bottom like tiles. Whether the PV array can be walked on for maintenance and repair depends upon the modules used (see manufacturer's specifications). In all cases, it is important to ensure that no stones or metal chippings are trapped in the soles of shoes in order to avoid scratching the module surface.

MANUFACTURERS OF ON-ROOF MOUNTING SYSTEMS

Alfasolar, Altec Solartechnik, Alustand (Switzerland), Beck Solartechnik, Biohaus, Conergy, Corus Bausysteme, Creaton, Deger Energie, Donauer, Ecosolar, Edelstahl Büchele, Elektro Spiegler, Energiebau Solarstromsysteme, EWS, HaWi Energietechnik, IBC, Krauss, MHH, MP-Tec, Modersohn, NAPS, natürlich Zenkel & Lauterbach, Nelskamp, Osmer, Phönix Sonnenstrom, Ralos, RegTec, Ricom, Schletter, Schüco, SE-Consulting, Siblik, Solara, SolarMarkt, Solarstocc, Solarworld, Soltech, Solvis, Sotec, Sunset, Temtec, Total Energie, Ubbink Econergy, VM Edelstahltechnik, Wagner & Co.

6.3.2 In-roof systems

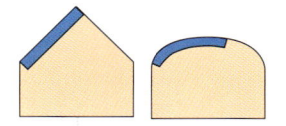

Figure 6.66
In-roof or roof-integrated system mounting
Source: D. Wunderlich

Section systems for standard modules | **Solar roof elements** | **Combined systems photovoltaic/solar thermal**

For in-roof mounting, the modules lie in the plane of, and replace, the conventional roof covering. Entire roof surfaces or only partial areas of the roof can be covered with modules. The PV array takes on a dual function: power generation and weatherproofing. The mounting system therefore needs to be rain proof between the modules and at the edges of the array. As for conventional cold roofs, at least one layer of sarking is required. If laying at less than the specified minimum roof slope, or with increased requirements, the structure is no longer considered to be sufficiently rainproof. This means that additional measures, such as a sub-roof, are required. In order to avoid moisture damage to the roof from condensation forming on the rear of the modules, sufficient ventilation behind the modules must be guaranteed. Although the cold roof construction usually used with sloping roofs generally permits rear ventilation, the ventilation space is generally smaller than with roof-mounted systems.

Figure 6.67
Nieuwland 1MW housing estate in Amersfoort, The Netherlands
Source: REMU

SECTION SYSTEMS FOR STANDARD MODULES

Figure 6.68
Sections for roof-integrated system
Source: Schüco

Section systems for framed and frameless laminates consist of a frame structure that is not self-supporting and that is fixed to the existing roof substructure. This often requires additional roof battens to enable the sections to be mounted independently of the existing battens and the rafter spacing. The modules are placed in this frame structure and secured either at points or linearly. Weather protection is achieved by overlapping the laminates in clapboard fashion by using rubber seals between the laminates or by drainage channels in the sections below the laminates that collect and drain any water that enters. With some systems, sarking is laid beneath the modules, which are laid with open joints. The modules are ventilated in the same way as conventional roof coverings in the plane of the roof substructure and the channels. For large PV array surface areas, a sufficiently large ventilation cross-section must be ensured in order to guarantee that the air draught will work. The closure at the verge, ridge and eaves, and the connection to the conventional roof covering, are made using special closing strips and connecting sheets. The use of standard modules facilitates rational and cost-effective mounting, particularly with large array surfaces. Special or custom-made modules are used only in individual cases.

MANUFACTURERS OF IN-ROOF SECTION SYSTEMS

Biohaus, Conergy, Ernst Schweizer Metallbau, Heisterkamp, IBC, Osmer Elektrotechnik, Schletter, Schüco, Soltech, Sunset, 3S Swiss Sustainable Systems, Ubbink Econergy, Wagner & Co.

Figure 6.69
Installation example
Source: Schüco

SOLAR ROOF ELEMENTS

SPECIAL MODULES WITH ROOF COVERING PROPERTIES

Solar roof elements, popularly known as solar tiles, are produced in two different versions. These are either special modules that are modified in form and function to be comparable to conventional roof coverings; they have large sizes so that they can replace several roof tiles and panels and the amount of electrical wiring is reduced. A normal roof tile overlaps on two sides with the tiles above and next to it so that rainwater can flow over it. The form of the overlap (groove and slot) ensures that flying snow and heavy rain do not penetrate below the roof covering and that no water runs under the tiles. In modifying the solar modules, an attempt has been made to make use of this principle and special overlapping frames have been developed. These can generally be fitted directly to the existing roof battens. Unlike conventional roof tiles, the relatively large weight of which generally ensures sufficient security in a storm, the lightweight solar modules still need to be mechanically fastened. The elimination of labour-intensive frame systems and the significantly smaller module sizes simplify mounting on the roof. Module outputs are typically around 50Wp.

*Figure 6.70
Nieuwland 1MW housing estate near
Amersfoort, The Netherlands: In the
foreground are houses fitted with the Braas
PV 700 roof system
Source: REMU*

ROOF COVERINGS WITH INTEGRATED PV MODULE

The other version includes roof covering elements that are fitted with a solar module in the factory. The roofing material is used as a mechanical support for the integrated PV module and performs weatherproofing. Impermeability is therefore comparable with conventional roof covering. The modules are attached to the roof covering elements on the rear side of the modules, in most cases by gluing. The absence of structural elements above the PV array plane favours self-cleaning of the modules. It is possible to remove individual elements according to the type of roof covering. The modules often have a very small size, which has the advantage that they can also be fitted on small and complicated roof areas, or on roofs of listed buildings. However, this is associated with higher wiring expense and increased surface area requirements. For large-scale use, metal roof systems with individual outputs of greater than 100Wp are particularly interesting in this area; these entail little additional work and expenses relative to conventional profiled sheets without PV integration.

Module ventilation takes place in the plane of the roof substructure. If the modules are covered with fixing cassettes, frames, etc. on the rear side, additional ventilation channels are important in order to ensure sufficient cooling of the solar cells. If only covering part of a roof with the modules, it should be noted that not every system can be combined with every roof covering and every roof slope. As with conventional roof coverings, it must be ensured that the minimum allowable roof slope is maintained to ensure that the roof remains rain proof. With reduced tilt angles, additional measures must be taken in accordance with roofer regulations.

*Figure 6.71
Kalzip framed panels with integrated
thin-film modules
Source: Corus Bausysteme*

MANUFACTURERS OF SOLAR ROOF ELEMENTS

Atlantis Energy, Corus Bausysteme, Eternit, Hoesch Contecna, ICP, Imerys Dachprodukte, Lafarge, Pfleiderer, Rathscheck, Rheinzink, RWE Schott Solar, SED, Star Unity, United Solar Ovonic.

COMBINED SYSTEMS FOR PHOTOVOLTAICS AND SOLAR THERMAL

The combined systems are often based on proven systems for sky lights. For this reason, window manufacturers are also becoming engaged in this area and offer sky lights, PV modules and thermal collectors with the same dimensions and with the same cover frames. These are mostly complete systems. In addition, modules and collectors for new buildings or complete roof refurbishments can already be integrated into prefabricated roofs in the factory.

Figure 6.72 Roof-integrated PV system on the parsonage, built around 1900, belonging to the Protestant church of Vilmnitz/Rügen, Germany

MANUFACTURERS PRODUCING COMBINED SYSTEMS

Buschbeck Solartechnik, Flumroc, Roto Frank, Schüco, Viessmann, Westfa, w+d Holzsysteme Schuh.

6.4 Flat roofs

6.4.1 On-roof systems for flat roofs

As with roof-mounted systems on pitched roofs, with roof-mounted systems on flat roofs the modules are mounted on a metal framework above the existing roof skin. The modules are generally tilted at a favourable angle using a support frame. PV

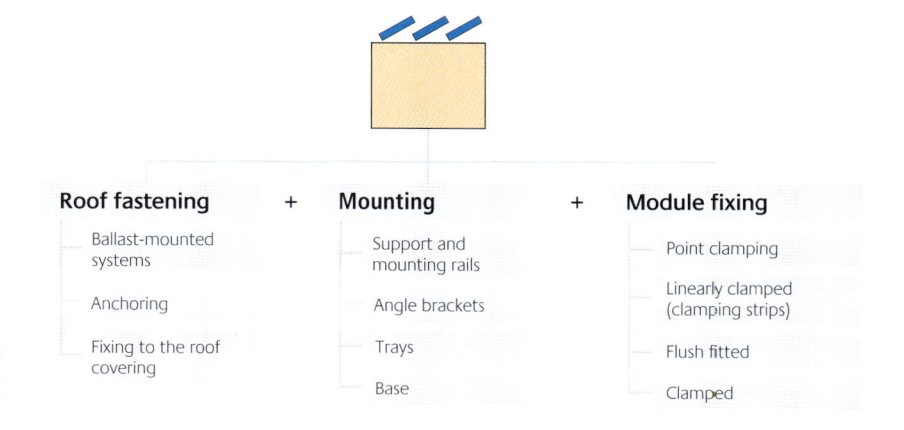

Roof fastening	+	Mounting	+	Module fixing
Ballast-mounted systems		Support and mounting rails		Point clamping
Anchoring		Angle brackets		Linearly clamped (clamping strips)
Fixing to the roof covering		Trays		Flush fitted
		Base		Clamped

Figure 6.73 On-roof system mounting Source: D. Wunderlich

systems generally restrict access to the roof. It should therefore be ensured before installing the PV system that the roof's ability to function will be maintained throughout the service life of the PV array. One advantage, however, is that the module shading reduces the thermal loading on the roof and thus prolongs its ability to function. In terms of the metal components, the same corrosion protection should be ensured as described in the earlier section on 'Corrosion'.

ROOF FASTENING

Considerable importance is attached to the method of securing mounts to flat roofs. Because PV arrays have large exposed areas, considerable wind forces must be taken into account when securing arrays. The choice of fixing depends upon the structure of the roof. Can the roof accept greater loads? This determines whether the system can be free-standing (ballast-mounted systems) or must be fixed to the roof (anchoring). With profiled sheet roofs, the frame is fixed directly to the roof covering.

BALLAST-MOUNTED SYSTEMS (FREE-STANDING INSTALLATIONS)

With ballast-mounted systems, the flat roof mounts are anchored without penetrating the roof. Concrete blocks, slabs or plinths are placed on the flat roof without any further fixing and the support frames are secured to these with screw anchors. For the concrete elements, it is possible to use standard building materials such as curbs, paving slabs or specially made foundation slabs. If necessary, matting should be laid beneath to protect the roof skin from sharp edges. Alternatively, the concrete weights can be inserted in channels on the support frame.

Figure 6.74
Mounting on foundation slabs
Source: Energiebiss

Figure 6.75
Mounting on strip foundations
Source: Energiebau

Figure 6.76
Anchoring with concrete slabs that are inserted in the frame channels
Source: (left) MHH; (right) Osmer

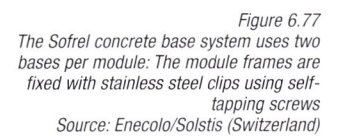

Concrete bases with different flange lengths are also used, which also act as support frames. These include standard L-blocks from local building suppliers or concrete bases that have been specially developed as a PV mounting system. Retaining clamps are required to fix the modules to the bases directly or via support rails.

Figure 6.77
The Sofrel concrete base system uses two bases per module: The module frames are fixed with stainless steel clips using self-tapping screws
Source: Enecolo/Solstis (Switzerland)

With roofs on which concrete slabs or gravel have functioned as loading, it is sometimes possible to use these as ballast. Gravel coverings can weigh 90kg/m^2 and more, and concrete slabs can weigh up to 125kg/m^2. The support frame is then attached directly to the slabs, or the slabs – or gravel – are used as ballast. This method is particularly interesting for tray systems in which trays made of UV-resistant plastic or fibre cement are laid on the roof and filled with gravel or pavement slabs to

Figure 6.78
L-block fixing system
Source: Phönix Sonnenstrom

Figure 6.79
ConSole plastic tray system
Source: Ubbink Econergy

Figure 6.80
Solmax universal tray system with flexible plastic walls for different module sizes: The module frames are secured with POP rivets to the aluminium channel on the tray system
Source: Solstis (Switzerland),

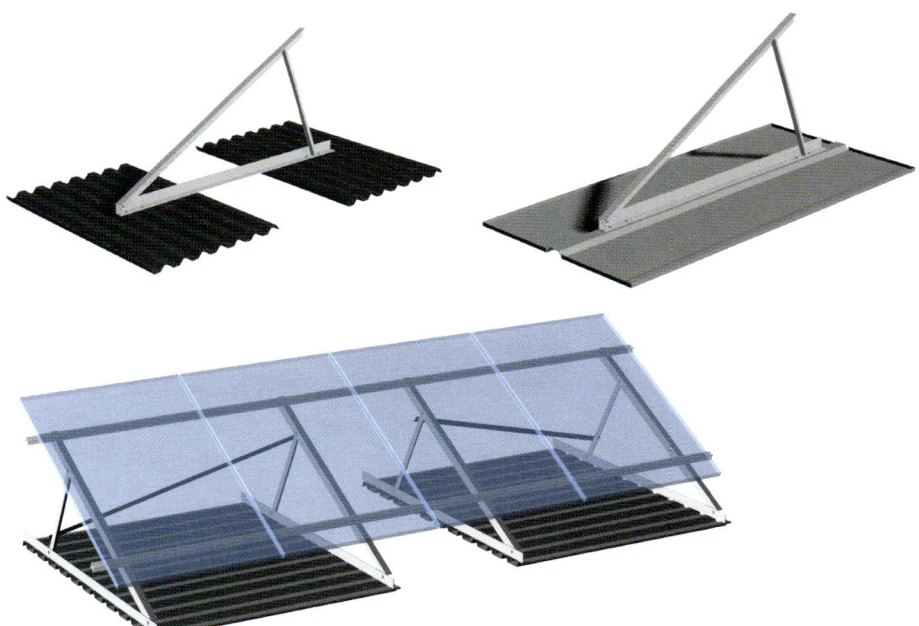

Figure 6.81
Plastic panel, aluminium tray and trapezoidal sheet for gravel ballast
Source: (left) Schletter; (right) Altec

achieve the necessary weight. The modules are fixed on the filled trays either via support rails or directly with the module frames. The advantage of this system is that it is easy to transport the equipment onto the roof because the ballast is already there.

Figure 6.82
Solgreen grass roof system: The plastic panel is covered with the layer of vegetation; the module is fixed to the stainless steel frame, which extends upwards
Source: Enecolo (Switzerland)

This is also the advantage of systems where the frame is anchored to sheets or plastic panels. The panels are laid on the roof skin or root protection layer and covered with gravel or, in the case of grass roofs, with the plant substrate.

If plants are to be placed on the roof, they should grow no higher than the lower edge of the modules. The cooling effect of a grass roof increases efficiency in summer since the surface temperature of a grass roof on a hot summer day can be up to 45°C less than with an unprotected roof surface.

Figure 6.83
Solarbasis SB 200 grass roof system: Plastic panel with water reservoir, fill troughs and canal system on the lower side
Source: ZinCo

The major advantage of ballast-mounted systems is that it is not necessary to penetrate the roof. However, the PV-array and ballast must be heavy enough to ensure that the installation remains firmly fixed even with the maximum expected wind load. The necessary weight depends upon the height of the building, its location and the nature of the substructure (frictional coefficient between the frame and the roof skin). Here, it is absolutely necessary to follow applicable building codes and regulations. As part of the structural specifications for their products, many mounting system suppliers also provide tables from which the necessary ballast can be derived. The ballast elements are, however, not included in the scope of delivery of the mounting systems, but must be obtained locally. It is possible that the necessary weight (guidance value: approximately 100kg/m^2) exceeds the structural loading capacity of the roof. This can be determined from any existing construction plans used for calculating the roof structure. In case of doubt, a structural engineer should be consulted.

ANCHORING (FIXED SYSTEMS)

Figure 6.84
Subsequently sealed anchoring
Source: Osmer

If it is not possible to use ballast-mounted systems for structural reasons, the PV array must be rigidly anchored to the roof construction. Here, the supporting frames are mounted on crossbeams that are secured either to the roof itself or to the roof parapet. Where the roof's waterproofing is penetrated, the anchorage points must be carefully sealed. When designing the layout, the number of penetration points should be reduced to a minimum. When refurbishing flat roofs, the anchoring can be particularly easily realized since the pressure points of the solar substructure can be sealed at the same time.

Figure 6.85
German president's office, Berlin:
The roof anchoring was specially
designed by the architect
Source: Solon

The PV array on the German president's office was intended to be flush with the parapet. The crossbeams desired by the architects for aesthetic reasons led to shading of the adjoining modules. However, by using an optimum module layout, it was possible to reduce the resulting yield losses to less than 10 per cent (see Figure 6.85).

The PV array on the Berliner Bank (see Figure 6.86) rests on base supports that are screwed to the concrete roof and support a structure of standard steel sections. The roof area beneath the modules is extensively greened.

Figure 6.86
Berliner Bank roof anchoring
(custom-made design), Germany
Source: Energiebiss

Although having a very gently sloping roof surface, the PV array on the golf course at Berlin Wannsee (see Figure 6.87) is designed to be easily seen across its length. Increasing the height of the PV array made it necessary to transfer the correspondingly greater loads (e.g. increased wind loads) to the steel structure beneath the roof surface.

Figure 6.87
Berlin-Wannsee Golf Course, Germany:
Special design by Poburski Solartechnik
Source: Poburski

FIXING TO THE ROOF COVERING

Figure 6.88
Supports directly mounted on trapezoidal sheet
Source: Schletter

Figure 6.89
Support frames run parallel to the trapezoidal sheet with a lower rail system for transferring loads
Source: Schletter

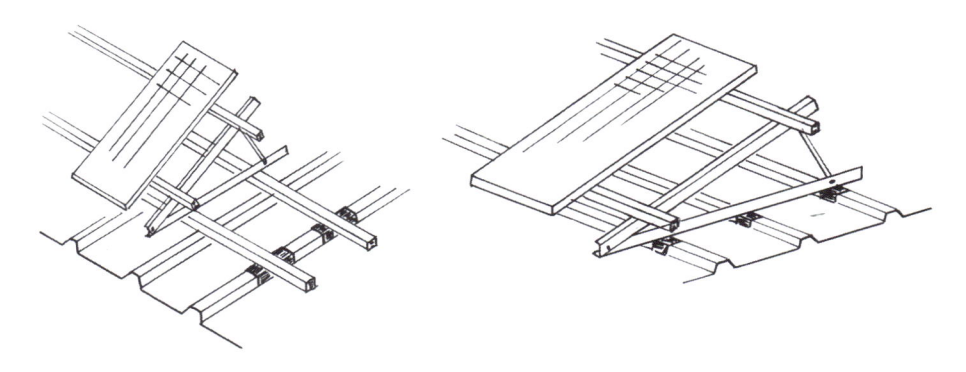

Strictly speaking, metal roofs are not considered to be flat roofs from a structural point of view as the profiled sheets provide a rainproof but not absolutely waterproof roof covering. They are often used, however, with gentle roof slopes. The seams or vertical creases can be employed for directly fixing to the profiled sheets, where the clamps and clips used for mounting on sloping roofs are also used. If the profiled metal decking and its fixings on the roof are able to withstand the additional wind loads caused when modules are racked, the modules can be tilted at the desired angle using a correspondingly larger number of roof fixing points and a flat roof mount.

Figure 6.90
New Trade Fair Centre, Munich, Germany
Source: Shell Solar

MOUNTING

Numerous mounting frames for flat roofs are available on the market. Often, the rails used for sloping roof systems can be deployed with customized support systems. Besides the common fixed support frames, there are also individual systems that can track seasonal solar altitude. There are flat roof mounts with low overall heights that only support lying modules, and flat roof mounts with greater overall heights that raise modules to a steeper angle or support several module rows on top of one another. In some cases, low-level supports use more space but have the advantage that they distribute the structural loads on the roof better. Low-lying supports have the added benefit that the modules cannot be seen from the street. The row spacing on the roof is chosen in accordance with the construction height to prevent module rows from shading one another (see the section on '3.7 Shading with free-standing/rack-mounted PV arrays' in Chapter 3).

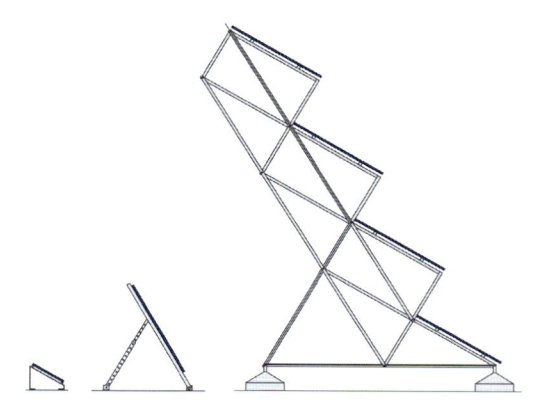

Figure 6.91
Support frames with different construction heights
Source: F. Berger

All kinds of building joints should be kept free of any structures so that no force is applied to them. In snowy regions, the modules should be mounted with sufficient spacing between the modules' lower edges and the roof surface in accordance with the winter snow level. This allows snow to slide off so that the modules are not shaded.

ADDITIONAL WIND LOAD

To minimize the wind load, it is also advisable to ensure that there is sufficient ground clearance so that the wind can flow freely around the module rows. In addition, there should be a minimum distance to the roof edge of 1.2m along the long side of the building and around 1.5m along the building's narrow side.

In aerodynamic terms, tilted modules create a slab-like surface that provides resistance against air currents. The air builds up on the windward side and creates a high-pressure area. On the leeward side, the recirculation current creates low pressure. The pressure difference between the front and rear sides creates a force that the mounting system must be able to withstand.

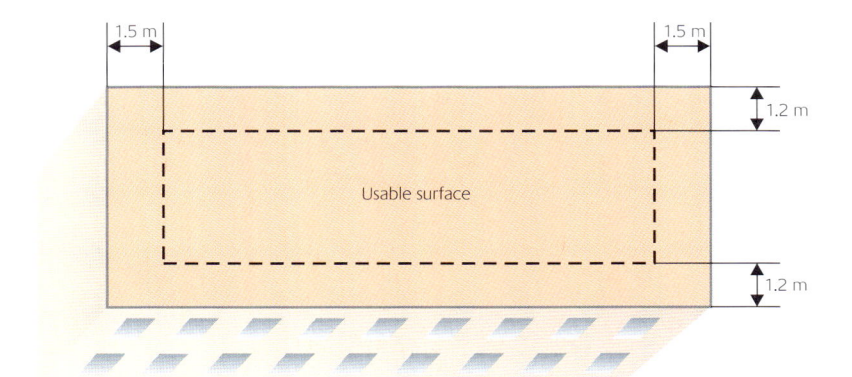

Figure 6.92
Recommended distances from the roof edge for arrays on flat roofs

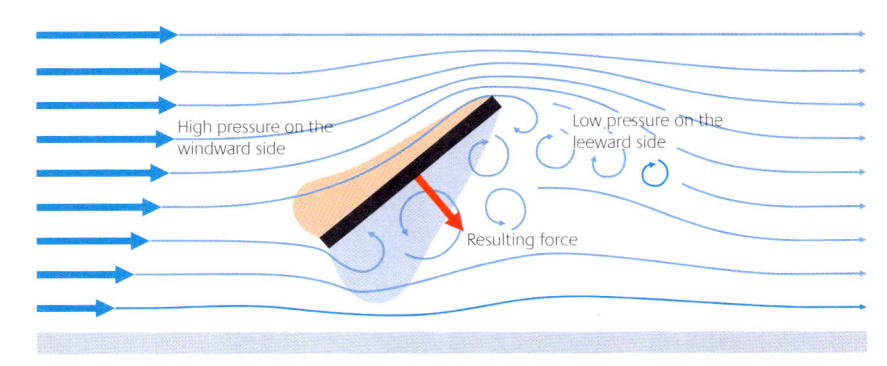

Figure 6.93
Schematic showing the wind load acting on the front of a module

The resistance and shear forces that act tangentially on the modules are much less than the pressure forces. As a result, these generally do not have to be taken into account as long as the surface area over which the air flows is not too great, as is the case with a single-row module arrangement. According to the German DIN 1055-4 standard, the wind load depends upon the shape of the components (= module row) and the wind's velocity pressure. The total wind force that acts on a component is calculated using the formula:

$$F_W = c_f \times q \times A_{ref}$$

Here, c_f refers to the aerodynamic force coefficient for the reference surface A_{ref}, which takes into account the shape of the components, as well as the effects of pressure, suction and resistance. For air flowing at an angle against slab-like surfaces such as solar arrays, DIN 1055-4 does not provide any force coefficients, so that similar load cases must be determined. If the force coefficient is used for components with a rectangular cross section where the air flows perpendicular to the cross-sectional side, this results in a force coefficient $c_f = 1.26$ for modules that are mounted vertically. However, this value does not take into account either the angular flow or the resistance of the building on which the modules are mounted. When using other coefficients based on wind tunnel experiments from recognized sources, the value may be lower.

Erfurth and Partner Beratende Ingenieure GmbH (2001) recommends a force coefficient of +1.2 for pressure and –1.8 for suction when maintaining a distance of 1.5m from the edge of the roof.

MOUNTING SYSTEMS: VARIATIONS

Most flat roof mounts are mounted on triangular supports at the standard module tilt angle of 30°. Here, the triangular supports, consisting of separate rail sections with lengths that vary according to the module size and tilt angle, are assembled on site or are supplied in standard sizes. The triangular supports are fixed to selected roof fastenings and, if required, to one another in order to distribute the loads across the roof surface.

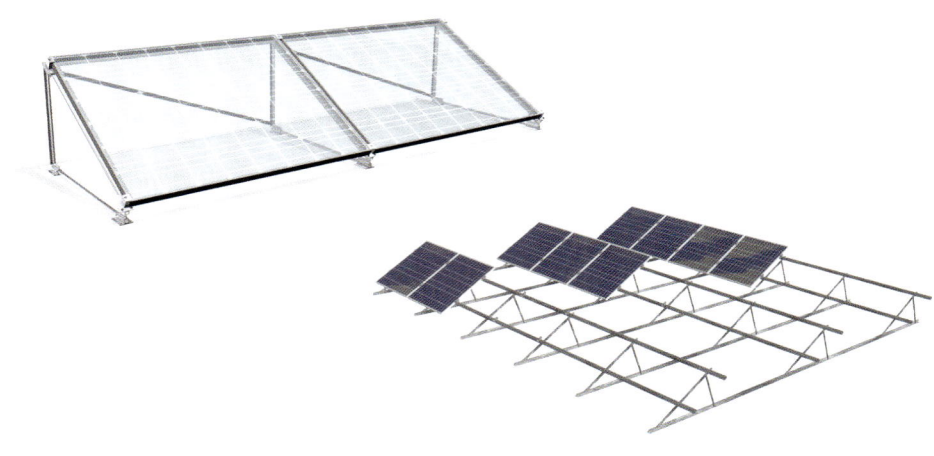

Figure 6.94
Triangular supports made of rail sections
Source: (left) Conergy AG; (right) Schletter

Figure 6.95
Angular and curved supports with a fixed tilt angle
Source: (left) Solar-Fabrik; (right) Donauer

There are also angular supports made of curved tubes or flat material, or bent or cut metal sheeting, with a fixed mounting angle.

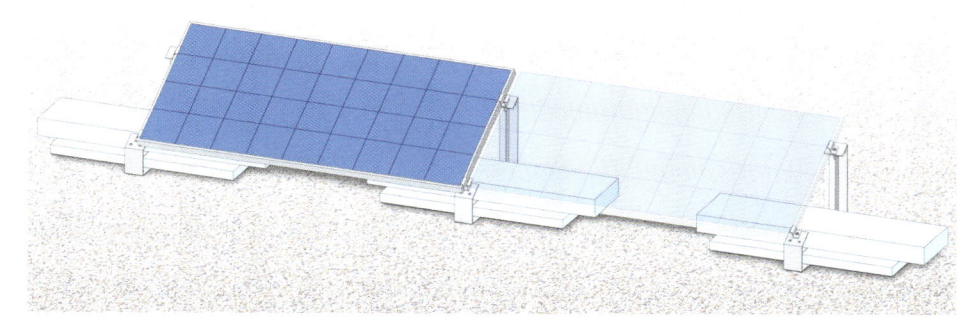

Figure 6.96
Angular supports for directly mounting modules without rail sections
Source: Schüco

Figure 6.97
Mounting with brackets
Source: AluStand

If trays or bases are used to provide ballast, they already provide a specific tilt angle as a result of their shape.

MODULE FIXING

The modules are fixed analogous to pitched roof mounting using module support rails. Here, they are not fixed to roof hooks but to triangular brackets, supports, trays or plinths. In some cases, the modules can be clamped directly to the support frames without using rails.

MANUFACTURERS OF FLAT ROOF MOUNTING SYSTEMS

Alfasolar, Altec Solartechnik, Biohaus, Conergy AG, Donauer, Edelstahl Büchele, Energiebau Solarstromsysteme, EWS, HaWi Energietechnik, IBC, MHH, MP-Tec, natürlich Zenkel & Lauterbach, Osmer, Phönix Sonnenstrom AG, Ralos, RegTec, Schletter, Schoenau, Schüco, SE-Consulting, Solara, Solar-Fabrik, SolarMarkt, Solarstocc, Solarworld, Soltech, Solvis, Sunset, Total Energie, Ubbink Econergy, VM Edelstahltechnik, Wagner & Co, ZinCo.

MOUNTING SOLUTIONS FOR LARGE-SCALE AND LIGHTWEIGHT ROOFS

Figure 6.98
Mounting on existing sarking with the Solrif roof-integrated system
Source: Ernst Schweizer

In terms of optimizing costs, large-scale roofs provide cost-effective sites for PV systems, particularly for large communal and jointly owned solar plants. Factories, schools and administrative buildings, multi-storey car parks, sports and entertainment venues, and (last but not least) agricultural buildings often offer large contiguous surface areas, but not the structural requirements for fixing standard mounting systems for PV modules. They require fixing systems with a low mass per unit area, a low number of anchoring points and, accordingly, wide spacing in order to span the often very large distances between the structural roof elements (rafters, purlins and trusses) and to allow rational mounting on the roof. Here, standard systems for sloping and flat roofs are mostly further elaborated upon (e.g. with stronger support rails or with angular mounts for gently sloping roofs) and are adapted to the respective roof. This means, however, that individual structural tests are indispensable here.

MANUFACTURERS OF SYSTEMS FOR LARGE-SCALE ROOFS

Energiebau Solarstromsysteme, Goldbeck Solar GmbH, Schletter, Solarmarkt, Sunset.

TRACKING SYSTEMS FOR FLAT ROOFS

Tracking flat roof systems range from manually adjusted systems that allow one to achieve a seasonal tilt angle to automatic single- and dual-axis trackers.

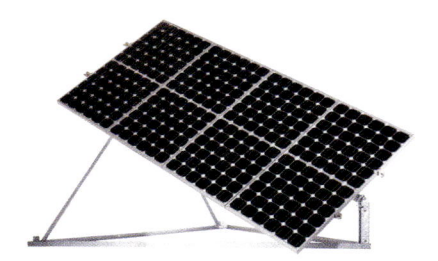

Figure 6.99
Tracking frame with triangular base for simple roof mounting
Source: Deger

Figure 6.100
Sincro Sun System 3 S tracking system
Source: Elettropiemme (Italy)

Figure 6.101
Flat roof support that can be tilted by hand between an angle of 10° and 60°
Source: Schletter

MANUFACTURERS OF TRACKING SYSTEMS

Concept Engineering, C.W. P. GmbH, Deger Energie, Elettropiemme (I), Kroschl Solartechik, Schletter, SPT Solar Power Tower GmbH, Solar-Trak GmbH.

6.4.2 Roof-integrated systems

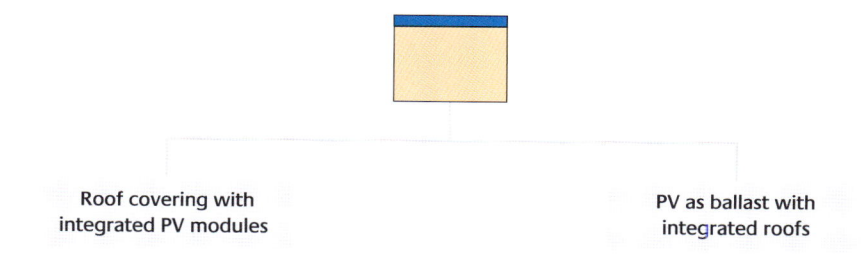

Figure 6.102
In-roof or roof integrated system mounting

- Roof covering with integrated PV modules
- PV as ballast with integrated roofs

A consequence of integrated roof mounting on flat roofs is that the PV arrays usually have very flat tilt angles and higher module temperatures. Compared with the optimum tilt and orientation, this means less solar irradiance and lower yields. On the other hand, thin-film cells can make best use of their advantages (see the section on '2.1.10 Electrical characteristics of thin-film modules' in Chapter 2). Furthermore, the poorer self-cleaning ability can cause the modules to become dirty, which means that it might be necessary to regularly clean them. There is also greater use of the roof surface: horizontal mounting enables greater power outputs to be achieved in Wp per square metre of roof area and the modules can be laid parallel to the roof edges regardless of the orientation of the building. Since it is not necessary to use support frames, costs can also be saved relative to roof-mounted systems.

MANUFACTURERS OF INTEGRATED ROOF SYSTEMS FOR FLAT ROOFS

Alwitra, Powerlight, Solar Integrated Technologies/Sarnafil.

Figure 6.103
Plastic roof membrane with integrated PV modules
Source: Mencke & Tegtmeyer

Figure 6.104
PV integrated roof
Source: Powerlight

6.5 Façade basics

The façade is what provides our initial impression of the building. Enormous attention is paid to providing the façade with a representative external appearance that conveys the style and philosophy of architects and builders. Current tastes, regional traditions and the latest technology are also reflected in the design of façades. Within this context, photovoltaic modules can enrich the design repertoire when treated as façade elements. In modern buildings, transparent glass façades provide a link to the outside world. Innovative solar cells can be simultaneously integrated within the glass sheets used. Thus, the façade element can also become a solar module.

6.5.1 External wall structure

The façade forms either the exterior wall itself as external rain-screen cladding fixed to a structural frame or, as the external skin, the outer component in a load-bearing,

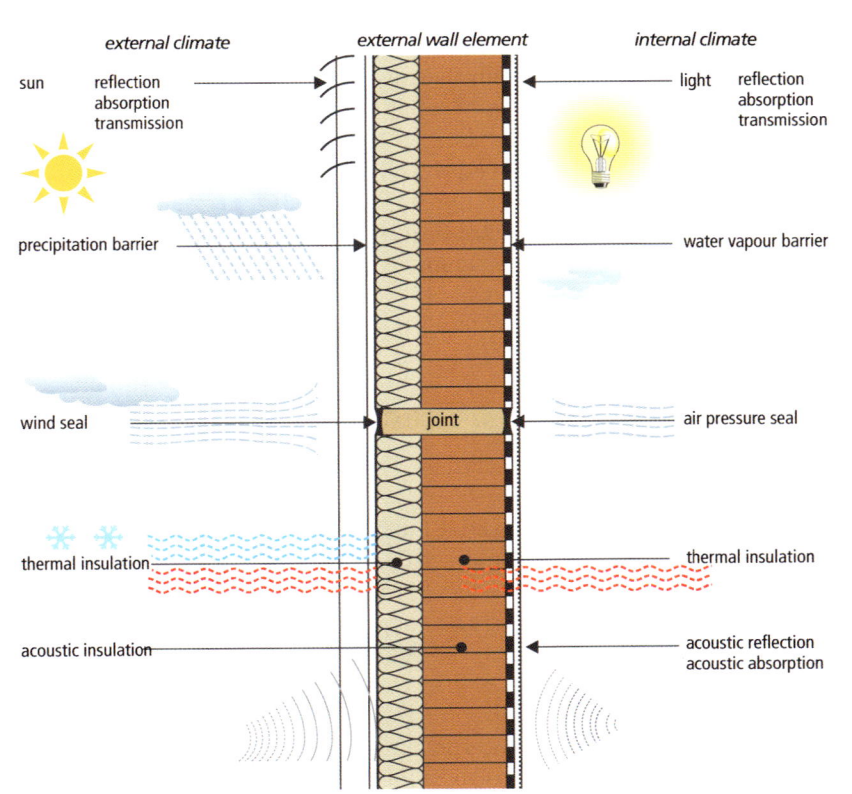

Figure 6.105
Functions of the individual layers in an exterior wall
Source: Solarpraxis

exterior-wall structure. Modern external walls consist of several individual layers that are combined together to fulfil their separate functions. The façade carries out the following functions:

- external delineation of the building and visual protection;
- separates the effects of the external climate from the internal climate (heat, humidity, sound and fire protection, electromagnetic shielding);
- utilization of daylight and solar protection (glare and overheating);
- determination of the building's appearance and its impact on the townscape;
- conversion of thermal and electrical energy.

The façade supports only itself and wind loads; the load-bearing structure (solid walls or framework) supports the entire building loads (roof, floors and dead load). Over the course of time, structures have developed from conventional load-bearing wall construction (single-leaf, non-ventilated construction) to multilayered ventilated construction. Whereas with load-bearing wall construction, the building material used takes on several functions, in multilayered construction the various functions are carried out by individual specific layers.

LOAD-BEARING WALL CONSTRUCTION

Traditionally, in Central European climatic conditions external walls were built of load-bearing masonry, first using stone and, later, brick. Here, mortar is used to hold together and seal the stones. To protect the mortar joints against precipitation, render is often applied. Windows are installed or integrated within the load-bearing exterior walls either as single 'punched' windows or combined to form continuous bands of windows. Besides traditional masonry, these days load-bearing walls are also manufactured from concrete. With the increasing industrialization of building, prefabricated construction methods have prevailed, as is exemplified by system-built housing using large-scale, self-supporting wall and roof panels.

Particularly in damp climates, moisture damage on sides exposed to heavy rain eventually led to the introduction of cavity wall masonry. Here, the penetration of moisture from outside towards the inside is interrupted by an air gap between the outer and inner skin. At the same time, the ventilation facilitates drying out. It is also no longer necessary to have external render. Although cavity wall construction is more complicated, it provides improved heating, humidity and sound protection. Additional heating insulation in the air gap (not to be completely filled!) further increases the construction quality.

The same effect can be achieved with different types of cladding as the external leaf. Slate, tiles and timber cladding were used originally; these days, stone or plastic panels, fibre boards, metal sheeting or coloured glass sheets are used, as well as (ever more frequently) PV modules. This reduces the wall depth compared with solid external leafs. In addition, weathered or outdated cladding can be easily replaced. In structural terms this is a single-leaf solid wall with ventilated cladding known as a rain screen (Haferland, 1987).

FRAMED STRUCTURES

The use of framed structures has long been practised in timber construction. High-rise buildings are also constructed using framed structures, as are many industrial and administrative buildings. Instead of solid walls, framed structures use a skeleton of steel, reinforced concrete or timber to carry the building loads. The frame consists of columns, beams, trusses and the roof elements. Here, the internal spaces are enclosed by non-structural façades that act as an envelope surrounding the shell construction and, as an enclosed skin, protect the building from the effects of weather and other external influences. The lightweight wall elements, also known as façade elements, only support their own dead weight while resisting wind loads, preventing water penetration and providing heat insulation. Glazing, individual windows and bands of windows can be integrated within the elements. This type of construction is known as a curtain wall. Compared with load-bearing wall construction, this results in a much reduced wall thickness and construction weight for the same heat insulation effect and bearing capacity. It is the load-bearing capacity of steel and concrete that makes it at all possible to build skyscrapers.

The considerable degree of prefabrication provides numerous advantages. Thus, both ventilated rain-screen façades and framed structures with curtain walling offer quick weather-independent, cost-effective and precise construction and open up diverse design possibilities. Modular construction principles allow different construction elements and materials to be combined with one another.

6.5.2 Façade types

COLD FAÇADES

Cold façades are cavity wall structures. The outer leaf, consisting of cladding or a masonry facing, provides the weather protection and gives the building its external architectural appearance. The load-bearing exterior wall situated behind provides the structural support and is thermally insulated. Between both leaves is an air gap that can disperse moisture and water vapour. All parts of the outer façade construction are built without any thermal insulation as there is no connection to the building's warm areas. In structural terms, these are mostly ventilated rain-screen façades.

The rear ventilation means that the cold façade is very good for integrating PV elements. Here, opaque laminates or glass–glass modules are used, whose junction boxes are situated on the rear side of the modules where they cannot be seen. The wiring is routed in cable ducts that are fixed to the load-bearing wall in the proximity of the thermal insulation.

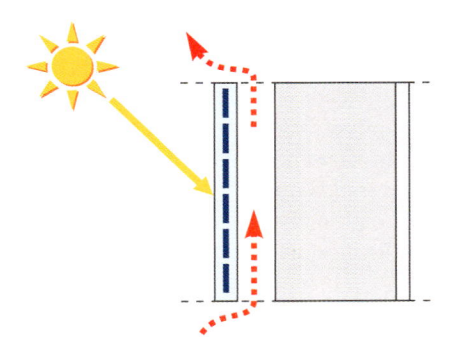

Figure 6.106
Cold façade
Source: Solarpraxis

WARM FAÇADES

Warm façades are façades that provide the functions of weather and acoustic protection and thermal insulation. They also sometimes provide structural support. Warm façades are not ventilated. Here, sections with thermal insulation are used. The façade elements must have lower U-values. These can either be opaque insulating panels or transparent/semi-transparent insulating glass. Warm façades are generally constructed as curtain walls using mullion-transom stick systems, or using unitized construction or spandrel panel construction. The most important technical features of curtain walls are defined in product standard EN 13830.

In warm façades it is possible to replace conventional insulating glass with PV modules. Insulating glass modules are used for the transparent or semi-transparent areas. Laminates or glass–glass modules can be used instead of the opaque spandrel glazing used for insulating panels. When using glass–glass modules, the rear glass sheet should be made opaque (e.g. by screen printing) or have very narrow cell spacing to prevent the thermal insulation behind from being seen.

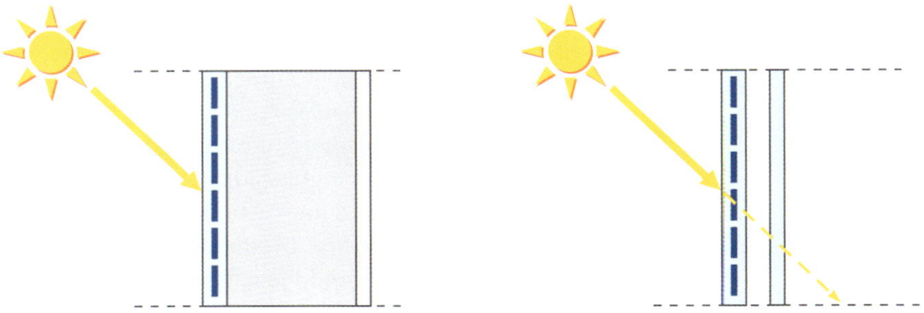

Figure 6.107
Warm façade, opaque
Source: Solarpraxis

Figure 6.108
Warm façade, transparent
Source: Solarpraxis

Generally, the electrical cabling is not routed out via a conventional rear junction box but sideways, and is sometimes protected in a conduit. In order to accommodate the bypass diodes, which are particularly important in the façade area, a very small junction box can be fixed to the façade profile that has the same thickness as the modules. This is only possible, however, for a limited module power output or number of diodes. Alternatively, with somewhat more expenditure it is possible to apply an external, easily accessible module junction box or install the bypass diodes in the PV array combiner/junction box.

With warm façades the cables are incorporated within the profiles. Because of the holes that have to be drilled through the profile sections, particular care should be taken to ensure that the regulated vapour pressure equalization between inside and outside is maintained to avoid interstitial condensation building up in the façade.

DOUBLE-SKIN FAÇADES (DOUBLE FAÇADES)

With this type of façade, an additional transparent glass envelope is constructed in front of an existing complete façade to improve the building climate and/or the sound insulation. Between the heat-insulated inner façade and the outer skin is an unheated thermal buffer zone, which is ventilated if required and can incorporate solar shading devices. Double-skin façades are designed to adapt to ambient conditions and balance out seasonal climate fluctuations. Thus heat, coldness, light and wind are regulated to attain optimum comfort without any complex technology or use of energy. Sometimes the heat energy that builds up in the cavity is used not just passively but also actively. The outer façade is extremely suitable for integrating photovoltaics since it consists of single glazing and the modules can also provide solar shading.

Figure 6.109
Office block using a passive building design with a double-skin façade
Notes: PV modules with three difference cell types are integrated within the outer skin. The heat gain is exploited by means of a heat recovery system.
Source: Biohaus

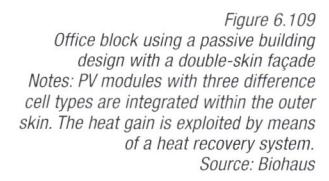

Figure 6.110
Refurbishing a façade by applying a double skin
Notes: PV modules have been integrated within the outer unitized façade. Together with the cable connections for the solar shading, the PV connections have been concealed in the spandrel area.
Source: Scheuten Solar

6.5.3 Façade structures and construction methods

VENTILATED RAIN-SCREEN FAÇADES

A ventilated rain screen refers to ventilated cladding used for exterior walls. Rain-screen façades are hung from load-bearing exterior walls and consist of a substructure

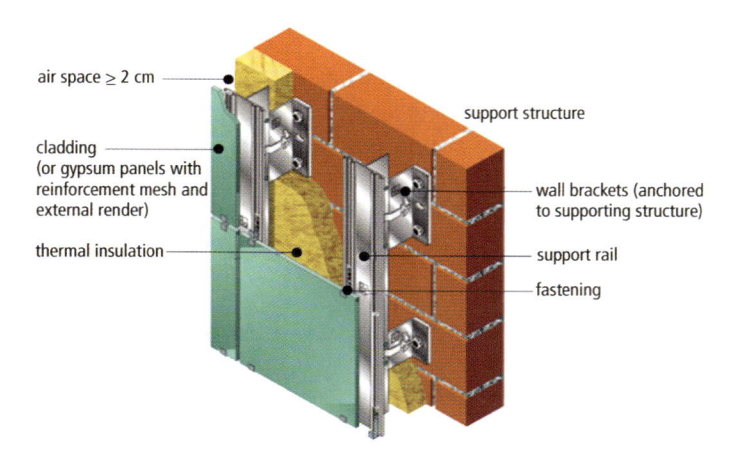

MOUNTING SYSTEMS AND BUILDING INTEGRATION   239

Figure 6.111
Structure of a ventilated rain-screen façade
Source: BWM Dübel & Montagetechnik

and its anchoring, as well as thermal insulation, the ventilation cavity, the cladding and its fixing elements.

FAÇADE CLADDING

As weather protection and as a design element, façade cladding provides the outer skin and is mostly secured to the load-bearing wall construction with a support structure. With cladding, a distinction is made between small and large elements. Small cladding panels of slate, fibre cement or timber are laid similarly to roof coverings and secured with nails, screws or clips. Photovoltaic modules are treated as large cladding panels. Large cladding panels and sheets are available in various forms and materials, such as fibre cement, laminates, ceramics, glass, timber, metal, plastic or stone. Flat, bent or profiled elements can be attached with open, closed or overlapping joints. Every cladding element has to be individually fixed.

Figure 6.112
Üstra public transport company in Hanover-Leinhausen, Germany: Façade design with various types of cladding – mono-crystalline solar modules, ceramic panels and profiled aluminium panels
Source: Solon AG; Gerhard Zwickert

Building cladding and façades are covered by national building codes and regulations. These need to be referred to and observed. Planning approvals will also usually be required.

CONNECTION AND FIXING ELEMENTS

Metal connection and cladding elements connect the cladding together and to the substructure. These fixings need to be of a type that is compliant with building codes and regulations.

THERMAL INSULATION

Only standardized or approved materials may be used for the thermal insulation. The insulation panels are generally fixed back to the exterior wall with mechanical fixings or bonded on.

SUPPORTING STRUCTURE

The substructure forms the structural connection between the load-bearing exterior wall and the cladding. Therefore, it needs to be specifically dimensioned in relation to the building and proved that it is structurally stable. It can consist of timber or metal (steel or aluminium) and must be able to level out deviations and unevenness in the shell structure. At the same time, the support structure carries the thermal insulation and creates the ventilation space. The continuous air space must be at least 20mm except with slate covering (10mm) and profiled metal sheet cladding that can be laid as strips. At the head and foot fixings, ventilation openings of 50cm^2 per metre must be arranged along the length of the façade that can be secured with insect mesh. The choice of supporting structure is determined by the type of cladding and the shell construction. There are, therefore, many systems on the market. Substructures must comply with the technical regulations stipulated in national building codes and regulations.

ANCHORING PARTS

Wall fasteners (brackets) or wall sections anchor the substructure to the wall or, if there is no substructure, fix the cladding directly to the wall. The necessary fixings are either already integrated within the wall construction (e.g. embedded anchor plates or rails) or dowel systems are subsequently drilled in. Thermal separation needs to be provided to avoid the occurrence of thermal bridges.

MULLION-TRANSOM STICK SYSTEM

With this façade system, the supporting sections are connected with screws and socket joints to form a frame construction. The vertical posts (mullions) are continuous and are secured to the floor slabs or, less frequently, to load-bearing parapets or columns forming part of the building's structural frame. The horizontal members (transoms) are inserted between the mullions. All intersections and boreholes can be powder coated in the factory after processing to prevent corrosion. Frequently the infill elements are mounted on the mullion-transom profiles with screwed-on pressure plates. The glass sheets, façade panels or PV modules effectively float in rebates that are furnished with water vapour equalization and drainage systems to discharge interstitial condensation to the outside. Continuous silicone or EPDM sealants prevent water penetration. Mullion-transom curtain walls can be used both as cold and warm façades.

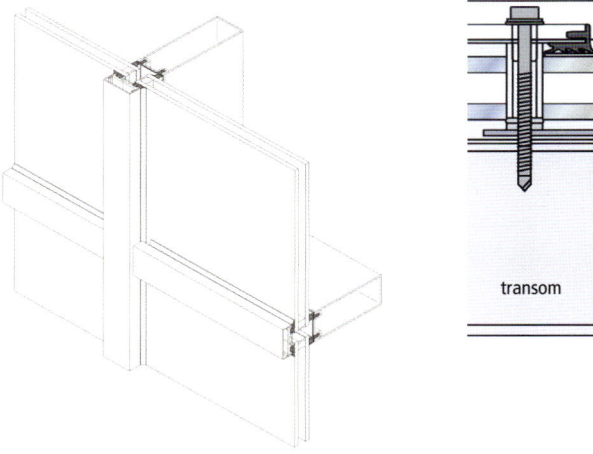

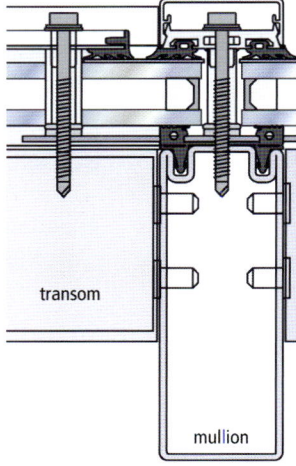

Figure 6.113
Steel mullion-transom stick system with narrow face width: Here the glass panes are mounted on the mullion and transom profiles with pressure plates
Source: (profiles) rp technick

UNITIZED FAÇADES

Unitized façades enable most of the assembly work to be transferred from the building site to the manufacturer's factory. The complete one- or more storey-high wall elements, which are already furnished with windows, spandrel panels, etc., only need to be mounted to the shell structure onsite by securing them to anchors.

Figure 6.114
Benrather Karree building, Düsseldorf, Germany: Mounting the façade elements
Source: Josef Gartner GmbH

SPANDREL PANEL CONSTRUCTION

As with unitized façades, with spandrel panel construction the wall elements are also prefabricated, including the infill panels; but the panel sections are not storey high.

LIGHTWEIGHT STRUCTURAL GLAZING SYSTEMS

To provide glass façades with as much lightness and transparency as possible, ever-lighter structural glazing systems are being developed where the distribution of the loads enables the supporting structure to be reduced to a minimum.

The first stage of development was provided by the suspended glazing systems first used during the 1960s. Suspended glazing systems enable all-glass façades to be constructed without mullion-transom profiles. Here, large glass sheets are suspended from anchor connectors fixed to the upper floor slab and the vertical joints are sealed with silicone. This enables large areas of glazing higher than 10m to be supported. It is also possible to suspend several sheets above one another with clamping plates. Like a coat of chain mail, the upper sheets support the entire façade.

Suspended glazing systems are often braced internally and restrained with glass fins. Here, the sheets are connected to one another via flush point fixings and stiffened with a network of tension cables and rods. Instead of the upper floor slab, brackets or springs can be used to absorb the tensile forces.

Modern façade structures designed to create even more transparency use tensioned steel cables or glass to create the stiffening. Here, the sheets are connected

Figure 6.115
Atrium of the Düsseldorf 'City Gate' building, Germany: Suspended glazing system with point fixings
Note: Structure: the vertical tension rods keep the horizontal steel lattice beams in position and transfer the loads downwards. They are suspended from a steel girder at the top and held in tension at the bottom with a double spring. Structural engineering: Ove Arup and Partners/Lavis Stahlbau Offenbach.
Source: Werkfoto Gartner

at all four corners with fixings to the cable net. The high tensile forces from the horizontal cables are absorbed by the adjacent building walls.

Such lightweight façades have not yet been implemented with integrated solar cells. Their high transparency and non-shaded glass surfaces provide interesting potential for photovoltaics.

Figure 6.116
Atrium of the German Foreign Office in Berlin: Cable-suspended glass curtain wall
Note: Structure: vertically and horizontally tensioned stainless steel cables are connected with spacers as in a tennis racket. Glazing: 2.7m x 1.8m laminated safety glass. The glass strips on the inside are purely for decorative reasons. They are dichroitically coated and cast coloured, letting southern sunlight into the atrium.
Engineer: Schlaich Bergermann and Partners.
Source: Müller Reimann Architekten

6.5.4 Fastenings

Fastenings refer to the parts that secure the façade elements or cladding to the façade or supporting structure. Fastenings must have sufficient stability and protection against rust and galvanic corrosion and be easy to mount. The method of fixing depends upon the façade system and needs to be matched with the elements being fixed. Glass in curtain wall façades is linearly supported whereas, for ventilated exterior wall panelling, point fixing and bolt-fixed support systems are usual. With panels, cassettes or overlapping cladding, it is fairly easy to design concealed fixings.

LINEAR-SUPPORTED FIXING

Here, a distinction is made between two-sided, three-sided and four-sided support. The perimeters of the elements are screwed to the substructure via a frame of pressure plates that clamp the edges into position. With four-sided supports, the glass can have smaller thicknesses than, for example, with two-sided supports or point fixing. Timber structures have the advantage that the clamping screws can be used at any point, whereas metal systems require prefabricated threads, such as in the mullions and transoms. If it is intended to have concealed fixing, the façade elements can be secured by bonding. The usual methods of fixing photovoltaic modules using glazing beads, pressure plates and bonding with structural sealant glazing systems are described in section '6.8 Solar protection devices'.

POINT FIXINGS

With point fixings the cladding is secured with clamps, patches, rivets, hooks, clips or bolts. The fixing elements are positioned either in the joints or in boreholes. Drilled panels are secured with screwed clamping plates or countersunk screws. For fixing with concealed undercut anchors (see the section on '6.8.1 Module fixing'), the panels are not completely drilled through but furnished on the rear side with mushroom-shaped sockets in which the dowels are embedded. This enables the fixings to be concealed. In PV façades, point fixings are mostly fixed along the edges of modules and less often on the rear side of modules or in boreholes (see section '6.8.2 Fixed solar shading').

Figure 6.117
Kaiser clothing store in Freiburg, Germany: Horizontal louvers; façade glazing and solar shading louvres with point fixing in boreholes
Source: Solon, AG

Figure 6.118
Ventilated rain-screen façade with PV modules point fixed in the joints with concealed fixings used for the ceramic panels
Source: 3S

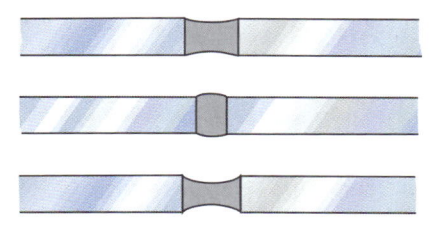

6.5.5 Joints and joint sealing

Figure 6.119
Adhesive sealing with permanently elastic sealant (silicone): Material compresses and stretches to allow movement
Source: Solarpraxis

In order to join façade elements together as precisely as possible and to accommodate deviations in size, joints are formed between adjacent components. They also accommodate mechanical movements (e.g. changes in length caused by fluctuating humidity, temperature and ground deformations) in order to prevent cracks from occurring.

Joints must fulfil the same requirements as the adjacent components in terms of rain and wind resistance, UV resistance, thermal insulation, fire protection, sound proofing and air tightness. For this reason, joints, as a potential weak point, must be reliably sealed. In façade construction there are various solutions:

- overlapping (not suitable for glass elements – that is, PV modules);
- contact sealing made of EPDM (neoprene) as permanently elastic sealing profiles, which (with sufficient contact pressure and clean glass surfaces) create a flush connection via flanges or lips, or as partly pre-compressed sealing strips that expand to the width of the joints after being inserted; contact seals are also used with overlapping;
- adhesive sealing with permanently elastic sealant (e.g. silicone).

SILICONE JOINTING WITH PV MODULES

As with laminated safety glass, with silicone joints between PV modules it should be ensured that the silicone cannot come into contact and react with the acrylic spacers of glass–glass modules or with laminate films (EVA/PVB). For this reason silicone profiles must be inserted into the joints and used as spacers. The joints can then be sealed with liquid silicone. When using insulating glass modules, even more care should be taken with the joints as the cabling along the glass edges must be routed out of the modules and through the joints into the supporting construction. The cable insulation should not come into direct contact with the silicone.

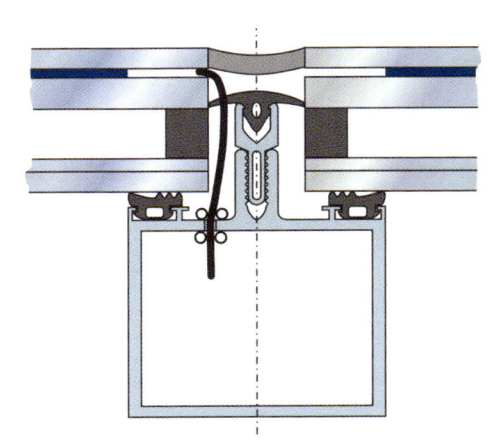

Figure 6.120
Silicone wet seal with silicone profile
Source: Solarpraxis

If the shell construction is insulated against wind, rain and heat, ventilated rain-screen façades with larger sheets and panels can also be designed with open joints. With small-sized panels it is necessary to provide additional protection against heavy rain (e.g. rebates or stepped joints). With some module fixing systems (e.g. glazing beads and pressure plate systems or structural sealant glazing), the fixing elements also act at the joints.

6.6 Photovoltaic façades

Photovoltaics can be used in front of or in façades. Although compared with surfaces sloping at an optimum tilt angle, the respective irradiance (see Chapter 1) and the expected yields are lower, façades offer other advantages. If expensive façade elements such as stone panels or stainless steel are replaced with photovoltaic elements, this leads to only slight additional costs, making the system very interesting from an economic point of view. The possible added prestige value of photovoltaic elements should also not be forgotten.

Modules provide enormous design possibilities. They can be manufactured in any form and size and furnished with all the visual and functional attributes of normal glazing (see the section on '2.1.4 Design options for PV modules' in Chapter 2). The photovoltaic elements are also fixed like conventional glazing. Not only is one or multi-sided support possible, but also point fixing or adhesive bonding in structural glazing systems. With PV modules, it should also be ensured that the cabling, electrical junctions and bypass diodes are easily accessible should they have to be replaced in the event of failure.

*Figure 6.121
Display window photovoltaics:
Zara clothing store in Cologne, Germany
Source: (modules) Solon; Constantin Meyer
Photography, Cologne*

6.6.1 Mounting modules on existing façades

Photovoltaic modules can be very easily secured to existing façades. Fire walls or windowless walls in large factory and industrial estates provide much potential. If no special demands are made in terms of the shape and size of modules, it is possible to use standard modules. As the modules do not have to provide any weatherproofing, they can be freely combined (e.g. as lettering and logos for advertising purposes or as pattern-like 'solar art'). Modules must be compliant with building codes and regulations.

*Figure 6.122
Energieforum Berlin, Germany, with
PV modules in front of the spandrels:
Framed laminates, anthracite
background foil – BP Solar;
fixing system – Osmer Solar*

*Figure 6.123
Shell solar factory in Gelsenkirchen,
Germany: Framed laminates – Scheuten
Solar; fixing system: Conergy
Note: The modules were mounted on the
flanges of the metal roof with special
Kalzip clips. No holes had to be punched
through them at any point.*

Figure 6.124
Future and Technology Centre in Herten, Germany: Glass–glass modules – Photowatt; fixing system – Aluhit
Source: Kramm and Strigl Architects

Figure 6.125
Meyer & Meyer shipping company in Osnabrück, Germany: Framed laminates – BP; fixing system – Schüco
Source: Meyer & Meyer

6.6.2 Façades with integrated modules

With façade integration the modules replace the façade elements and take on their functions in cold or warm façades (for a description of the façade types see section 6.5.2). In cold façades they replace the external cladding; in warm façades and in front of unheated areas they even replace the complete external skin. Modules can cover sections of the façade or entire areas. The PV array fulfils three functions: electricity generation, external envelope (weather protection, possibly also heating insulation, etc.; see section '6.5 Façade basics') and as a marketing instrument. The PV modules must fulfil the same constructional, structural and legislative requirements, as well as the requirements concerning corrosion protection and durability as conventional façade elements. The product standards for façades have already been explained in section 6.5.

Building codes and regulations need to be referred to regarding structural issues, heat and moisture protection, thermal insulation and building energy consumption, and fire protection. Structurally, façades have to cope with their own dead weight, temperature and wind loads.

MODULE FIXING

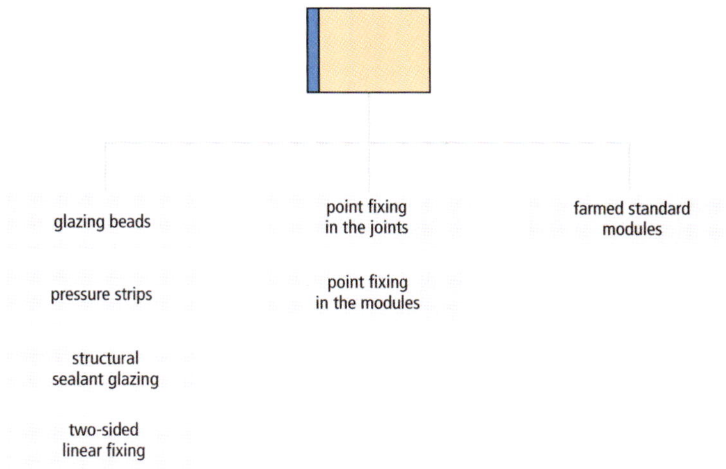

```
                         ┌────────────────────┐
                         │   module fixing    │
                         └──────────┬─────────┘
         ┌────────────────┬─────────┼──────────┬──────────────────┐
   glazing beads    point fixing            framed standard
                    in the joints              modules

   pressure strips  point fixing
                    in the modules

   structural
   sealant glazing

   two-sided
   linear fixing
```

Figure 6.126
Fixing systems for PV modules in the façade
Source: D. Wunderlich

The following examples of photovoltaic warm and cold façades are classified according to the fixing systems generally used in glass façade construction: glazing beads; pressure plates; structural sealant glazing; two-sided linear-supported fixing; and point fixing in the joints or modules, as well as façades with framed standard modules. The respective relevant provisions must be observed.

GLAZING BEADS

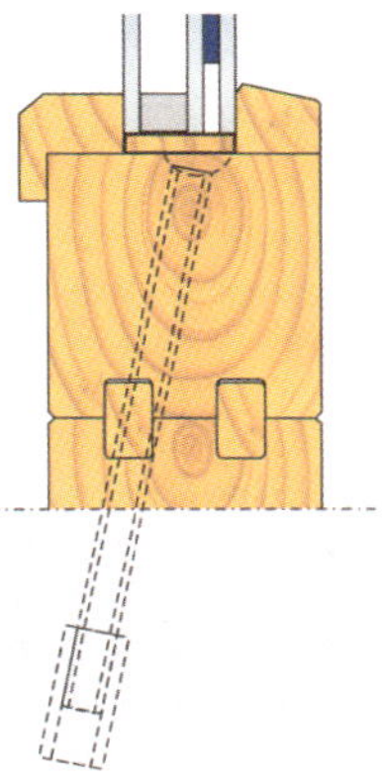

Figure 6.127
Cable ducting through the timber frame (section)
Source: Solarpraxis

Glass fixing with glazing beads within frames built into masonry walls is the most common form of window construction. The weight of the sheets is transferred via supporting blocks. The glazing beads provide the mechanical fixing for the sheets and the sealing. If PV elements are used instead of normal single or insulating glass, holes must be first drilled into the timber frames for the cabling.

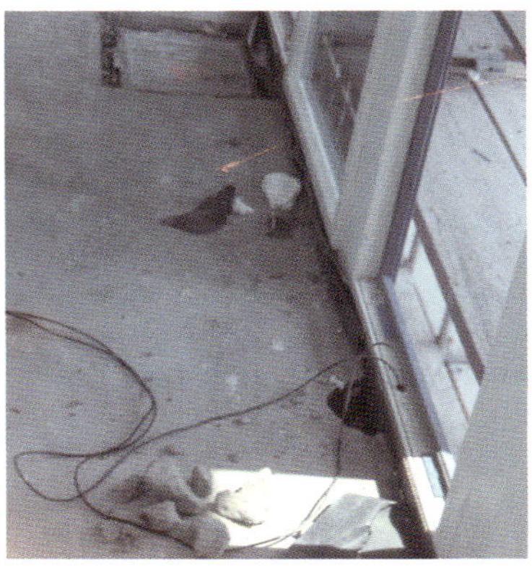

Figure 6.128
Karl Philipp Moritz House, Berlin, Germany: Warm façade – transparent; insulating glass modules – Solarwatt; window sections – Hübner

Figure 6.129
Cabling through the pre-drilled timber frame using a feed coil

Figure 6.132
German Federal Ministry of Economics, Berlin: Sloping warm façade – transparent and opaque; glass–glass and insulating glass modules – Scheuten Solar; mullion-transom stick system – Hartmann
Notes: (left) Insulating glass modules: view from the inside.
(right) Insulating panel with glass–glass module.

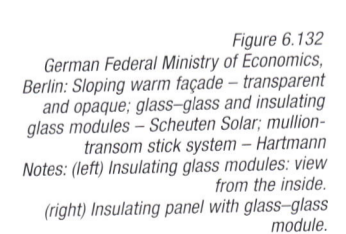

The modules in the lower area use a composite glass unit construction that provides additional protection against breakage.

Figure 6.133
Hastra (Hanover Municipal Utilities), Germany: Sloping warm façade – transparent; insulating glass modules – Scheuten Solar; mullion-transom stick system – Gartner
Source: Gartner

PRESSURE PLATES

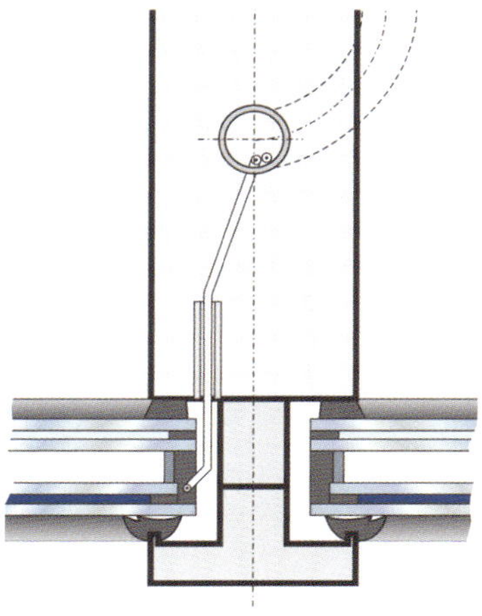

*Figure 6.130
Mullion-transom stick system
with cabling in the profiles
Source: (profile) Gartner*

Glazed 'pressure wall' framing systems are a further development of the construction described above, using glass beadings designed for curtain wall construction. Here, sections are secured from the outside that exert linear pressure on two adjacent glass sheets and the supporting frame. The weight of each sheet is absorbed via a setting block. Inserted sealing strips ensure water tightness. Since with glazed pressure wall systems the supporting structure is behind the glass plane (in the picture this is a mullion-transom construction), the face width and the cover strips can be relatively slim. This is important when using PV elements. They must be narrow and flat to ensure that the cells along the edges are not covered or shaded. With sloping façades, the horizontal sections should also be bevelled so as not to impede snow from sliding off. In order to fix insulating glazing, pressure plates must be thermally separated from the substructure.

Any electrical contact of current carrying cables with the metal façade must be avoided. The cabling must be routed along the shortest path possible away from the glass rebates and without any mechanical loading (e.g. in a conduit). If modules with rear junction boxes are installed, it must be ascertained during planning that the junction boxes will not conflict with the mullion sections.

*Figure 6.131
Stawag utility company Aachen, Germany:
Warm façade – transparent; insulating
glass modules – Scheuten Solar;
mullion-transom stick system – Gartner
Source: STAWAG*

As part of the refurbishment of the Stawag building (Figure 6.131), insulating glass modules have been installed in the glass façade. The glass panels on the inside have been designed to scatter light in order to prevent strongly contrasting light from occurring in the staircase.

With the transparent glass–glass modules used in the insulating panels, the cell spacing is so narrow that the thermal insulation cannot be seen.

*Figure 6.135
Rembrandt College In Veenendaal, The
Netherlands: Cold façade; glass–glass
modules – Saint Gobain; mullion-transom
stick system – Schüco
Notes: The modules follow the line of the
sloping façade with specially cut cells.
The string interconnection was adapted
to the shadow patterns created by the
visually animated façade.
Source: Schüco*

*Figure 6.136
Lift tower at the Constance Cultural Centre,
Switzerland: Warm façade – transparent;
insulating glass modules – Saint Gobain;
mullion-transom stick system – Schüco
Notes: Here, silver-coloured solar cells
have been used without an anti-reflective
coating.
Source: Schüco*

*Figure 6.137
Amica Aachen, Germany: Warm façade –
opaque; glass–glass modules – Scheuten*

MOUNTING SYSTEMS AND BUILDING INTEGRATION   251

Figure 6.138
Exhibition Hall 7 at Expo 2000 in Hanover, Germany: Façade designed as a curtain wall; construction – Poburski Solartechnik; amorphous modules – BP Solar
Notes: The modules are fixed twice with pressure plates running linearly along the horizontal module edges.
Source: Poburski Solartechnik

Figure 6.139
Berufsgenossenschaft Holz building in Munich, Germany: Cold and warm façade; glass–glass and insulating glass modules – Saint Gobain; mullion-transom stick system – Schüco
Notes: The insulating glass modules (to the right in the first photo) are examples of partly used modules with large transparent areas.
Source: SEV Bayern

Figure 6.140
Missawa Home Kinki office building in Kobe City, Japan; EFG laminates – Solarwatt; façade system – Kobe Steel Ltd
Source: Solarwatt

Figure 6.141
Staircase tower of the Ceramique building in Maastricht, The Netherlands: Sloping warm façade – transparent; insulating glass modules – Saint Gobain; mullion-transom stick system – Schüco
Source: Schüco

STRUCTURAL SEALANT GLAZING

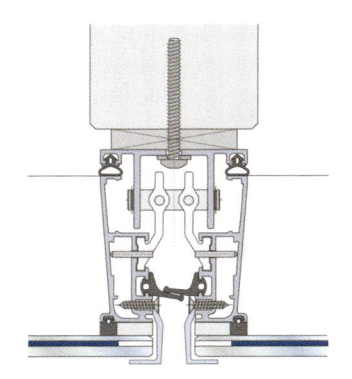

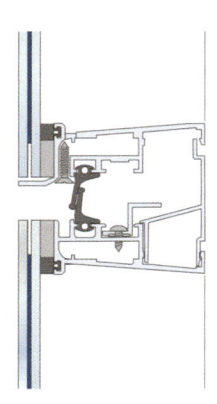

Figure 6.142
SSG façade with glass–glass module
Source: (profile) Wicona

With structural sealant glazing (SSG) the glass element is adhered directly to a carrier frame. These steel or aluminium frames are secured to the support structure (usually mullion-transom stick systems). This creates façade surfaces that from the outside appear to be frameless and without support. The bonding is generally carried out in a factory certified for SSG bonding. Carrier frames and glass are produced as elements and installed in the support structure on the building site. The adhesive supports the dead load and the wind suction forces, while also providing the sealing. In Germany, for buildings higher than 8m, the glass weight must also be secured mechanically (e.g. using support brackets). Structural sealant glazing is suitable for both cold and warm façades.

Structural sealant glazing is extremely suitable for photovoltaic mounting since there are no external frames that could shade the module edges. The prefabrication in the factory facilitates the wiring of the modules and reduces the risk of damaging expensive modules on the construction site. Laminates, glass–glass and insulating glass modules can be used in SSG façades. When using laminates, the rear Tedlar layer must be milled along the edges after lamination to ensure secure bonding with the aluminium frame construction. SSG façades may require approval from building authorities in each individual case.

Figure 6.143
Training Academy of the North Rhine-Westphalia Ministry of the Interior in Herne, Germany: Cold façade – transparent; glass–glass modules – Scheuten Solar; SSG system – Wicona
Source: Flabeg

Figure 6.144
Bayerische Landesbank in Munich, Germany: Warm façade – opaque; glass–glass modules – Saint Gobain; SSG system – Schüco
Source: Tobias Grau

Figure 6.145
Tobias Grau lighting manufacturers in Hamburg, Germany; Warm façade – transparent; insulating glass modules – Saint Gobain; SSG system – Schüco
Source: Tobias Grau

Figure 6.146
Lamy writing instrument manufacturers in Heidelberg, Germany: Warm façade – opaque; laminates – Siemens; SSG system – Rinaldi
Source: BFK Architects, Stuttgart

TWO-SIDED LINEAR-SUPPORTED FIXING

With two-sided, linear-supported fixing, frameless laminates or glass–glass modules are supported on the upper and lower edges with glazing bars. These are screwed to sections that, in turn, are linear supported or point fixed to the façade-supporting structure. The free unsupported edges can be flushed joined with silicone sealant (see section '6.5.5 Joints and joint sealing').

For ventilated rain-screen façades, DIN 18516-4 specifies in Germany that the depth of coverage of the glass by the frames must, depending upon the glass thickness and the span, be at least 15mm. With four-sided support, 10mm is sufficient.

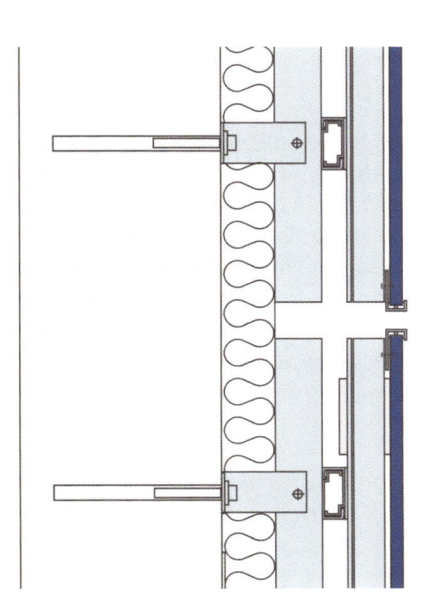

Figure 6.147
Residential tower block at Helene-Weigel-Platz, Berlin, Germany: Section through façade
Source: Solarpraxis

Figure 6.148
Residential tower block at Helene-Weigel-Platz in Berlin, Germany: Cold façade; glass–glass modules – Saint Gobain; construction – Ikarus
Source: Ikarus

Figure 6.149 Aluhit clip fixing system on the TüArena sports hall in Tübingen and a plate-fixing system from Fassadentechnik Schmidt on a residential tower block in Freiburg, Germany
Source: SunTechnics, Solar-Fabrik

In Helene Weigel Platz Berlin the glazing bars were screwed to vertical aluminium sections with hanger fixings that enable the glazing to be hung to the façade substructure with horizontal hanger rails.

POINT FIXING ALONG THE EDGES

Point fixings are less visible than linear fixings. Small fixings in the joints carry, via narrow wedges, the loads in line with the glass panes, whereas clamping plates carry the forces perpendicular to the panes. In Germany, according to DIN 18516-4, with point fixing of glass sheets, the clamping surface covering the glass must have an area of at least 1000mm^2 and the depth of glass coverage must be at least 25mm. If the fixings are arranged in the corners, asymmetrical clamping areas are required.

POINT FIXING ON THE MODULE BACK

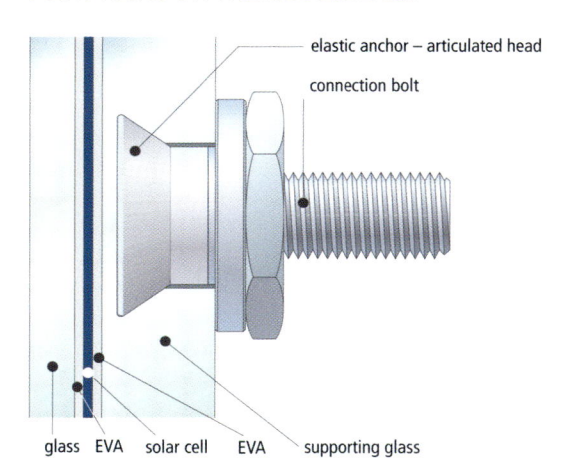

Figure 6.150 Module anchor in undercut borehole
Source: Fischer ACT

With the fischer Zykon panel anchor for glass (FZP-G) made by fischer ACT, the glass–glass modules are supported with undercut anchors on the back. This enables a concealed fixing which, in contrast to bonded SSG façades, is mechanical. Here, undercut recesses are drilled into the back of the supporting glass (10mm or 12mm thick heat strengthened safety glass (HSG) glass or safety glass) before tempering. FZP anchors with metal dowels and plastic caps are inserted into the mushroom-shaped cavities so that they grip tightly. They support the modules without penetrating the glass and are secured to the support structure with the outwardly extending bolts. The fastening remains invisible from the front and does not require any frames.

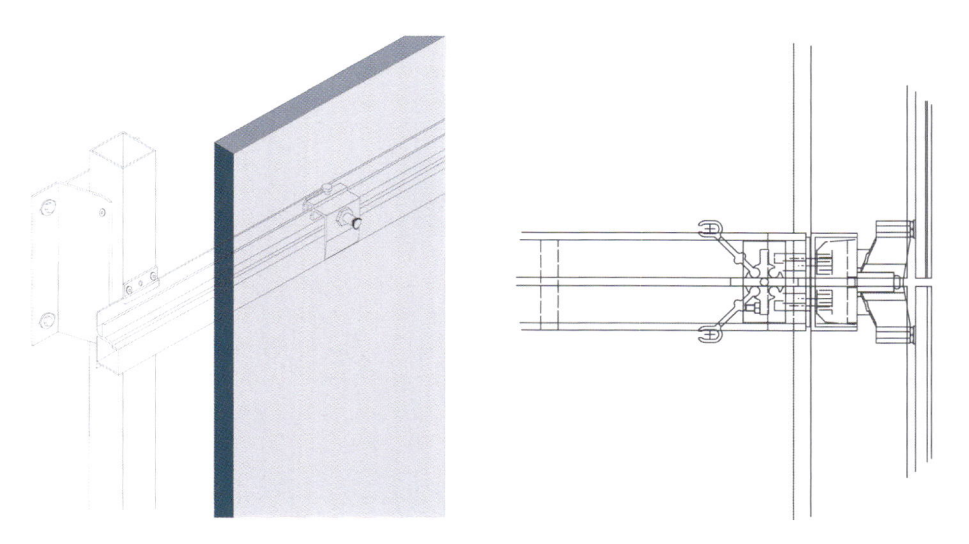

MOUNTING SYSTEMS AND BUILDING INTEGRATION   255

*Figure 6.151
(left) Substructure with hanger solution;
(right) substructure with star-shaped
two- or four-point solution
Source: Fischer ACT*

The system is only suitable for integrating modules in cold façades. As a substructure, it is possible to use a hanger or star-shaped fixing.

POINT FIXING THROUGH BOREHOLES

*Figure 6.152
Plastic-covered plate anchor with drilled
façade module
Source: Fischer ACT*

*Figure 6.153
Fischer factory in Denzlingen, Germany,
with modules directly point fixed to
the substructure
Source: Fischer ACT*

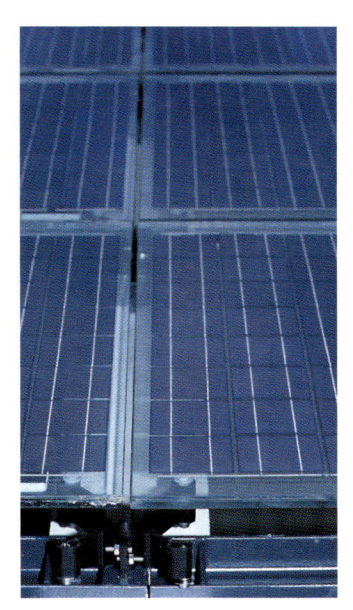

*Figure 6.154
Kriegerhornbahn in Lech/Arlberg, Germany:
PV façade with four-point fixings at an
altitude of 2176m
Source: Fischer ACT*

This type of fixing is rarely used in PV façades since the point fixing creates shadows on the front side when the sun is shining high in the sky or from the side. This area should not be covered with cells.

Figure 6.155
The façade modules on the mountain and valley stations of the Piz Nair cable car system in St Moritz, Switzerland, are mounted to the vertical U-sections of the substructure with rubber-cushioned hole-fixing elements: Glass–Tedlar laminates – Creaglas
Source: SunTechnics

CASSETTE FIXING

Figure 6.156
Isometric drawing and prototype with crystalline modules
Source: Sykon

The Biosol solar façade has been developed by Biohaus in collaboration with Sykon, which produces windows, doors and façades. The façade modules, based on amorphous standard laminates (glass–Tedlar), are inserted in a special aluminium frame and can be hung in the façade substructure as cassettes. The substructure comprises vertical members with locating bolts in the U-sections. The modules are fixed with safety screws to prevent them from rattling. All frame parts are based on a cold façade system according to the German specification DIN 18516. The Biosol façade can also be used with semi-transparent modules as an outer skin of a double façade.

Figure 6.157
Special application with a thermal insulation composite system: The façade system was matched to the timber framing; amorphous module – M-54 façade
Source: Biohaus

6.7 Glass roofs

Glass roof structures are used in building areas that are to be top lit by daylight. Here, the same materials and mullion-transom sections can be used as with glass façades. However, special structural measures are required because of the high thermal loading and different mechanical stresses. The drainage system must also be adapted to the inclination. The horizontal pressure plates are levelled off to improve the discharge of precipitation. Alternatively, it is possible to use light weight roof structures (e.g. curved trussed roofs; see Figure 6.169 of the Lehrter Stadtbahnhof railway station in Berlin, Germany).

Glass roofs are frequently equipped with additional solar protection devices to prevent overheating or glare affecting the areas below. Here it is possible to use PV elements to provide shading and glare protection. Particularly suitable are translucent roofs above unheated areas (e.g. staircases and atria) and open spaces (e.g. railway platforms and carports) as the lower module temperatures mean that higher yields can be expected.

The modules used must comply with all safety requirements and building codes for glazing in overhead areas. The type of fixing (two, three or four sided) and the span determine the required thickness of the glass composite. In addition, only specific glass types may be used.

In Germany, PV modules are only permitted if they either have building regulations approval in accordance with the German *Technical Regulations for Linear-Supported Glazing* (TRLV) (to date, there is only one such product on the market) if they are installed in insulating glass structures with laminated safety glass (LSG) as the lower pane or if the rear sides of the modules are made of LSG. In the latter case, the modules would have to consist of three panes. This assembly is only possible, however, with modules using resin-encapsulated cells since the PVB interlayer used in laminated safety glass would become soft when laminating the cells and the front pane with EVA. In addition, 'suitable measures' must be taken to prevent larger glass parts from falling down (e.g. installing nets or grills). This can be particularly visually intrusive with semi-transparent glass roofs. Another reason why such a solution is less satisfactory is because the overall thickness of the glass and the weight of the panes become considerably greater; thus, the price of the modules and the supporting structure also increases.

As a result, photovoltaic elements in overhead glazing are mostly implemented by using insulating glass construction if the areas below are heated (warm roofs) or, in the case of cold roofs, by gaining approval in individual cases or certification regarding the remaining structural capacity.

OVERHEAD GLAZING ABOVE HEATED SPACES

Figure 6.158
Epiphanias Baptistery in Hanover, Germany: Four modules in the shape of a cross were covered with only half the usual number of cells; insulating glass module with LSG as the inner pane; Solarnova mullion-transom stick system made of wood
Source: Solarnova

Figure 6.159
Winter garden with semi-transparent thin-film modules used in the overhead area: Amorphous raw modules – RWE Schott Solar; insulating glass modules – Glaswerke Arnold
Source: Glaswerke Arnold

Figure 6.160
Halle Utility Company, Germany: Warm roof; insulating glass modules: Scheuten Solar; mullion-transom profiles: Schüco
Source: Schüco

Figure 6.161
Nursing home in Strassen, Luxembourg: Warm roof; insulating glass modules – Saint Gobain; mullion-transom stick system – Schüco
Source: Schüco

In the nursing home in Strassen (Figure 6.161), because the span was greater than 4m, it was necessary to divide the glass area into two. In order to facilitate the flow of water off the very gently sloping skylight, the joint between the modules has been sealed with silicone. In addition, the module has been designed with stepped insulating glass panes so that the water in the eaves area can also flow away easily.

Figure 6.162
Bonn-Rhein-Sieg University of Applied Sciences, St Augustin, Germany: Warm roof; insulating glass modules – Saint Gobain; timber framework with steel support beams and aluminium pressure plate – Schüco
Source: Schüco

Figure 6.163
IPS Pressevertrieb GmbH in Meckenheim, Germany: Warm roof; insulating glass modules – Saint Gobain; mullion-transom stick system – Schüco
Note: The roof, which spans 270°, uses only three different types of modules.
Source: Schüco

Figure 6.164
Steinhuder Meer Bathing Island in Lower Saxony, Germany: Cold roof of the service pavilion; glass–glass modules – Solon
Note: Timber post construction: clamping profiles support modules and webbed panels on all sides.
Source: Solon AG; aerial photo: Gerhard Zwickert

OVERHEAD GLAZING ABOVE UNHEATED SPACES

Figure 6.165
Training Academy of the North Rhine–Westphalia Ministry of the Interior in Herne, Germany: Cold roof creating a protected climate zone; glass–glass modules (made from heat strengthened glass) – Scheuten Solar; saw-tooth roof construction – Wicona
Source: Scheuten

Figure 6.166
Information Centre for Ecological Building in Boxtel, The Netherlands: Cold roof above a passageway; glass–glass modules – GSS
Source: GSS

Figure 6.167
Bayerische Landesbank in Munich, Germany: Cold roof of a glass extension; glass–glass modules (with laminated safety glass) – Saint Gobain; mullion-transom stick system – Schüco
Source: Schüco

Figure 6.168
Semi-transparent atrium glazing at the Barsinghausen Educational Centre Lower Saxony: Overhead glazing with individual approvals; glass–glass modules – Solarnova
Source: Solar Engineering

OVERHEAD GLAZING ABOVE OPEN SPACES

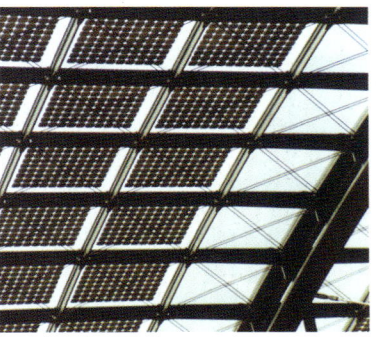

Figure 6.169
Lehrter Bahnhof railway station in Berlin, Germany: Glass–glass modules (made from heat strengthened glass) – Scheuten Solar
Notes: Steel-glass construction: cable-tensioned arched trusses and cable-tensioned grid of vertical and horizontal cross-members that provide support for the glazing. All modules have different sizes (depending upon their position). They lie on silicone profiles and are screwed to the grid nodal points with pressure plates in the corners.
Source: Scheuten

Figure 6.170
Platform roof at the Adlergestell suspension railway station in Wuppertal, Germany: Glass–glass modules (made from heat strengthened glass) – Scheuten Solar
Notes: SSG system: the 1m x 3.8m glass–glass modules made of heat strengthened glass are bonded to a stainless steel frame.
Source: Scheuten

Figure 6.171
Car park roof cover: Kochi Zoological Garden, Japan
Source: NEDO

Figure 6.172
Platform roof at the railway station in Morges, Switzerland
Source: EPFL-LESO

Figure 6.173
Carport
Source: NEDO

Figure 6.174
Bus stop
Notes: Acrylic plastic modules: Sunovation.
Source: Sunovation

Figure 6.175
House entrance
Notes: Acrylic plastic modules: Sunovation.
Source: Sunovation

Figure 6.176
Church spire: Kirchsteigfeld in Potsdam, Germany
Notes: Glass–glass laminates: GSS.
Source: GSS

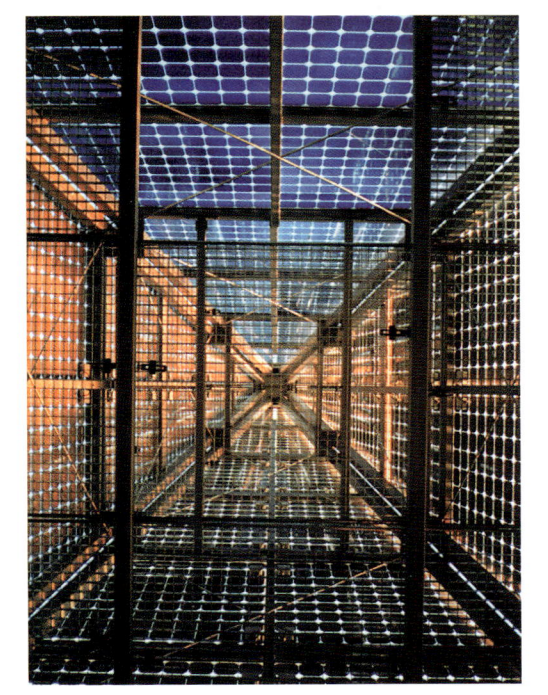

Figure 6.177
Interior
Source: Akut, Berlin

Figure 6.178
Deck access roof for a block of flats in Vauban, Freiburg, Germany
Notes: Glass–glass modules: Saint Gobain.
Source: Saint Gobain

Figure 6.179
Platform roof: Hundertwasser Railway Station, Uelzen, Germany
Notes: Modules: RWE Schott Solar.
Source: SMA

Figure 6.180
Curved solar roof on the Stillwell Avenue Metro Station in Brooklyn, New York, US
Notes: Amorphous raw modules – RWE Schott Solar; overhead glazing – Glaswerke Arnold.
Source: Glaswerke Arnold

SKYLIGHTS ON LISTED BUILDINGS

When converting the listed bus depot into the Berlin Arena (Figure 6.181), insulating glass modules were integrated within the skylights. In order to conform to the listed building requirements, textured glass was used on the rear side.

Figure 6.181 Arena in Berlin, Germany: Saw-tooth roof – warm; insulating glass modules – Saint Gobain

*Figure 6.182
SBB Locomotive Depot in Bern, Switzerland: Saw-tooth roof – cold; laminates – former Atlantis company; mullion-transom stick system – Eberspächer
Source: Eberspächer*

6.8 Solar protection devices

The extensive glazing used in façades and roofs in modern buildings (e.g. in all-glass façades) has a considerable effect on the building climate. Whereas incident solar radiation in winter can be used to provide passive heat gain, south-facing glazing in summer can lead to unwanted heat build-up. To prevent the cooling loads from being unnecessarily high when using large-scale external glazing, it is necessary to have a solar protection concept with custom-made shading elements. Here, external solar protection devices such as external louvres, canopies and blinds are generally more effective than elements on the inside of the glass (e.g. internal louvres) since they do not even allow the solar radiation to enter the building where it would then be converted into heat.

Whereas solar shading devices provide protection from solar radiation, photovoltaics need the sun. These seemingly conflicting functions can be combined as both elements require optimum orientation to the sun. Combining external solar shading devices and photovoltaic electricity generation has many advantages since solar shading devices are cost-intensive components, with the structure and technology used for tracking systems being particularly expensive. If, however, the glass or metal elements that provide the shading are replaced by photovoltaic elements, this does not substantially increase the overall costs. In addition, the optimum tilt angle and the good ventilation guarantee high yields. For this reason, such photovoltaic solar shading devices can be very interesting from an economic point of view.

*Figure 6.183
Clinical and Molecular Biology Research Centre, University of Erlangen, Germany: Fixed solar canopy in the upper part of the building and photovoltaic louvres tracking along a single axis to provide solar shading for the lower floors
Source: Solon AG; Wolfram Murr*

If they project into the street space, solar shading devices must accord with the safety requirements for overhead glazing. Here glass–glass modules and laminates using heat strengthened glass (double-sheet construction) or laminated safety glass (triple-sheet construction) are used. The former always require individual approval. Individual approval has already been granted for glass–Tedlar laminates with heat strengthened glass.

There are fixed and tracking solar shading devices. Tracking enables both the solar shading effect and the energy yield to be optimized, allowing energy yield to be increased by up to 30 per cent. The following examples are listed according to fixed solar shading devices (e.g. canopies) and moving solar shading devices (e.g. louvres).

6.8.1 Module fixing

Generally, the same types of module fixings are used for solar shading devices as in façades. In addition, point fixing through boreholes is also frequently used.

POINT FIXING THROUGH BOREHOLES

The louvres are fixed via boreholes in glass. In extreme cases, it is even possible to drill through cells, which causes the cells to be deactivated. For modules and glazing, point fixing through boreholes always requires individual approval.

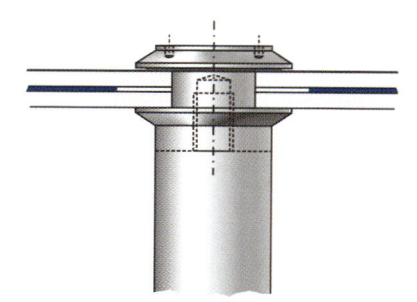

Figure 6.184
Point fixing through borehole: Manet
Source: Manet

POINT FIXING WITH MODULE CLAMPS

Here, the modules are supported with special clamps and elastomeric seals. This prevents the glass or the module frames from being damaged by drilling. This type of fixing also requires individual approval.

Figure 6.185
Point fixing with module clamps (section):
Colt-Shadovoltaik
Source: Colt

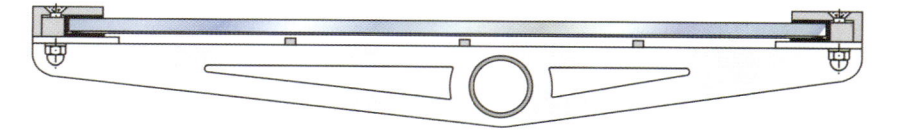

LINEAR SUPPORT

In Germany, systems with two-sided linear support are easier to gain individual approval since two-sided linear support is regulated in accordance with construction type lists. Systems with four-sided linear support provide the least difficulties when seeking approval. They entail, however, considerable expenditure in terms of materials and often look cluttered.

6.8.2 Fixed solar shading

CANOPIES

Although fixed canopies are the simplest form of solar shading device used on façades, they can be very effective if carefully positioned. PV modules in laminated safety glass assemblies can be secured with fixing channels and brackets to cantilevered roof supports to provide optimally tilted solar canopies. With the right size and correct distance to the glazing to be shaded, the canopies can block out the high summer sun while, in winter, letting low-lying incident and diffuse solar radiation penetrate under the canopy deep into the interior of the building.

Figure 6.186
Bayerische Landesbank in Munich, Germany: Point fixing through borehole – Schüco; glass–glass modules – Saint Gobain
Source: Saint Gobain

Figure 6.187
Rembrandt College in Veenendaal, The Netherlands: Canopy linearly supported on four sides – Schüco; glass–glass modules – Saint Gobain
Source: Saitn Gobain

Figure 6.188
Town hall in the town of Monthey, Switzerland: Canopy linearly supported on two sides; glass–glass modules – Scheuten Solar
Source: Scheuten

In order to achieve sufficient transparency in Monthey town hall (Figure 6.188), cells were spaced at a distance of 3cm to 4cm.

Figure 6.189
University of Applied Sciences, St Augustin, Germany: Canopy linearly supported on two sides, adjustable angle – Schüco; glass–glass modules – Saint Gobain
Source: Schüco

Figure 6.190
Clinical and Molecular Biology Research Centre, University of Erlangen, Germany: Point fixing through borehole – Manet/Austria; glass–glass laminate (made from heat strengthened glass) – Solon
Source: Solon AG; Wolfram Murr

OTHER SOLAR SHADING DEVICES

Roof surfaces can also be protected in this way with fixed solar shading. Besides the frequently used horizontal arrangement, the shading elements can also be applied vertically. In all cases, however, fixed solar shading can only be positioned very roughly. Apart from the solar altitude, the required incident light can only be controlled by the degree of transparency of the elements (see section '2.1.4 Design options for PV modules' in Chapter 2).

Figure 6.191
Apartment building in Probsteiggasse in Cologne, Germany: Solar shading above a roof terrace; glass–glass modules – Scheuten Solar
Source: Ulrich Böttger

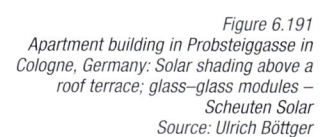

Figure 6.192
Constance Utility Company, Switzerland: Scale-like solar shading above a mono-pitch roof; glass–glass modules – Scheuten Solar
Source: Scheuten

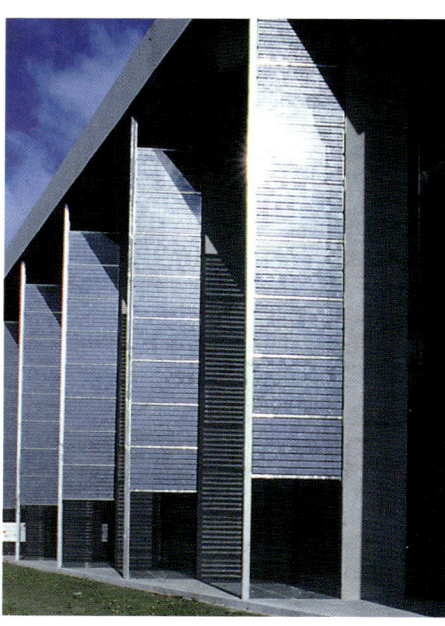

Figure 6.193
Terraced house in Cologne, Germany: Solar shading above a roof terrace
Source: Wilhelm Schulte

Figure 6.194
Environmental Technology Centre, Adlershof, Germany: Solar vertical shading panels; glass–glass laminates – Solon
Source: Solon AG

Figure 6.195
Zero emission plant at Solar-Fabrik, Freiburg, Germany: Glass–film modules – Solar-Fabrik

6.8.3 Moveable solar shading

Moveable solar shading elements include louvres, blinds and extendable canopies. They can be adjusted vertically or horizontally and allow the amount of incident solar irradiance to be tailored to meet the requirements of every daylight situation. Generally, they only tilt around a single axis (i.e. either along the vertical axis in accordance with the solar path or horizontally in accordance with the solar altitude).

Figure 6.196
Audi AG Museum in Ingolstadt, Germany: Two solar shading panels, adapted to the circular building, track the entire course of the sun
Notes: The louvre-type modules are split into two (PV above, metal panels below) and integrated within the moving circular segments on the roof. Linear support in mullion-transom profiles. Glass–glass modules: Solon.
Source: Solon AG

SOLAR SHADING LOUVRES

Solar cells are particularly suited for integration within moveable solar shading louvres. When arranged above one another, only the lower parts of the louvres are usually covered with cells to ensure that, whatever the angle, they are not shaded by the louvres above. In order to make sure that the shading function is, nevertheless, fulfilled, the rear glazing can be tinted or screen printed (fritted). The louvres' semi-transparency enables the view to the outside to be maintained, while the natural weather and sky conditions can still be seen.

The frameless louvres are point or linear supported and can be mounted on a cantilevered fixing or a continuous supporting tube. The loads are transferred either directly via the window and façade mullions or via cantilevered brackets. A rod and lever system enables all of the louvres to be adjusted simultaneously. The louvre tracking corresponds with the current solar attitude and daylight requirements, and is powered using a joint control system, mostly with a 230V or 240V servomotor per row, or using an autonomous thermo-hydraulic control system. Brightness sensors allow louvres to be moved to a predetermined position in order to gain the maximum possible light transmission when there is diffuse irradiance. A diverse range of other control functions can also be implemented.

Figure 6.197
Paul-Löbe-Haus in Berlin, Germany: Vertical louvres above the glass saw-tooth roof of the central foyer; linear support using clamp fixings along the short edges; amorphous semi-transparent (10 per cent) modules using laminated safety glass (triple-sheet construction) – Solon
Source: Solon AG; Wolfgang Reithebuch

Figure 6.198 Wesertal Technology Tower, Ohrberg in Hameln-Emmerthal, Lower Saxony, Germany: Horizontal louvres with module clamps; laminates (from heat strengthened glass) – GSS; Shadovoltaik – Colt
Notes: The louvres fixed on the front façade tilt via a single-axis tracking system. Two wing structures with louvres are also connected to the building edges and enable dual-axis tracking via a central vertical driveshaft. The changing positions of the wings enable the time of the day to be read like a sundial.
Source: Niederwörmeier, Darmstadt

Figure 6.199
Clinical and Molecular Biology Research Centre, University of Erlangen, Bavaria, Germany: Rear view of free-standing louvres; point fixing through borehole; glass–glass laminates – Solon; Voltaikwings: Ado Solar
Source: Solon AG; Wolfram Murr

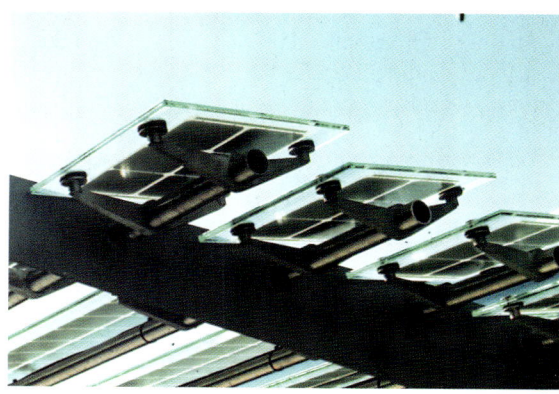

Figure 6.200
Sparkasse, Fürstenfeldbruck, Bavaria, Germany: Horizontal louvres; point fixing through borehole; assembly on continuous support tube; glass–glass laminates – former Atlantis company
Source: Atlantis

Figure 6.201
Kaiser clothes store in Freiburg, Germany: Horizontal louvers; point fixing through borehole; glass–glass laminates – Solon; Voltaikwings – Ado Solar
Source: Solon

MOUNTING SYSTEMS AND BUILDING INTEGRATION 269

Figure 6.202
Residential complex at the IGA Stuttgart, Germany
Source: Hegger Architekten

Figure 6.203
Office building in Tokyo, Japan: Vertical louvres with point fixing through borehole; glass–glass laminates – former Atlantis company
Source: Atlantis

Figure 6.204
Training centre for the Heating, Plumbing and Ventilation Guild in Cologne, Germany: Horizontal louvres with holographic optical elements (HOE), single-axis tracking; glass–glass modules – Saint Gobain; HOE – Institute for Light and Building Technology, Cologne
Source: Saint Gobain

Figure 6.205
Netherlands Government Building Agency in Delft, The Netherlands: Point fixing with module clamps along the edges; glass–film modules – Solarwatt
Source: Solarwatt

Figure 6.206
Zollern-Alb-Kurier building in Albstadt-Ebingen, Germany: Solar shading louvres with transparent solar cells connected to a solar tracking system; glass–glass modules – Scheuten Solar
Source: Sunways

6.9 Mounting systems for free-standing installations

*Figure 6.207
Geiseltalsee Solar Park, Germany
Source: BP Solar*

Just as with PV modules fixed to buildings, free-standing installations require sturdy and weather-resistant support structures. The selection and usability of mounts and foundations depend upon the quality, the load and the pH value of the ground. There are also other special circumstances that may have to be taken into account, such as shallow topsoil layers, if the PV installations are to be built on former landfill sites.

Stone or concrete strips or slabs are frequently used as pad foundations, which are either precast or made *in situ*. Compared with these, timber post or steel screw foundations reduce ground compaction and sealing, and are easier to remove and dispose of. The heat radiation caused with stone or concrete foundations does not occur. No earthworks are required to mount them and the foundations are able to carry loads immediately. However, they are not suitable for all types of ground and require sufficient depth.

Frames can be made of both timber and metal. The good energy balance and easy disposal of untreated timber needs to be carefully weighed against the cost benefits and quick mounting of prefabricated metal frames made from standard and mass-produced components. With large-scale contiguous module surface areas, it should be ensured that drainage is not hindered – for example, by providing sufficient space between the modules so that rain can run off the site. The frames should be high enough to allow mechanical mowing or to allow farm animals to graze.

*Figure 6.208
Steel screw foundation
Source: Krinner*

*Figure 6.209
Aluminium frame on concrete pad foundations
Source: Conergy*

*Figure 6.210
Galvanized steel frame on Gabion foundations
Source: GWU Solar*

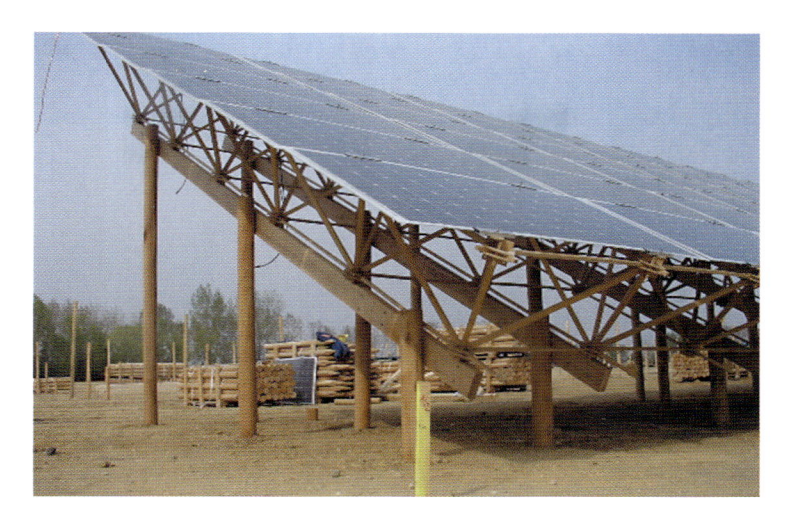

MOUNTING SYSTEMS AND BUILDING INTEGRATION 271

Figure 6.211
Timber frame on post foundations

Figure 6.212
Timber frame on concrete pad foundations: Structurally, the timber posts are separated from the foundations and are thus protected from rising damp

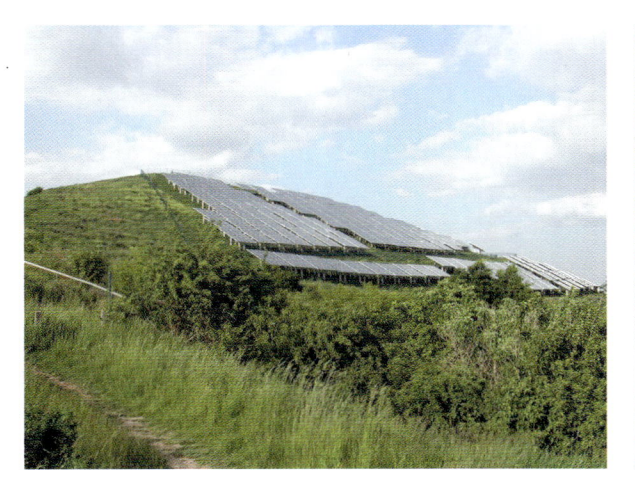

Figure 6.213
Fürth-Atzenhof landfill site, Germany: The four concrete posts supporting each module table are connected with a concrete slab beneath the humus layer

7 Installing, Commissioning and Operating Grid-Connected Photovoltaic Systems

7.1 General installation notes

PV arrays are mounted externally. For this reason, the specifications for external mounting (IP protection category, UV and weathering resistance) must be observed for the components used (module junction boxes, PV combiner/junction boxes, etc.). These are considered in more detail in Chapter 2.

In contrast to conventional voltage sources (public electricity grid), PV modules display a significantly different operating behaviour. Practices common with AC installations can cause faults with DC installation. Therefore, particular attention should be paid to the differences between DC and AC installation.

7.1.1 Notes on DC installation

- The modules are live when mounted and installed. They cannot be switched off. During the day the PV modules provide full nominal voltage. It is recommended that modules without touch-proof plug connectors should be covered with a light-proof material during electrical installation.
- The level of DC current is proportional to irradiance. The nominal voltage, on the other hand, is reached even when there is low irradiance.
- PV arrays are current sources whose short-circuit current is only around 20 per cent above the nominal current. This should be taken into consideration when designing protection technology (fuses, circuit breakers, etc.).
- The PV current is DC, which means that if there is an insulation fault this can cause a permanent arc. For this reason, the installation (except with voltages < 50V) must be earth/ground fault and short-circuit proof and the cable connections must be carefully executed. Only circuit breakers that have a proven DC switching capacity may be used.
- When connecting the DC main cable, the PV combiner/junction box must not be live. This is achieved by opening the isolation terminals in the combiner/junction box. Otherwise there is a very high risk of an arc occurring as the entire power of the PV array can be present.
- Because there is no PV array combiner/junction box with systems using string inverters, isolation is achieved by isolating the string at a module cable. Plug connectors should also not be isolated under current since there is also a risk of arcing. The DC main disconnect/isolator switch is used for load switching.
- When connecting circuit breakers or circuit breaker devices, care needs to be taken to ensure correct polarity and energy flow direction.

7.1.2 Notes on module mounting

- The module manufacturer's assembly and installation instructions should be carefully followed. This applies, in particular, to the type of mounting or clamp and the points on the module provided for this purpose as only the specified mounting system has been stress tested. The most secure mounting is achieved using the pre-

drilled holes in the module frame. In addition, the maximum permitted surface loads resulting from local wind and snow load effects must not be exceeded.
- No additional holes may be drilled in the module frame, otherwise the warranty may be voided.
- Installation should only be carried out in dry weather conditions using dry tools.
- Modules should not be stepped on during installation, and no heavy or sharp-edged objects should be placed on them.
- The accessibility of flat roofs must be ensured after installing the modules for maintenance and inspection purposes. Skylights and roof access must be kept clear. The roof skin is not designed to be walked on frequently.
- Frameless modules are at extreme risk of breakage during transport and installation and must be handled with great care. The corners and edges are particularly sensitive.
- In large systems or for roof areas that are difficult to access, it may make sense to have a crane to supply preassembled and prewired module assemblies.

Figure 7.1
Roof installation of module fields with a mobile crane
Source: Wagner & Co

Figure 7.2
Roof installation of module using scaffolding
Source: Agit GmbH, Berlin

7.1.3 Notes on module interconnection

- With modules with larger power tolerances (> 5 per cent), it is recommended that the modules are individually measured before installation to ensure that modules with similar MPP currents are interconnected in a string. This avoids losses owing to mismatching.
- Only modules of the same type should be used in the same system.
- For connecting the modules together, module types that have connecting cables with single-pole touch-proof plug connectors are quicker and easier to connect to each other.
- Pay attention to the polarity of the cables when connecting the modules to each other and in the PV array combiner/ junction box. If the polarity is reversed, bypass diodes and the inverter's input stage may be damaged.
- Keep in mind that the modules produce power during the day: do not disconnect the plug connectors under load. If it is necessary to disconnect them after installation, switch off the inverter and trip the DC circuit breaker (if fitted). The plug connectors can be disconnected under open-circuit voltage.
- In modules without preassembled module connection cables:
  – Strip insulation on connecting lead to approximately 16mm.
  – Connect firmly in spring clamp terminals without metal end sleeves.
  – Remember strain relief and correctly implement water-proof cable feed-through.
  – Form drip loop before the cable entry point into the module junction box.
  – Seal box cover so that it is watertight.
- Measure the open-circuit voltage per string before connecting together (minimum requirement).

In addition, measurements of the short-circuit current and insulation resistance per string provide security for correct installation.

7.1.4 Notes on cable laying

- Earth/ground fault-proof and short-circuit proof cabling: separate laying of positive and negative cables wherever possible; double insulation.
- Pay attention to the cable's permitted bending radius.
- Cables should be handled carefully in winter temperatures: the cable insulation is more easily damaged.
- Do not lay cables on the roof covering. Fix them to the support frame instead. Install all cables using suitable fastenings.
- Rainwater run-off should not be impeded.
- Cables should be laid in shaded areas, if possible.
- Cable ties must be weather resistant.
- Avoid extensive looped circuits (see section '4.7 Lightning protection, earthing/grounding and surge protection' in Chapter 4).
- Cables should be laid as far as possible from lightning conductors or the lightning conductor system (avoid crossovers).
- Avoid sharp edges and mechanical damage.
- Minimize the overall length of the module cabling (see the section on '4.5 Selecting and sizing cables for grid-tied PV systems' in Chapter 4).
- The cables should be installed so that they are protected from children, rodents and pets.
- Pay attention to the polarity of the cables and connectors when connecting them together.
- The DC cables may not be run through spaces or parts of spaces in which highly inflammable materials are stored or in which an explosive atmosphere can form.
- Label the DC cables when bunching cables with different electricity types (DC and AC).

7.2 Example installation of a grid-connected PV system

In this case study, the aim is to install a grid-connected PV system with a power of 10kWp on the existing sloping roof of a private home. In our example, 80 modules are to be installed of the type KC-125-2 produced by Kyocera. Each has a power of 125Wp. The PV array is divided into three strings with 27, 27 and 26 modules in series, each of which is connected to a string inverter of type Sunways 3.02 with 3kW nominal power rating. For on-roof installation the MHH-alutegra SD sloping roof base frame is used. This mounting system is suitable for framed modules that are mounted in vertical orientation. Two horizontal mounting rails are required for each module row. All the materials used for assembling the PV array substructure (rails, roof hooks, etc.) and all clamps and fixing materials (timber screws, rail connectors, angle brackets, etc.) are corrosion resistant and compatible with one another as materials. Before beginning work, carefully read through the installation instructions for the modules and the mounting system.

7.2.1 Preparation

For the preliminary planning of the arrangement of the modules on the roof, it is helpful to produce an exact plan of the roof that contains the dimensions of the roof surface, the size, height and position of existing roof fixtures or superstructures, and the spacing and position of the rafters. The individual modules should then be drawn on this roof plan. The modules should be arranged so that the entire PV array is free of shading on the shortest day of the year from 9.00 am to 3.00 pm. A shading analysis may be necessary (see Chapter 3). As well as an appealing visual appearance, access to the system and space for expansion or a solar thermal system may be important considerations. In order to reduce the wind load, the array should be a sufficient distance from the edges of the roof (rule of thumb: five times the distance between the modules and the roof surface). The minimum distance from chimneys is 60cm. In our example the regions around the chimney and above the antenna are excluded. The modules are arranged in five rows of 5 and 9e modules in the upper rows and 22 modules each in the bottom three rows.

Figure 7.3
Existing roof surface for installation
of the PV system
Source: MHH Solartechnik GmbH

So that the modules can be mounted securely on the roof structure, the required number and position of the roof hooks and the required screw sizes need first to be calculated. The easiest way to do this is using load tables or a project-based structural calculation supplied by the manufacturer of the mounting system. The wind and snow conditions at the site, the altitude above sea level, the building dimensions, the slope of the roof and the modules used are all taken into account in the structural design of the mounting system. Using these parameters, the required number of roof hooks per square metre and the maximum spacing between them (depending upon the load-bearing strength of the rails) can be read off from the load tables. The rafter spacing and the spacing of the module support rails determine the actual spacings. Now the distribution of the roof hooks and the mounting rails can be entered in the roof plan. At the edges of the roof and when going round skylights or similar features, an increased number of roof hooks is generally required.

For our example array, the mounting rails are fixed at a spacing of approximately 70cm (two tile heights, corresponding to roughly half the module height). With a rafter spacing of 60cm, this results in a horizontal spacing of 1.2m since a roof hook is positioned on every second rafter.

As well as the roof plan, an electrical wiring diagram with the string wiring, the position of the inverter and, where necessary, the PV combiner/junction box, the cabling between the modules and the inverter, and the approximate wiring distances will speed up the subsequent installation.

7.2.2 System installation: Step by step

ATTACHING THE ROOF HOOKS

To assist orientation, the position of the modules can be drawn on the roof using chalk. Where the roof hooks are going to be fitted, the roof tiles must be removed so that the rafters of the roof structure are visible at the designated points. The roof hooks are positioned so that the leg lies over the wave trough of the roof tile below and the mounting plate is located across the full width of the rafter. If the roof hook does not clear the tile surface by at least 5mm, it must be shimmed. Most manufacturers provide suitable shim plates. The roof hooks are then fixed to the rafters using two timber screws (minimum screw diameter of 8mm, with a length of 80mm). Pre-drilling the holes in the rafters and lubricating the screws makes it easier to screw them in and helps to prevent them from shearing off. The screws should be screwed into the rafters to a depth of at least 60mm to 80mm. If insulation is located over the rafters, correspondingly longer screws must be used.

Note that instead of roof hooks, it is also possible to use fixing tiles (see the section on '6.3.1 On-roof systems' in Chapter 6).

Figure 7.4
Pre-drilling the holes and screwing the roof hooks to the rafter: A shim plate is required only if the roof hook clears the roof tile by less than 5mm
Source: Schletter

TILE CUTTING

The tiles that were lifted must lie flush on the roof tiles below and at the side when they are put back in their original places. The leg of the roof hook will prevent this with tiles that are grooved at the top and bottom. The roofer or installation engineer needs to cut or abrade these tiles so that they fit together cleanly again. Depending upon the roof tile, only the top tile or possibly both tiles will need to be adapted. Then the tiles that were removed can be replaced again. Next, the roof cover is sealed again and the roof is protected against weathering.

Note that the roof hooks should not alter the position of the tiles since this could otherwise lead to roof leakages.

Figure 7.5
Depending upon the roof tile, it may be necessary to cut sections away from the tiles so that after fitting the roof hooks they lie flush again and the roof hooks do not touch them (this is best done using an angle grinder with a suitable cutting disk)
Source: (left) Schletter; (right) MHH Solartechnik GmbH

Unfortunately, the installer in Figure 7.5 is not adhering to health and safety regulations. Protective gloves and eye protection must be worn when cutting roof tiles.

FITTING THE FIXING RAILS

The cross-member rails are previously cut to size or supplied in the required sizes, and then fastened to each roof hook. In our example, the rails are secured from below through the elongated holes using a hexagon socket screw, a washer, a spring washer and a nut. If using T-head bolts and slot nuts or threaded plates, ensure that they are inserted correctly into the rail groove. Unevenness in the roof can be compensated for via the elongated holes in the roof hooks and rails and, if necessary, by using spacers (such as flat washers) This is important in order to achieve a level array surface later on. Across the width of the roof, multiple rails need to be joined together. They are

Figure 7.6
Sawing the rails to length and fixing to the roof hooks: To extend the rails horizontally, sections are connected together using flat connectors
Source: MHH Solartechnik GmbH

connected together using screwed flat connectors with a gap remaining to allow for linear expansion. Once the rails are in vertical alignment (a plumb line is helpful here), the screw fasteners are tightened using a torque wrench to the torque specified by the manufacturer.

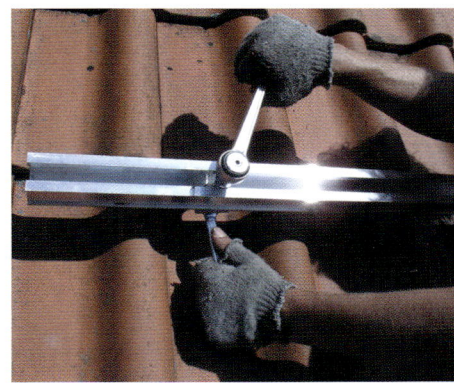

Figure 7.7
Vertical alignment of rails and tightening of bolts
Source: MHH Solartechnik GmbH

EQUIPOTENTIAL BONDING AND EARTHING/GROUNDING OF ARRAY SUPPORT STRUCTURE

Since transformerless inverters are used, the metal array support frame generally needs to be equipotentially bonded to the building. The capacitive discharge currents resulting from the system must be safely conducted to earth/ground (personal protection). Earthing/grounding and equipotential bonding codes and regulations differ from country to country. These must be consulted and observed.

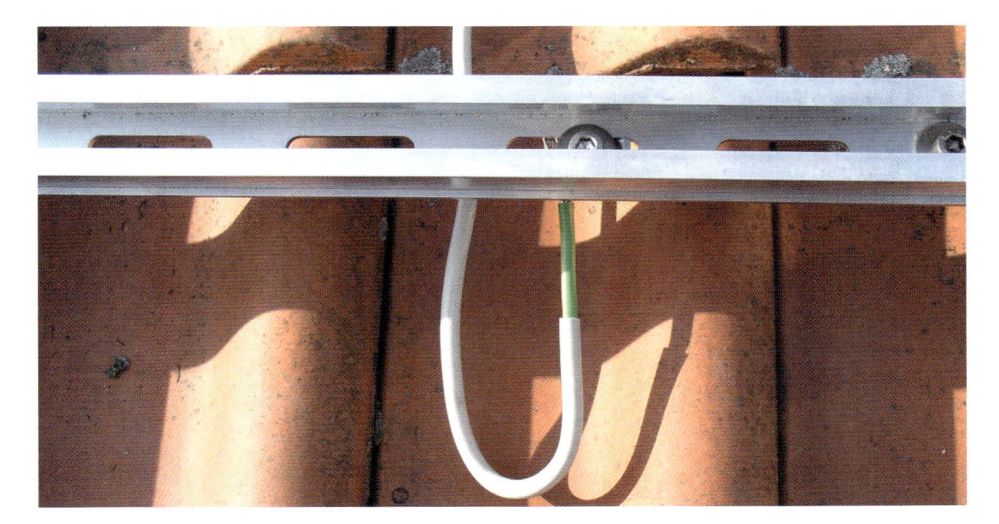

Figure 7.8
Connecting the module mounting frame to the building's equipotential bonding system
Notes: Earthing/grounding and equipotential bonding codes and regulations differ from country to country. These must be consulted and observed.
Source: MHH Solartechnik GmbH

MOUNTING THE MODULES

To prevent slippage, bolts are placed in the mounting holes on the module frame with the shaft outwards and secured in place with nuts. The threaded part of the bolt projects from the back and can be used to hang the modules in the upper horizontal rail during installation. Before the individual modules are finally secured in place, they are electrically connected to each other. The module leads that are already fitted with plug connectors are simply plugged together. For modules without plugs, the module junction box has to be opened and the connections wired up inside it. The cables are best placed and secured in the transverse rails (e.g. using UV-resistant cable ties). This ensures that rainwater is not prevented from running off the roof and that no snowmelt can build up in the area of the array as a result of cables lying on the roof surface. It ensures that no drip water can run into the plug connectors or module junction boxes. The cables must be laid so that no mechanical damage can occur to the insulation through sharp edges, pointed objects, etc. (short-circuit and earth/ground-fault proof wiring!).

Note: if metal cable conduits are used, insulating edge protection must be employed, where necessary.

In the system in this example, the easiest way of installing the modules is in rows from top to bottom. At the start of the row, the first module is clamped to the rails by its outside long edge using two preassembled end clamps. Preassembled middle fastenings are inserted into the rails with a laterally positioned rail nut and pushed up to the module. When rotated, the nut engages in the rail. The next module is positioned flush next to it and the bolt of the middle fastening is tightened using a cordless screwdriver. The row end is, in turn, concluded with end clamps. The uniform torque can be established either on the first tightening of the individual fasteners or afterwards at all clamps.

Figure 7.9
Mounting the modules: An anti-slippage precaution prevents modules that have not been finally fixed into place from sliding off the roof
Source: MHH Solartechnik GmbH

The modules are attached to the ends of the module fixing channels using angle brackets screwed to them. Thin weather-proof spacers (e.g. neoprene) are inserted between the angle brackets and the module frames. The brackets enable sufficient mechanical tension to be generated in the module rows so that no rattling or vibration sounds can be created by the module frames.

RUNNING THE STRING CABLES THROUGH THE ROOF

The string cables are run in protective conduits through the roof's inner cladding, thermal insulation and vapour-proof barrier at a centrally defined point to the outside. The cable laying must not adversely affect the roof's vapour barrier or thermal insulation. It must also be ensured here that the cabling is short-circuit and earth/ground-fault proof.

The protective conduits are first inserted through the previously made openings and fixed to prevent them from sliding out. The cables are then drawn through them; with long distances, for example, this can be done with the help of a feed coil. It is also possible to draw the cables through the conduits in advance to enable the protective conduits and cables to be installed simultaneously. Running the cables through the protective conduits ensures a high level of operating safety and a long service life for the cables. The protective conduits should be inserted through the vapour barrier at the overlapping points of the sheeting. This ensures that it can be easily sealed again after installation.

Note that protective conduits must be UV resistant and rated for use in external areas.

Finally, the string cables are run through the opening of a ventilation tile onto the roof. This is inserted at an appropriate point in the roof tiling and ensures that the roof remains impermeable to leakages at the lead-through point. For aesthetic reasons, this tile should be situated beneath the modules and be invisible from outside. The string cables are attached to the mounting frame and connected to the corresponding modules (first and last modules in a string).

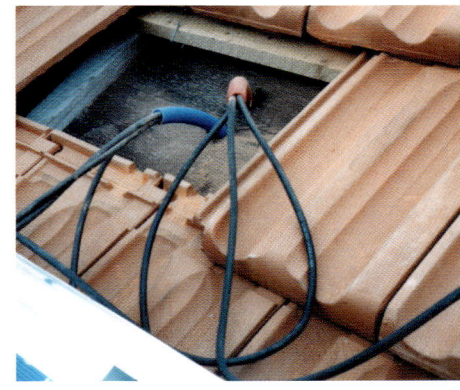

Figure 7.10
Feeding cabling through vent tile
Source: Solon

Figure 7.11
Running the string cables through the roof
Source: agitsol

The assembly and installation of the PV array on the house roof is now complete. During the assembly of the array, the individual module strings are measured electrically (open-circuit voltage, short-circuit current and insulation resistance) and the results recorded. This ensures that all array strings work without problems and that the work on the roof is, in fact, finished.

Figure 7.12
PV array installation complete
Source: MHH Solartechnik GmbH

STRING WIRING INSTALLATION INSIDE THE BUILDING

The string wiring is routed inside the building along the shortest possible route to the DC main disconnect/isolator switch (or to the PV array combiner/junction box, if present). Here, strict attention should be paid to earth/ground-fault proof and short-circuit proof installation when laying the wires. Because these wires carry direct current, they should be marked as such, especially if they are routed together with other wires in the building. There are often existing wiring routes or conduits that can be used. The string cables are connected to the terminals of the DC main disconnect/isolator switch terminals or of the PV combiner/junction box (caution is required with voltages > 120V DC). Surge voltage protectors and string fuses ensure the appropriate operational safety, while the two-pole DC main disconnect/isolator switch ensures safe system switch-off under load (e.g. for servicing and maintenance).

In the example system, the module strings are connected to the respective inverter via a two-pole DC circuit breaker.

Note: for DC voltages > 50V, the two-pole DC main disconnect/isolator switch must have at least a 5mm gap between the contacts in order to enable reliable, safe isolation. The manufacturer's specifications on the type plate of the switch should expressly state the suitability for switching DC at the required voltage level.

Figure 7.13
DC main disconnect/isolator switch with overvoltage protectors and inverter for a string
Source: MHH Solartechnik GmbH

INVERTER INSTALLATION

Connections to the respective string inverters are made from the DC main disconnect/isolator switches (or PV combiner/junction boxes) to the respective string inverters' DC input terminals. The inverters must be installed in a place where faultless operation is guaranteed. Factors to be considered include the ambient temperature, the heat dissipation capability (e.g. for installation in a cupboard), the relative humidity and the noise emissions. For service and maintenance purposes, the inverters should be easily accessible. The manufacturer's instructions must be followed.

If there is a large distance between the PV combiner/junction box and the inverter, an additional DC main disconnect/isolator switch should be installed before the inverter. This enables safe isolation of the DC main cable from the inverter, even under load.

Figure 7.14
Inverter room with one DC main disconnect/isolator switch and inverter per string along with the PV sub-distribution system
Source: MHH Solartechnik GmbH

Note: a data line enables the transfer of system parameters to a PC. This is of particular interest for remote monitoring.

INSTALLING THE MAINS CONNECTION

The AC inverter outputs are connected to the mains grid via protective equipment (e.g. fuses and line circuit breakers) and via the distribution network operator's feed meter, in the meter cupboard. In our example, the existing meter cupboard with supply meter is expanded by one meter housing (observing connection conditions), which receives the feed meter. In this case, no problems were encountered in the expansion as the main power supply was new and could remain unchanged.

Figure 7.15
Meter cupboard with feed meter: Metering arrangements will differ from country to county; this is one possible configuration that is suitable in Germany, but may not be suitable elsewhere
Source: agitsol

The commissioning of the PV system starts with setting up the meters. For this, all relevant measurements are taken and entered in the commissioning log. The mains voltage is switched on, the DC voltage connected and, with this, the inverter operation is started. The display on the inverter enables the relevant operating states to be read off, allowing conclusions to be made as to whether the system is functioning properly.

7.3 Guarantee

The installation engineer and planner, if involved, provide a guarantee for the PV system. In addition, any warranty obligations that are still effective from previous work carried out should be taken into consideration when penetrating the roof or installing electrical equipment. The basic guarantee period in Germany is two years and some installation firms voluntarily offer longer guarantee periods. Within the guarantee period, the installer must remedy any defects in the installation or defects caused by improper installation at his own expense. In addition to electrical safety and system safety, the weather-tightness and structure of the roof and the structural integrity of the PV array are relevant with respect to liability law.

7.4 Breakdowns, typical faults and maintenance for PV systems

Total failures of PV systems are extremely rare. As a rule, PV systems operate without fault. Most systems work for many years without developing costly faults or needing repairs. Between 1990 and 1995, more than 2000 grid-connected photovoltaic systems were installed on the roofs of private detached and two-family homes in a PV support programme in Germany. These were then monitored over several years in an accompanying survey programme. The operating behaviour was intensively evaluated by various institutes and organizations and the faults and problems that occurred were analysed (Fraunhofer IES, 1994–1998). In the 100,000 Roofs programme, which ran from 1999 to 2003, questionnaires on system faults and repair costs were evaluated (KfW, 2002).

Both of these investigations, as well as our own experience and analyses, provided the following results.

The most reliable component was the PV array with the modules, the DC cabling and the PV combiner/junction box. But when faults did occur in the PV array, rather than the solar modules, it was mostly the bypass or string diodes that were responsible, and these failed, for example, as a result of electrical storm activity. Today, string diodes are only rarely used (see the section on '2.2 PV array combiner/junction boxes, string diodes and fuses' in Chapter 2). In some cases, module strings were inactive owing to bad wire connections or connections that had become detached. During recent years, the quality of module connections has improved significantly since the use of plug connectors is now widespread. The use of cable ties or wiring that is not UV or temperature resistant was highly problematic. The insulation also needs to withstand the mechanical loads. All insulation ages in the course of time. For electrical power supplies, the physical operating life of power cables is generally specified as 45 years. The insulation can also be damaged by UV radiation, overvoltages or mechanically. In some areas, house martins chewed the insulation of the module cables. Suitable cable protection is readily available on the market.

Any insulation fault – however it is caused – on the DC side can result in arcing and can cause fires. As a result, the wiring should be periodically checked for any mechanical or thermal damage: the best way to do this is to measure the insulation resistance. Automatic insulation monitoring, as is performed by many inverters, is therefore a very useful feature. An insulation fault is then signalled and the inverter isolates the system from the grid. However, the illuminated PV array will still supply direct current to feed the arc. Consequently the fault cannot be isolated by the inverter. If an insulation fault is indicated, the cause of the fault should be traced as quickly as possible. In a system with one or two strings, wiring faults can be detected by checking the inverter.

Figure 7.16
Storm damage to an insufficiently secured PV array
Source: Schletter

In some cases, modules were distorted or put under stress when they were installed on the roof in order to form a flat array surface mechanically. Under the influence of temperature and wind, or over the course of time, the module glass may shatter. More rarely, expansion joints were not provided between the modules. In some cases, the wind load was not taken into account and too few roof hooks were used. Some systems showed signs of corrosion on the mounting frame. This was due to the wrong choice of materials. Here, it should be ensured that compatible metals are used. For example, brass screws must not be used on galvanized mounting systems.

Most frequently, there were faults with the inverter, even if these faults no longer occur as often as they did in systems in the 1000 Roofs programme (see Figure 7.17). Of around 3600 PV systems that were commissioned in 2000 as part of the 100,000 Roofs programme, inverter faults occurred 301 times. However, only 14 units – corresponding to 0.39 per cent – had to be repaired (KfW, 2002).

A common source of faults was incorrect dimensioning and/or cable or voltage matching with the PV array. Most installation firms have now mastered this, and software such as simulation programs or design tools from inverter manufacturers provides support here. Other sources of inverter trouble were surge voltage effects resulting from electrical storms or grid switching, ageing or thermal overload. Some failures were simply due to device faults. Extended warranties or service agreements are recommended and can be agreed with the inverter manufacturer. Sometimes harmonic effects on the system can trip the ENS/MSD device without there being a grid failure or a fault caused by the inverter. In this case, the power network operator should be contacted and suitable measures agreed (e.g. adjust the ENS/MSD impedance threshold or similar).

As well as fuse failure, other faults included faults in the meter cupboard (e.g. as a result of circuit breakers tripping).

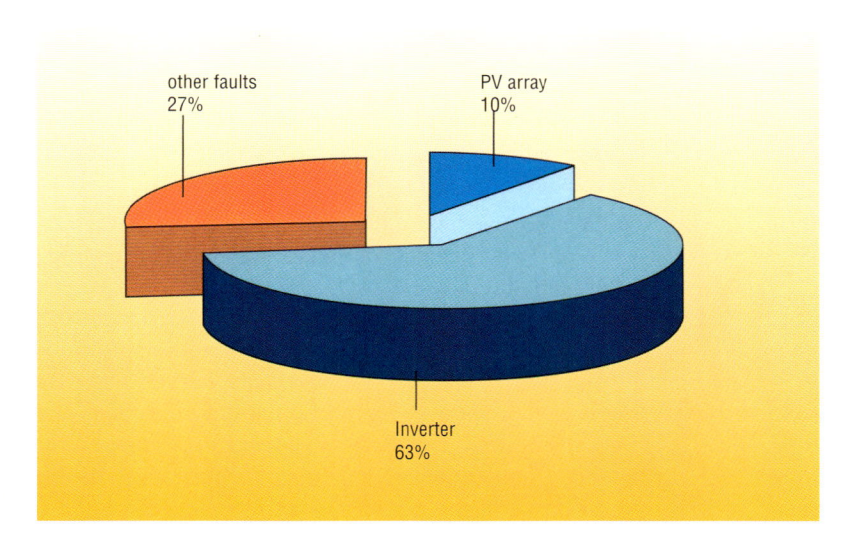

Figure 7.17
Percentage breakdown of faults among PV system components according to analysis of the 1000 roofs programme
Source: Fraunhofer ISE, 1997

7.4.1 Maintenance

Photovoltaic systems have low maintenance requirements. However, periodic maintenance routines carried out by the system operator or the installation firm help to avoid faults and long downtimes, which, in turn, optimizes the yield. In order to carry out servicing and maintenance, it is necessary to have the operating instructions (especially for the inverter) and good system documentation that should contain recommended maintenance items. The inverter's fault display should be checked on a daily basis, if at all possible. In parallel, the operating results should be read off, noted down and checked once a month. Systems for automatic fault and operating data monitoring with a notification function make the system operator's task easier in this respect.

7.4.2 Maintenance and upkeep checklist

Table 7.1 Maintenance and upkeep checklist

		Operating without any fault display?
Daily	Inverter	
Monthly	Yield check	Log the meter readings regularly (not necessary in systems with automatic recording and evaluation of operating data)
	PV array surface area	Heavy soiling? Leaves, bird droppings, air pollution or other types of soiling? Clean with copious amounts of water (use a water hose) and a gentle cleaning implement (a sponge), without using detergents Do not brush, or wipe the modules with a dry cleaning implement to avoid scratching the surface Are all modules still correctly fixed? Is the generator surface area subject to any mechanical stress? (e.g. as a result of a warped roof structure)
Every six months	PV combiner/junction box (if present)	Are there any insects/is there humidity in the device? (if mounted outdoors) If possible, check fuses
	Surge arresters	*Check after thunderstorms as well* Surge voltage arrester intact (window white or red)?
	Cables	Look for charred spots, broken insulation and other kinds of damage (e.g. cables damaged by animals) Check the fixing points
Every three to four years	Repeat the measurements as during commissioning	Only to be carried out by a trained professional
	Inverters in outdoor applications	Humidity may penetrate in spite of suitability for outdoor applications Only to be controlled by a trained professional
If suspected	Modules PV combiner/junction box AC protective equipment	Peak output measurement by a trained professional Check string fuses Line circuit breakers, AC fuses and RCDs

7.5 Troubleshooting

PV arrays are exposed to the weather. Within their service life span of 25 to 30 years, various faults can develop. The method of correcting the fault depends upon the type of fault and the type of PV system.

First, customers should be asked when and how they became aware of the fault. Circuit diagrams and a technical description of the system are very helpful. Before taking measurements, a visual check of the PV system should be carried out – in particular, of the PV array. Look out for mechanical damage and soiling.

The wiring and electrical connections should be checked. The following faults can occur that do not trip any fuse and, hence, lead to dangerous touch voltages or to arcing:

- poor or loose wire connections;
- earth/ground leakage through insulation faults;
- short circuit through insulation faults.

The measurements required to find faults in grid-connected systems are essentially the same as those required for commissioning. Today, increasingly, remote diagnostics via a modem and PC are also possible with more modern inverters.

The following causes of system faults or failures may exist, arranged according to their frequency:

- fault at the inverter;
- loose wiring connection;
- defective string fuses;

- defect with a module and, hence, partial or total failure of a string (bypass diodes or contacts of individual cells in the module);
- defective surge voltage protector;
- insulation fault.

Beginning at the feed-in point or the meter housing, the measurement check of the inverter and the PV combiner/junction box should start with the respective connecting wires. At the inverter, test the operating data by checking the LED or error code, or using remote software and a laptop. The inverter's operating data record (e.g. feed-in power, V_{MPP} and I_{MPP}) can provide important clues as to the fault. For the measurement check, test the AC side and then the DC side at the inverter. If there is no mains voltage, it is possible that the ENS/MSD has tripped (e.g. system impedance too high). Afterwards, check the DC cable and the DC main disconnect/isolator switch. When measuring the insulation resistance, the resistance to the earth/ground potential should be at least 2Mø. In the PV combiner/junction box, check the string fuses, surge voltage protector and, where applicable, the string diodes. Then, at the PV combiner/junction box, check whether a faulty string can be found.

At the string fuses and, if applicable, at the string diodes, it is possible to measure the voltage during operation in parallel by using a voltmeter. If excessive differences are present in the individual string voltages and/or string short-circuit currents, this is an indication either of too high mismatching in the generator or of an electrical fault in one or more strings. Following this, it may be necessary to make individual measurements at the modules of the corresponding string. Here, for longer strings, divide the string in half and find out which is the faulty half of the string. Then use the same method on the faulty half of the string so that you get closer to the faulty module. The module connections and bypass diodes should also be tested.

As well as measuring the open-circuit voltage V_O, the short-circuit current I_{SC} can also be ascertained. It should be borne in mind that this depends proportionally upon the irradiance.

For troubleshooting earth/ground faults or short-circuit faults in a multi-string system, the strings need to be separated and measured individually. To do this, first switch off the inverter and, if present, switch off the DC switch or DC switches. Then one module per string should be completely darkened by covering it over. Now the strings can be separated without the danger of arcing and measurement can begin.

Measuring devices have been developed for detailed fault investigation on modules or the photovoltaic array on site. These can gauge the complete I–V curve, taking into account irradiance and temperature. This enables an assessment of the performance of the modules, of a string or of the photovoltaic generator, as a whole. Peak power output meters produced by PV-Engineering in Germany even enable the modules' nominal power to be found with an accuracy of 5 per cent. The effect of shading can also be deduced from the characteristic curves. In some cases, insulation resistance or series resistance measurement is integrated at the same time. This allows installation engineers, for example, to locate damage in the wiring insulation or bad contacts. A PC interface on the devices and software for evaluating the measured data enable detailed fault investigation. The results can be documented in measurement reports.

Table 7.2 lists fault types and shows which checks and measurements can be used to detect these.

Before remedying the fault, it should be assessed whether fixing the fault is covered by the installation engineer's or planner's warranty, and whether equipment manufacturers' warranties can be called on (product liability). If this is not the case, a cost estimate for remedying the fault can be made. An inspection log should be produced for the re-commissioning.

	Visual inspection	Multimeter measurement	Earthing resistance measurement	Input/output check	Insulation resistance measurement	Over/under voltage check	I–V curves	Inverter data readout check	Test the AC circuit	Grid analysis
Fault type										
PV modules										
Soiling	✗									
Delamination	✗	✗					✗			
Bypass diodes		✗						(✗)		
Contact points		✗		✗			✗	(✗)		
Moisture	✗	✗			✗		✗			
Defective modules	✗	✗			✗		✗	(✗)		
Inverter										
Efficiency				✗				✗	✗	✗
Control characteristics				✗		✗		✗	✗	✗
Harmonic content									✗	✗
Line voltage disturbances								✗	✗	✗
Installation										
Faulty fuse	✗	✗		✗						
Defective string diode		✗		✗				✗		
Short circuit/earth leakage	✗				✗					
Defective surge voltage protectors	✗	✗			✗	✗				
Increased earthing resistance			✗							

Table 7.2 Fault types and the checks and measures to detect them

7.6 Monitoring operating data and presentation

A comprehensive operating data monitoring system ensures that failures or faults are signalled and quickly detected. This allows the system owner to instigate measures to remedy the fault and to minimize lost payments if the electricty is being sold onto the grid. Although photovoltaic systems generally operate without problems, when faults or breakdowns do occur these may not be detected for several months if no effective operation monitoring system is in place. This can result, in certain circumstances, in a significantly reduced payment from the feed-in tariff at the end of the year. Correct and trouble-free operation can be seen from the inverter's display or by observing how the meter is running. But experience shows that regular manual checks on the yield are not conducted over longer periods. Apart from this, a partial system failure (e.g. bypass diode short circuit and string failure) can only be detected with effort and a high degree of experience and technical knowledge. Equally, reading the output/feed-in meter once a year is, in most cases, not sufficient since any defect that might have occurred will only be discovered months too late. And a cursory glance at the annual feed statement may fail to detect even large yield reductions.

Figure 7.18 Operating data capture device and data logger Source: Tritec

Figure 7.19
Wireless operating data capture device with display
Source: Otronic

Almost all inverter manufacturers offer operating data that measure functions directly integrated within their devices or as an optional add-on. Many inverters record the main operating data and therefore enable rudimentary monitoring of the operation of the photovoltaic system. This enables conspicuous faults in the system to be recorded and displayed. The data can be read off from a display and/or sent to a PC (see also section '2.3.4 Characteristics, characteristic curves and properties of grid-connected inverters' in Chapter 2). For larger volumes of data, an external data logger or a separate data collection device is required, or a link to a server PC or the internet. In some cases, the inverter or data capture systems automatically carry out a system check. But this will only detect and signal any noticeable faults (such as total failure or earth/ground leakages). Then acoustic alarm signals may be triggered, for example, or messages sent by fax, email, text message or the internet. Or the system operator can maintain, back up and process the operating data on a home computer using special software supplied by inverter manufacturers and measuring system providers – this is a very involved activity to carry out on a routine basis. With internet-based system monitoring, a service provider takes over the evaluation of the operating data (see below). The operating data can also be displayed via external display units located in the living room, for example, or on a TV screen.

Figure 7.20
Photovoltaic system operating-data display unit for the living room
Source: SMA

Figure 7.21
Operating data displayed on a TV screen
Source: SMA

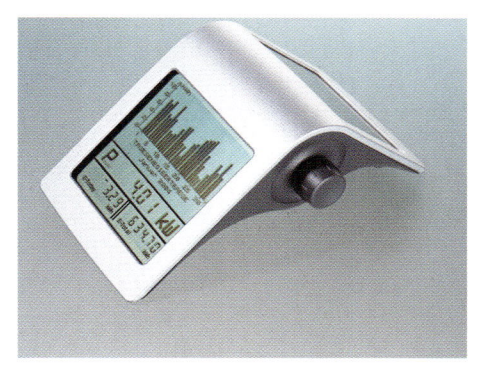

If measurements are not constantly taken, it is not easy to tell whether a system is delivering optimum yields. Depending upon the weather, the currents, voltages and feed power change constantly. A precise examination of operating data is only possible by comparing this with weather data measured at the site. A calibrated PV sensor and a temperature sensor are fitted in the module plane and measure the irradiance and module temperature. An irradiance sensor that uses the same cell technology as the modules will supply good comparison values. With larger systems, sometimes weather monitoring equipment also measures horizontal global radiation, relative humidity and wind speed. The measured values are used to calculate the expected yield. By comparing the expected yield with the actual yield, the performance of the system can be assessed. An acceptable deviation from the expected yield (e.g. as a result of partial shading) can be taken into account. If the acceptable discrepancy is exceeded, the system operator and, if necessary, the installation engineer are automatically informed (e.g. by e-mail). In some larger PV systems, even the individual strings are monitored with current monitoring devices. If significant deviations or string failure occurs, this is signalled to the operation monitoring system.

Figure 7.22
Energy meter with pulse output
Source: Conergy

Figure 7.23
Temperature and irradiance sensor
Source: Conergy

Without an irradiance sensor the comparison must be carried out based on weather data from weather stations in the local area. With suitable weather data, a yield simulation is conducted for the system. It is possible to make an additional approximate check on the monthly operating results over the internet by comparing the yields of one's own system with the yields of other systems. As well as these websites, there is also a substantial amount of PV system data online; however, in general, only a few systems are displayed or analysed at each time. The disadvantage of a system check without a radiation sensor is that reduced yields can only be detected after several months or at the end of a year.

In the PVSAT project sponsored by the EU, the project partners (University of Oldenburg, Meteocontrol, Enecolo, Utrecht University, Fraunhofer ISE and Hochschule Magdeburg) developed a method that enables the expected system yield to be calculated based on irradiance data from satellite images from the Meteosat satellite. As a result, there is no need for an irradiance sensor on the PV system. The principle of the method is shown in Figure 7.24. The hourly energy yields are sent over the internet once a day via a simple piece of hardware to the PVSAT server. There, the energy yield data are compared with the expected yields calculated from the irradiance data and the system simulation results. The average accuracy of the method based on monthly average values is 7 per cent. Possible faults or malfunctions can be detected, and clues are provided to assist in tracing the fault. For more information, see www.pvsat.com.

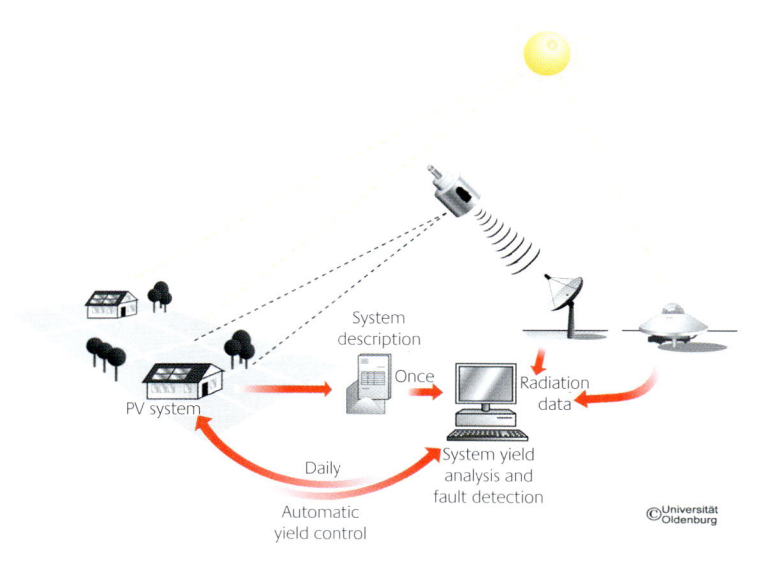

Figure 7.24
Principle of yield checking in the PVSAT project
Source: Oldenburg University

7.6.1 Internet-based system evaluation

In a local operating data analysis system, the system operator maintains, backs up and processes the operating data. Inverter manufacturers and measuring system providers offer analysis software with their devices. Even with the analysis software, local operating data analysis makes a number of requirements of the system operator – for example, fault notification, data backup, software problems or updates, to name but a few.

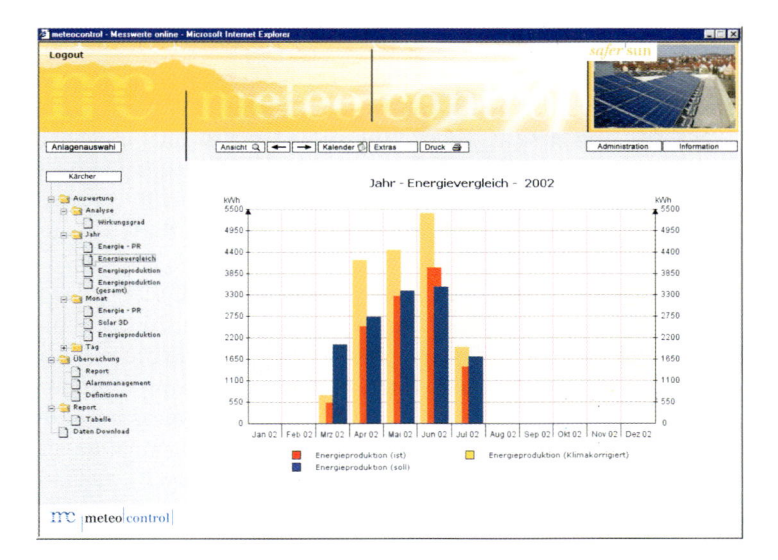

Figure 7.25
Web-based yield-checking and operating data display
Source: Meteocontrol

With internet-based system analysis, an external provider performs these tasks and the system operator is automatically informed if there is a fault in the system. Via a modem, the data logger automatically transfers the operating data to a server over an existing analogue or Integrated Services Digital Network (ISDN) telephone line at intervals that can be set by the system operator. The server stores, graphically processes, analyses and outputs the data according to the customer's wishes. An information system gives operators the facility to view the current yield data for their photovoltaic system at any time over the internet. To do this, users log onto the website over the internet using their personal password. The irradiance data, yield data and module temperature graphs, for example, are displayed graphically. Graphs are generated with resolutions ranging from hours to years, providing analysis capabilities.

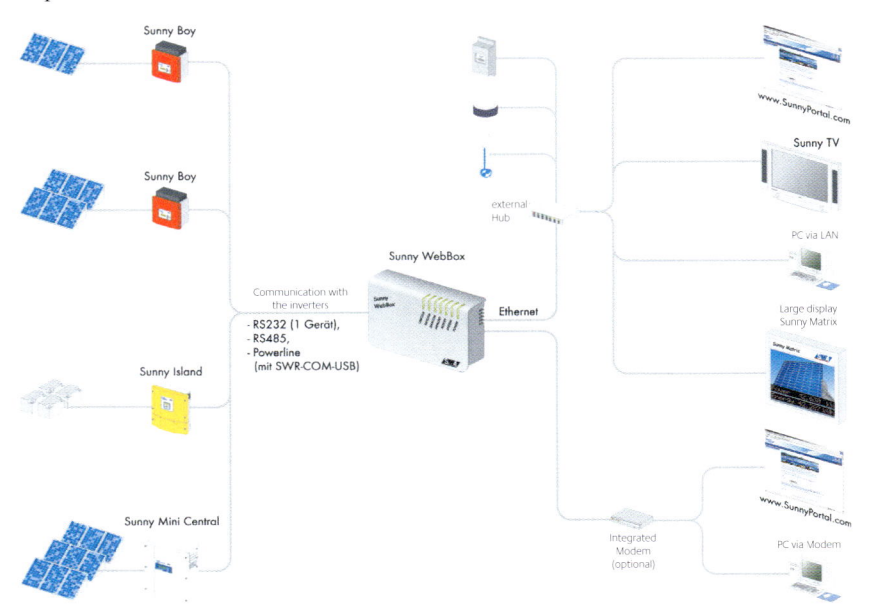

Figure 7.26
Communication scheme via the Sunny-WebBox for SMA inverters
Source: SMA

7.6.2 Web-based data transmission and evaluation

Manufacturers should follow the usual web standards for transferring and analysing operating data over the internet. Extensible mark-up language (XML) is becoming established as the format for internet data files. By using this mark-up language to define document types, it is possible to exchange data and data structures between different platforms and applications, regardless of the language. Today, there are already extensions for practically all widely used programming languages that can be work with XML data. For the central provision of data, relational database systems are primarily used. The client–server principle enables these systems to serve distributed clients on different platforms. The standardized query language used for this is structured query language (SQL). The relational database system MySQL, for example, is an open source programme that is available as freeware and is widely used. The XML format achieves high compatibility and a widely employed relational database system ensures future compatibility. In a world where software and the internet are developing at such a speed, these are qualities that should not be overlooked.

The draft international standard IEC 62350 'Power Systems Management and Associated Information Exchange' from 2005 sets out specifications for data communication ('logical nodes') between the various decentralized power sources. Section 5.4 deals with photovoltaic systems. It proposes standardized data terms for system information and characteristics, measured values and status information. This allows for the worldwide exchange of information and, in future, will enable optimized energy management of the electricity grid by regulating decentralized power generators.

PROVIDERS OF SYSTEMS FOR OPERATIONAL DATA MONITORING AND YIELD CHECKS

Conergy Electronics, Meteocontrol, Otronic, Papendorf Software Engineering, Shell Solar, Skytron, Tritec Energie, and various inverter manufacturers.

7.6.3 Presentation and visualization

As well as monitoring operation, it is also possible to visualize the operating data. Photovoltaic systems operate silently and are often located unobtrusively on roofs. System operators increasingly want to draw attention to their environmentally friendly power generation systems by making a feature of the operating data, graphs and other information on display boards, screens or terminals. In most such cases, the instantaneous power, daily yield, total energy yield and instantaneous solar irradiation are displayed. Specially designed display boards ensure that visitors notice the solar energy system. Photographs, schematic depictions and system diagrams increase the visual impact and informational value, especially since visitors are often unable to see the system's actual location on the roof.

Figure 7.27
Indoor display terminal
Source: SMA

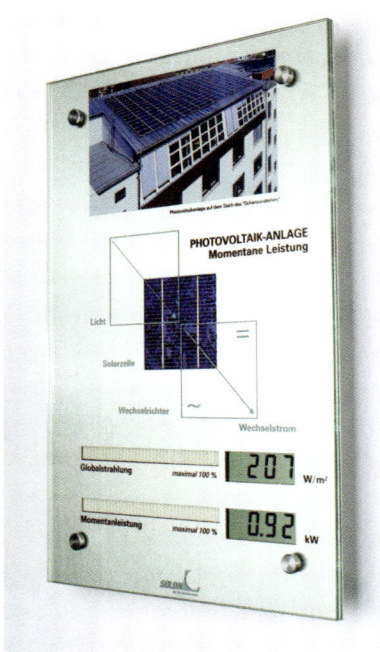

Figure 7.28
Display terminal for outdoor use
Source: Skytron

Figure 7.29
Indoor display panel
Source: Skytron

7.7 Long-term experience and quality

7.7.1 Long-term behaviour of PV modules

Photovoltaic systems will operate trouble free over many years and deliver a reliable energy yield, provided they are well planned and carefully installed, and that high-quality components are used. Many quality criteria have already been discussed in Chapter 2. The modules are the most durable components in a PV system. Some crystalline modules have now been in service for more than 25 years and show hardly any signs of ageing. Over the decades, the sun's ultraviolet light causes ageing and results in bleaching of the cells and slight degradation. Degradation here means a drop in the electrical output power of a solar module over its life span. In rare cases, weather ageing causes damage to the plastic cell encapsulation, which leads to corrosion. However, tests on 25-year-old modules showed that even modules with visible signs of ageing (such as browning effects and cell corrosion) were still supplying, on average, 75 per cent of their original output power (Quaschning, 1998). A performance drop of approximately 8 per cent was found in two modules in an 18-year-old grid-connected photovoltaic system at the Swiss LEEE-TISO institute in Lugano, while all the other modules had no discernible drop in output at all (http://leee.dat.supsi.ch). Research and investigations by TÜV Rheinland have found the average annual degradation to be less than 0.5 per cent (Vaaßen, 2004).

In long-term investigations on a ten-year-old grid-connected photovoltaic system, the Fraunhofer ISE found no relevant loss of module power (Kiefer, 2004). However, the phenomenon called initial degradation does occur. This begins when the modules are first exposed to light over a significant period of time after production. With crystalline cells, this can be up to 2 per cent. After that, the degradation is very slight (Vaaßen et al, 2003). For as long as the laminate remains sealed, the cells hardly age at all. If the laminate is mechanically damaged or comes apart due to ageing effects, air and moisture can penetrate the module. This then results in accelerated module ageing. The key factor in ageing is the long-term behaviour of the plastic that was used for encapsulation and its compatibility with other chemical substances (e.g. the anti-reflective coating). All plastics age. Some plastics (e.g. EVA) start to discolour when exposed to sunlight over long periods. This process is seen on the PV modules as a browning effect (yellow or brown discolouration). The plastic (e.g. EVA or resin) can become detached from the cell: this process is called delamination. It occurs more frequently in modules that are exposed to high temperatures and humidity. The ageing of the encapsulating plastic can result in follow-on effects such as cell bleaching, bubble formation or cell corrosion as the result of water vapour penetration. In overall terms, the long-term degradation of crystalline silicon modules is slight, amounting to approximately 0.2 per cent per year.

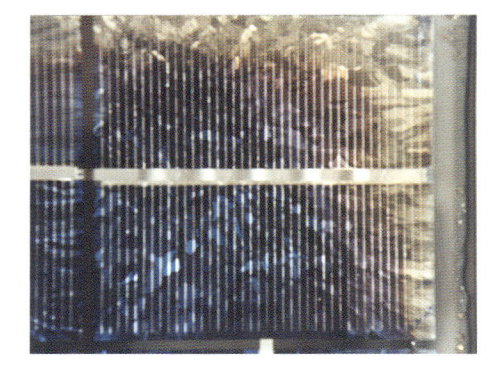

Figure 7.30
Cell discolouration (browning) caused by discolouration of the embedding plastic at the edges of 28-year-old PV modules on the roof of the Berlin Technical University, Germany
Source: TU Berlin

Thin-film modules experience degradation to a greater extent. With amorphous silicon (a-Si), in particular, degradation of the initial power output by 10 per cent to 15 per cent occurs during the first 1000 hours as a result of light degradation. After this, the power output remains almost constant. However, this effect is taken into account in the power specifications of amorphous modules, which means that the stated nominal power is 10 per cent to 15 per cent lower than the initial power of brand new modules. Over the subsequent course of their life, a-Si thin-film modules degrade further, but at a much slower rate. Unlike amorphous silicon, cadmium telluride (CdTe) and CIS cells are not affected by light ageing. But because of the thin layer thickness there is the risk of internal short circuits at the grain boundaries of the finely crystalline semiconductor material and as a result of small defects in the semiconductor layer. In thin-film modules, warping due to temperature changes can, in some cases, damage the thin metal contact that connects the cells together, resulting in high transfer resistance. As a consequence, the module's output power falls. In addition, as with crystalline modules, ageing phenomena can occur in the encapsulation material. Performance degradation of 0.25 per cent to 0.5 per cent per year can be assumed for thin-film modules.

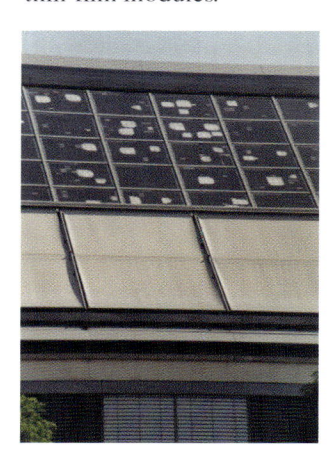

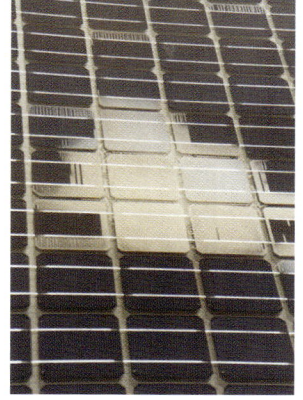

Figure 7.31
Delamination in a glass-glass module

7.7.2 Quality and reliability of inverters

Most inverters on the European market have CE approval and therefore meet electrical safety standards. The inverters are the most vulnerable component in a photovoltaic system. Long-term experience with many different units has shown that the average period for trouble-free operation is five to eight years. On average, costly repairs or a complete replacement are required after ten years of operation (Häberlin and Renken, 2003). Manufacturers now offer optional extended guarantees for up to 20 years. A service agreement with the inverter manufacturer that includes a replacement service minimizes the yield losses resulting from device failure and repairs. Serious yield losses often result from the system operator being slow to discover an inverter fault. Faults are often caused by the failure of fuses in the device or – in the event of storms or grid switching – by faulty varistors (surge protectors). Dimensioning the inverter's power too low or increased ambient temperatures can result in failure through overloading the electronic components. Improved cooling increases the device's life span.

Continuous monitoring during operation and regular maintenance can optimize the operation of the whole system. A frequent cause of inverter failure is faulty electrolytic condensers, which are used in the DC input to store energy while the polarity is switched. Reducing the number of electrolytic condensers is one way in which manufacturers can increase the durability of the devices.

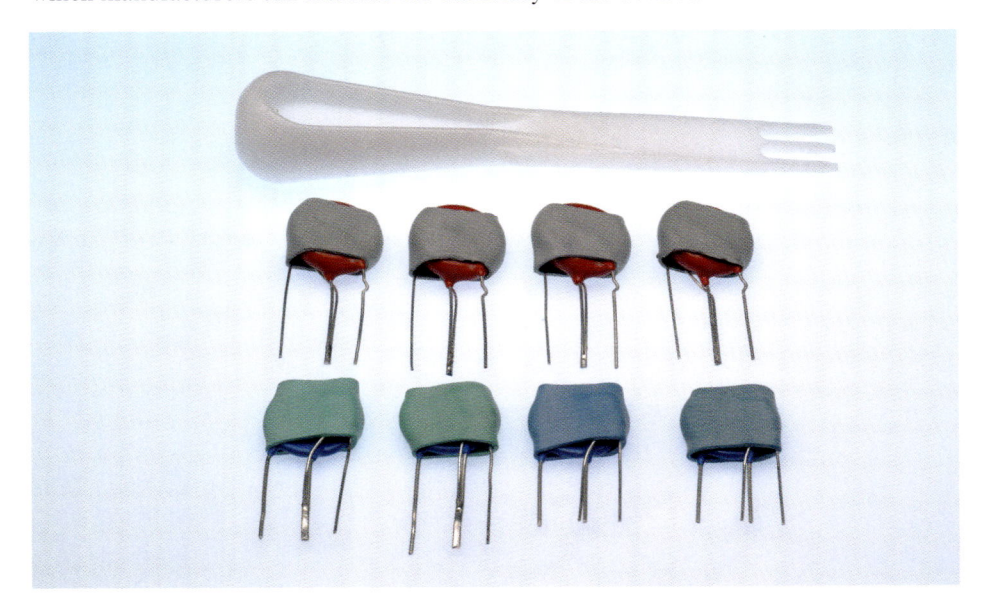

Figure 7.32
A relatively common fault in inverters is varistor failure: An engineer can easily fix this by replacing the faulty varistors
Source: SMA

LORENZ®

The right solution for every roof

As an experienced systems provider, we offer you PV modules and power inverters from leading brand producers as well as professional module mounting systems, behind which stands our total PV installation expertise of 25 years:
The **LORENZ**®- mounting system

Its advantages:

- Short installation time
- Simple and secure module attachment
- Easy replacement of individual modules
- Simple adjustment for uneven roofs

The LORENZ®- additional value:

- High stability through cross-braced frame construction
- Attractive design and use of high-grade materials
- Highest corrosion resistance

energiebau
solarstromsysteme gmbh

www.energiebau.de

8 Stand-alone Photovoltaic Systems

8.1 Introduction

Stand-alone PV systems are systems that are not connected to the public electricity grid. They are generally much smaller than grid-connected systems, and because they are very often in rural areas, the PV modules are frequently ground mounted as space is usually not a problem. The three main categories are:

1. systems providing DC power only;
2. systems providing AC power through an inverter;
3. hybrid systems: diesel, wind or hydro.

The main applications are:

- rural electrification in the developing world – small lighting systems to larger systems for clinics, hospitals and schools;
- telecom applications;
- street and furniture in urban areas (e.g. bus shelter lights and traffic sign lights etc.), usually where extending the grid would be more expensive than installing a small stand-alone system;
- power back up systems where power outages can be of a long duration;
- remote dwellings and other buildings;
- solar water pumping.

This chapter describes the main components of stand-alone PV systems (except modules, which are described in Chapter 2) and outlines a procedure for designing/sizing stand-alone systems.

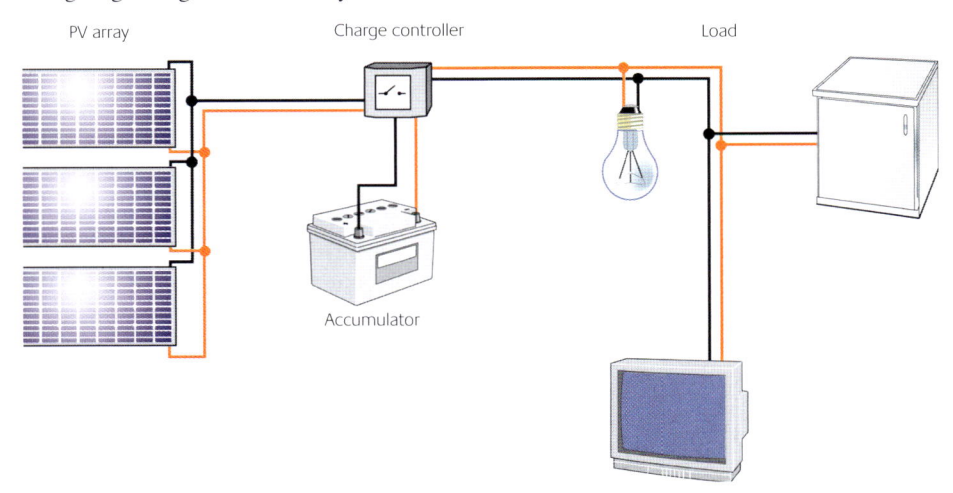

Figure 8.1
A stand-alone PV system: DC only
Source: V. Quaschning

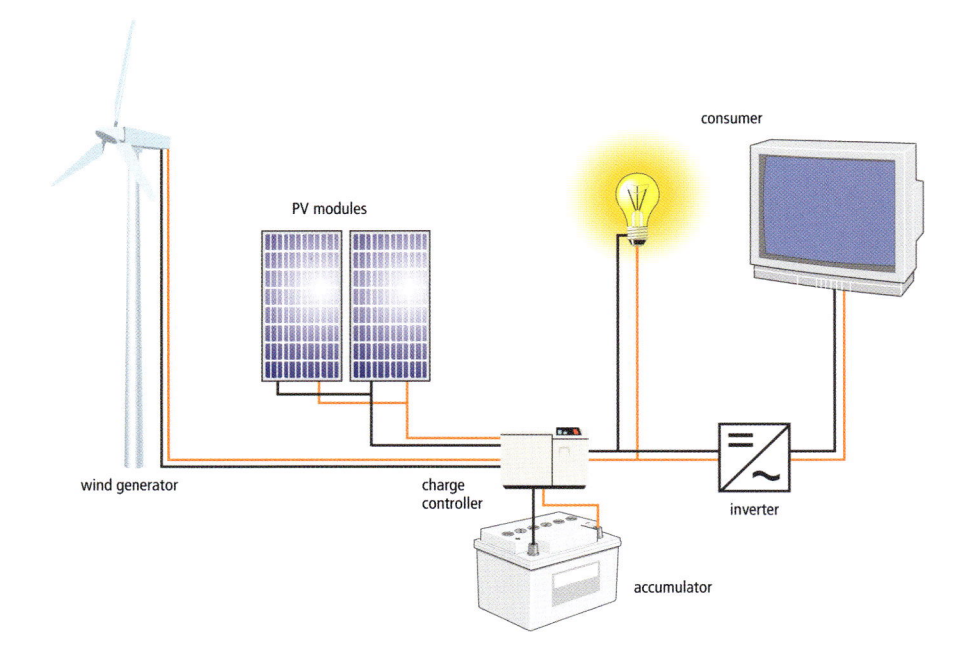

Figure 8.2
A stand-alone PV–wind hybrid system
Notes: More common in northern latitudes, energy is supplied from the PV modules during summer, and from the wind turbines during winter. Two controllers, one for wind and one for PV, is more common. Another common configuration is the PV–diesel hybrid system. PV–hydro systems are rarer.
Source: V. Quaschning

8.2 Modules in stand-alone PV systems

The modules in PV arrays in stand-alone systems are usually configured to give nominal DC voltages of 12V, 24V and, in larger systems, 48V. This means that the modules are usually connected in series. In order to facilitate easier interconnection of the modules, modules with junction boxes should be used rather than modules with plug-in leads (as used to configure the long strings used in grid-connected arrays). Most stand-alone systems are 12V.

8.3 Batteries in stand-alone PV systems

Energy storage is required in most stand-alone systems, as energy generation and consumption do not generally coincide. The solar power generated during the day is very often not required until the evening and therefore has to be temporarily stored. Longer periods of overcast weather also have to be catered for. Most stand-alone solar systems have batteries, an exception being solar water pumping systems: the water is pumped when sufficient sunlight is available and stored.

The most common type of battery found in stand-alone solar systems comprises rechargeable lead-acid-batteries. These are the most cost-effective and can handle large and small charging currents with high efficiency. In PV systems, the storage capacities are generally in the range of 0.1kWh to 100 kWh, although a few systems in the MWh range have already been implemented (see section '8.3 Batteries in stand-alone PV systems'). Other commercially available types of rechargeable batteries are nickel-cadmium, nickel metal hydride and lithium ion batteries. These are used mainly in small appliances such as radios, clocks, torches and laptops.

8.3.1 How lead-acid batteries work: Construction and operating principles

Lead-acid batteries consist of multiple individual cells, each with a nominal voltage of 2V. When built as a block assembly, the cells are accommodated inside a shared housing and wired in series internally (e.g. six cells for a 12V block). In large battery banks, separate individual cells are used on account of their size and weight. These then need to be connected together when they are installed. In this case, by

Figure 8.3
Range of batteries designed for use in stand-alone solar systems
Source: Sonnenschein brand, Deutsche Exide Standby GmbH

connecting the cells in series or parallel, different system voltages and capacities can be created.

A battery cell is a container filled with electrolyte made from diluted sulphuric acid (H_2SO_4), in which two plate units of different polarity are mounted. The plates serve as electrodes and consist of a grid-shaped lead carrier and the active material. The porous active material is the actual energy store, its sponge-like structure providing sufficient surface area for the electrochemical reaction. In the charged state, the active mass in the negative electrode is lead (Pb) and in the positive electrode lead dioxide (PbO_2). To isolate the positive from the negative electrodes, 'separators' are used.

When electricity is drawn, electrons flow from the negative to the positive pole via the loads, causing a chemical reaction between the plates and the sulphuric acid. This causes lead sulphate ($PbSO_4$) to form on the surfaces of both sets of plates, as sulphur from the acid binds to the active material. The electrolyte is therefore used up as the battery is discharged. This reduces the concentration of the acid. This change can be measured extremely well with a hydrometer, which makes it possible to check the state of charge of a battery cell.

When recharging the battery using the PV module, with a voltage that is higher than the current battery voltage, the electrons move in the other direction (i.e. from the positive pole to the negative pole). This reverses the chemical change that took place during discharging. The process is not completely reversible. Small quantities of lead sulphate do not dissolve again (sulphation). The battery's capacity has lessened slightly as a result of the discharging/charging process. This capacity loss is greater the deeper the discharge was. If only part of the battery capacity is used, the decrease in capacity is relatively small. Thus, if used in applications with small discharge capacities, the service life of the battery – that is, the number of cycles – increases considerably. Figure 8.5 shows the relationship between discharge depth and number of cycles. Battery life in terms of cycles is normally specified for cycles with a given discharge depth and with a fully charged battery in each cycle. The cycle life of a battery is defined as the attainable number of cycles before the capacity in the charged state falls to 80 per cent of the rated capacity. After that point, the battery can still be used; but the available capacity falls continuously and the risk of sudden failure increases, particularly as a result of a short circuit (see section '8.3.4 Ageing effects').

MANUFACTURERS OF LEAD-ACID BATTERIES SUITABLE FOR USE IN PV SYSTEMS

Akku Gesellschaft, BAE, Bäären, Bayern Batterie, Deutsche Exide, Hawker, Hoppecke, Mastervolt, Moll, Swisssolar, Varta.

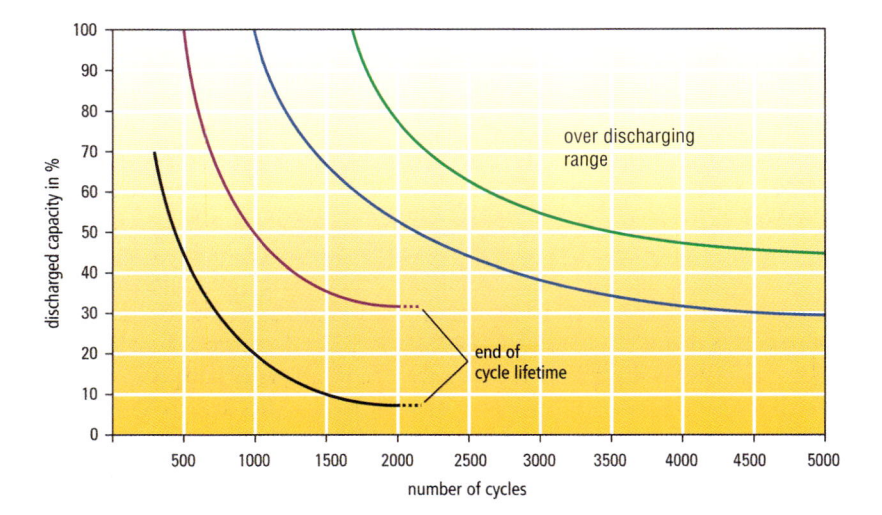

Figure 8.4
Cycle life of lead-acid batteries: The number of cycles depends upon the battery type and the discharge depth
Notes: Graphs such as these are provided by battery manufacturers so that the life of a battery in a given system can be estimated. Occasionally, the information is given in table form.

8.3.2 Types and designs of lead-acid batteries

Lead-acid batteries can be divided into different types according to the plate technology and the type of electrolyte that they use. In solar installations, usually lead-acid grid plate batteries with fluid electrolyte (known as *solar batteries* or modified starter batteries), gel batteries, tubular plate batteries and block batteries (individual cells) are used.

LEAD-ACID GRID-PLATE BATTERIES WITH FLUID ELECTROLYTE (WET CELLS)

The most common battery type is the lead-acid battery with grid plates and fluid electrolyte. Because of its use as a starter battery in vehicles, it is manufactured in large numbers and thus referred to as the starting–lighting–ignition (SLI) battery.

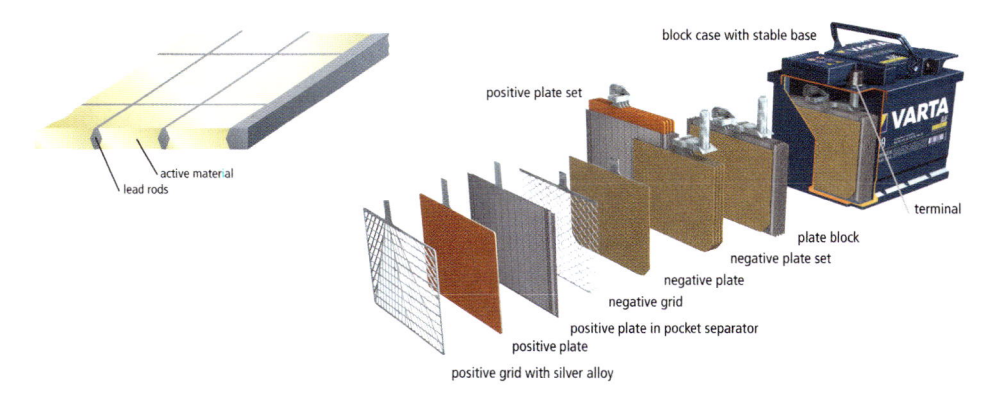

Figure 8.5
Cross section through grid plate
Source: C. Hemmerle

Figure 8.6
Construction of a lead-acid grid-plate battery with fluid electrolyte (SLI battery)
Source: Varta

Both the positive and the negative electrodes are grid plates. Since the active material can simply be spread onto the grid as a paste, grid plates can be produced cheaply. Modified versions of these batteries are sometimes known as so-called solar batteries. Here we will briefly examine the question of whether a vehicle starter battery can be used in a stand-alone solar system. A vehicle starter battery is built with a large number of relatively thin plates. This results in a large active surface area. As a result, a high starting current can be provided for a short period. This flows for only a few seconds, which decreases the battery's capacity only by a very small fraction. To favour high performance over short periods (thin plates), cycle resilience is reduced. Therefore, a starter battery in normal cyclical solar use with up to 50 per cent discharge depth would soon become unusable after just a few days. Its capacity would be reduced to a small fraction of its initial capacity. So, generally, vehicle starter batteries are not suitable and should only be used in stand-alone solar systems if they are the only type of battery available (such as in many areas of the developing world where truck batteries are used in small systems – giving service lives of up to two years). And when they are used, average daily discharge should be kept low by sizing the battery appropriately and, occasionally, by adjusting the low voltage cut-off of the charge controller downwards, where this is possible.

In contrast to vehicle starter batteries, so-called *solar batteries* are built with thicker plates to increase cycle resilience, and the lead grids are hardened with an antimony additive. In addition, the electrolyte has a somewhat lower acid content to reduce its own corrosion and thereby increase the service life. The capacity of a battery depends both upon the strength of the current with which it is discharged and upon the temperature. As temperatures fall, the capacity grows smaller; with rising temperatures, it increases. For example, if the temperature falls from 20°C to 0°C, the available capacity decreases by approximately 25 per cent.

The lead-acid battery characteristic curve in Figure 8.4 shows the cycle resilience of a *solar battery* (made by Varta). At 70 per cent discharge depth, this means that when 70 per cent of its capacity is regularly discharged, it attains a service life of only 200 cycles. At 50 per cent discharge depth, it increases significantly to somewhere over 400 cycles, and at 20 per cent, it is 1000 cycles. This type of battery is well suited to sporadic applications (e.g. in camper vans, boats and weekend homes). Provided the battery is sufficiently well dimensioned, a 50 per cent discharge will only happen rarely, with the result that the service life will be lengthened dramatically. So that the discharge level should not, if possible, fall below 50 per cent, the use of a charge controller is required. This then disconnects the loads in good time to protect the battery.

LEAD-ACID GEL BATTERIES

A more advanced version of the traditional lead-acid battery with grid plates is the gel battery. In these, the acid is thickened into a gel by the addition of additives. The particular advantages of this are:
- no problematic acid stratification; reduced sulphation;
- higher cycle resilience;
- no gassing, enabling use even in poor ventilation conditions;
- fully sealed leak-proof housing; installation in any location (e.g. boats and camper vans);
- maintenance free since there is no need to top up electrolyte during the service life.

Batteries with fluid electrolyte are manufactured in an enclosed design as an inseparable unit with housing. Only the sealing plugs can be opened to check and top up the electrolyte level. In contrast, gel batteries do not require any sealing plugs since under normal operating conditions there is no need to add water. As a result, they are built as enclosed batteries and are fitted with safety valves, which allow any gas to escape that results from overcharging. Gel batteries are therefore maintenance free.

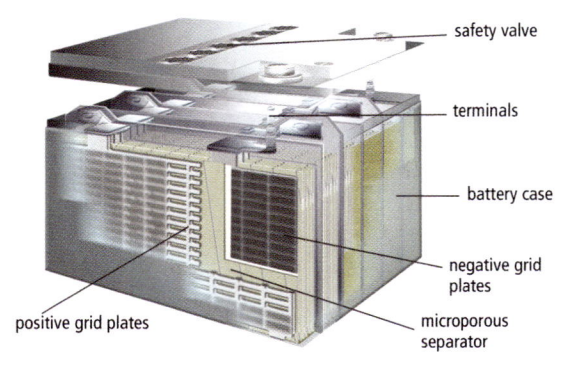

Figure 8.7
Components of a lead-acid gel battery
Source: Deutsche Exide Standby GmbH

It should be noted that this battery type requires a charge controller suited to its properties as gel batteries are highly sensitive to overcharging. The charge cut-off voltage must be strictly adhered to so that gassing does not occur, which dries out the battery. Because of the enclosed design, it is not possible to check the state of charge by measuring the acid concentration. The only way to gain approximate information on the state of charge is to measure the battery voltage.

The characteristic curve of the gel battery in Figure 8.7 shows that a gel battery (Sonnenschein brand) is good for 1000 cycles with a discharge depth of 50 per cent. The number of cycles is therefore more than twice that of the battery with fluid electrolyte. Sufficient dimensioning, as is typical in practice for a discharge depth of 30 per cent, results in 2000 cycles for the gel battery, but only 700 cycles for the battery with fluid electrolyte. Gel batteries have a longer service life, but are more expensive than lead-acid batteries with fluid electrolyte. The application field for these batteries clearly opens in the direction of all-year-round use with several years of service life.

STATIONARY TUBULAR PLATE BATTERIES (TYPES OPZS AND OPZV)

For heavy-duty long-term operation over 15 to 20 years and all-year-round use in larger-sized and large stationary power supply installations (e.g. in a building supplied in stand-alone operation), the use of deep-cycle tubular plate batteries is recommended. Greater weight and volume, and more expense at the installation site (they need to be installed in a special battery room and a support structure is necessary), as well as two to three times higher procurement costs, are involved. Tubular plate batteries are available as enclosed OPzS-type batteries (from the German *Ortsfeste Panzerplatte Spezial* – Stationary Tubular Plate Special), which have fluid electrolyte and special separators, or as sealed OPzV batteries (Ortsfeste Panzerplatte Verschlossen – Stationary Tubular Plate Sealed), which have gel electrolyte. They are standard in emergency power systems and not specially developed for the solar industry. These products are mature, have been around for decades, and are an excellent choice for stand-alone solar applications.

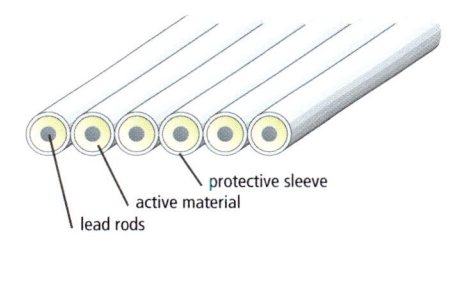

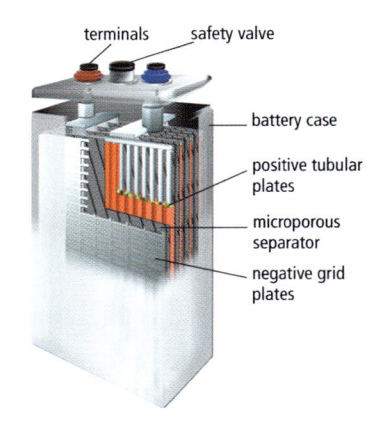

Figure 8.8
Cross section through a tubular plate
Source: Deutsche Exide Standby GmbH

Figure 8.9
Components of an OPzV battery with positive tubular plates and negative grid plates
Source: Deutsche Exide Standby GmbH

Stationary tubular plate batteries differ from solar and starter batteries in the design of the positive electrodes, which are formed as tubular plates. In these plates, the rods are surrounded by permeable tubes through which the electrolyte can pass. The protective sleeve ('tube') mechanically secures the active material in the space inside and also limits sedimentation (when fine particles of the active material fall down to the bottom of the battery case). Tubular plates are particularly stable and ensure that the battery has a high cycle resilience. The cycle resilience of OPzS and OPzV batteries (Figures 8.8 and 8.9) (Sonnenschein brand) is significantly higher than other battery types. At 50 per cent discharge depth they have a life of around 3500 cycles, reaching 5000 cycles if only 45 per cent of the nominal capacity is discharged. OPzS batteries require maintenance every 0.5 to 3 years; OPzV batteries are maintenance free.

BLOCK BATTERIES WITH POSITIVE FLAT PLATES (OGI BLOCK)

OGi (from German Ortsfeste Gitterplatten – 'stationary grid plate') block batteries with fluid electrolyte are also among the stationary battery types. The positive electrodes in this case are flat plates, a compromise between grid plates and tubular plates. The rods are not individually encased, but are enclosed by a common protective sleeve. This enables flat plates to be manufactured more easily and more cheaply than tubular plates; yet they still have a longer life than grid plates. The negative electrodes of the block battery are once again grid plates.

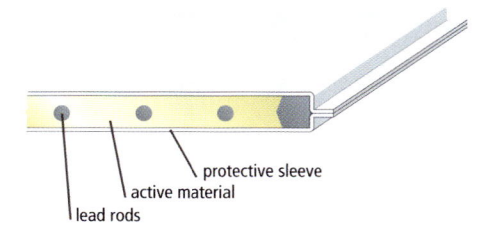

Figure 8.10
Cross-section through a flat plate
Notes: DIN name: OGi block (stationary grid plates).
Source: C. Hemmerle

Block batteries feature high current reliability and high cycle resilience. They reach 1300 cycles at 75 per cent discharge depth, and 4500 cycles at 30 per cent discharge ('Vb' range, Varta bloc). Because of the large reserve of acid in the housing, maintenance is required only about every three to five years. This battery type is often used in PV systems since even very small currents are useful for recharging and very good charging efficiency of approximately 95 per cent to 98 per cent is achieved.

8.3.3 Operating behaviour and characteristics of lead-acid batteries

BATTERY CAPACITY

The capacity C of an accumulator is the quantity of electricity that can be discharged under the respective discharge conditions until the battery is fully discharged. The nominal capacity is the product of the constant discharge current I_n and the discharge time t_n:

$$C_n = I_n \times t_n$$

The nominal capacity is determined by the geometry and number of parallel-wired cells. However, this is not a constant value. It depends upon the temperature, the discharge cut-off voltage and, above all, upon the discharge current. With a low current discharge, the depositing of sulphur in the plates takes place slowly. This achieves a greater penetration depth than with a greater current discharge. With high current discharge, the sulphur molecules deposited at the beginning block the rapid penetration of the following molecules. The result is that more power can be discharged from a battery by discharging it slowly at low currents than by discharging it rapidly at high currents. Hence, the manufacturer specifies a rated capacity (i.e. the electricity that a battery can supply when fully charged) under defined conditions. The rated capacity always stands in relationship to the associated discharge current or – as it is usually represented in the trade – the discharge time that the capacity relates to.

If the total power of a battery is discharged in ten hours, a much greater current flows than if it were discharged over 100 hours. Figure 8.11 shows a typical example of this relationship: for a 100-hour discharge, the battery has a capacity of $C_{100} = 100$Ah (i.e. it can be discharged for 100 hours at a current of 1A). If the same battery is discharged at 8A, the discharge cut-off voltage is reached after ten hours and only 80Ah can be discharged. The battery's C_{10} capacity is 80Ah. The usual rated capacities are C_{10} for stationary, C_{20} for starter and C_{100} for so-called *solar batteries*.

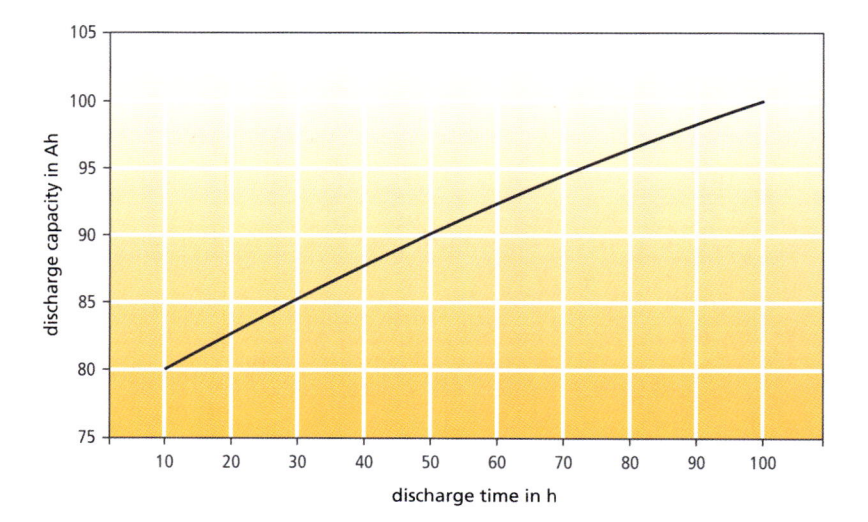

Figure 8.11
Relationship between discharge time and dischargeable capacity with the example of a lead-acid battery with grid plates
Source: C. Hemmerle

CURRENT

Like the capacity, the current is specified depending upon the charging or discharging period. While the charge current is produced by the PV module/s, the discharge current depends upon the load. Typical currents for a battery in a solar system are as follows.

Maximum charge current:

$$I_{20} = C_{20} / 20\,h$$

Medium charge current:

$$I_{50} = C_{50} / 50\,h$$

Medium discharge current:

$$I_{120} = C_{120} / 120\,h$$

VOLTAGE

The nominal voltage of a lead-acid battery is 2V per cell. Conventionally, 12V batteries have 6 cells, and 24V batteries 12 cells in series. OPzS and OPzV batteries are generally offered as single cells. In operation, the voltage at the electrodes

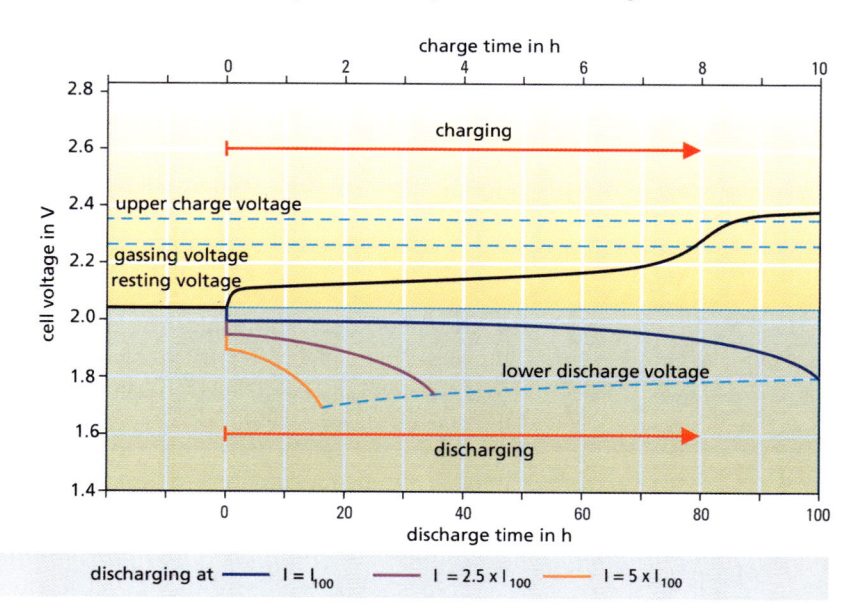

Figure 8.12
Voltage progression during charging (top) and discharging (bottom) with constant current, with the example of a 100Ah battery (C100)
Source: C. Hemmerle

fluctuates according to the operating conditions. To protect the battery, two limit values are defined, which must be adhered to. During charging, the charge cut-off voltage is the upper limit. For discharging, the discharge cut-off voltage is the permissible lower limit. Another parameter is the gassing voltage, above which a battery starts gassing significantly in the charging process. The change in voltage during charging and discharging is explained in more detail below.

The open-circuit or no-load voltage of a battery is called the resting voltage and cannot be measured immediately after charging or discharging since a thermodynamic equilibrium has to set in first. The resting voltage relies upon the electrolyte concentration. Depending upon the state of charge and the battery type, it fluctuates somewhere between 1.96V and 2.12V per cell (see Table 8.1). In practice, for example, it is somewhere between 12V and 12.7V for a 12V battery.

CHARGING AND DISCHARGING

During the charging process, the battery voltage gradually increases. At some point it reaches a value at which slight gas formation begins (water separates into hydrogen and oxygen). The escaping gas mixture is highly explosive (oxy-hydrogen gas). Close to this gassing voltage, the manufacturer defines a charge cut-off voltage for the specific battery. This voltage must be controlled with a charge controller so that the battery does not sustain any damage through lack of fluid and an excessive acid concentration, and so that the risk from oxy-hydrogen gas is minimized. Since the charge cut-off voltage depends upon the temperature, the charge controller also needs to measure the temperature and include this in its charge algorithm.

So-called *solar batteries* are usually subjected to a daily cycle characterized by charging in the daytime and discharging during the night. A typical daily discharge may range from 2 per cent to 20 per cent of the battery capacity. In addition, there is a seasonal cycle. During the winter months, the low solar irradiance results in a low level of energy generation. Depending upon the discharge depth permitted by the charge controller, the battery's state of charge may fall to 20 per cent of the rated capacity. During longer periods of time with low solar irradiance, it is possible that the energy produced by the PV array will not be sufficient to recharge the battery fully and the cycles take place at a low state of charge. In contrast, with high levels of irradiance in summer, the battery is operated with a high state of charge, typically between 80 per cent and 100 per cent. Here, there is a risk of overcharging. Between the opposing requirements of restoring the maximum state of charge as soon as possible in the seasonally variable charging phase, but without overcharging the battery, the planner needs to set the maximum permissible battery voltage at the charge controller. A typical threshold value for the maximum cell voltage for lead-acid batteries is 2.4V for each cell. Some charge controllers enable this voltage to be exceeded for short periods (e.g. equalization charging or fast charging).

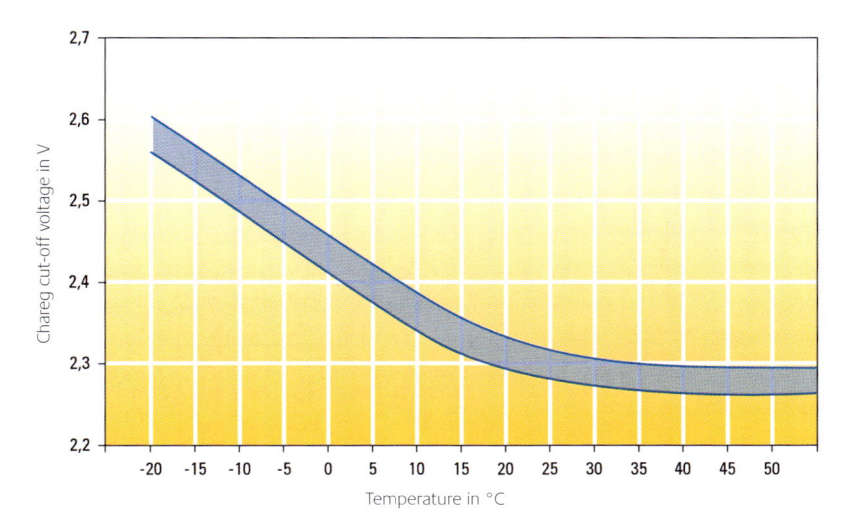

Figure 8.13
Temperature dependence of the charge cut-off voltage with the example of a lead-acid grid plate battery
Source: C. Hemmerle

As Figure 8.13 shows, the charge cut-off voltage of a battery at 20°C is around 2.3V to 2.35V per cell. It falls at higher temperatures (+ 40°C –> 2.25V to 2.3V). At lower temperatures it rises (0°C –> 2.4V to 2.45V). The ageing state of the battery also has an influence on the charge cut-off voltage. It is generally reached somewhat earlier in older batteries than in new batteries. Intelligent charge controllers are able to allow for this.

Despite the formation of oxy-hydrogen gas and loss of fluid, a certain occasional gassing of the battery is advantageous because the rising gas bubbles mix the acid. This increases the service life and performance of the battery. However, the gassing increases corrosion and reduces efficiency. In addition, this results in requirements for maintenance (lost water must be replaced) and for installation in a separate ventilated room. Heavy overcharging does not increase the energy stored in the battery. It results in capacity loss and/or overheating.

As the battery is discharged, the output voltage changes. After an initial jump reduction owing to ohmic losses, it falls continuously, decreasing more strongly towards the end of the discharging process until the discharge cut-out voltage is reached. If the battery is discharged further and the voltage falls below the discharge cut-out voltage (deep discharge), the acid concentration undergoes a strong reduction and sulphation occurs. Deep discharging should therefore be avoided under all circumstances. As can be seen in Figure 8.13, the level of the discharge current determines the change in voltage over time and the value of the discharge cut-out voltage. With a higher discharge current (i.e. with a faster discharge), it is true that the discharge cut-out voltage falls; but overall, less energy can be discharged.

STATE OF CHARGE

The currently available portion of the battery capacity is its state of charge. Especially in autonomous power supplies, it is important to know how much energy there is left in the 'tank'. Measuring this exactly in PV systems turns out to be extremely difficult owing to the irregular cyclical operation with rare full charges. Two relatively simple methods make it possible to determine the state of charge.

For unsealed batteries (with fluid electrolyte), the acid density can be measured using a hydrometer or a special sensor. Here it should be ensured that the electrolyte is well mixed. Since the density is roughly proportional to the acid concentration, it is possible to infer the state of charge directly from the acid density. The limit values depend upon the battery type. Typical values are listed in Table 8.1.

With sealed batteries (with immobilized electrolyte), the acid density cannot be measured. Here the charge status can only be determined by measuring the voltage. However, for an estimate of the state of charge, according to Table 8.1, cell voltage first has to level out at the resting voltage. Therefore, the battery should be in an open-circuit condition for at least four hours prior to measurement. Even this method does not supply exact results as the acid stratification and ageing condition have an effect on the battery voltage. In larger stand-alone systems, it is recommended that a precision voltmeter be permanently installed for measuring the battery voltage, as well as additional ammeters for regularly checking the charge and discharge current.

Even if no current is discharged from the battery, chemical reactions are constantly taking place at the electrodes. These slowly discharge the battery. Self-discharging in the case of batteries used in solar systems should not exceed 3 per cent per month.

Table 8.1
Determining the state of charge by measuring the cell voltage or the acid density
Source: Varta Batterie AG (1986); Exide

Charge state	Lead grid plate battery acid density ρ_{in}	Resting voltage V_0	OPzS battery acid density ρ
0%	1.10g/cm³	1.96V	1.10g/cm³
20%	1.13g/cm³	1.99V	1.13g/cm³
40%	1.16g/cm³	2.01V	1.16g/cm³
60%	1.20g/cm³	2.05V	1.19g/cm³
80%	1.24g/cm³	2.08V	1.22g/cm³
100%	1.28g/cm³	2.12V	1.24g/cm³

CHARGE FACTOR, CHARGE EFFICIENCY AND ENERGY EFFICIENCY

The charge factor designates the ratio of supplied charge (the supplied amount of current in Ah) to the dischargeable charge. The ideal charge factor would be 1. However, because of conversion losses, in practice, it is anywhere between 1.02 and 1.2, depending upon the discharge depth and the battery. The inverse value of the charge factor is the charge efficiency, which can be anywhere between 83 per cent and 98 per cent. If no manufacturer's data is available, the charge efficiency of lead-acid batteries can be assumed from Table 8.2.

Table 8.2
Charge efficiency of lead-acid batteries at different states of charge at a reference temperature and a cycle depth of less than 20 per cent of the rated capacity
Source: E DIN IEC 61427:2002-04

Charge state	90%	75%	< 50%
Ah efficiency	> 85%	> 90%	> 95%

By contrast, energy efficiency also takes account of storage losses, and for a new battery is between 70 per cent and 85 per cent, depending upon the charging method. Accordingly, it is calculated from the ratio between the discharged energy (in Wh) and the energy supplied to the battery.

8.3.4 Ageing effects

The major drawback to lead-acid grid plate batteries is their short service life. With 100 to 800 full cycles, they work for three to eight years. Stationary batteries reach a service life of 10 to 15 years. The causes for this are the various reversible or irreversible ageing processes, some of which mutually influence and amplify each other.

Acid stratification (reversible). Because the 'heavy' acid falls to the bottom, a greater acid density prevails in the bottom part of the cell than at the top. This results in a potential difference (see Table 8.1), which leads to a discharge in the lower part. An occasional gassing charge to mix the acid is helpful at each maintenance interval.

Sulphation (irreversible). If the battery is not sufficiently recharged after being discharged, sulphate crystals grow and can no longer be completely transformed back into lead or lead oxide. As a result, in the course of the operating period, the active mass decreases and the dischargeable capacity falls. The lower part of the cell is particularly badly affected by sulphation since a full charge is rarely attained here.

Corrosion (irreversible). Corrosion on the lead grid of the positive pole is caused by the positive potential being high. This leads to increased grid resistance and occurs increasingly at cell voltages of over 2.4V or under 2V. Large-area corrosion scaling that falls from the positive conductor onto the electrodes can cause short-circuiting.

Sludging (irreversible). The changing volume during charging and discharging causes a loosening of the active material. If there is gas formation, it can be flushed out and collects on the bottom of the battery as lead sludge. If the sludge space on the bottom is too small, short circuits can occur.

Drying out (irreversible). If gassing occurs during charging, the lost water must be replaced. If the battery is not topped up with distilled water in time, it dries out and no longer functions (Sauer, 2002).

8.3.5 Selection criteria

In the individual case, the selection of the battery depends upon very many factors and will be influenced by system management and climatic conditions. The special requirements made of the battery in operation can be broadly classified according to the operating time per year, the type of loads (high or low power drain) and the number of cycles per week. However, it is still difficult to make generalizations about which battery type is best for which typical applications as the basic conditions (such as cost, housing capabilities, maintenance capabilities and reliability requirements) can be the deciding factors.

Batteries for use in a PV stand-alone system should have the following features:

- good price/performance ratio;
- low maintenance requirements;
- sufficiently long service life;
- low self-discharging and high energy efficiency;
- can be charged with small charge currents;
- high energy and power density (space requirements and weight);
- vibration-resistant (mobile use or for transportation);
- protection against health and environmental hazards; recyclable.

No storage type fulfils all the stated requirements to the same extent. It has to be decided which are the most important properties according to the respective application.

In sporadically used stationary systems (holiday home, weekend home and summer house), a simple so-called *solar battery* with fluid electrolyte meets the requirements for the PV accumulator and is an excellent choice. It is cost-effective, has low maintenance requirements (all that is required is to check the electrolyte level occasionally and top up with distilled water) and achieves a long service life if it is generously dimensioned. If it can be housed in a protected location, it is safe against external damage and spillage of acid.

Even safer, maintenance free, with a longer life and low cost is the simple gel battery. It fulfils all the requirements well. The gel battery has proven itself in mobile applications as no acid escapes if damage occurs in transit. It is completely held in the gel. Because no gassing occurs, the requirements for the battery accommodation are less stringent. On the other hand, batteries with immobilized electrolyte are very sensitive to overcharging.

In the case of systems that are used all year round, the service life of simple so-called *solar* and gel batteries is generally not sufficient. They need to be replaced every few years. In this case, the use of stationary batteries of block battery, OPzS or OPzV type can be recommended. The OPzV design with immobilized electrolyte (gel) is preferable when a well-ventilated location is not available and when true zero maintenance is desired.

When purchasing batteries, it should also be ensured that the manufacturer's specifications, listed below, are available:

- *Capacity specification with the associated discharge times.* Since manufacturers base the nominal capacities on different discharge times, at the very least the C_{10} and C_{100} values for the 10- and 100-hour power discharge should be specified in order to be able to compare various products.
- *Nominal acid density, acid volume or weight.*
- *A graph or table showing the cycle resilience or the expected service life in years, in relation to the discharge depth* (similar to Figure 8.5). With the aid of this information it is possible to tell which is the most cost-effective product. Bottom-end products only have 400 cycles at 30 per cent discharge depth, whereas a product 50 per cent more expensive achieves double the number of cycles. The cycle resilience relates to the number of full cycles. A full cycle comprises the discharge of the nominal capacity and a subsequent full recharge. The number of cycles with partial discharge (e.g. to 30 per cent discharge depth) can be converted to full cycles by multiplying with the discharge depth (e.g. 400 partial cycles x 30 per cent discharge depth = 120 full cycles). Products for which no cycle resilience is specified should be considered with scepticism.

In general, transparent housings should be preferred as they enable better visual checking with regard to acid level, sedimentation, corrosion of the terminals and possible ice formation. Sealed batteries, however, are not available with transparent housings.

Figure 8.14
Battery system at the Mindelheimer Hütte, Germany: The acid level can be easily checked from the markings on the transparent housing
Source: Fischer Energy Systems

8.3.6 Battery safety and maintenance

In the interests of a long service life and high supply reliability, the accumulator banks should undergo a thorough servicing every six months. This maintenance work comprises:

- keeping the tops of the batteries clean to avoid possible short-circuits resulting from moist dust and grime;
- checking that the terminals are firmly seated;
- checking the electrolyte level, and opening the vent caps and topping up with distilled water where necessary;
- measuring all cell and block voltages and all cell acid densities (with a proper full charge in zero-current state or with very small currents);
- intensive full charge to 2.4V per cell with a gassing phase over several hours (equalization charge) to mix the electrolyte solution (not for gel batteries).

In addition, a thorough visual check should be carried out. The results of regular maintenance make it easier to assess the ageing of the batteries and enable the detection of individual cell failures. Complete documentation of the service work is also helpful here.

When installing and operating battery systems, national codes and regulations should be observed, and the installation, commissioning and maintenance specifications of battery manufacturers should be rigorously adhered to. Batteries represent the greatest source of danger in stand-alone systems. Safety measures are necessary both in respect of electrical safety and for handling corrosive sulphuric acid, and because of the formation of explosive gases (oxy-hydrogen gas). Sulphuric acid is hazardous to humans and the environment, and causes severe burns on contact with the skin, the respiratory organs or the eyes. Gel batteries are significantly less critical in safety terms, as here gassing is a factor of 100 lower, and the housing is sealed.

From the electrical viewpoint, a main fuse is required in all cases between the battery and all other devices (load, controller, PV generator). This applies even for the smallest systems with only one or two PV modules.

Special requirements exist for the installation location, particularly for batteries with fluid electrolyte. This location must be clearly signed from the outside as a 'Battery Room' and on the inside with signs forbidding smoking and open flames. If the nominal voltage exceeds 65V or the nominal power of the charging equipment overall exceeds 2kW, then the battery room is considered an 'electrical plant', and at 220V or more a 'self-contained electrical plant'. The battery room must be sufficiently ventilated.

8.3.7 Recycling

The greatest environmental impact from lead-acid batteries results from heavy metal lead. To prevent harmful substances from polluting the environment, used batteries should not be disposed of with domestic refuse. Local authority collection points and vendors will take back old batteries and see that they are recycled. In continental Europe, lead-acid batteries are 95 per cent recycled and the recovered lead is reprocessed for new batteries. This material cycle enables the minimization of environmental impacts associated with energy storage using lead-acid batteries.

8.4 Charge controllers

In stand-alone systems, the system voltage of the PV array should be matched to that of the batteries; the usual system voltages are 12V, 24V and 48V. The charge voltage must be higher than the battery voltage. For example, with a 12V battery, it can be up to 14.4V. Crystalline standard modules with 36 to 40 solar cells supply a nominal voltage of 15V to 18V. The nominal voltage must be higher than the batteries' charge voltage so that the MPP voltage at higher temperatures is sufficient to charge the batteries. Furthermore, voltage losses occur through the cables and the line diode, which are usually limited to around 1 per cent to 2 per cent. At low temperatures, the modules' MPP voltage is approximately 21V and the open-circuit voltage up to 25V, with the result that the charge cut-out voltage of the battery can be exceeded. The charge controller therefore measures the battery voltage and protects the battery against overcharging. This is achieved by:

- switching out the PV array when the charge cut-out voltage is exceeded, as happens with series controllers; or
- short-circuiting the PV array with a shunt controller; or
- adjusting the voltage with an MPP charge controller.

At low irradiances, the PV voltage breaks down, with the result that the battery discharges via the array. To prevent this, a reverse current diode is used. This is usually integrated within the charge controller.

Optimized operation of batteries over a long service life requires charge controllers to be flexible. The charge cut-out and discharge cut-out voltages are dependent upon the state of charge of the accumulator. In addition, the properties of different battery types (batteries with fluid electrolyte and gel batteries), the temperature and, quite possibly, the age of the battery all need to be taken into consideration. To monitor the battery temperature, a temperature sensor is used, which is connected to the charge controller. High-quality charge controllers can make a big difference in extending the service life of the batteries.

The fundamental tasks of a modern charge controller are:

- optimum charge to the batteries;
- overcharge protection;
- preventing unwanted discharging;
- deep discharge protection; and
- information on the state of charge of the batteries.

*Figure 8.15
Charge controller for outdoor applications with state of charge indicator
Source: Steca*

*Figure 8.16
Charge controller with liquid crystal display (LCD)
Source: Sunware*

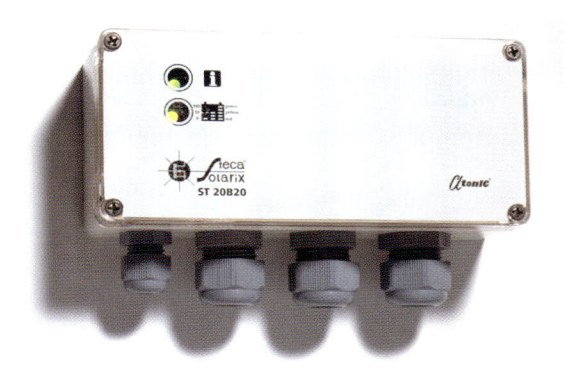

*Figure 8.17
Smallest charge controller 'NANO'
Source: Solarwatt*

MANUFACTURERS OF CHARGE CONTROLLERS

ATT TBB, Heliotrope, Mastervolt, Meyer Solar Technologie, Morningstar, Phocos, Reusolar, Schams Electronic, Solarwatt, Steca, SunSelector, SunWare, Trace, Uhlmann Solarelectronic.

8.4.1 Series controllers

When the charge cut-out voltage is reached, a series controller interrupts the module power using relay or power conductor S_1 and switches it back on after a defined voltage drop. This creates an oscillating state of constant switching operations around the charge cut-out voltage, as well as permanent forward losses. On account of these disadvantages, a charge controller with continuous regulation was developed.

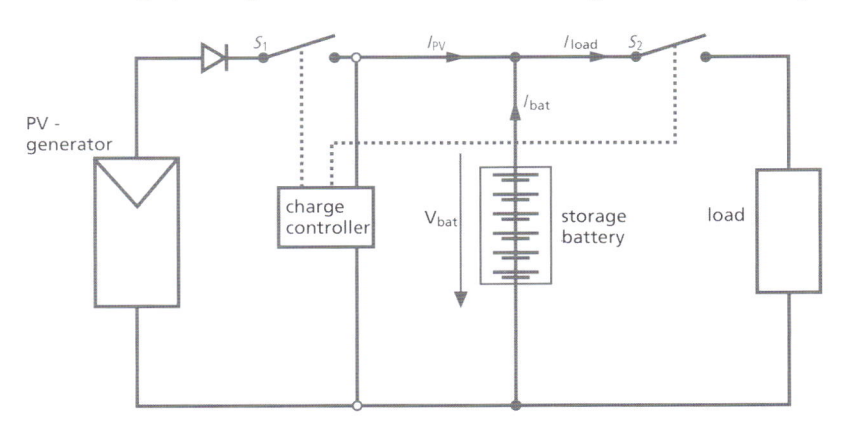

*Figure 8.18
Principle of a series charge controller
Source: V. Quaschning*

8.4.2 Shunt controllers (parallel controllers)

A shunt controller continuously reduces the module power when the charge cut-out voltage is reached. But since the module continues producing power, the unneeded part of the module's power is used as a short-circuit current in the module (i.e. it is converted into heat). The modules can take the short-circuit current without any problems – it merely leads to a slight additional warming. This method is ideal for the battery as charging is safe and swift.

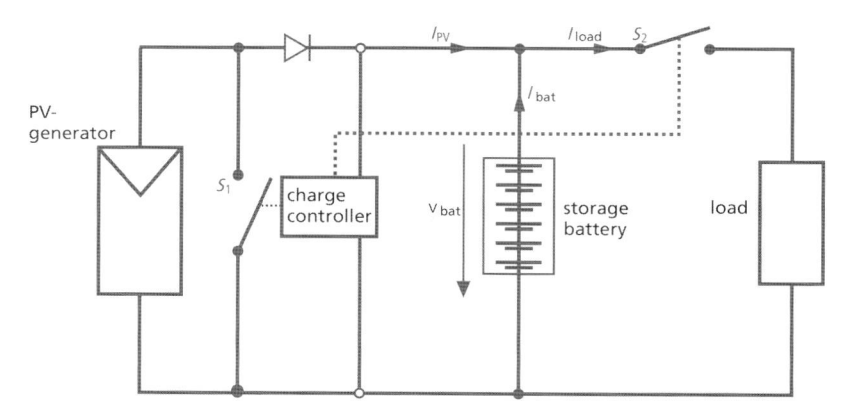

Figure 8.19
Principle of a shunt charge controller
Source: V. Quaschning

8.4.3 Deep discharge protection

A deep discharge protector with relays to isolate the loads from the battery is integrated within most charge controllers (S_2 in Figures 8.18 and 8.19 and S_1 in Figure 8.20). It is useful if there is a signal, in good time before the switch-off point is reached, that the power needs to be switched off. A built-in reset button is helpful in this case. For example, if the battery voltage has fallen below the discharge cut-off voltage because of a brief high current discharge (a refrigerator comes on) and the controller has isolated the load, as a result, after a short recovery break the battery voltage is generally high enough again to be able to power small loads (e.g. a lamp, for a few more hours without the battery voltage falling below the discharge cut-out voltage again). The reset button then enables further power discharge.

The latest charge controllers have a built-in temperature sensor that measures the ambient temperature. Normally the charge controller is sited in direct proximity to the battery, so the temperatures of both devices are very similar. Only in the event that the battery is extremely heavily loaded will it heat up and therefore have a higher temperature than the charge controller. This eventuality can be ruled out by sufficiently dimensioning the batteries. An external temperature sensor mounted on the battery can prove useful here too. A readout display for the most important electrical variables, such as state of charge, battery voltage, charge current and load current, is integrated within many devices. An integrated circuit under the brand name Atonic, which is integrated within many charge controllers, offers additional benefits. This has a self-learning algorithm that enables it to record the battery ageing state and make appropriate allowances for it.

Charge controllers can only accommodate a limited current, both on the module side and on the load side. To protect the sensitive power electronics, an appropriate fuse is built in. Generally, the same limit values are set for the maximum permitted module current and the maximum permitted load current, which means a common fuse is built in for both currents. In commercially available devices, the standard values for the maximum current absorption range from around 5A to 30A. In larger systems with even higher currents, either special models have to be used or the system is split up into a number of groups so that the whole system does not fail if breakdowns occur. On the battery side, the currents are brought together again – otherwise there is the risk that the battery groups will not be under equal loads.

8.4.4 MPP charge controllers

Since the battery voltage determines the operating point on the PV characteristic curve and, therefore, the PV array is often not operating at the MPP point, series and shunt charge controllers do not always make optimum use of the available solar energy. These power losses can amount to between 10 per cent and 40 per cent, depending upon battery voltage, irradiance and temperature. This can be avoided by using an MPP tracker. An MPP tracker essentially consists of a regulated DC/DC converter. Regulation is performed by an MPP tracker, which approximately every five minutes passes along the current/voltage characteristic curve of the PV array and determines the MPP power. Then the DC/DC converter is set so that it takes the optimum power from the PV array and adjusts it to the charge voltage of the battery.

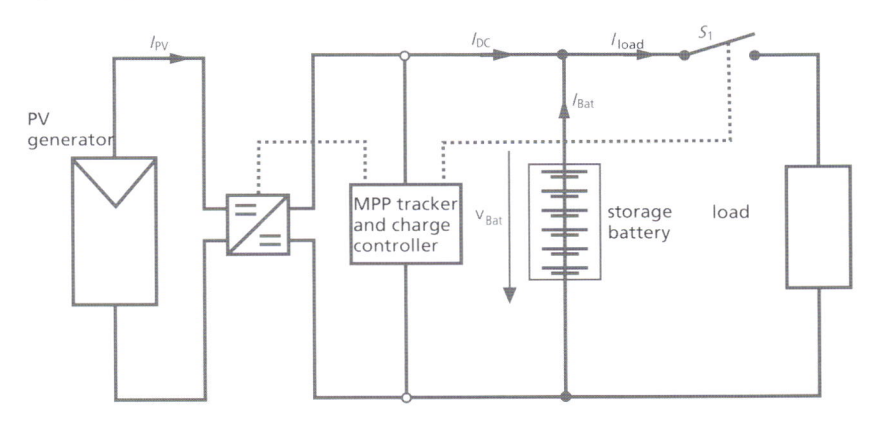

Figure 8.20
Principle of an MPP charge controller
Source: R. Haselhuhn

The efficiency of the DC/DC converter is around 90 per cent to 96 per cent. The use of an MPP tracker only makes sense with PV arrays of 200Wp and above. With lower-powered arrays, the converter's conversion losses are generally higher than the gain from the controller. Since the circuitry involved is more complicated, the price of the MPP charge controller is somewhat higher. This means that for reasons of economy, at present, MPP charge controllers are often only used with PV arrays of 500Wp and more.

Figure 8.21
MPP charge controller
Source: Meyer Solar Technologie

Figure 8.22
MPP charge controller
Source: Schams-Elektronic

8.5 Stand-alone inverters

In a PV stand-alone system, the storage is realized by the batteries and the operation of numerous loads using direct current. To be able to use conventional 230V AC loads, as well, on a DC grid, stand-alone inverters are used. Some inverters even have integrated charge controllers.

The object of a stand-alone inverter is to enable the operation of a large range of loads. These range from rugged construction tools through domestic appliances to sensitive electronic devices in communications technology.

The following requirements may be made of a stand-alone inverter:

- alternating current that is as sinusoidal as possible with a stable voltage and stable frequency;
- very good conversion efficiency, even in partial load range;
- high overload capability for switch-on and starting sequences;
- tolerance against battery voltage fluctuations;
- economical standby state with automatic load detection;
- protection against short-circuit damage on the output side;
- high electromagnetic compatibility (good EMI suppression);
- low harmonic content;
- surge voltage protection;
- bidirectional operation (i.e. conversion from AC to DC is also possible so that batteries can be charged from AC generators, if necessary).

Three different types of inverter concepts dominate the market: sine-wave, 'modified sine-wave' and the square-wave inverters.

MANUFACTURERS OF STAND-ALONE INVERTERS

ACR, ASP, FEG, Fronius/Steca, Ingeteam, IVT, Mastervolt, Microtherm, Outback, Panelectron, PDA, Power Master Technology, ProWatt, Siemens, SMA, Studer, SunPower, Trace, Victron Energy, Wagan, Xantrex.

Figure 8.23
Stand-alone inverter with optional charge controller function
Source: Steca/Fronius

Figure 8.24
Bidirectional stand-alone inverter with integrated charge controller, also known as inverter-chargers
Notes: These can be 'connected to the grid' but are not to be confused with the inverters used in grid-connected systems – inverter-chargers can take power from the grid to charge a battery bank, but cannot put power onto the grid.
Source: SMA

Figure 8.25
Sine-wave inverters suitable for PV stand-alone systems
Source: ASP; SOLON

8.5.1 Sine-wave inverters

The requirements listed above are best fulfilled by sine-wave inverters. These devices work on the principle of pulse width modulation. They are suitable even for operating sensitive electronic equipment. Compared to square-wave inverters, sine-wave inverters are higher in price as a result of the greater complexity of their circuitry.

8.5.2 'Modified sine-wave' inverters

A quality 'modified sine-wave' inverter will meet most of the above requirements, but not all. As the complex electronics required for sine-wave inverters has matured, many manufacturers are now phasing these out and replacing them with sine-wave inverters. If using a 'modified sine-wave' inverter, one should check its suitability for a particular application with the manufacturer unless the application is specified in the inverter manual.

8.5.3 Square-wave inverters

Square-wave inverters are very common and cheaper. The direct current is chopped into a 50Hz alternating current with square characteristics and stepped up using a transformer to a voltage of 230V. They can be very inefficient and are not recommended. Sensitive equipment can be damaged.

8.5.4 Application criteria for inverters in stand-alone systems

12V or 24V DC systems quickly come up against their limits when large loads are to be supplied and long cable runs are unavoidable. Either no DC model of the desired load is available, or it is disproportionately expensive, or transferring the substantial amounts of power results in extreme cable cross sections in the low voltage range, which also pushes costs up. Often it is advantageous to lay a DC grid for the small loads (light, etc.) and assign the large loads to an inverter. Converting the whole supply to 230V AC using inverters is not the ideal solution since high currents are generally only drawn for short periods. Most of the time, therefore, the inverter would be working in the lower partial load range with inevitably high conversion losses. Separate DC and AC systems make it possible to select a smaller and, therefore, lower-cost inverter and to utilize it better.

When selecting the power class of the inverter, the nominal power ratings of the loads can be taken as a guide:

$$\text{Nominal power of inverter} = \text{sum of the nominal powers of all AC loads} + \text{capacity margin}$$

The safety reserve is dimensioned according to how many loads with high start-up power will go into operation at the same time and how big, conversely, the short-term overload capability of the inverter is. In large systems with many loads, it may be necessary to use a load management system to ensure that only one or two loads can be started up at the same time. This saves inverter capacity and costs to a considerable extent.

Inverters with low power consumption can be connected to the charge controller as a load, provided the maximum power, together with the power of the other connected loads, does not exceed the maximum permissible value for the charge controller. Generally, however, inverters are connected directly to the battery as the currents they draw are too high (especially when starting up a larger 230V load). This would place too great a demand on the charge controller, causing the fuse in the charge controller to blow. However, direct connection to the battery means that the inverter must, in all cases, have an integrated deep discharge monitoring system.

8.6 Planning and designing stand-alone systems

This section describes a simplified procedure for designing stand-alone systems. It does not achieve the precision of a simulation program; however, the correct use of simulation programs by the user requires fundamental knowledge of PV stand-alone systems. They owe their precision to a considerable amount of complex input data – which, in practice, can only be obtained with great difficulty – and to the considerable expertise of the user in handling the data. In this respect, the design route described here should be seen as a first, but sensible, step on the way to producing a practicable design.

The most important task when planning stand-alone systems is to balance the energy consumption with the supply. Since solar energy is a limited and fluctuating resource, the daily electricity consumption, radiation level and battery or battery bank capacity must be realistically calculated and balanced with one another. The situation becomes difficult when the intention is to supply electricity throughout the year. At Central European latitudes, the available radiation amount is about six times greater in summer than in winter. The electricity demand, however, is generally greater in winter than in summer. Designing the PV array for winter requirements can therefore lead to a considerable unused surplus during the rest of the year. This would make it particularly uneconomic. The situation is also unsatisfactory from an ecological point of view. A PV array that can only be used for a fraction of its service life leads to very long energy payback times. Here, it might be considerably more sensible to use a hybrid system consisting of a PV array and an auxiliary combustion-engine generator and/or a wind generator. In some locations, sun and wind complement each other, with solar energy available in summer and wind energy available in winter. A diagram of a simple PV stand-alone system is illustrated in Figure 1.15 in Chapter 1.

8.6.1 Direct coupling of PV array, battery and loads

In order to achieve a greater understanding of the interaction between batteries and PV modules, this section briefly describes the electrical process within a directly coupled stand-alone system. Directly coupled stand-alone systems are used mostly for small-scale systems.

A small PV module is connected with a reverse diode with a battery (see Figure 8.26). At night, when the PV module does not provide any voltage, the battery must be prevented from discharging itself via the inner resistance of the PV module. The reverse diode ensures that the battery does not drive any reverse current through the unlit PV module. Otherwise it would become a load and would convert the energy to heat. The reverse diode is required to avoid these battery losses and to protect the PV module from thermal overloading. When the PV module is illuminated, the reverse diode causes an additional voltage loss of around 0.5V to 0.7V.

Figure 8.26 illustrates the I–V curve of the PV module and the dynamic characteristic curve of the battery. It can be seen that the battery voltage fluctuates around the open-circuit voltage of around 12.5V. It can drop below 11V before reaching the discharge cut-off voltage or increase to a charge cut-off voltage of more than 14V. Connecting the PV module to the battery shifts the battery characteristic curve by the value of the PV current from the module I–V curve. Point B on the battery characteristic curve would otherwise lie directly on the voltage axis.

In our example, three different loads can be connected to the battery via a change-over switch. Different operating conditions occur depending upon which load (resistance R_A, R_B or R_C) is connected. In case B, the resistance of the load is so large that the current for load B is completely delivered by the PV array. The resistive load line cuts the PV and the battery characteristic curve precisely at their point of intersection. Load A requires greater current. The PV module provides only part of the required electricity; the rest is provided by the battery. Thus, a discharge current flows and the battery is discharged. If load C is connected, a smaller current develops because of the higher resistance R_C. The PV current is greater than the current

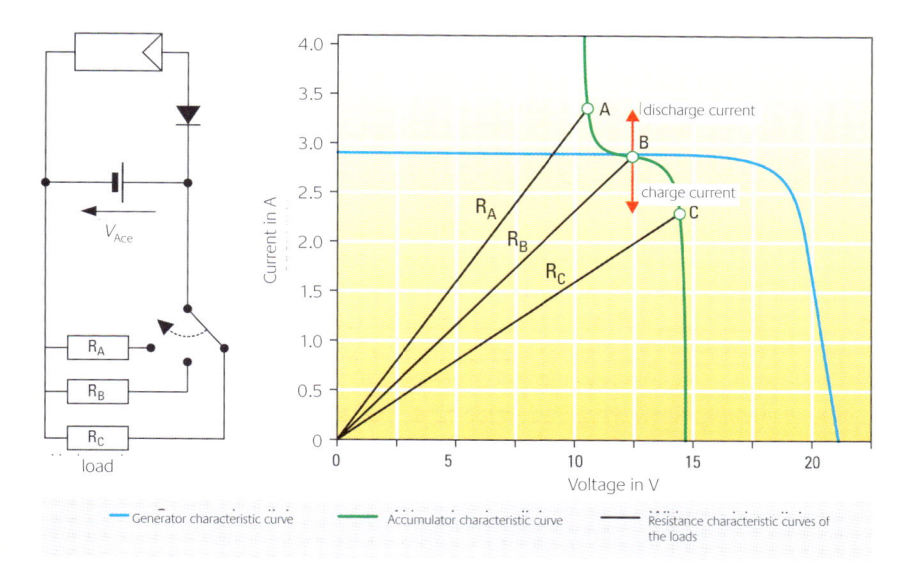

*Figure 8.26
Directly coupled PV stand-alone system
and PV module/battery I–V curve
Source: R. Haselhuhn*

— Generator characteristic curve   — Accumulator characteristic curve   — Resistance characteristic curves of the loads

required by the load; in addition, a charging current of around 0.9A flows. The PV module can therefore supply load C and charge the battery, as well. This system will operate reliably and with little need for maintenance if, taking into consideration the voltage drops of the reverse diode and the PV cables, the system is sized so that the charge cut-off voltage is not exceeded and the battery capacity is so tailored to the load needs that the discharge cut-off voltage is not undercut.

Charge controllers considerably improve the functioning of stand-alone systems. The charge controller essentially ensures that the battery is not over-discharged or overcharged (see section '8.9.1 Charge controller cable'). An MPP tracker, if incorporated, optimizes exploitation of solar energy.

The following section describes the design of a stand-alone system using the example of a small holiday home, in which despite the considerable summer–winter fluctuations, it is intended to achieve a year-round solar energy supply.

8.7 Measuring electricity consumption

The most important and complex stage in sizing a stand-alone PV system is providing a carefully worked out breakdown of the daily electricity consumption. This is listed in Table 8.3 using the example of a small holiday home.

The considerable radiation variation during the course of the year in northern latitudes needs to be taken into consideration when designing a system (this variation decreases the further south one is). This makes it necessary to take the level of solar radiation into consideration for the different months and seasons, or at least between the extremes of summer and winter. In the example below, the period from May to August was taken for the summer season and the period from September to April for the low radiation winter season. The calculation of the radiation energy is based each time on the weakest month, taking into account the location, inclination and temperature. The radiation calculation will be dealt with in detail in Chapter 9.

First of all we need to estimate the consumption of all the individual electric loads. All intended electric loads and their respective power consumption are listed with their probable daily operating times and their daily consumption amounts.

Table 8.3 shows the three main loads: refrigerator, water pump and television set. The refrigerator requires the most electricity. It will only be realistic to have a year-round solar supply if the refrigerator is switched off in winter, otherwise, winter consumption would more than double the size of the solar array. The water pump is an ideal load for a solar thermal system because, in winter, it requires only one tenth of the summer consumption. The rate of consumption is, in other words, synchronous with the radiation level of the sun.

Table 8.3
Power consumption table for a small holiday home
Notes: 1 Here it is planned to have a high-quality, energy-saving refrigerator. Modern refrigerators are always supplied with information on daily standard use. Since we are interested in this numerical value, information on the daily running time of the cooling unit is not necessary for our calculation.
2 To save energy, the refrigerator will not be used in winter. As we will see with the following calculation of the PV array, use of the refrigerator in winter would lead to considerable enlargement of the array and, thus, to much higher system costs.
3 In summer, it is planned to irrigate the garden with around 2000 litres of water per day. In winter, only water for domestic use will be needed, at around 200 litres per day. The pump supplies around 600 litres per hour. This results in the listed operating periods of 3.33 and 0.33 hours per day.
Source: R. Morsch

Load	Nominal power P_n in watts (W)	Daily operating time in hours		Daily consumption in watt hours (Wh)	
		Summer	Winter	Summer	Winter
Three lamps in the living area	3 x 12 = 36	1	3	36	108
One lamp in the sleeping area	12	0.5	0.5	6	6
Two reading lamps in the sleeping area	2 x 7 = 14	1	1	14	14
Refrigerator	50	Unknown[1]	Switched off[2]	300	Switched off
TV set	50	2	2	100	100
Water pump[3]	60	3.33	3.33	200	20
Totals	222			656	248

If the customer intends to have a comprehensive supply, then it is necessary to consider providing electricity reserves by increasing the size of the generator or by using an auxiliary device (hybrid system). Ensure that there are only high-quality, energy-saving loads. Their higher procurement costs are certainly worthwhile because this enables savings to be made on module costs that would otherwise be considerably higher than the additional costs for such high-quality, energy-saving appliances.

8.8 Sizing the PV array

After the daily electricity demand has been ascertained, the correct size of the PV array needs to be determined. There are different approaches to determining the yields of the diverse solar module types available on the market. The most sensible procedure would be to base this on the nominal power of a module at STC (see section '8.2 Modules in stand-alone PV systems').

8.8.1 Model for calculating the yield of a PV array

The following explanation is intended for readers who would like to take a more in-depth look at the material. Those who are more interested in the result than its derivation can skip this section and turn to the calculation examples in section '8.8.4 Brief summary of the calculation method for designing a PV array, taking the example of the small holiday home'. The example is for Berlin, Germany.

Table 8.4
Factor Z_2, average daily global radiation on horizontal surfaces in kWh/m²/day

	Jan	Feb	Mar	Apr	May	Jun	Jul	Aug	Sep	Oct	Nov	Dec
Station												
Birmingham, UK	0.65	1.18	2.00	3.47	4.35	4.53	4.42	3.87	2.67	1.48	0.83	0.45
Brisbane, Australia	6.35	5.71	4.81	3.70	2.90	2.43	2.90	3.61	4.93	5.45	6.33	6.32
Chicago, US	1.84	2.64	3.52	4.57	5.71	6.33	6.13	5.42	4.23	3.03	1.83	1.45
Dublin, Ireland	0.65	1.18	2.26	3.60	4.65	4.77	4.77	3.68	2.77	1.58	0.77	0.45
Glasgow, UK	0.45	1.04	1.94	3.40	4.48	4.70	4.35	3.48	2.33	1.26	0.60	0.32
Berlin	0.61	1.14	2.44	3.49	4.77	5.44	5.26	4.58	3.05	1.59	0.76	0.46
Houston, US	2.65	3.43	4.23	5.03	5.61	6.03	5.94	5.61	4.87	4.19	3.07	2.48
Johannesburg, South Africa	6.94	6.61	5.90	4.80	4.35	3.97	4.26	5.10	6.13	6.45	6.57	7.03
London, UK	0.65	1.21	2.26	3.43	4.45	4.87	4.58	4.00	2.93	1.68	0.87	0.48
Los Angeles, US	2.84	3.64	4.77	6.07	6.45	6.67	7.29	6.71	5.37	4.16	3.13	2.61
Melbourne, Australia	7.13	6.54	4.94	3.20	2.13	1.93	2.00	2.71	3.87	5.26	6.10	6.68
New York, US	1.87	2.71	3.74	4.73	5.68	6.00	5.84	5.39	4.33	3.19	1.87	1.48
Philadelphia, US	1.94	2.75	3.81	4.80	5.55	6.10	5.94	5.42	4.37	3.23	2.13	1.68
Phoenix, US	3.29	4.36	5.61	7.23	8.00	8.17	7.39	6.87	5.97	4.84	3.57	2.97
Sydney, Australia	6.03	5.54	4.23	3.07	2.61	2.33	2.55	3.55	4.63	5.87	6.50	6.13
Toronto, Canada	1.58	2.54	3.55	4.63	5.77	6.30	6.29	5.45	4.03	2.68	1.37	1.16
Vancouver, Canada	0.84	1.75	3.00	4.27	6.03	6.50	6.52	5.42	3.80	2.06	1.03	0.65

Table 8.4 lists the monthly mean totals for the daily global radiation on a horizontal plane Z_2 for a range of global locations. If the values are normalized to the irradiance at STC, then the factor Z_2 has the unit hours per day. For tilted areas, these values must be converted with the factor Z_3. The necessary factors for the location are provided in Table 8.5. Sizing and design software can provide all of this information. Module and system suppliers can also often provide information for monthly radiation values for both horizontal and tilted surfaces. In addition, there are also numerous sources on the internet:

- The PVGIS EU Joint Research Centre provides data for Europe, the Middle East and the whole of Africa at http://re.jrc.ec.europa.eu/pvgis/index.htm.
- Radiation data for the US can be obtained from the National Renewable Energy Laboratory (NREL) at http://rredc.nrel.gov.
- Data for all locations can be found at the NASA Surface Meteorology and Solar Energy Data Set at http://eosweb.larc.nasa.gov/sse.

Let's consider an example: the horizontal radiation value in May in Berlin and the surrounding area amounts to 4.77kWh/m²/day). In other words, during the course of an average May day the sun supplies a radiation energy of 4.77kWh per square metre on a horizontal plane (the time between sunrise and sunset in May is about 15 hours; we would receive the same amount of energy by compressing the radiation occurrence to 4.77 hours and having the standard irradiance of 1000W/m² impact upon a surface area of 1m²). A 50W module, for example, laid flat on the ground (horizontal irradiance) would then be radiated for 4.77 hours with the standard illumination of 1kW/m², and thus produce a daily yield of 4.77 hours × 50W = 238.5Wh (provided that its cell temperature remained the same during this period at 25°C).

It is possible to make a simple yield calculation as follows: by interpreting the numerical values from Table 8.4 as hours per day with a standard illumination of 1kW/m² we only have to multiply them with the nominal power of the modules used. For instance, a horizontal solar array of 0.5kWp in May supplies 0.5kW × 4.77 hours per day = 2.385kWh per day.

If we align this solar array to the south at a tilt angle of 45° relative to the horizon, we need to adjust the yield by using a tilt correction factor in accordance with Table 8.5. In May the factor Z_3 for the tilt is 0.94. The yield for our PV array would then be 2.385kWh per day × 0.94 = 2.242kWh per day.

Table 8.5
Factor Z_3 for deviation from the horizontal
Note: The data in this table is for Berlin; it was derived from sizing and design software.
Source: R. Morsch

Module orientation tilt angle	Jan	Feb	Mar	Apr	May	Jun	Jul	Aug	Sep	Oct	Nov	Dec
South, 30°	1.44	1.40	1.17	1.08	1.00	0.96	0.97	1.03	1.17	1.30	1.47	1.42
South, 45°	1.57	1.50	1.19	1.05	0.94	0.90	0.91	1.00	1.18	1.37	1.61	1.55
South, 60°	1.63	1.54	1.15	0.98	0.85	0.81	0.83	0.92	1.14	1.38	1.68	1.61
Southwest/southeast, 30°	1.37	1.33	1.15	1.07	1.00	0.97	0.98	1.03	1.15	1.25	1.40	1.36
Southwest/southeast, 45°	1.48	1.42	1.16	1.05	0.95	0.91	0.92	1.00	1.16	1.31	1.51	1.46
Southwest/southeast, 60°	1.52	1.43	1.12	0.99	0.88	0.82	0.84	0.93	1.12	1.30	1.55	1.49
West/east, 30°	1.01	1.01	0.99	0.98	0.97	0.96	0.96	0.97	0.99	1.00	1.01	1.00
West/east, 45°	0.99	1.00	0.96	0.95	0.93	0.92	0.92	0.94	0.96	0.98	1.00	0.98
West/east, 60°	0.95	0.96	0.91	0.89	0.88	0.86	0.86	0.88	0.92	0.94	0.96	0.94

We must also take into account the deviation of the cell temperature from the standard value. With the exception of the winter months, this is, on average, over 25°C, which reduces the power. The corresponding factors Z_4 are listed in Table 8.6. In May the factor is 0.88. This reduces the yield of our solar array to 2.242kWh/day × 0.88 = 1.973 kWh/day.

	Jan	Feb	Mar	Apr	May	Jun	Jul	Aug	Sep	Oct	Nov	Dec
	1.02	1.01	0.95	0.91	0.88	0.87	0.86	0.86	0.89	0.98	1.00	1.02

Table 8.6
Factor Z_4 for taking the cell temperature into account
Note: The data in this table is for Berlin; it was derived from sizing and design software.
Source: R. Morsch

SUMMARY

The ideal energy yield for the PV array is derived from the product of the nominal power of the PV array and the factors. We can summarize the yield calculation for a PV array in the following equation:

$$E_{ideal} = P_{PV} \times Z_2 \times Z_3 \times Z_4$$

Table 8.7
General parameters for a PV generator
Source: R. Morsch

Parameters	Symbol	Unit
Ideal energy output of the PV array	E_{ideal}	kWh/day
Nominal power of the PV array	P_w	kW_p
Factor from Table 8.4 for taking into account the location and month	Z_2	hours/day
Factor from Table 8.5 for deviation from the horizontal	Z_3	–
Factor from Table 8.6 for temperature correction	Z_4	

8.8.2 Cable, conversion and adjustment losses

With the data for daily electricity consumption of the intended loads, we are able to calculate the daily yield of a solar array in accordance with the geographical location, the season and the orientation of the array. The PV array, load and battery are connected with electrical wiring. In particular, the cables and battery reduce the electricity yield. Voltage losses occur in the cables, and the battery causes conversion and adjustment losses. It is only when we take these losses into account will we know the quantity of electrical energy that is actually available to power the loads.

CABLE LOSSES

When sizing the cables (see section '8.9 Sizing of the cable cross sections'), we will ensure that the losses are restricted to about 3 per cent per cable: this is quite an ambitious goal with low-voltage systems. On the other hand, solar electricity is a valuable commodity that should not be wasted. Only very rarely does it make economic sense to accept high cable losses and to compensate for them by procuring additional modules. In fact, the total loss will be 6 per cent: 3 per cent in the array to charge controller cable and 3 per cent in the battery to charge controller. Therefore, we will apply 6 per cent for the cable losses. This means that we must reduce the array yield by a factor of $V_L = 0.94$.

CONVERSION LOSSES IN THE BATTERY

The conversion of electrical energy into chemical energy and back again into electrical energy, which takes place in the battery, is a process that is difficult to calculate in energy terms since this involves construction details, age, temperature, depth of discharge, and the charge and discharge amperage. Here, it is only possible to make use of experience values. In practice, an average loss of 10 per cent is acceptable. This reduces the PV array yield by another factor of $V_u = 0.9$.

MISMATCHING LOSSES

These losses result from changes in the voltage level during operation: the battery presents the solar array with different voltage levels that are determined by its respective state of charge, the actual voltage dependent upon this and its temperature. This means that the PV array is frequently operating outside its maximum power point. At the same time, it relies upon radiation and temperature. This drifting of the voltage, known as mismatching, must be estimated for with an average energy loss of 10 per cent:

Additional loss factor: $V = 0.9$.

To reduce these losses, it is also possible to use MPP tracking. MPP trackers in the form of DC/DC converters can be deployed very effectively. Charge controllers with MPP trackers are also available. However, the higher associated investment costs must be taken into consideration.

8.8.3 Summary of the design outcome

The PV array is designed around the assumed daily energy consumption W in kWh/day. Depending upon requirements, the average value for summer or winter can be used.

Array design:

$$P_{PV} = \frac{W}{Z_2 \times Z_3 \times Z_4 \times V}$$

with:

$$V = V_L \times V_a \times V_u$$

The amount of solar power still actually available for the electrical loads used can now be reduced to the summarizing formula:

$$E_{real} = E_{ideal} \times V$$

or also:

$$E_{real} = P_{PV} \times Z_2 \times Z_3 \times Z_4 \times V$$

Corresponding to the statements above, the results for the overall losses are:

$$V = 0.94 \times 0.9 \times 0.9 = 0.76.$$

Table 8.8
Summary of design overview
Source: R. Morsch

Parameters	Symbol	Unit
Average daily energy consumption	W	kWh/day
Real energy yield of the PV array (available useful energy)	E_{real}	kWh/day
Ideal energy output of the PV array	E_{ideal}	kWh/day
Nominal power of the PV array	P_{PV}	kWp
Factor from Table 8.4 for taking into account the location and month	Z_2	hours/day
Factor from Table 8.5 for deviation from the horizontal	Z_3	–
Factor from Table 8.6 for temperature correction	Z_4	–
Total of losses from cables, conversion, mismatching	V	–
Summer excess	SE	kWh/day
Winter reserve	WR	kWh/day

DETERMINING THE SUMMER EXCESS AND THE WINTER RESERVE OF THE PV SYSTEM

In Germany, the design of a typical stand-alone installation generally leads to a summer energy excess. For the energetic and economic optimization of the system, the summer excess and the winter reserve are calculated. The level of both values should be agreed with the system's user. The values then determine the load situation. For this, the average monthly daily values of representative months are used for the energy yield of the PV system and energy consumption. The summer excess SE results from the difference between the average solar daily yield and the daily energy consumption W_{summer} in summer:

$$SE = E_{\text{real summer}} - W_{\text{summer}} \text{ in kWh/day.}$$

The winter reserve WR is calculated from the difference between the average daily yield and the daily energy consumption W_{winter} in winter:

$$WR = E_{\text{real winter}} - W_{\text{winter}} \text{ in kWh/day.}$$

Example. In the example above we had, somewhere in the Berlin area, a solar array of 0.5kWp with an angle of incidence of 45°, facing south. In the summer season, represented by the month of May, we can draw a useful energy of 0.5kW × 4.77 hours/day × 0.94 × 0.88 × 0.76 = 1.5kWh/day.

In the winter season, represented by December, we only have 0.5kW × 0.46 hours/day × 1.55 × 1.02 × 0.76 = 0.276kWh/day available.

Hence, the winter requirements of 248W/day for the small holiday home as calculated in Table 8.3 can easily be covered. However, the reserve would only be 11 per cent. In this case, the customer should be consulted on whether they want to try out the 0.5kW PV array first of all, somewhat expanding the system later if required. Allowance could be made for this when building the module base frame, in the choice of charge controller and in the cable cross section so that subsequent expansion is easy to implement.

During summer, there would be a considerable surplus: 0.656kWh/day are consumed, 1.5kWh/day are available – a surplus of 130 per cent. Additional loads can be supplied, and/or longer operating times would be possible.

8.8.4 Brief summary of the calculation method for designing a PV array, taking the example of the small holiday home

REQUIRED DATA

1. Daily energy consumption from Table 8.3:
   In the summer season (May to August): 0.656kWh/day
   In the winter season (September to April): 0.248kWh/day
2. Factor Z_2 from Table 8.4 for the horizontal radiation at the relevant location. Select the month with the lowest radiation in each case (in the summer season this is May; in the winter season, December).
   For the summer season (Berlin, May): $Z_2 = 4.77$ hours/day.
   For the winter season (Berlin, December): $Z_2 = 0.46$ hours/day.
3. Factor Z_3 from Table 8.5 for the orientation of the PV array (less than 45° tilt angle, facing south):
   For the summer season (May): $Z_3 = 0.94$.
   For the winter season (December): $Z_3 = 1.55$.
4. Factor Z_4 from Table 8.6 for taking the cell temperature into account:
   For the summer season (May): $Z_4 = 0.88$.
   For the winter season (December): $Z_4 = 1.02$.
5. Overall factor V for cable, conversion and mismatch losses:
   $V = 0.76$.

The equation for calculating the required array power P_{PV} is as follows:

$$P_{\text{PV}} = \frac{W}{Z_2 \times Z_3 \times Z_4 \times V}$$

Table 8.9 Summary of calculation for designing a PV array using the example of the small holiday home
Source: R. Morsch

Season	Daily energy consumption in kWh/day	Z_2	Z_3	Z_4	V	PV array output in kWp
Summer (May)	0.656	4.77	0.94	0.88	0.76	0.22
Winter (December)	0.248	0.46	1.55	1.02	0.76	0.45

To ensure winter operation with some degree of certainty, we therefore need to choose an array of approximately 0.5kW, even if this is considerably oversized for summer use.

8.9 Sizing of the cable cross sections

For the sizing of the cable cross sections, we point to the extensive discussion of cable sizing in section '4.5 Selecting and sizing cables for grid-tied PV systems' in Chapter 4. The current carrying capacity of the cables and the sizing of fuses must follow national codes and regulations. The method used here follows the German standard VDE 0298 Part 4.

The cable cross section can be calculated according to the following equations:

$$A = \frac{L \times P}{3\% \times V^2 \times \kappa} = \frac{L \times P}{0.03 \times V^2 \times \kappa}$$

Table 8.10
Electrical parameters for the sizing of cable cross-sections

Electrical parameters	Symbol	Unit
Cable length (positive cable + negative cable)	L	m
Power transferred in the cable	P	W
Cable cross section	A	mm²
Electrical conductivity (copper κ_{CU} = 56; aluminium κ_{AL} = 34)	$\kappa\Omega \times$ mm²	
Percentage cable loss (generally 3%)		%
System voltage	V	V

For the design of the PV array in the previous section, we assumed a voltage loss in the cables of 3 per cent. To allow for this condition, sufficient cable cross sections must be selected. The method for doing this is clarified in Figure 8.27.

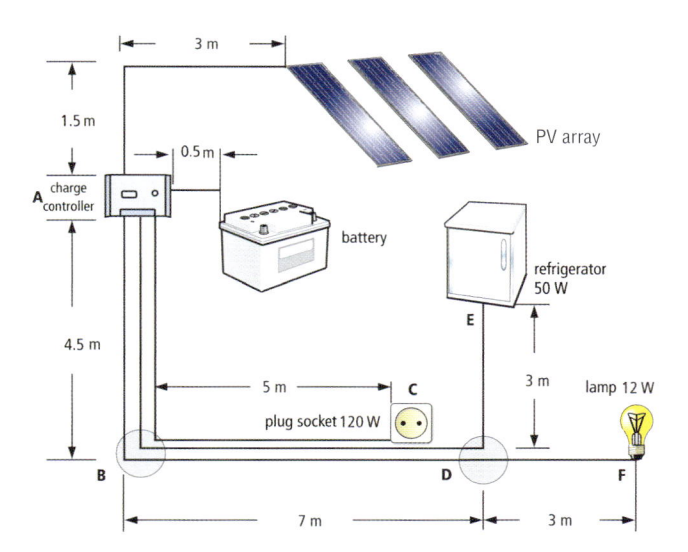

Figure 8.27
Determining the cable lengths
Source: R. Morsch

By way of example, a complete solar power system with PV array, charge controller, battery and three different loads is shown. First, each load has its own cable calculated. If the installation is carried out in this way, each load will receive its own fuse in a power circuit distributor directly after the charge controller.

However, it would also be possible to combine parallel routed cables between the distributor stations A, B and D. In this case, the cross sections of the parallel-running cables must be added. The main cable fuse and the power circuit fuses are selected accordingly. The required cross sections can be calculated by either using the equation above or by through Figures 8.28 and 8.29 and compiled Tables 8.11 and 8.12.

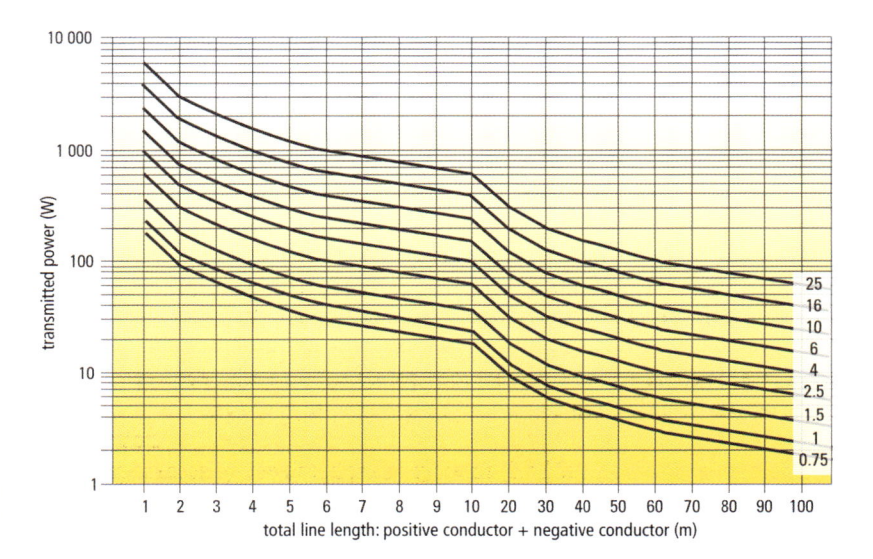

Figure 8.28
Recommended cable cross sections for a 12V system: Cable losses are 3 per cent

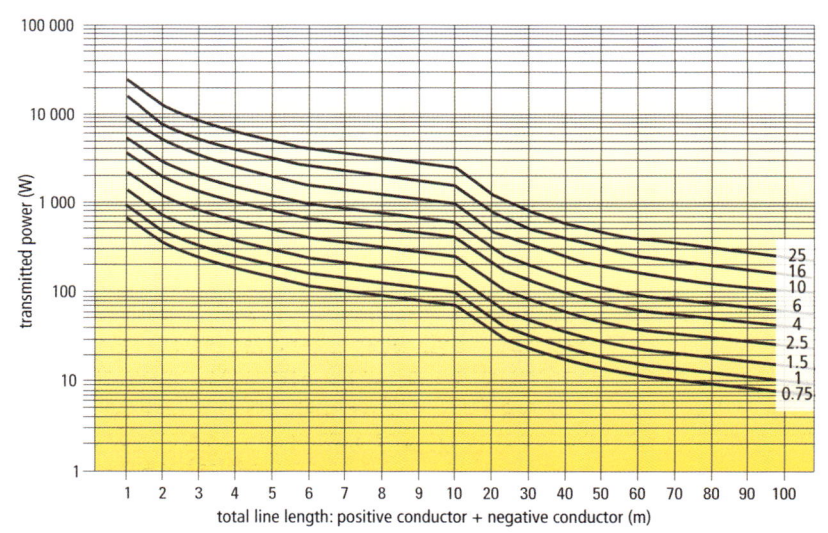

Figure 8.29
Recommended cable cross sections for a 24V system: Cable losses are 3 per cent

Table 8.11
Determining the cable cross sections for a system voltage of 12V
Source: R. Morsch

Cable between the following	Length in m (positive cable + negative cable)	Component	Connected load in W	Calculated ø in mm²	Selected ø in mm²
PV array – battery	10	PV array	500	20.7	25
Battery – A, B, C	20	Plug socket 10A	120	9.9	10
Battery – A, B, D, E	30	Refrigerator	50	6.2	6
Battery – A, B, D, F	30	Lamp	12	1.5	1.5
Combination of strings A, B and B, D (optional)					
A, B	10			17.6	16
B, D	14			7.7	10

Table 8.12
Determining the cable cross sections
for a system voltage of 24V
Source: R. Morsch

Cable between the following	Length in m (positive cable + negative cable)	Component	Connected load in W	Calculated ø in mm²	Selected ø in mm²
PV array – battery	10	PV array	500	5.2	6
Battery – A, B, C	20	Plug socket 10A	120	2.5	2.5
Battery – A, B, D, E	30	Refrigerator	50	1.6	1.5
Battery – A, B, D, F	30	Lamp	12	0.4	0.75[1]
Combination of strings A, B and B, D (optional)					
A, B	10			4.5	4
B, D	14			2.0	2.5

a With fixed wiring cross sectional area ≥ 1.5mm².

The current carrying capacity in accordance with VDE 0298 Part 4 is achieved with separated laying and ambient temperature ≤ 30°C using the calculated cross-section sizes. Because the system voltage for calculating the cable cross section is in the denominator as a quadratic value, doubling the voltage results in a quartering of the cross section. Wiring costs can be reduced by setting up the entire installation as a 24V system. On the other hand, however, there is the difficulty of finding suitable 24V loads. As a result of their scarcity, they are significantly more expensive than 12V loads. The voltage that is ultimately selected depends upon each individual case.

8.9.1 Charge controller cable

This cable is under load both during charging and during discharging. The cable cross section is determined according to which is the greater load. The maximum load occurs either when charging the battery at maximum array power or when discharging the battery through the loads with the greatest permissible simultaneity factor in the absence of irradiance. There are application scenarios (e.g. in systems that are only used at weekends) in which the array's sizing can be significantly reduced as it is able to recharge the battery on weekdays so that there is enough power available at the weekend (one example of this is presented below). In this case, the discharge power can be greater than the charge power. The cable cross section must then be based on the discharge power with the relevant simultaneity factor (i.e. be equivalent to the sum of all load cross sections).

8.10 Battery sizing

The battery's task is to compensate for the non-simultaneity of energy supply and energy consumption. It is designed to supply energy for 2 to 15 days under specified conditions without or with only minimal solar irradiance. In northern latitudes, considerable radiation fluctuations depending upon the time of year must be taken into account. It therefore makes sense to provide a reserve of at least two to three days for the summer months, and of at least three to five days for the winter months (time of autonomy). The battery capacity is stated in amp hours (Ah). So far, we have stated the energy consumption in watt hours (Wh). To be able to relate the values Ah and Wh to each other, we convert the consumption figures into Ah by dividing the Wh by the system voltage (e.g. 12V).

Example. 656Wh correspond to 656VAh/12V = 54.7Ah.
In order to gain a sufficiently long service life with fluid-electrolyte batteries, we may not plan for the total capacity as specified by the manufacturer. Instead, we can only plan with 50 per cent of this value. We therefore need to select the battery capacity C_n to be twice as large as was calculated from the consumption values. The following empirical equation can be used to determine the battery capacity:

$$C_n = \frac{2 \times W \times F}{V_n}$$

Here, F is a factor for the reserve days, W the respective average daily consumption and V_n the system voltage output by the battery (see Table 8.13).

Table 8.13 Average value of reserve days in summer and winter
Source: R. Morsch

Factor for the reserve days	Season	Days	Average value
F	Summer	2 to 3 days	2.5
F	Winter	3 to 5 days	4

The actual factor used here for the autonomy days should be discussed with the user and has to be increased in relation to the demands on supply reliability.

The results for the 'small holiday home' example is presented in Table 8.14.

Table 8.14 Determining battery capacity
Source: R. Morsch

Calculation of the battery capacity for a 12V system					
Daily consumption		Reserve		Battery capacity = double the reserve	
Summer	Winter	Summer (2.5 days)	Winter (4 days)	Summer	Winter
656 Wh/day	248Wh/day	1640Wh	992Wh		
Converted to Ah:		137Ah	83Ah	274Ah	166Ah
Calculation of the battery capacity for a 24V system					
Daily consumption		Reserve		Battery capacity = double the reserve	
Summer	Winter	Summer (2.5 days)	Winter (4 days)	Summer	Winter
656Wh/day	248 Wh/day	1640Wh	992Wh		
Converted to Ah		68Ah	41Ah	136Ah	82Ah

Of course, we need to select the greater of the two values – the summer value – since a sufficient reserve must also be available during summer.

On first appearances, in the 24V system the procurement costs are only half. But on closer reflection it becomes clear that to create 24V we need twice as many battery cells (we need 6 cells in the 12V system and 12 in the 24V system). In principle, we have the same costs. Although, in the 24V system, we only need cells to be half as big, we need twice as many of them!

8.11 Use of an inverter

Not all desired loads are available in 12V or 24V versions. Sometimes it is very expensive to purchase a new DC appliance, or a high-value AC appliance is already available and it is more practical to continue using it. In all cases, it can be of interest to convert the direct current into alternating current at the level of 230V (see section '8.5 Stand-alone inverters'). In our small holiday home example, it could make sense to operate the refrigerator via an inverter because an energy-saving 230V fridge might be available, and buying a new 12V fridge is more expensive than buying an inverter. In addition, during the summer season with its high current surpluses, the inverter would open up new possible uses, such as operating electric power tools.

It would now be possible to either purchase a smaller inverter with, for example, 400W and 50 per cent overload capability (the price for a high-end sine-wave inverter of this capacity is currently approximately 350 Euros). In this case, we would be restricted in the use of electric power tools to smaller equipment, or we could opt for a significantly larger product in order to expand possible uses. Thus, one or two 230V plug sockets would be installed after the inverter, and for the other loads (lighting, television and water pump), a 12V or 24V grid would be laid.

It makes little sense to convert all loads to 230V and have a correspondingly large inverter. The inverter would be very expensive and for 80 per cent to 90 per cent of its operating time would be running under extremely unfavourable partial load conditions, with poor conversion efficiency.

With long cable lengths and in larger stand-alone systems, the use of a central stand-alone inverter is recommended.

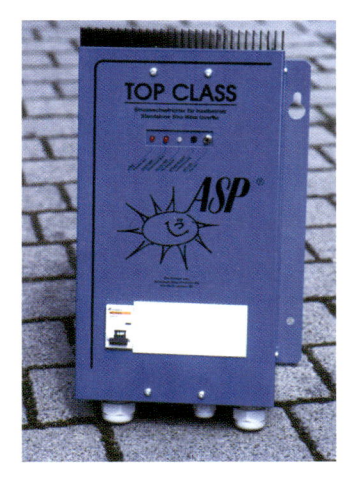

Figure 8.30
Stand-alone inverter
Source: ASP

Figure 8.31
Stand-alone system on a small holiday home
Source: R. Morsch

Figure 8.32
Small stand-alone system for a chalet in the mountains
Source: R. Morsch

Figure 8.33
Camper van with PV stand-alone system
Source: Solara

8.12 Photovoltaics in decentral electricity grids/mini-grids

Worldwide, there are more than 3 billion people who do not have access to a public electricity grid. In Europe, this occurs primarily in small islands or mountainous regions that are difficult to access. In developing and emerging countries, in contrast, this applies to large parts of rural areas. Especially in countries that have a large land mass, remote regions will not be connected to supply grids in the foreseeable future even if there is a significant expansion of power stations and grid capacity. In some of these situations, expandable and 'networkable' stand-alone systems could become the nuclei of an extensive supply structure. However, any such development cannot be realized with the system technology in use today, such as is described in the previous sections. Instead of custom solutions on a small scale, which are designed for specific conditions, what is required is the standardization of communication and system technology. This is the only way that stand-alone systems can be networked to create virtual power stations that can be expanded limitlessly and allow greater efficiency, supply reliability and cost-effectiveness than individual systems.

During the past, new concepts and a modular system technology were developed for stand-alone systems. These have been used in pilot systems for rural electrification. In the future, standardization and experience will facilitate the integration of a considerable percentage of renewable energy sources within the traditional supply networks in industrialized regions. Especially in Southern European countries, the grid-stabilizing properties of battery systems should be of interest.

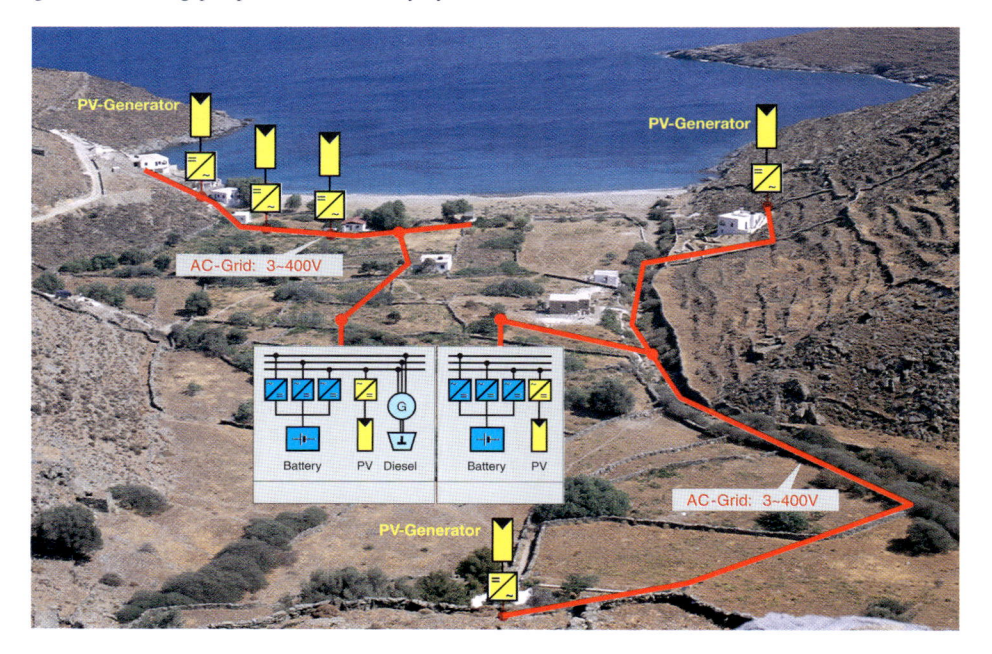

Figure 8.34
Decentral energy supply of the Greek island of Kythnos: A total of 600kW of wind turbines and 100kW of PV systems are integrated within the weak supply network along with a large battery bank
Notes: The fully automatic management and control system selects the optimum system configuration depending upon energy consumption and the available renewable energy, thus enabling up to 100 per cent renewable coverage. The grid quality on the island has been significantly improved as a result of this system.
Source: EC

Modular system technology has to be based on harmonized standards for the transmission of energy and information between the components, and to allow for an 'overseeing' management system. Different kinds of energy generators and storage units could then be combined, such as building blocks, to create a power supply with capacities ranging from a few kilowatts to large systems with several hundred kilowatts. These could power varying loads. This not only makes planning easier, it also facilitates maintenance, spare part provision and training for installation engineers and users. Two fundamentally different system versions are available: DC-coupled systems and a new concept: AC-coupled systems. In both cases, the central component is the stand-alone inverter. This can be synchronized with a parallel low-voltage mains grid or can itself function to form a grid. When supplying different kinds of loads, it keeps grid variables such as voltage and frequency constant. It also regulates the real and reactive power.

8.12.1 DC-coupled systems

In stand-alone systems of up to approximately 50kW, it can still make sense to couple the arrays and loads on the DC side. The PV array strings are connected to the DC distribution system via charge controllers in a master–slave system. DC loads can be supplied directly from the DC bus. A three-phase stand-alone inverter (alternatively, three single-phase stand-alone inverters) (also known as inverter-chargers) connects the AC loads to the DC bus. The main (master) charge controller controls the state of charge of the battery bank and automatically switches on the backup power source (e.g. a diesel generator) when the state of charge falls below a predefined level (e.g. 30 per cent). This then charges the battery via a bidirectional combination inverter–battery charger unit and at the same time supplies power to the load. As soon as bidirectional three-phase inverters with integrated charge controller are available, only one device will be necessary between the DC and AC sides.

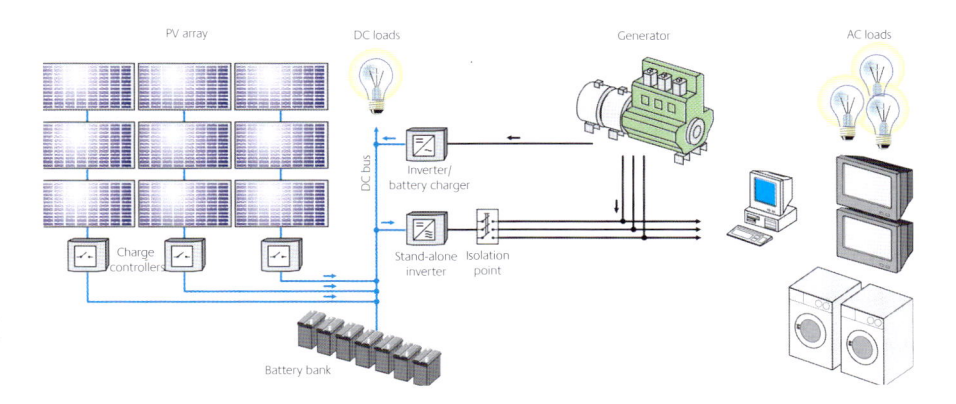

*Figure 8.35
Example of a DC-coupled stand-alone system with an additional generator and DC and AC loads: The system can also be constructed with a single-phase AC side and functions without a backup generator
Source: adapted from Meyer, 2002 and Steca*

Advantages of DC-coupled systems are as follows:

- DC loads can be used.
- Higher efficiency exists for battery charge and DC loads.
- Grid control is simple.
- The system is relatively robust.

Disadvantages include the following:

- It is difficult to expand.
- The DC voltage level is not standardized.
- High DC losses are experienced with long wiring routes.
- Stand-alone inverters and PV arrays must be designed for maximum load.
- DC wiring is laborious.

8.12.2 AC-coupled systems

For larger systems, by contrast, AC-side coupling of PV arrays and loads offers better connectivity with the low voltage mains grid and other stand-alone systems. The solar power is converted into AC power using conventional string inverters as used in grid-connected systems before being supplied to the stand-alone grid. The required numbers of strings, all with a uniform configuration, are connected in parallel to the AC distribution system. First the solar power is used to cover consumption. Any surplus is converted back into DC power via the stand-alone inverters and stored in the batteries, where it can be fed back into the stand-alone grid if insufficient solar energy is available. Depending upon size, the AC network can be single phase or three

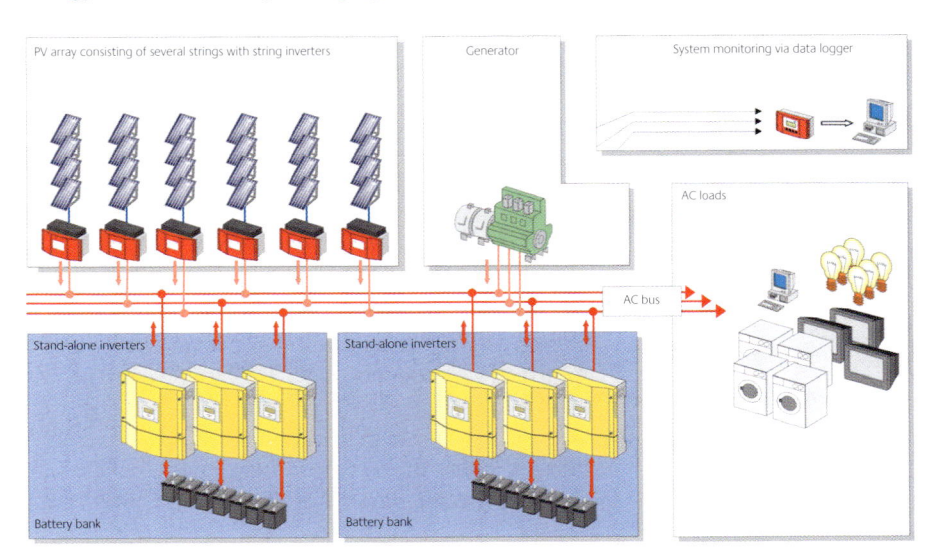

*Figure 8.36
Wiring diagram for an AC-coupled stand-alone system with backup power supply (e.g. diesel generator) as a three-phase supply: Simpler versions with a pure PV supply and single-phase systems are also possible
Source: SMA*

phase. The stand-alone inverters (also all the same type of unit) are wired in parallel in single-phase networks and in three-phase networks, and are shared equally between the three phases. The system monitoring facility enables communication with the inverters, functional control and operating data analysis depending upon the measuring equipment used. Transformerless string inverters enable systems up into the megawatt range.

Advantages of AC-coupled systems are as follows:

- Higher peak power output and more cost-effective components with lower maximum power are experienced.
- Large-volume series production components as used in grid-parallel systems are employed.
- Expansion capability is good.
- A DC distribution system is avoided.
- There is easy coupling to additional power sources.
- Optimization of operations is straightforward.

Disadvantages include the following:

- Grid control is complex.
- There are higher system costs for very small systems.
- Additional energy conversion stage for PV power involves losses.

Combined devices with integrated charge controller are best suited to this type of application. These are able to create a bidirectional link between the AC and the DC sides. With an integrated management system, the stand-alone inverter enables optimum coordination of the feed from various generators and uses the battery as a buffer. The integrated charge controller manages the use of the batteries to extend their life and, if needed, can cut off low-priority loads from the power system.

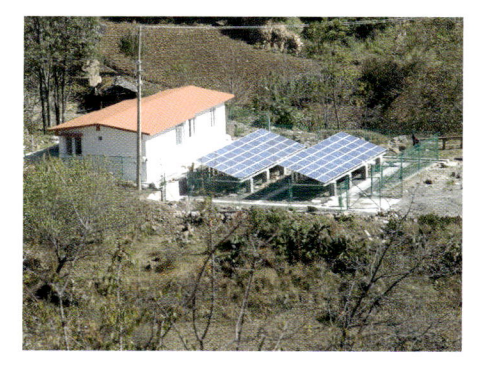

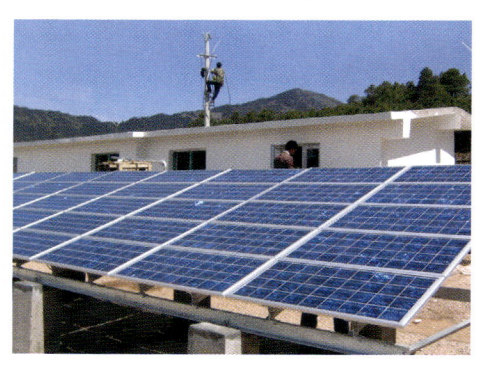

Figures 8.37 and 8.38
AC-coupled PV stand-alone systems in Yunnan Province, China
Source: SMA

AC-coupled systems of this kind have been, and are being, installed, for example, as part of a major rural electrification programme operated by the German Federal Ministry for Economic Cooperation and Development (BMZ), which has built renewable electricity systems in Yunnan and Xinjiang provinces in China with capacities of 9kWp to 27kWp. A total of 170 stand-alone PV systems are to be built as part of the project. As well as health stations and village schools, every household will also obtain electricity. The AC coupling means that the systems can be expanded as energy requirements grow, and existing conventional power sources (e.g. diesel generators) can be integrated. Smaller systems have been built as single-phase systems, while larger villages have gained three-phase networks with a diesel generator as backup for the PV system (Meyer, 2002).

9 Economics and Environmental Issues

9.1 Cost trends

Today, applications where PV systems are economically competitive in comparison with conventional power conversion systems are generally small applications or at locations remote from the mains grid. However, where attractive tariffs are paid for electricity fed into the grid by grid-connected PV systems, PV systems can become economically viable for individual customers. Good feed-in tariffs were first offered in Germany on a large scale in 2000. These led to a boom in the PV sector. Recently, productive feed-in tariffs, based on the German model, have also been introduced in Spain, France, Austria, the Czech Republic, Italy and Greece.

The extent to which even smaller grid-connected photovoltaic systems in Germany and, globally speaking, grid-connected PV systems, in general, will come to form a significant proportion of the power supply depends above all upon the future development of costs.

Figure 9.1 shows that, above all, it is the production volume that has a heavy influence on the costs. Thus, since the start of the 1980s, system prices in Germany have fallen by more than 70 per cent. This trend was reversed in 2004. In Germany, an increase of approximately 10 per cent was recorded (Stryi-Hipp, 2005) due to demand for modules exceeding supply. But at the beginning of 2007, with expanding production capacities for solar silicon, prices were going down again.

More than 95 per cent of the modules manufactured in the world are silicon-based. The PV industry requires in the region of 10 to 12 tonnes of silicon (Si) per megawatt. Growth is possible only by developing Si production and making savings on resources. In the medium term, increased expansion of Si production capacities is necessary to avoid further shortages.

Economies of scale are crucial: a 100 per cent production increase means a 20 per cent reduction in costs (Hoffmann et al, 2004). However, with increasing mass production the material costs make up an increasingly large proportion of the total cost of a module, with the result that, in the future, cost reductions of only 15 per cent to 18 per cent per 100 per cent increase in production can be expected. In the November 2004 Sarasin study (Fawer-Wasser, 2004), an average annual worldwide growth rate in cell production of 13 per cent was assumed until 2020. This means 5.8GWp production in 2020. Hoffmann et al (2004) forecast a production volume of more than 15GWp for 2020 based on annual growth rates of 20 per cent to 25 per cent.

The German Bundestag's Sustainable Energy Supply Commission anticipates a cost reduction in the production costs of small PV systems (2.6kWp) of 43 per cent by 2010 compared to the costs in 1999. Between 2010 and 2020, the costs are expected to fall by another 50 per cent (see Figure 9.1). Similar reduction rates are expected for large-scale systems, but with much lower absolute costs.

In areas with high solar irradiance, such as southern Spain, power production costs could fall to significantly less than 0.20 Euros per kWh as soon as 2010. This means that solar power would be competitive with peak load power.

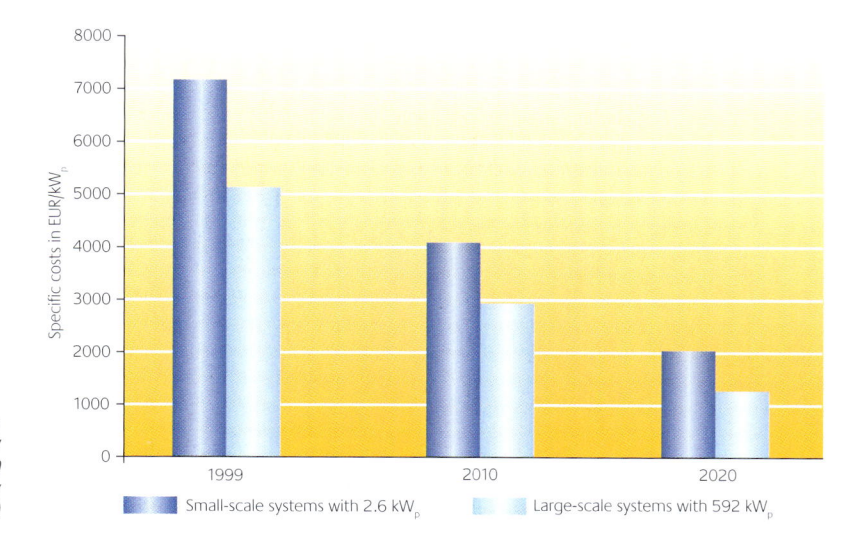

Figure 9.1
Cost forecast for large and small PV systems in Germany up to 2020
Source: Sustainable Energy Commission, German Bundestag (2002)

■ Small-scale systems with 2.6 kW$_p$ ■ Large-scale systems with 592 kW$_p$

9.2 Technological trends

At present, the main starting material used for solar cells is crystalline silicon, with a market share of 94 per cent (see Figure 9.2). Thin-film cells have a market share of only 6 per cent (see Figure 9.2), but amorphous silicon cells are mainly used in mini- and micro-applications for the consumer market.

Figure 9.2
Worldwide market share of various materials in solar cells as of 2004
Source: Photon 4/2005

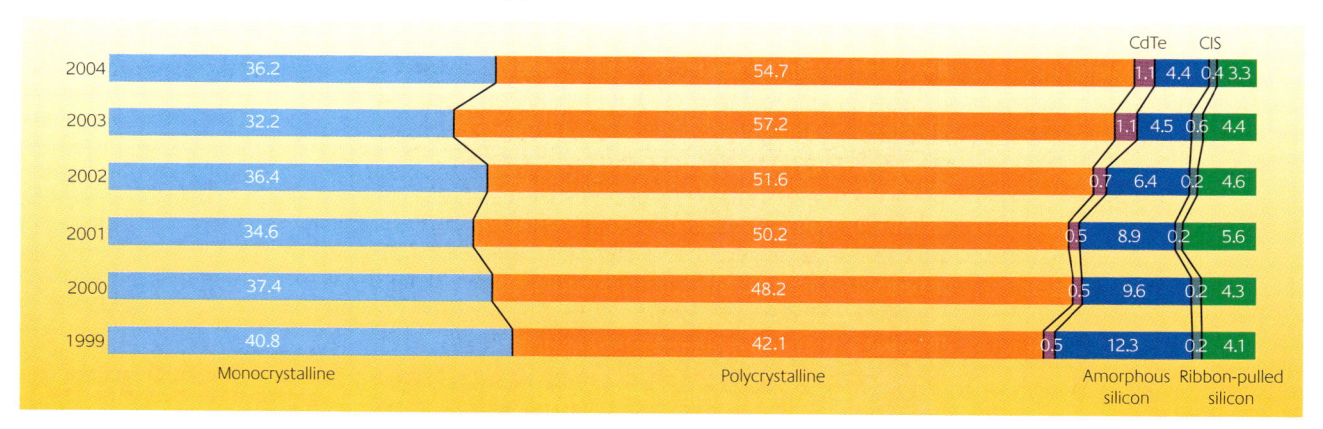

Year	Monocrystalline	Polycrystalline	CdTe	Amorphous silicon	CIS	Ribbon-pulled silicon
2004	36.2	54.7	1.1	4.4	0.4	3.3
2003	32.2	57.2	1.1	4.5	0.6	4.4
2002	36.4	51.6	0.7	6.4	0.2	4.6
2001	34.6	50.2	0.5	8.9	0.2	5.6
2000	37.4	48.2	0.5	9.6	0.2	4.3
1999	40.8	42.1	0.5	12.3	0.2	4.1

The market share of polycrystalline silicon cells has increased during the last three years from 52 per cent to 55 per cent, while the share of mono-crystalline silicon cells has remained constant at around 36 per cent. The market share of ribbon silicon has fallen slightly from 4.6 per cent in 2002 to 3.3 per cent in 2004. In the medium term, there will be no substantial changes to these market shares since the production capacities for modules with crystalline Si have been (and are being) rapidly expanded. However, in the long term it is anticipated that the market share of thin-film cells will increase strongly. The lower material usage in production and more efficient manufacturing processes permit expectations of large reductions in cost.

Almost all of the large manufacturers are currently working feverishly on further developing thin-film cells and on introducing new material-saving technologies in the crystalline field (see the section on '1.4 Solar cell types' in Chapter 1).

The standardization of modules and the trend towards modules with higher output will lead to additional cost reductions, particularly for fitting costs.

In the past, there have already been significant efficiency and reliability improvements in inverters. The requirements relating to the impacts on grid quality will become tougher in future because (with an increasing number of photovoltaic systems) the influence of inverters on the mains grid will be greater. As well as the development of cost-effective devices, a trend towards greater flexibility and modularity can be seen in inverters. At the same time, systems for monitoring operating data are gaining ever greater importance.

9.3 Economic assessment

9.3.1 Power production costs

In a photovoltaic system, the investment costs essentially determine the production costs of the generated solar power (see the section on '4.8 Yield forecast' in Chapter 4) since no costs are incurred for fuels and the actual running costs (insurance, maintenance, etc.) are low. Via the power production costs, the photovoltaic system can be compared with other energy systems and the break-even point can be calculated for the payment received for power fed into the grid.

The desired return on capital has a decisive influence on the calculation of the power production costs. The calculation, excluding interest on capital, is fairly uncomplicated. Here, the overall costs that are incurred over the lifetime of the system are determined. As well as the investment costs A_0, the operating costs A_{Op} also need to be taken into account, such as costs for maintenance, repairs and insurance premiums.

If photovoltaic arrays are taken into account in planning from the outset, rather than being integrated within or 'stuck onto' a building after it is built, photovoltaic systems integrated within the roof or façade can result in considerable savings A_S in construction materials. In addition, the energy quantity generated annually by the photovoltaic system, E_a, must be included in the calculation.

With a useful life n for the system – generally a useful life of 20 to 30 years is assumed – the electricity production costs c_E are calculated as follows:

$$c_E = \frac{A_0 + A_{Op} - A_S}{n \times E_a}$$

Example. In the case of a photovoltaic system installed on an existing building ($A_S = 0$) with a power of 3kWp, we calculate the investment costs A_0 to be 14,138 Euros. The operating costs A_{Op} are 2000 Euros (i.e. 100 Euros per year), the useful life n is 20 years and the energy quantity generated each year E_a is 875kWh. The power production costs are therefore as follows:

$$\frac{14{,}138 \text{ Euros} + 2000 \text{ Euros} - 0 C_E \text{ Euros}}{20 \times 2625 \text{kWh}} = 0.307 \text{ Euros} / \text{kWh}$$

If interest on capital is taken into account (as is usually the case), we use a dynamic method to factor in the interest-on-interest effect over the service life or the amortization period of the system. This annuity method enables the investment costs and all other costs incurred to be converted into capital costs that stay the same, year on year.

The equation is:

$$a = \frac{p}{1 - (1 + p)^n}$$

The annuity factors for various useful life spans and interest rates are provided in Table 9.1. As can be seen, the annual costs are strongly dependent upon the interest rate and the amortization period. For example, if existing private funds are used to build the photovoltaic system, a considerably lower interest rate can be assumed than if a bank loan has to be taken out.

It is also possible to determine the present ('discounted') value of future payments that will be incurred for operating costs (A_{Opp}). To do this, all the payments A_i in the various operating years i of the system are worked out. Then, with the interest rate p, this results in the present (i.e. the 'discounted') cost of the future operating costs:

$$A_{\text{Opp}} = \sum_{i=1}^{n} \frac{A_i}{(1+p)^i}$$

This is then used in the subsequent calculations.

Table 9.1
Annuity factors a for various life spans n (years) and interest rates p

Interest rate p									
n	1%	2%	3%	4%	5%	6%	7%	8%	9%	10%
10	0.1056	0.1113	0.1172	0.1233	0.1295	0.1359	0.1424	0.1490	0.1558	0.1627
15	0.0721	0.0778	0.0838	0.0899	0.0963	0.1030	0.1098	0.1168	0.1241	0.1315
20	0.0554	0.0612	0.0672	0.0736	0.0802	0.0872	0.0944	0.1019	0.1095	0.1175
25	0.0454	0.0512	0.0574	0.0640	0.0710	0.0782	0.0858	0.0937	0.1018	0.1102
30	0.0387	0.0446	0.0510	0.0578	0.0651	0.0726	0.0806	0.0888	0.0973	0.1061

The power production costs c_E with an annuity factor a are then calculated as follows:

$$c_E = \frac{(A_0 + A_{\text{Op}} - A_S) \times a}{E_a}$$

Example. In the photovoltaic system from the previous example, interest on capital of $p = 6$ per cent $= 0.06$ is assumed. For the sake of simplicity, the operating costs in this example are assumed to be 2000 Euros for repair costs, payable after 15 years. The discounted operating costs are therefore:

$$A_{\text{Opp}} = \frac{2000 \text{ Euros}}{(1 + 0.06)^{15}}$$

With investment costs $A_0 = 14{,}138$ Euros, no cost savings ($A_E = 0$), a useful life of $n = 20$ years and the energy quantity generated each year $E_a = 2625\,\text{kWh}$, the power production costs with the annuity factor $a = 0.0872$ are calculated as follows:

$$c_E = \frac{(14{,}138.00 \text{ Euros} + 834.33 \text{ Euros}) \times 0.0872}{2625 \text{ kWh}} = 0.497 \text{ Euros/kWh}$$

Figure 9.3 shows the dependency of power production costs on the interest rate and the annual energy yield for small systems with investment costs of 5000 Euros/kWp and for large systems with investment costs of 4000 Euros/kWp. The operating costs are 1 per cent of the investment costs with a variable interest rate.

Of course, the assumed interest rate for the capital tied up in the system plays a role in calculating the power production costs, as in the example calculation above. A lower assumed interest rate results in lower power production costs.

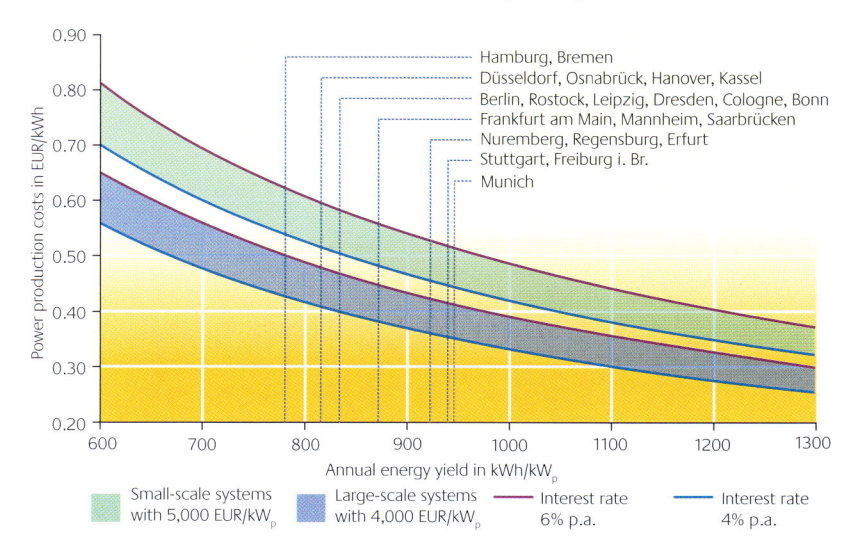

Figure 9.3
Power production costs depending upon location and interest rate
Source: yield data are derived from www.pv-ertraege.de: annual yields from 2000 to 2004 are relayed by postcode regions, Germany

Figure 9.3 therefore shows the power production costs with interest rates of 4 per cent and 6 per cent and a useful life of 20 years. The figure also shows the great influence of energy yield on power production costs. The energy yield, in turn, is dependent upon the irradiance conditions at the system's installation site. A system built in the Hamburg region of Germany will generate an average annual yield of 780kWh per kWp. By contrast, an average annual yield of 944kWh per kWp can be expected for systems located in the Munich area. About twice as much, again, can be expected from a system in southern Spain.

9.4 Environmental impact

Photovoltaic systems require no fuels when operating and do not release any harmful emissions. Nevertheless, questions regarding the energy used during manufacture, the implemented material streams and the possibilities for recycling the modules must be addressed.

9.4.1 Energy payback and harvest factor

Energy is required in order to manufacture a photovoltaic system. The cell technology used is the deciding factor here. Crystalline cells have the highest energy requirements, which is due mainly to the high temperatures required in silicon and wafer production. Manufacturing raw silicon is very energy intensive. It currently takes around 10,000kWh per kWp for the crystallization of solar silicon. This figure does not take into account the amount of energy in the silicon waste from the semiconductor industry that is generated through the production of mono-crystals for electronics applications. With rising demand, as soon as there is separate silicon production for photovoltaics, energy consumption could be further reduced. The production of thin-film modules requires roughly only half the amount of energy. The frames and mounting systems for thin-film modules are somewhat more complex, which means that the advantage compared to crystalline modules is reduced. It takes approximately 1000kWh per kWp just to produce the aluminium frame for mono-crystalline modules, and approximately 2000kWh per kWp for amorphous silicon modules. However, more frameless modules or plastic frames could be used in the future. With crystalline silicon modules, the mounting system for pitched roofs requires approximately 1500kWh per kWp. With amorphous silicon modules, this value almost doubles due to the increased space requirements. The inverter accounts for approximately 280kWh per kWp. Figure 9.4 shows the cumulative primary energy requirements for grid-connected photovoltaic systems based on different module technologies.

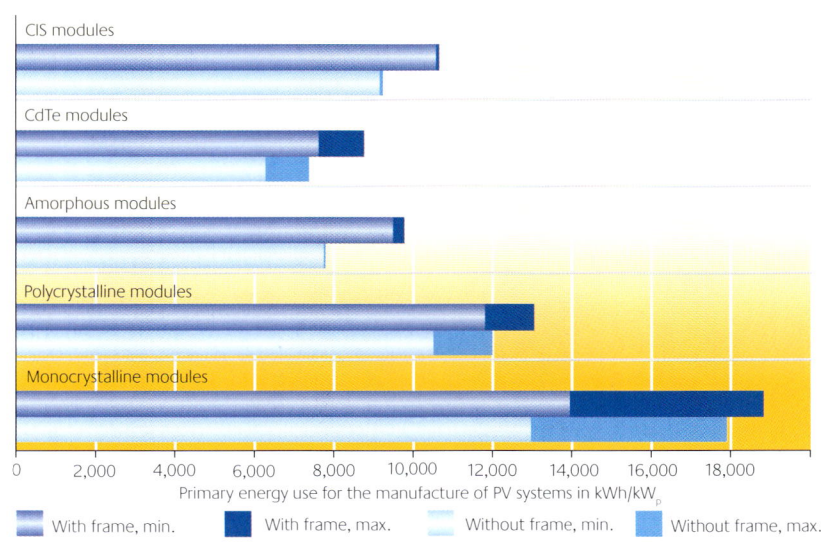

Figure 9.4
Primary energy use for the entire production chain for PV systems with different cell technologies: Minimum and maximum values from various ecological balance studies
Source: Alsema, 2000; Pehnt et al, 2003; Jungbluth and Frischknecht, 2000; Knapp and Jester, 2000

The energy used to manufacture photovoltaic systems can be seen as a kind of credit that is paid back with solar power. How long does this solar repayment take? If we assume a system output of 900kWh per kWp per year (Central European location) and compare it with the electrical energy delivered by a conventional power station (assuming 35 per cent power station efficiency), then there is an annual primary energy saving of 2571kWh per kWp. In other words, a conventional power station would require a 2571kWh of primary energy to produce the equivalent amount of electrical energy that the PV array would produce (900kWh). So, in one year of the operation of the PV array, 2571kWh of primary energy (fossil fuel) does not have to be consumed at the power station. Viewed this way, energy amortization periods of between 2.7 and 6.2 years are achieved (see Figure 9.5). With double the solar irradiance (e.g. in Southern Europe or North Africa), approximately twice as much solar power is generated. The energy amortization periods become shorter accordingly. In California or Spain, it would be twice as short – between 1.5 and 3 years.

Particularly with thin-film cells, considerable reductions in the energy payback period can be expected in the future due to improved production technology. Initial estimates suggest that periods of less than a year can be achieved. But the energy payback periods will also fall with silicon-based modules since many solar silicon manufacturers are currently developing and testing new processes. These processes will use much less energy overall than the normal Siemens process. Examples are the fluidized bed reactor process, the tube reactor process and the vapour-to-liquid deposition process.

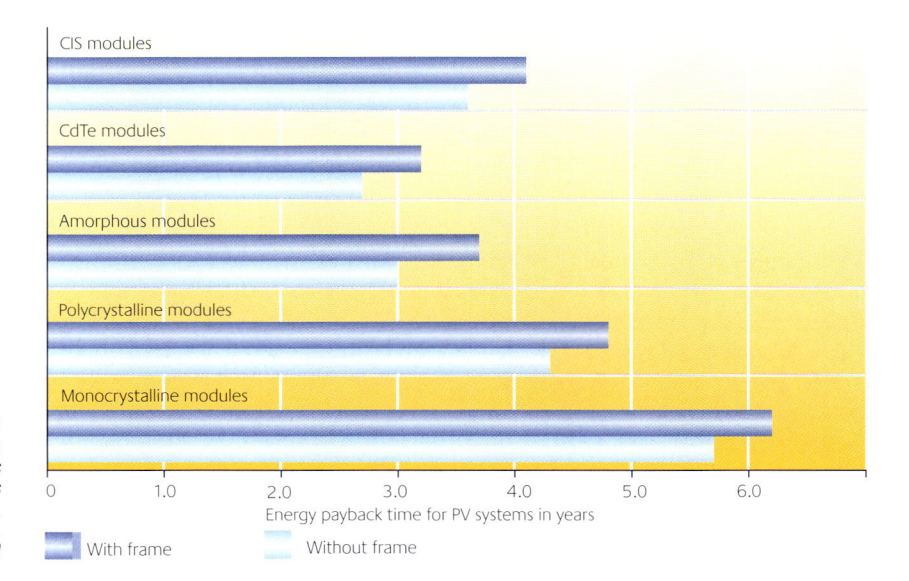

Figure 9.5
Energy payback time for PV systems in Germany based on the average values from various studies
Source: Alsema, 2000; Pehnt et al, 2003; Jungbluth and Frischknecht, 2000; Knapp and Jester, 2000

As well as the energy payback time, the 'harvest factor' is often stated. This factor specifies how much more energy the photovoltaic system generates than was required to produce it, over the predicted service life of the system. With an assumed service life of 30 years, 5.3 to 11 times as much energy is generated as is required to manufacture the system. In buildings with integrated systems, on the one hand, the energy yield is not as high; but the energy from the construction material savings can be taken into account in the calculation (Haselhuhn, 2005).

9.4.2 Pollutants in the production process

Environmentally harmful substances comparable to those in the semiconductor industry occur in the production of silicon solar cells; but they stay within closed cycles. The finished silicon solar module contains no environmentally damaging components apart from the soldered joints. In the soldered connections, lead solder containing about 40 per cent lead is frequently still used. As a result, PV systems can have a lead content of up to 300 grams per kilowatt (http://Global.mitsubishiElectric.com/bu/solar). RWE Schott Solar and Mitsubishi Electric Corporation have been using lead-free solder in their modules for a number of years. These alternative tin/silver compounds are not yet used by other module manufacturers owing to the higher solder temperature and the associated technological requirements.

Cadmium telluride (CdTe) and copper indium diselenide (CIS) modules currently have a global market share of less than 1 per cent. CdTe modules, in particular, contain up to 170g/kWp or 900mg/kg of cadmium (Giese et al, 2002). However, this only occurs in compounds that are non-toxic. Despite this, old CdTe modules are considered to be hazardous waste. In normal operation, CdTe modules do not pose any environmental risk. But in a fire – and only in a fire – poisonous gases can be produced. In tests on cracked modules, the levels of cadmium that leached into rainwater washing over the modules were below the permitted levels for drinking water (Steinburger, 1995). CIS modules generally have thin layers of cadmium sulphide. The amount of cadmium in these is less than 1 per cent of that in CdTe modules.

Table 9.2
Composition of Si standard module and thin-film modules
Source: Sander et al, 2004

Weighted Si standard module (proportions of materials in percentage)						
Glass	Frame	EVA	Cells	Rear film	Contact box	Mass/power output
62.7%	22%	7.5%	4%	2.5%	1.2%	103.6kg/kWp

Thin-film modules (proportions of materials in percentage)						
Glass	Frame	EVA	Contact box	Chemical elements	Rear film	Mass/power output
74,53% No rear side because glass–glass module	20.4%	3.5%	1.1%	0.1%	0.0%	285.2kg/kWp

Another critical substance in CIS modules is selenium; but it is used only in very small quantities. CIS modules are considered less potentially hazardous than CdTe modules. The second generation of CIS modules will enter the market in the next few years. These modules use sulphide instead of selenium.

9.4.3 Module recycling concepts

In the light of the EU Waste Electrical and Electronic Equipment (WEEE) directive and the Restriction of the Use of Certain Hazardous Substances (RoHS) directive in electrical and electronic equipment, manufacturers are working on recycling and avoidance strategies. On 1 July 2006, the RoHS directive banned the use of hazardous substances such as cadmium, lead and mercury in electrical and electronic equipment. Even though solar modules do not yet come under this directive, inclusion within the scope of the directives would suggest itself, in principle, in order to support the attainment of the environmental and waste objectives with regard to PV modules (Sander et al, 2004).

Factory recycling (i.e. the partial reuse of discarded raw material in production) has already been introduced by almost all manufacturers. Relevant volumes of waste are not expected until 2030, at the earliest, as installed modules reach the end of their service life. It has estimated that the total mass of scrap available for recycling will be 13,300 tonnes in 2030 and 33,500 tonnes in 2040 (Schlenker and Wambach, 2005).

But for energy conservation reasons, it already makes sense to recycle the materials in faulty solar modules or those that have reached the end of their life. As well as the silicon cells, the glass and aluminium components can also be reused.

After many major module manufacturers had carried out various studies, in the summer of 2003 Deutsche Solar AG opened a pilot Si solar module recycling plant in Freiberg, Saxony. The process enables recycling of raw materials, silicon wafers or complete solar cells. Modules are broken down into their constituent parts in a special furnace at a temperature of 500°C. This burns away the EVA embedding material and the backing film. The dismantled cells can then be processed and reused. However, the recycled solar cells have a somewhat lower efficiency than the originals.

Figure 9.6
Faulty modules for recycling
Source: Deutsche Solar AG

Recycling saves up to 80 per cent of production energy. The energy payback time for recycled crystalline silicon modules is therefore reduced to about one year. Thin-film amorphous silicon modules can also be recycled at the plant; but only the glass and metal are recovered. In principle, module recycling is also possible for CdTe and CIS modules. Work on a solution is in progress.

The research project Recovery of Solar Valuable Materials, Enrichment and Decontamination (RESOLVED) should be mentioned here. This is run by the German Federal Institute for Materials Research and Testing (Bundesanstalt für Materialprüfung, or BAM) in conjunction with Deutsche Solar AG in Freiberg and other partners. The project is funded by the EU's LIFE–Environment demonstration programme and its purpose is to recover semiconductors from thin-film modules and to decontaminate waste materials. Particular attention is focused on ecological balance, process sustainability, resource availability and socio-economic effects.

Expected outcomes include:

- pilot demonstration of pretreatment and wet mechanical processing of photo semiconductors from thin-film modules for high-efficiency materials recycling;
- demonstration of the efficiency of material separation and recovery by producing and testing prototype cells from recovered material;
- demonstration of quality control in various process stages with instrumental large-scale analysis.

The individual process steps are:

- The cover glass and the substrate plate (normally also glass, coated with the recyclable semiconductor material) are mechanically or thermally separated.
- The semiconductor layer is removed from the carrier substrate by sandblasting. The cleaned glass substrate and the cover glass go to a glass recycling plant.
- The mixture of semiconductor material, blasting material and glass particles is separated using a wet mechanical method. The separated and concentrated semiconductor material is further purified in a flotation process.
- This clean semiconductor material is used to produce a new thin-film module.

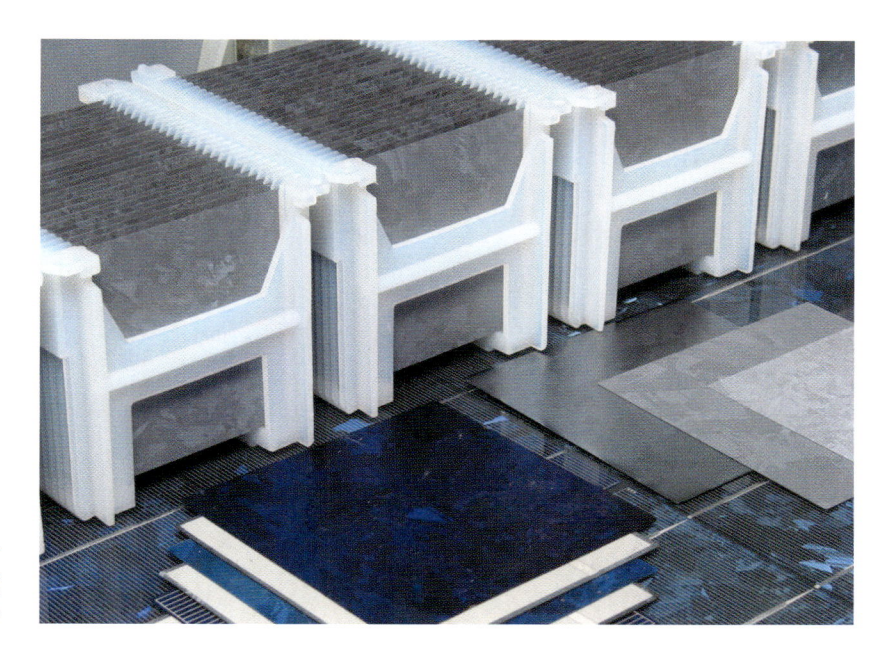

*Figure 9.7
Recycled wafers from the Deutsche
Solar AG pilot plant
Source: Deutsche Solar AG*

10 Marketing and Promotion

One of the first things anyone entering the solar industry needs to do is get to know the market in the country they are working in. The market for grid-connected PV differs from country to country and is largely determined by national support schemes and the level of feed-in tariffs, although even where these are not very attractive, many systems are being installed for prestige reasons and as a statement of an organization's commitment to the environment. The market for stand-alone systems is different again, and is not only limited to rural areas of developing countries and buildings off the grid in developed countries, but also lends itself to small systems such as signage lighting and street furniture in urban centres where it can prove cheaper than extending the grid. PV can also be integrated within power back up systems, especially where power outages are frequent and can last several hours. The PV global market, at the time of writing, is significantly more than 2000 MWp annually (modules produced) and growing at a rate of about 25 per cent per annum.

It is important when marketing PV systems, or any other renewable energy technology, to find out right at the beginning why a customer is interested in the technology. This may be because an organization wants to make an environmental statement, because of attractive feed-in tariffs or providing energy security, or – in the case of stand-alone systems – it might simply be the simplest and easiest way of providing an electricity supply. This will influence the final offer to the customer and the overall system design concept. This chapter explores marketing techniques and concepts.

10.1 Marketing PV: The basics

10.1.1 Customer orientation: The central theme

What does marketing actually mean? A good working definition is as follows: marketing is the totality of measures that make it easy for your customers to decide on you and your products. Here the topic has been divided into three sections: the basics, systematic marketing planning and sales talk, which is the most direct contact with the customer.

Solar energy represents the future and a respectful attitude to the natural and human environment. When you sell a PV system you are always also selling part of this idea. This works most persuasively when you are customer oriented: the benefits to the customer become the central theme of your marketing.

As a reader you are my customer, so what benefits do you gain from this chapter? You'll find that acquiring customers and sales are more fun – and you'll experience greater success with less effort.

What do you gain from selling solar energy systems? Answers to this question include:

- promoting a good cause (what benefits the environment and society also benefits you);
- advising your customer (what benefits your customer, benefits you too);
- customer loyalty (additional product(s); higher name recognition);
- gain in expertise (practice makes perfect, both technically and when it comes to giving advice);
- image boost;
- personal satisfaction (emotional self-interest);
- job security (securing your own job; future opportunities).

10.1.2 The iceberg principle

Just as only one seventh of an iceberg is above the water, only one seventh of our communication happens at a conscious level – by far the greater part of human communication takes place at a subconscious level that is not visible on the surface. Yet, these hidden six-sevenths of communication affect the other person's behaviour and decisions. So when you want to build up trust and a long-term customer relationship that benefit both parties, remember that people generally make on-the-spot decisions by using their emotions (90 per cent) and then their reason (10 per cent).

In order to acquire customers, systematic, rational planning is needed. But for lasting success, your mental attitude, your feelings and your customer's feelings are important – as also are the clothes you wear and the way in which you present yourself. You are not just selling a product, you are also selling what you represent as a person: are you believable? Are you good company? Do you get along well with the customer?

The iceberg also means that a customer will be able to tell very quickly whether you are just faking it or whether you seriously mean what you say (e.g. when you are being customer oriented). So it's worth your while considering fundamental questions such as: what is the aim of your work? What impression do you want to make on your customers? It always gives you and your employees a mental boost if you live out your mission statement.

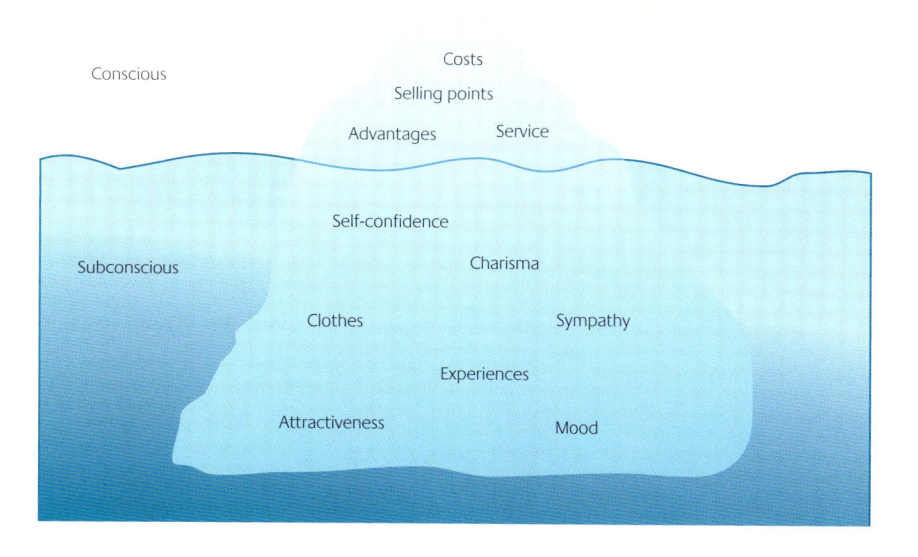

Figure 10.1
The iceberg principle:
Only one seventh of communication takes place at the conscious level
Source: Binder-Kissel

10.1.3 The pull concept

Someone who wants to make more sales starts by feeling a pressure – the pressure to sell stuff. According to the iceberg principle, the seller will transfer this pressure onto the customer, who will react to this pressure in the same way that you react – by resisting it! You stand a much better chance if you see yourself as a kind of 'genie of the lamp': you read your customer's mind, see their wishes in their eyes and can summon up all kinds of excellent products and services that will benefit your customer. This way you create a 'pull' force that attracts your buyer.

You want customers to buy because they are convinced of the usefulness of the product. In the pull concept, the subconscious and instincts play a particularly important role. As well as factual information, you also convey an image of your company and the product. You will gain customers if you make the product and your company so attractive that customers want to have something from you (and not from anybody else). Table 10.1 depicts the major differences between push and pull.

Table 10.1
The difference between push and pull
Source: Binder-Kissel

Push	Pull
Exploit every conversation to 'hit on' potential customers Aggressive acquisition	In general, conversations build up trust first, then develop approaches to working together
Sell at any price	Only sell if the customer genuinely benefits from it
One sales talk = one sale	Honest, genuine advice Accept 'no' for an answer
Company image is secondary	Image is a key factor
Sale = success factor	Word of mouth: a recommendation is also a success
Critical customers who felt pushed into buying High complaint rate Price sensitivity	Satisfied customers who 'come willingly' Fewer complaints Lower price sensitivity

Modern marketing works according to the pull concept. There are fewer and fewer differences between competitors' products. The importance of relationships between the company and its customers has never been greater. As a result, marketing doesn't stop with the sale. In the sale, it sees the beginnings of a new sale. This may be another product or another customer – the important thing is your behaviour and how you generate a pull effect: the customer simply enjoys working with you, recommending you to others, buying something from you. If you make it easy for customers to see how they benefit, then they will find their way to you.

10.2 Greater success through systematic marketing

10.2.1 The benefits come first

You know and appreciate your product from the seller's point of view. But every customer will see a product differently. It depends upon how you look at it. The following tale illustrates the point:

> *A novice says to an older monk: 'I see that you're smoking while you perform your duties. The abbot told me I wasn't allowed to smoke while I prayed.' To which the older monk replies: 'It depends what you say to the abbot. I asked him if it was alright to pray while I smoked.'*

Customers don't want to know that photovoltaic modules have an anti-reflective coating as much as they want to know how high the system's yield is. Technicians and installation engineers generally focus on the technical data as being the essential features of their product. Customers are primarily interested in what they gain from the product – this can be as different for men and for women as it is for private customers and business customers!

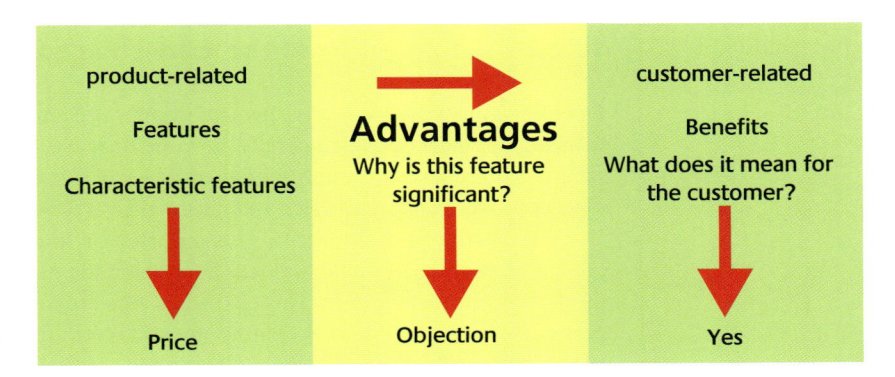

Figure 10.2
From feature to benefit
Source: Binder-Kissel

product-related		customer-related
Features Characteristic features ↓ Price	Advantages Why is this feature significant? ↓ Objection	Benefits What does it mean for the customer? ↓ Yes

Figure 10.2 shows how features, advantages and benefits have different effects on the outcome of the sales talk. Think about your products and services and list the benefits to the customer. Table 10.2 provides some examples.

Table 10.2 From feature to benefit

Feature	Advantage	Benefits (for customer)
Special glass	Hail resistance	Long service life
		No trouble or repair costs
Roof-mounted modules	Better ventilation	Higher yields
Cells made of mono-crystalline silicon	High efficiency	Low space requirements
Thin-film cells	Shading tolerance	Good yields even with shading

You will notice that more than one benefit can be found for each feature. Try to identify every possible benefit from the customer's point of view. The following list of benefit areas should help you in your investigation:

- time savings;
- problem solutions;
- cost savings;
- environmental protection;
- safety, reliability and security;
- prestige;
- enjoyment;
- convenience and ease of use;
- health;
- information;
- entertainment;
- public relations (for business customers);
- employee motivation (for business customers).

Figure 10.3
The bait has to attract the fish, not the angler!

10.2.2 The four pillars of the marketing concept

Develop a marketing concept. Instead of operating at random, you will then be working the market systematically. Just one or two days of concentrated planning can save you a great deal of time, money and frayed nerves. You can selectively fill up low-order periods to increase sales and improve cash flow.

A marketing concept like this is supported on four pillars: analysis of your company; analysis of your products; analysis of the market; and analysis of your marketing implementation. This chapter is concerned mainly with the latter two points. However, the two other areas are still worth investigating. Here are a few suggestions and questions to get you started.

THE COMPANY

WHAT TARGETS ARE YOU PURSUING IN YOUR WORK? WHAT IS IMPORTANT TO YOU?
You want your marketing to acquire the orders for you that serve your objectives. The way you see yourself is also important here. For example, do you want to offer the lowest possible prices or is quality your overriding priority? How important are ecological or social concerns to you? Make your company's image fit what you decide here.

WHAT ARE YOU OR YOUR COMPANY'S PARTICULAR STRENGTHS?
For example, are you a problem-solver who enjoys designing complex systems? Or are you particularly good at handling standardized projects? The more you concentrate on your core competences, the more clearly recognisable you are to everyone else. Customers are then better able to distinguish you, for example, as 'the flat roof specialist'.

WHAT ARE YOU OR YOUR COMPANY'S WEAKNESSES?
What kind of customer support do you offer? How long does it take you to respond to customer enquiries? Have the main contact people in your company received training in how to make contact with customers? Also think about resources with regard to your marketing. If your acquisition of new customers is successful, will you be able to manage the orders and still maintain your usual quality standards?

THE PRODUCTS

WHAT ARE THE GREATEST BENEFITS THAT YOU OFFER YOUR CUSTOMERS (STRENGTHS)?
This concerns the services (quality and planning services) and the equipment you sell (modules, inverters, etc.). Are you up to date? Do you have the right products for your target group?

WHERE DOES YOUR PRODUCT HAVE WEAKNESSES COMPARED TO WHAT YOUR COMPETITORS OFFER?
What have prospective customers repeatedly said when they decided not to buy your product? What do customers complain about who have experience with your products? Perhaps you could include an alternative product in your range in the medium term, or, for example, compensate for a higher price by including additional services, such as a free check-up after one year.

WHAT PRODUCT IS THE MOST LUCRATIVE? WHAT DO YOU ENJOY THE MOST?
These questions could have consequences for your marketing, your calculations or the general direction of your company.

MARKETING IMPLEMENTATION

WHO DOES YOUR MARKETING?
The best answer is: everyone! From your receptionist to the customer service engineer, every employee should know how they can contribute to winning and keeping customers. Are your sales representatives motivated? Is there a need for training?

WHAT MARKETING RESOURCES DO YOU HAVE?
These resources range from literature, brochures and model letters to standardized procedures. How do I handle a customer who is making a complaint? How should I behave on the telephone? How are enquiries processed?

WHAT EXPERIENCE IN MARKETING DO YOU HAVE, TO DATE?
Rate your experience in terms of time, money and enjoyment, on the one hand, and success, on the other. Were there any unexpected effects that you could build on and utilize? What have you never tried before? Don't be satisfied with quick answers. For example, if you weren't happy with the outcome of an exhibition, how much of that was the exhibition's fault, and how much was your fault? Are there other exhibitions

where you could reach your target audience more effectively? How could you prepare better for the exhibition? Did you attend to all the prospective customers/visitors?

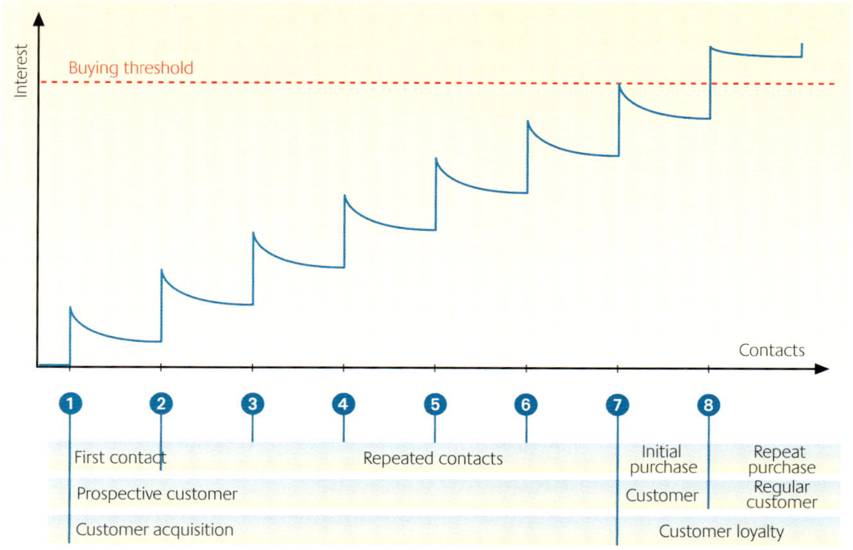

Figure 10.4
Phases of customer acquisition
Source: Wittbrodt

STRATEGIC CONSIDERATIONS: THE 7C STRATEGY

Before taking a closer look at the range of available marketing tools, this section introduces the 7C strategy (C for contacts). This will provide the basis for a more focused selection of marketing tools, which is what turns marketing at random into a marketing concept.

Figure 10.4 shows the phases of customer acquisition. It is natural enough that your customers might forget about you after a while. Don't take it personally; do refresh their memory with regular activities. This pre-sales phase requires some mental staying power. As soon as you detect an interest, it becomes easier.

There is a rule of thumb that says you need to make seven contacts before you acquire a new customer. This may take years to achieve. Don't let yourself be discouraged – the first contact almost never leads to success. As Winston Churchill stated: 'Success is not final, failure is not fatal: it is the courage to continue that counts.'

Essential marketing tools are presented in the following section. Your choice of communication media depends partly upon the target group, partly upon the message, and partly upon the advertising budget. As well as the placement costs (for advertising) and media distribution costs (e.g. postage costs for mail shots), production costs can vary dramatically. Production for print advertising, for example, comprises only a fraction of the production costs for TV commercials.

MARKETING REACH

This list is presented in order of increasing reach. At the same time, the form of address moves from direct (personal) to indirect (e.g. an advert), and the absolute costs increase (not the costs per contact):

- face-to-face contact;
- telephone;
- event;
- mail shot;
- newspaper/magazine/TV/radio.

Rule of thumb: the smaller the target group, the more directly the target group should be addressed.

10.2.3 Range of marketing options

DIRECT MARKETING

Face-to-face contact

The iceberg principle (see section 10.1.2) applies for face-to-face contact. Here, your whole personality comes into play to build up trust and create a personal bond. Face-to-face contact is generally used for extensive consultations with prospective customers and in a sales talk is preparation for the sale. This is dealt with in detail in section '10.3 A good sales talk is fun'.

Telephone

A telephone conversation is an intensive form of personal contact in which all components of the iceberg play a part. The telephone creates a personal closeness to the customer. You can hear the undertones: how genuine is the interest? Is this an anxious customer who needs security, or is he well informed and needs clear facts? The telephone is a good instrument for helping hesitant prospects across the contact threshold to the next, more concrete, step.

Statutory regulations can restrict the use of the telephone for the first contact with private customers. Incidentally, this also applies to fax and email. The telephone is generally used to take up an earlier expression of interest or to follow up on a mail shot. Calling customers on the phone is an art that you and your employees can learn. Some suggestions are provided in section 10.3 – dealing with this topic more intensively will pay dividends.

Mail shot

The mail shot is one of the most commonly used marketing tools. It is a good compromise between direct and indirect contact and the financial outlay is reasonable. You can use mail shots for various purposes:

- generate attention (e.g. to create initial contact);
- provide up-to-date information to a wide target group, such as offers, events and news (e.g. as part of a 7C strategy);
- a regular reminder to customers that you are there as part of customer retention.

Generally, mail shots should aim to trigger a reaction: customers should be invited to get in contact with you.

There are three things to know about mail shots:

1. It is getting more and more difficult to achieve the desired attention. You are doing well if you get a response rate of 1 to 2 per cent.
2. There is little point in doing a mail shot without a telephone follow-up.
3. It follows from the first two points that you should put in a lot of effort and creativity.

Aims of the mail shot

Have you come across the acronym AIDA? You mail shot should aim to trigger your customers':

- attention;
- interest;
- desire;
- action.

Attention

The first aim of a mail shot is not to get thrown away! And this is also the hardest part when it comes to designing a mail shot. Concentrate all of your creativity on this task. Here are some ideas:

- *Evocative heading:* create a vision in your reader's mind.
- *Overall impression:* the graphical appearance determines the initial impression. Take time to make your text and image layout pleasing to the eye.

With regard to typical eye movement across the page, the following is generally true:

- When reading, our eyes move from left to right.
- Elements at or near the centre attract most of our attention.
- We see images before we understand text. For this reason, explanatory text should be positioned below the picture or to the right of it, otherwise this holds up the reading process.
- *Image:* if your product is photogenic or you have a suitable drawing, you should use the image as an eye-catcher and memory aid.

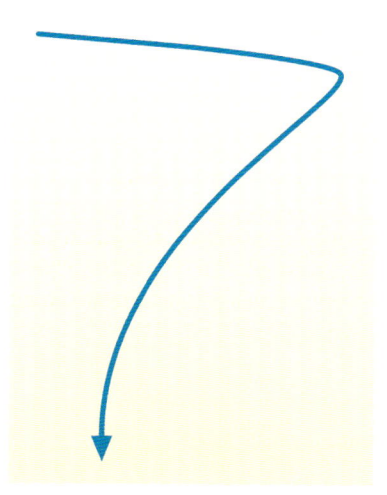

Figure 10.5 When reading, our eyes move from left to right Source: Weidler

- *Gimmick* (a playful note): unusual folding, embossed letter paper, a free gift, a fold-out cardboard sundial – the chances of the letter not getting thrown in the bin increase, although so does the cost. There must be balance between the product and the gimmick. The content of your message must stand up to the expectations created by the gimmick.

Interest
Interest means:

- understanding;
- recognizing benefits.

Put yourself in your target group's position. What benefits do they respond to? How can you put these across in a way that is easy to understand? Brevity is the key. A mail shot should not be longer than one page. It is not just the text, as a whole, that benefits from being short. For sentences and words, too, short is beautiful. The brain can quickly grasp the meaning of 12 syllables of text. Therefore, parts of the text that are especially important should not be longer than this.

Desire
If recognition of benefits is to turn into desire, the customer has to believe you. The less you put the reader under pressure, the less you come across like a market-stall trader and the higher your chances are. So it is probably better to go for restraint when designing the mail shot. In an age where everything is flashy and sparkly, with mega and giga bargains to lure in the punters, a solid, consciously modest style can hit the mark.

Action
You want your readers to do something. You should make this as easy as possible for them:

- *Tempo:* sentences near the end should be even shorter than at the beginning.
- *A concrete appeal to what to do next:* 'Fax us this letter'; 'Contact us by 3 July'.
- *Testimonials:* Let satisfied customers say good things about your product. This way, other people confirm what you say, lowering the inhibition threshold.
- *Free gift:* if the expense is moderate, you can offer a free gift – e.g. 'Early birds who respond by 10 August will receive a free solar alarm clock.' An early reply discount is also a kind of free gift.
- *Pre-printed reply form:* a fax form is the standard today. Fill out the sender part in advance. Also, it's often enough simply to have your mail shot faxed back to you. The personalized address lets you know who it comes from.

CHECKLIST

When you are happy with your mail shot design, give it to someone impartial and ask that person:

- Does the design appeal to you?
- In five seconds, is it clear to you what it is about?
- Is my phrasing clear and positive?
- Does this appeal to you?
- Is its message credible?
- How would you react to it?

Table 10.3
The three areas of an address record
Source: Binder-Kissel

Address data	Name of business and legal form, address, phone, fax, email, customer number, type of business, size of business and possibly turnover
Profile data	Last name, first name, job description and title, direct line and mobile number of contact Additional information could be included, such as date of birth, hobbies, etc.
Action dates	Date of first contact, contact history (conversation notes), action history, products ordered, order media, order value, as well as reminders and unpaid items

ORGANIZING THE MAIL SHOT

You need excellent organization to give the best possible support to your well-formulated, well-designed mail shot. Here are a few tips:

- Who will do the follow-up calls and when? You should give recipients time – at least a week or ten days.
- How will you ensure that each customer only deals with one employee?
- How will you indicate customer interest next to their details?
- What happens if they are interested? The customer should get an answer from you within two days.
- Can you actually handle the orders if there is a lot of interest? You might consider sending out the mail shot in stages.

MARKETING TIP

Give a topical reason for your mail shot that concerns the recipient personally in some way. This will help you to build a bridge and attract attention. It could range from the weather to new laws – for example, new energy saving regulations or a building refurbishment programme.

BROCHURES, SALES DOCUMENTATION

You need printed information for your marketing. As a starting point, a leaflet that provides information about your business (a promotional leaflet) is recommended. This should be professionally designed and printed. Like a special business card, it is a memory aid that you can hand out to prospective customers (e.g. at an exhibition stand).

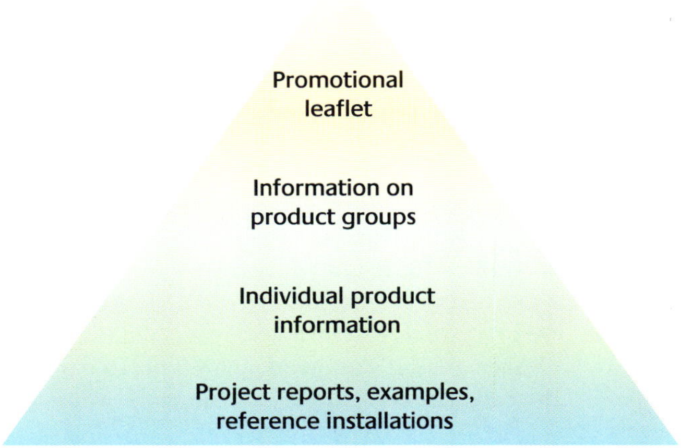
Promotional leaflet
Information on product groups
Individual product information
Project reports, examples, reference installations

Figure 10.6
The publication pyramid: The further down you go, the more simply the information can be presented and produced

Whether or not the whole range of information material is required will depend upon the individual case. You don't need a lavish four-colour brochure for everything! The point is that you should have something up to date ready to give to people, especially for the most frequently asked-for information. This could include information you've formatted and printed yourself (e.g. examples of installations with a photo). A standardized appearance will be worthwhile for someone who has a lot of demand for this kind of material.

EVENTS

Events have a particular charm of their own: you can come into contact with a lot of people in a short time. Preparing for and staging an event is a motivating experience for the whole firm. In addition, it will often get you a free mention in the local press. On the other hand, preparation involves a lot of work and the event itself is staff intensive. During an exhibition, the business can often run only at a limited pace. In addition to the costs, you will also be taking less money.

Some event ideas

Event ideas include the following:

- Participate in trade fairs and exhibitions.
- Put on your own exhibitions or open days.
- Take part in local campaigns or action weeks.
- Provide product presentations (e.g. in a restaurant or hotel).
- Conduct customer seminars.
- Organize talks/lectures.

Events are an excellent opportunity of collecting the addresses of potential customers. Don't take all your information material along with you: just offer to send it – that way you get the address. By doing this at exhibitions, you'll also prevent your expensive brochures from going straight into the trash.

Marketing tip

Try lateral thinking – for example, invite a school class in for a solar morning. Not only will you have the satisfaction of passing on your knowledge, you'll gain 20 to 30 sales helpers. No one educates adults as effectively as children with their parents!

MASS MEDIA

As soon as you want to reach large numbers of people, the mass media is the way to go. Some examples of the media include:

- internet;
- press;
- cinema;
- radio;
- TV.

Today, using the internet more or less goes without saying (see below). When it comes to the press, there are many options:

- advertisements;
- inserts;
- press releases;
- placements (a journalist writes about a topic that you suggest).

At least in their local area, even small businesses can do successful publicity work. Editors are always interested in news: perhaps you have installed a solar energy system on a prestigious building or a school class installed its own system with your help – call the local newspaper! Sometimes a good photo opportunity is all it takes: a mobile crane with a large module hanging from its hook or an architecturally appealing solar array. Ask journalists what they are interested in. Give them background information in which they have a general interest. Keep a clear distinction between public relations (PR) and advertising. If you just want to do advertising, then get in touch with the advertising department directly – otherwise a journalist is likely to feel put upon.

Contact via email and forms

The internet is the only advertising medium that allows direct contact to be made without switching media. This is an opportunity that should not be wasted. Hence, every website should include a contact form that the customer can easily fill out and send to the company directly by email.

There are a few simple rules for the structure of the form. It should have only a small number of fields that the user has to complete: normally their name and an email address or telephone number are sufficient. A free text field is also a good idea so that customers can make a specific enquiry. A 'call me back' field is also a nice gesture as it indicates to the potential customer that you are willing to pay for the cost of the call.

The speed with which you deal with enquiries makes a big difference to the effectiveness of online forms. Since users have a concrete interest in your company's products or services, it pays to get back to them. Many people expect a reaction on the same day or not later than the next working day. For this to happen, there is a need

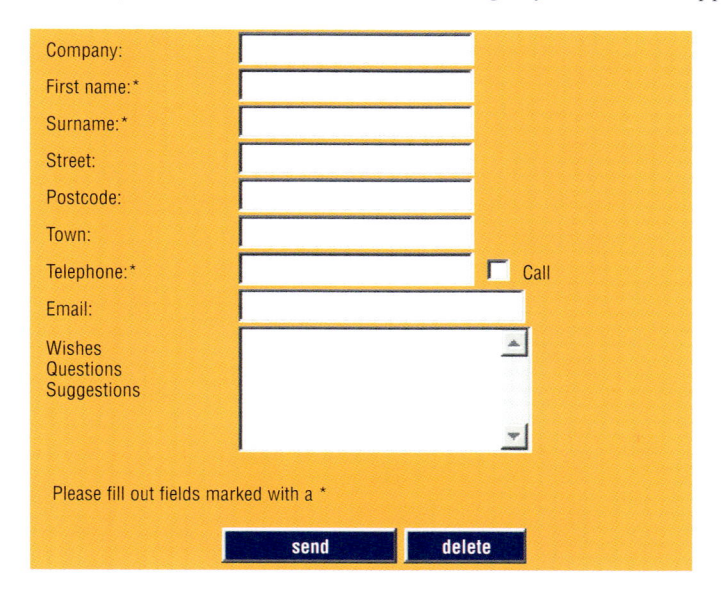

Figure 10.7
The contact form asks for only the essential information to establish contact with the customer

for clarity in your business regarding who reads and replies to emails. It is essential to have arrangements in place during holiday periods, as well, since there is little point in replying weeks later.

You should also publicize your business's email address as an alternative so that customers can write to you directly from their email programme. This also allows customers to send you digital photos.

MEASURING SUCCESS

Another special feature of the internet is that it makes it very easy to know how many visitors your website has generated. Web statistics tools for this purpose are offered by nearly all providers. The main statistics here are the number of page views and the number of visitors. Other variables, such as hits and data transfer, are useful for specialists but normally not of interest to you. A good website should have at least a few visitors each day; otherwise, it hardly justifies the expense of creating the internet presence.

Apart from the number of visitors, a website's success also depends upon the number of enquiries it generates for your business and, ultimately, upon how much it increases your sales. This factor is very hard to determine. While a direct online purchase may happen rarely, the internet is often a sales support tool.

A simple calculation serves to illustrate this: for a website to be worthwhile economically, the additional proceeds must be greater than the costs of the internet presence.

The costs of maintaining an internet presence can be calculated from the following variables:

$$C = S + L \times R$$

C = costs of the internet presence over the amortization period;
S = set-up costs (one off);
R = running costs per year;
L = lifetime of the website in the internet until the next major revision (amortization period).

The internet presence enables you to acquire new customers who bring additional net income and, hopefully, increased profit:

$$P = A \times T \times M \times L$$

P = acquired additional profit;
A = additional customers per year through the internet;
T = average sales (turnover) per new customer per year;
M = profit margin.

Example. The following values may apply for a large installation business:

- new customers: $A = 5$;
- average sales (turnover): $T = 5000$ Euros per year per new customer;
- profit margin: $M = 10$ per cent;
- website set-up costs: $S = 2500$ Euros;
- website running costs: $R = 800$ Euros per annum;
- amortization period (lifetime): $L = $ three years.

$$C = S + L \times R = 2500 \text{ Euros} + 3a \times 800 \text{ Euros per annum} = 4900 \text{ Euros.}$$

$$P = A \times T \times M \times L = 5 \times 5000 \text{ Euros per annum} \times 0.1 \times 3a = 7500 \text{ Euros.}$$

Additional profit = $P - C = 2600$ Euros.

This means that an internet presence that costs around 2500 Euros and generates running costs of around 800 Euros per year is worthwhile in this case since it generates an additional profit of 2600 Euros in three years.

IMPLEMENTATION COSTS

The internet presence comprises the following four cost factors:

1 web design;
2 web server;
3 web directories;
4 internet access.

Some guideline values are provided here, although these depend to a very large degree upon individual requirements and the country in which you live.

Web design. Web design work should be entrusted to a professional agency, otherwise there is no guarantee that the pages will work in the way they are supposed to. Pages that don't work properly are counterproductive! A well-designed website with ten pages and a contact form will cost around 2500 Euros.

Web server. Rent a web server from a provider – you don't need to have one in your company. A simple website doesn't require a high-performance server and will typically cost around 300 Euros per year.

Web directory. Any website needs to be advertised on the internet. The best way to do this is to appear on topic-related websites with large amounts of visitor traffic. Your internet advertising agency must also ensure that the sites are reliably found by search engines. Around 500 Euros per year need to be earmarked for this purpose.

Internet access. Your own internet access that you use for answering email from your website doesn't need to be included here since it is generally used for other purposes and its share in the costs is negligible compared to other items.

MAINTENANCE

Anyone who takes internet marketing seriously and wants to gain more than a handful of new customers each year needs to maintain their website regularly. This maintenance primarily consists of updating content rather than changing the website's appearance.

The principle for updates is 'a little but often'. It cannot be stressed frequently enough that there can only be a noticeable improvement in advertising effectiveness if there is continuity in the maintenance of content. No customer expects daily news from an installation firm; on the other hand, they do not expect to see the 'latest' Christmas offers still being advertised at Easter. In most cases, one hour per month is sufficient to add completed projects to reference lists, announce special dates and delete old information. Today this work can be taken care of by using low-cost editing systems that allow you to modify web pages without knowledge of HTML and without making undesired changes to the layout.

ACTIVE ADVERTISING IN THE INTERNET

Advertising on the internet is different from classic print media in that you first have to bring your website to your potential customers' attention. There are various ways of doing this – the best of these are provided by the internet itself and are described here.

10.2.4 Six steps to the target

THE BASIC IDEA

Acquiring customers for solar energy can take a lot of stamina. A systematic method involving six steps can be used known as the marketing cycle (see Figure 10.8). For large campaigns, it's best to go through this cycle twice: start off with a test run with a small number of target customers, then use the experience from this to improve the campaign for the main run.

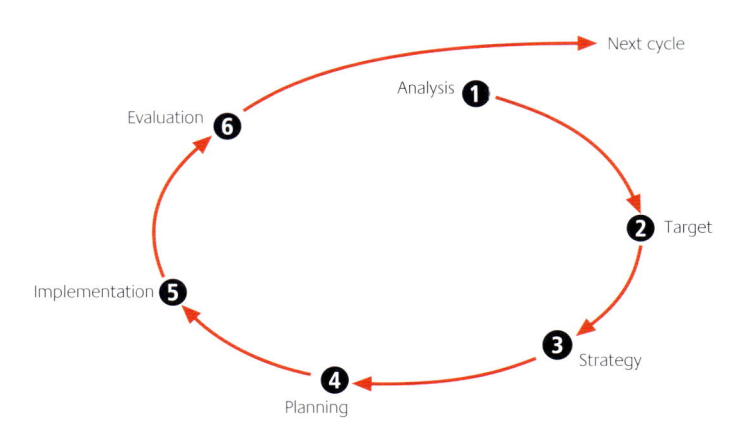

Figure 10.8
Six-stage marketing cycle

Now I can hear you saying: 'But I don't have time for that!' In this case, make the time! Systematic marketing will take less time in the long run than if you start campaigns in dribs and drabs that don't bring you the success you hoped for.

Set aside half a day a month, for example, to update your marketing plan. You should also take two days during a quiet period (e.g. over the Christmas/New Year break) to think through your marketing concept for the year ahead. You will also need time for the implementation (e.g. for an exhibition or a mail shot). Once you have a good plan, you can share the workload for these campaigns with others. It also strengthens the team spirit in your business if everyone works together on the same activity for a few days. And success in a well-planned campaign will stimulate further successes.

THE MARKETING CYCLE

The theory behind the six steps in the marketing cycle is discussed below, with a practical example.

ANALYSIS: STEP 1

You can find notes to help you in section '10.2.2 The four pillars of the marketing concept'. You should also:

- Assess the costs and benefits of your marketing activities to date.
- Assess your economic situation.
- Have an eye towards the future development of your business.

Case study. Chief Executive Mr Watt of Watt Solar Inc has learned through talks with competitors and association members that he is obviously profiting less than others from the solar boom. A business assessment shows he hasn't spent much time on marketing. So far his customers have always come to him. They wanted new outdoor lighting or their wiring needed replacing. Mr Watt realizes that this is no longer enough.

SETTING A TARGET: STEP 2

Having targets is what enables you to set priorities and selectively decide which actions to take. Targets also make it easier to keep track of things in hectic situations. They release extra energy as they activate your subconscious and help you to concentrate on the essentials.

So that you act on your good intentions, you need to work out your concrete targets. Put the results in writing and think SMART:

- Specific: what exactly?
- Measurable: how will you verify your success?
- Active: positive formulation instead of saying what you won't do.
- Realistic: ambitious but attainable.
- Timing: when do you want to have achieved it?

MARKETING AND PROMOTION 355

What do you think of this target: 'Next year I want to have more time for my family.' Yes, it's too vague. A better target would be: 'Starting from 1 March I will go home at 5.00 pm on Tuesdays and Thursdays.'

Case study. In the first few days of the new year, Mr Watt makes the following plans (among others):

- By 9 January I will draw up a marketing strategy for the current year.
- This year I want to actively acquire 20 new customers for solar energy systems.

STRATEGY: STEP 3

Work out a 7C strategy for your target group(s) and translate this into an approximate timetable. In this timetable, write in the marketing campaigns month for month, as well as fixed dates, such as exhibitions and holidays.

Here are a few more tips:

- As a general rule, plan between four and seven contacts per customer each year.
- Don't give your contacts information overload. Include selective material, not a whole bundle.
- Vary the media: send Christmas greetings.
- The form of contact is not as important as the fact that customer contact is sustained over a period of time.

Consider when is a good time or not such a good time. For example, take the following periods into account:

- mail shot: holiday periods, Christmas/New Year;
- telephone marketing: holiday periods, working hours;
- daily rhythm: lunch break, meetings, Friday afternoon;
- events: bank holidays, exhibitions, parallel events, weather.

Reserve half a day each month for updating your 7C strategy.

Case study. This is what an excerpt from the Watt Solar 7C strategy looks like:

- Implementation time frame: 1 January to 31 December 2006.
- Target: acquire 20 new solar customers.

Activities 2006

Month	Marketing activity	Notes
January	 Get promotional leaflet designed and printed	Participation in manufacturer's training for new module
February		
March	 Exhibition stand "all about us"	
April	 Send requested materials to prospective customers from exhibition	Easter holidays
May	Follow-up calls	
June		Whitsun holidays
July		Summer holidays
August		
September	 Send out invitation to in-house exhibition	
October	 In-house exhibition "New Photovoltaic Systems"	Half-term holidays
November		
December	 Christmas greetings	

Figure 10.9
The Watt Solar 7C strategy (excerpt)
Source: Watt Solar

PLANNING: STEP 4

Create a marketing plan for one to two years. Don't give up on the plan if success doesn't happen overnight. Persistence and long-term thinking are key qualities in customer acquisition.

The 7C strategy provides the rough time frame. For detailed planning, create an activity plan for the individual campaigns (e.g. a mail shot). In this you specify:

- who (responsibilities)?
- who does what (definition of tasks)?
- by when (timetable)?

Plan the requirements for preliminary and follow-up work:

- Produce the advertising material (e.g. a mail shot letter).
- Consider the advertising medium (e.g. editorial deadline for trade publications).
- Address ordering/provision (address research in your own database).
- Prepare your own involvement. You control the activities: success depends upon your availability. Do small amounts when you can. You don't have time to call 50 people in one day; but if you make two calls each day, you'll get the job done in about a month.
- Availability of the employees you need: a long time in advance, schedule a period for concerted action by all of your employees who are involved in customer acquisition. First, no one will have time for it otherwise; second, it creates a strong feeling of team spirit.
- Follow up on prospective customers: good organization pays dividends. You need to know who has received what, who your A-list customers are and when the next action is due. And you need to know when to call off your efforts or move an address from category A (very good prospects) to C (less good prospects).

Case study. For the first campaign in his 7C strategy (promotional leaflet), Mr Watt has drawn up an activity plan (see Figure 10.10)

who?	does what?	by when?
Team	Content brainstorming	10/1/2004
Watt	Choose marketing agency to take charge of text, graphics and production	10/1/2004
Watt	Decide on content, discuss with marketing agency (briefing)	17/1/2004
Müller	Sort out photo rights, source photos, forward to marketing agency together with logo and other material	24/1/2004
Marketing agency	Present draft designs and sample text	24/1/2004
Watt, marketing agency	1st revision cycle	7/2/2004
Watt, marketing agency	2nd revision cycle	14/2/2004
Müller	Compile addresses who to send leaflet to and where to display it	19/2/2004
Watt	Decide print run and clear for printing	19/2/2004
Watt	Check delivered leaflets	29/2/2004
Müller	Send out leaflets to external addresses and give to all staff	3/3/2004
	Retain copies for documentation and as reserves	
	Set aside copies for exhibitions and other events	
	Track print run and make note to reorder in good time (3 months in advance)	
Watt	Collect suggestions for improvements for next print run	from 29/2/2004

Figure 10.10 Planning promotional leaflet for Watt Solar Source: Watt Solar

IMPLEMENTATION: STEP 5

There's not much else to say here apart from do it! The biggest danger in marketing campaigns is that just when they're about to start, something else more urgent needs to be done. Stick to your 7C concept! Marketing is important. As the boss, you have to give it the urgency it deserves.

Mr Watt followed this recommendation. He had agreed the timing well in advance with his employees and during the campaign periods assigned one person to emergency customer contact duty. This allowed the rest of the team to give their full attention to the marketing tasks.

Addresses are central to marketing implementation. So how can you get hold of them?

SOURCING ADDRESSES

A common source is the firm's telephone book or address book. Members' directories or business directories are often useful, as well. A highly effective way for you to collect addresses yourself is via an exhibition stand or an information stand in town. Make it a rule to record the addresses and wishes of every prospective customer. Never take too much promotional material with you. Instead, send the selected information material to the prospective customer – and to do that you need the address.

There are two ways you can use the internet for sourcing addresses:

1. Website: a concise, informative website is enough – it doesn't have to include flashy graphics. A more important consideration is that your website is registered with search engines and that you keep checking that the search engines are actually finding your site. This is where you give prospective customers the opportunity to request material from you.
2. Portals: register in the major portals for your line of business. A basic entry is often free of charge; but, equally, it is often worth paying for a link to your own website.

Finally, there are many service companies who specialize in selling customer addresses. Some address providers focus on particular regions or industries. Hence, it's often worth comparing providers before making a commitment. The addresses you require can be searched for according to a range of criteria.

DATABASE MANAGEMENT

A well-maintained computerized customer database is a valuable resource for any business. Give this the attention it needs. Figure 10.11 shows a structural outline for an address record.

Address data	Name and title, address, telephone, fax, e-mail, customer number, industry, occupation
Profile data	Surname, first name, job, position, direct dial and mobile number. Possible additional information such as date of birth, hobbies, etc.
Action data	Date of first contact, contact history (discussion notes), action history, products ordered method of ordering, purchase value, reminders and open posts

Figure 10.11
The three areas of an address record
Source: Binder-Kissel

The following method has proven effective:

- Everyone involved should have the ability to view and modify addresses. Notebook computers are recommended for the sales force so that they can make notes immediately following talks with customers.
- Notebooks and the server database need to be synchronized automatically.
- Keep a contact history with the address record that documents important conversations.
- Keep an action history where you keep a record of all customer acquisition actions. What documentation does the customer already have? What invitations or greeting cards have they already received?
- As well as keeping records of all your contacts, also make a note of all other information that is gained during the customer acquisition phase (e.g. customer preferences).
- Enter 'search terms' that tell you what product(s) the customer is interested in. This will make it easier to compile address lists (e.g. for mail shots).

■ Place your customers into two or three categories: A-list customers are the ones you absolutely must keep up contact with; C-list customers are less interested in your products.

EVALUATION: STEP 6

Immediately after a campaign, evaluate the costs and benefits (e.g. in your monthly marketing planning). A yearly summary is also worthwhile. Only when you have a thorough evaluation will you get the full benefit of the fruits of your work. This will allow you to make the next round even more efficient and effective.

Case study. At Watt Solar, the marketing has paid off. After one year they have gained 23 new customers with an order volume of around 320,000 Euros. There were considerable costs: 26,000 Euros and 40 person days from the employees. However, much of what was produced can continue to be used in the year ahead, and the customers were so satisfied that the first order will not be the last.

10.3 A good sales talk is fun

10.3.1 What does 'successful selling' mean?

Good selling doesn't mean making the sale at any price. Customer satisfaction is the key focus. If you give good advice, customers will recommend you to others even if they don't make a purchase themselves. But customer satisfaction isn't the whole story – you want to be satisfied, too, and not invest time unnecessarily. This is why successful selling means achieving a good balance between the cost and benefit of selling.

Buying – and, hence, selling – is all about trust. This includes sincerity and respect. Good salespeople take customers seriously, even if their wishes sometimes seem strange at first or they express themselves awkwardly. Most products in a given category differ only slightly from each other. *You* make the difference as a salesperson. This is a demanding task – but it's also fun and lets you use your creativity!

This section provides you with some tools and a well thought-out guide to effective sales conversations in order to help you sell more. From finding out the customer's needs to closing the sale, your personal style is important here. Every salesperson needs to be authentic – customers will instantly see through hollow posturing.

PERSONAL REQUIREMENTS

Selling means dealing with people. Two points:

1 Your subconscious will make itself felt! Think back to the iceberg principle: only one seventh of communication is visible on the surface – the rest is invisible, but all the more effective for it. Your gestures and facial expressions are particularly expressive – your body language never lies. For this reason, 'tricks' are of only limited use. The important thing is that what you say and what your body says are consistent. It's important to realize that you can only convince your customer if you profoundly believe in your products and your own abilities yourself.
2 The thing you can change the quickest is yourself! You cannot transform difficult customers. But you can give yourself better protection against them through the way in which you lead the sales conversation.

Figure 10.12 Smiling (sometimes only inwardly) is a good selling aid

Good salespeople build up an intense personal relationship with their customers. They are interested in the customer, not just in the immediate order. They make it easy for the customer to get in contact by being friendly and reachable. They are useful to the customer – and give the customer energy saving tips while checking out the possibilities for a solar system installation in the cellar.

During the sales talk, the benefits to the customer are the central focus. But for the talk to be a success, the accompanying emotional components are equally important (see Figure 10.12). Even the greatest selling points on their own can seem dry and uninteresting without the accompanying components.

THE FOUR PHASES OF A SALES TALK

The following section looks at the four phases of a typical sales talk in detail:

1. Build a bridge.
2. Find out the need.
3. Offer a solution.
4. Achieve the result.

These phases may take place in several stages and be partially repeated. After successfully closing the sale and installing a solar energy system, the salesperson's job is not done. They remain the customer's partner and support the customer so that the customer helps them sell more – through their enthusiastic recommendations to others!

10.3.2 Build a bridge

Figure 10.13
Build a stable bridge to your customers
Source: Heidler

YOUR ATTITUDE

Here the iceberg comes into play again. If you want to talk to a customer, you need a positive attitude about your task and about the customer. A good exercise is to consider 'why am I looking forward to this talk?'

WHY AM I LOOKING FORWARD TO THIS TALK?

Before meeting a customer, stop for a moment. Consider what you will enjoy about the encounter. You could also think about the great evening last night or about your plans for the weekend. You could stock up on some inner sunshine.

MAKING CONTACT

Whether on the phone or in a face-to-face talk, when you make contact, your opposite number first needs to know to whom they are talking. It is particularly helpful here if you can refer to mutual acquaintances or activities. You're sure to have noticed how a cautious or reserved voice on the telephone has relaxed once you mention somebody they know so that the other person can 'place' you. This is called a bridge. Once you've found the bridge, you can cross it without getting wet!

THE CHECKLIST

Note the name of the person you are talking to and address him or her by name several times. Tell the other person:

- Who you are and which company you are from.
- For the first contact: why you are contacting the person at this time.

Find a positive introduction: ideally, things that relate to the customer, such as:

- 'I know you from the horticultural society.'
- 'I got your address from Mr Jones.'

For the follow-up contact:

- 'How are you?'
- 'Nice to see you (to talk to you).'
- Other possibilities include small talk to get started (weather, journey there, etc.).

Tell the customer why you have contacted him/her:

- 'We install solar energy systems.'

Don't automatically assume that the person you are talking to has time for you. You can ask a question to show that you don't mean to hassle the person and are leaving the choice up to them:

- 'Is this a good time to call?'
- 'Do you have time to talk right now?'

At this stage, don't talk about the details of what you offer. The reason for this is that making your presentation too early will mean your customer will start asking questions – which are bound to include questions about the price! But you've only just got the customer's interest: find out their requirements!

10.3.3 Find out the customer's requirements

Figure 10.14
You should find out the customer's requirements as if you are a detective investigating a case

This is the most important phase in a sales talk. Use it to find out your customer's needs.

QUESTIONING TECHNIQUE

If you can ask good questions, you're already well on the way to leading the conversation where you want it to go: 'He who asks, leads.' The iceberg principle applies yet again here: if you don't feel a genuine interest in the other person, they might feel that your questions are intrusive or pushy.

CLOSED QUESTIONS
Closed questions, to which there is only a 'yes' or a 'no' answer, are the most commonly used form of question. You can use them to confirm clear agreements:

- 'Can we meet on Friday?'
- 'Is 5 May all right with you to come and install the solar system?'

You can use semi-closed questions (the answer is usually one word: a fact) to get specific information at the beginning of a conversation. These are also known as 'W' questions: who, what, where, when (and how much):

- 'How many people live in this house?'
- 'When will the roof be put on?'

This type of question is particularly suitable for getting to the point when you want your customer to make a commitment. You should ensure here that you always link the closed question to a selling point. This makes it more difficult for the customer to say no:

- 'Would you like the anodized frame so that the modules match your roof better?'
- 'Would you like two modules more to increase the yield?'

OPEN QUESTIONS
Open questions make the other person think much more about their answer. As a result, they involve the other person in the conversation more, and you get a wealth of information – which may include things you hadn't expected. In the stress of everyday life, this unexpected information can be tiresome at first since it requires thought and flexibility on your part. However, it can often turn out to be a gold mine. Investigation is worthwhile here. Perhaps you don't sell any modules, but you might sell a new outdoor lighting system.

As long as you are in the 'find out the requirements' phase, you should primarily ask open questions. This will also help you to keep up the momentum in a hesitant conversation and get taciturn people to open up:

- 'What are your thoughts about supplying energy to your house in the medium term?'
- 'What is particularly important to you?'
- 'What difficulties did you encounter there?'

You can use open questions to direct the conversation and keep it on track. Open questions may seem innocuous, but they require good preparation and a certain amount of practice. Even experienced salespeople have a tendency to ask more closed questions than open ones. When under pressure, closed questions often come to mind more readily. This is why it is worth regularly formulating open questions beforehand, until you can produce them effortlessly.

PRACTICAL TIPS

Use your initiative! Imagine you are walking next to a high wall and want to know where there is a pond on the other side. How would you find this out? Throw a small stone over the wall every few steps – until you hear a splash. With this detective's instinct you'll find the information that matters to you.

BACKPACK TECHNIQUE

You should always have everything 'with you', as if you carried your whole range of products and services in a backpack. When you visit a customer about a rewiring job, you should keep your eyes and ears open for additional opportunities to do something for the customer. Make it a habit always to take a look at the roof and ask every customer: 'You have such a nice south-facing roof. What do you think of solar technology?'

COFFEE FILTER

Let your opposite number talk. Give them time to formulate complex ideas or to get things clear for themselves, just like a coffee filter that you only throw away when the last drop has passed through.

10.3.4 Offer solutions

Now you prepare to close the sale: you propose solutions and make an offer. You should only do this once you know exactly what the customer's needs are. This is because as soon as you offer solutions, the customer's need stops growing.

BENEFITS

The benefits are at the beginning of the road to the solution. You have investigated the customer's needs. Now identify the benefits of your products that meet these needs.

USE CUSTOMER-ORIENTATED PHRASES

Address the person to whom you are talking. Say 'you' at least five times as frequently as you say 'I' or 'we'. If you combine this with the other person's name, you will get their full attention. Everyone becomes attentive if they read or hear their own name.

Use the customer's phraseology. If they talk about a 'solar panel', you shouldn't say 'module'.

Pay attention to the way in which you speak (make a phraseology checklist) – you already know the technical stuff.

PRESENTATION

Nothing is as persuasive as a practical example. Take your customer with you to visit demonstration installations. If you can't do that, show them (good) photos or bring models or components with you. The more sensory channels you appeal to, the more you will fix your product in the customer's mind. You can use the same things that work in advertising to help you here: show letters from satisfied customers or, in your advertising materials, use testimonials from customers you have already impressed. When filling your reference folder, think of the popular selling points. Do the calculations for an example system.

The iceberg applies to the presentation, as well: feel how much you enjoy the work, how much you're looking forward to meeting the customer, before you ring their doorbell. Then you can really let loose with a barrage of good ideas.

10.3.5 Achieve the result

Now is the moment for the transition from the no-obligations talk to the act of commitment. Don't miss it – it is also possible to talk too much and to lose the sale.

THE TRANSITION

The buying signals checklist shows you how to recognize when you've reached this phase of the conversation.

BUYING SIGNALS CHECKLIST
- Expressions of assent:
  - 'That sounds good.'
  - 'That would mean that in the future ...'.
- Questions of detail:
  - 'So in that case should I buy a system with mono-crystalline or polycrystalline cells?'
- Buying-related questions:
  - 'What would the complete system cost?'
  - 'When could you do that?'
- Non-verbal signals:
  - The customer moves closer to you.
  - They go up to the product and touch it.
  - They nod their heads in agreement.

NOW SYSTEMATICALLY TAKE THE LEAD

Eliminate any final obstacles to buying:

- 'What else can I do for you?'
- 'Who's going to take the decision?'
- 'How's your schedule?'

Do you still remember the iceberg? In your mind, convince yourself that the conversation is going to end successfully. When you are firmly convinced that you're making a good offer, this transfers onto the customer as well.

You want the customer to say 'yes': so make it easy for them to say yes. This phase is suitable for closed questions that your customer can answer with a 'yes':

- 'Have you spoken with Mr Howson about his system?'
- 'Has he confirmed my data?'
- 'Have you looked at the grant application forms?'
- 'Do you like the cover frames?'

Nevertheless, take care not to seem overly confident of success. For customers, the purchase is a difficult decision since they have to decide against many other ways in which they could use their money. On the basis that 'a problem shared is a problem halved', it helps customers to see that you don't have it easy either. So don't mind if customers have another query that costs you effort to answer or if they try to get another concession out of you.

DEALING WITH OBJECTIONS

Objections are normal: you also consider the pros and cons of any decision. Unfortunately, it is also normal and can be disastrous to try to knock back objections just to prove yourself right. This can degenerate into bickering. It's better to take the other person's thoughts seriously and help them to see other points of view:

- 'It's true that the large system is more expensive. But you're using your roof space more effectively.'

It is important that the customer retains the choice over whether they want to be persuaded or not.

PRACTICAL TIPS

Echo technique
The echo technique is useful if the other person suddenly surprises you with objections and counter-arguments. You repeat what was said in your own words and ask a question in return:

- 'So you're concerned about the roof leaking? How do you mean?'

This gives you time to 'absorb' the question. Your opposite number feels acknowledged and that you take what they say seriously. They will explain what they mean. Then, you can either dispel your customer's concerns or you realize that they really want something else.

Let rubbish pass you by
If you notice that your customer is finding fault with things that are actually of no importance – perhaps because they're having a bad day – then let them. Like twigs and leaves in a swollen river, don't pay attention to these things at first. If there is genuine interest in there somewhere, it will surface again. This way you avoid getting into pointless arguments.

Support positive signs
If you notice that the customer is more interested in the inverter with a display than the modules on the roof, bring this up again in your closing conversation. Perhaps you'll even be able to sell a data logging unit.

Use the power of silence
When everything has been said, by being silent you help the customer to collect their thoughts.

THE PRICE QUESTION

It's not the price that sells, it's the product. If you are helping customers to make *their* wishes come true, the price is not the main focus of attention.

If possible, don't state the price until the end of the sales conversation, when it is clear what the customer is getting in return and how the price will be offset (e.g. through grants or savings). If the customer asks you about the price too early on, reply:

- 'Could you wait just a moment? The price depends upon a range of factors and I'd like to tell you about them first.'

If the customer persists in asking about the price, then tell them it – otherwise they may feel that you're giving them the run-around.

A solar energy system is a long-term investment. Make it clear that the purchase is effective over a long period of time (e.g. by stating the price per year or month over the lifetime of the system). Or include the grant and amortization period in the calculation straightaway:

- 'This system costs you 10,000 Euros. It pays for itself after about 15 years. For the rest of its total lifetime of 25 years, it earns money for you!'

If someone latches on to the price and says 'this is too expensive!', then you might be able to find out more about their reasons if you ask:

- 'In your view it's too expensive: too expensive compared to what?'

Don't give discounts. If customers reckon you're an easy target when it comes to price negotiations, they will always squeeze discounts out of you. If you want to make concessions, offer the customer more value for the same price (e.g. a better inverter or the interface for viewing power yields on a PC).

If the customer insists on getting a discount, then you could ask:

- 'What feature or benefit do you want to do without?'
- 'Let's assume we agree on this price. What else is important to you?'
- 'I can give you a discount if you buy a larger surface area.'

Perhaps there are some fine-tuning possibilities: adjust the payment method or payment terms, reduce the amount of options, or reduce risk surcharges. The main thing is that you should stick to your principle: no discount without something in return.

You have to be 100 per cent committed to your prices yourself so that you can appear confident in price negotiations. The customer will notice immediately if you think your own prices are excessive or if you are unsure of yourself. So do thorough calculations beforehand.

Bibliography

Alsema, E A, 2000, 'Energy payback time and CO_2-emissions of PV systems', *Progress in Photovoltaics: Research and Applications* 8, S. 17–25

Basore, P A, 2004, 'Simplified processing and improved efficiency of crystalline silicon on glass modules', in *Proceedings of the 19th European Photovoltaic Solar Energy Conference*, 7–11 June, Paris, France

Baumgartner, F, NTB, 2004, 'MPP voltage monitoring to optimise grid connected system design rules, Beitrag zur 19', European Photovoltaic Solar Energy Conference, June, Paris

Baumgartner, F, NTB, 2005, 'Euro Realo inverter efficiency: DC-Voltage Dependency, Beitrag zur 20', European Photovoltaic Solar Energy Conference, June, Barcelona

Becker, G, 2001, *Innovative gebäudeintegrierte Solarstromanlagen– Architekturwettbewerb des Solarenergiefördervereins Bayern e.V.*, Broschüre, Munich

Becker, G, 2002, *Solarstrom aus Fassaden – Architekturwettbewerb des Solarenergiefördervereins Bayern e.V.*, Vortrag, Munich

Becker, H, 1997, 'Blitz- und überspannungsschutz bei Photovoltaikanlagen', *Photon* 12

Bendel, C, Nestle D and Malcher S, 2005, *Dezentrale Energieeinspeisungen ins Niederspannungsnetz, Tagungsband des 20. Symposiums Photovoltaische Solarenergie*, Hrsg. OTTI-Kolleg

Bernreuter, J, 2005, 'Die Branche hat geschlafen, fachartikel in sonne wind and wärme', *Ausgabe 29. Jahrgang*, BVA Bielefelder Verlag GmbH & Co, KG

Bopp, G, 1999, 'Inwieweit tragen PV-Anlagen zum Elektrosmog bei?' *Beitrag zum 14. Symposium Photovoltaische Solarenergie*, OTTI-Kolleg

Bopp, G and Schätzle, R, 2002, 'Inwieweit tragen Photovoltaikanlagen zum Elektrosmog bei?', Beitrag zur EMV-Tagung des VDB, Hamm 2002, Fraunhofer ISE

Brösicke, W, 1995, *Vorlesungsskript Elektrische Energiewandler – Photovoltaik Teil 3*, FHTW, Berlin

Burger, B, Fraunhofer ISE, 2005, 'Auslegung und dimensionierung von wechselrichtern für netzgekoppelte anlagen', *Vortragsfolien für das 20. Symposium Photovoltaische Solarenergie*, OTTI-Kolleg, Veranstalter

Decker, B, 1998, 'Betriebserfahrungen mit PV-Fassaden', Beitrag zum DGS-Symposium Energiefassaden, Berlin

DGS, 1999, 'EUPOS – Schulungsunterlage Photovoltaik', Munich

DGS LV Berlin, Brandenburg, Haselhuhn, R. and Spitzmüller, P, 2002, *Daten aus dem Solarkataster Berlin*, DGS, Berlin

Dierks, K, 1997, *Baukonstruktion*, 4. Auflage, Werner Verlag, Dusseldorf

Dietze, G, 1957, *Einführung in die Optik der Atmosphäre, Akademische Verlagsgesellschaft*, Geest & Porting K.G., Leipzig

Domnik, K, 2005, 'Fachgerechte und normen- bzw. zulassungskonforme Integration von Photovoltaikelementen in vorgehängte hinterlüftete Fassaden (VHF) NACH DIN 18516-ff, Tagungsbeitrag zum 20', Symposium Photovoltaische Solarenergie, 10 May, OTTI Regensburg

Durisch, W, Bitnar, B, Shah, A and Meier, J, 2004, 'Impact of air mass and temperature on efficiency of three commercial thin film modules', 19th European Photovoltaic Energy Conference, Paris, Hrsg.: WIP

Eicker, U, 2002, *Multifunktionsfassade – Die öffentliche Bibliothek in Mataro, Beitrag in Gestalten mit Solarzellen*, S. Rexroth [Hrsg.], C.F.Müller Verlag, Heidelberg

Erban, C, 1999, *Planungsunterlagen Saint Gobain*, PV zur Gebäudeintegration, Aachen

Erfurth + Partner Beratende Ingenieure GmbH, Steinbeis-Transferzentrum Energie und Umwelttechnik Chemnitz, Solarpraxis, 2001, *Tragkonstruktionen für Solaranlagen. Planungshandbuch zur Aufständerung von Solarkollektoren*, Solarpraxis, Berlin

eta Energieberatung, 1998, *Genehmigung und Montage von PV*, Anlagen, Pfaffenhofen

Fawer-Wasser, M, 2004, 'Solarenergie – ungetrübter Sonnenschein?' *Sarasin Studie*, November

Fischedick, M, Nitsch, J et al, 2002, 'Langfristszenarien für eine nachhaltige Energienutzung in Deutschland', *Auftrag des Umweltbundesamtes*, June

Fraunhofer ISE, 1994–1998, Diverse Auswertungen des Bund-Länder-Breitentestprogramm für netzgekoppelte Photovoltaik-Anlagen, basierend auf Untersuchungen von verschiedenen Instituten und Organisationen: DGS LV Berlin Brandenburg, FhG-ISE Freiburg und Leipzig, FZR Rossendorf, ISFH Hammeln, IST Augsburg, TÜV Rheinland, WIP Munich

Fraunhofer ISE, 1997, '1000-Dächer Meß- und Auswerteprogramm Jahresjournal 1996', Fraunhofer ISE

Frei, R and Meier, C, 2004, '6 Dünnfilm-Technologien in 3 verschiedenen BIPV-Varianten - Erste Resultate des Performance-Tests unter realen Bedingungen, Beitrag zum 19', Symposium Photovoltaische Solarenergie, OTTI-Kolleg

Freitag, O and Weber, U A, 2000, 'Systematische Untersuchungen zur Abschattungsproblematik von PV-Generatoren mit ihren Auswirkungen auf den Energieertrag, Beitrag zum 12', Internationale Sonnenforum 2000 der DGS, Berlin

Genennig, B, 2002, 'Evaluierung des 100.000-Dächer-Solarstrom-Programms im Auftrag des BMWi, UIL und IE', Leipzig

Giese, L B et al, 2002, 'Wiederverwertung von Anlagenkomponenten zur Erzeugung erneuerbarer Energien – Forschungsbedarf in der Abfallbehandlung', Dokument der Bundesanstalt für Materialprüfung [BAM], Berlin

Götzberger, A, Voß, B and Knobloch, J, 1997, *Sonnenenergie: Photovoltaik. Physik und Technologie der Solarzelle, 2,* Auflage, B. G. Teubner Verlag, Stuttgart

Green, M A, 2004, 'Third generation photovoltaics: Theoretical and experimental progress', in *Proceedings of the 19th European Photovoltaic Solar Energy Conference*, 7–11 June, Paris, France

Häberlin, H and Renken, G, 2003, *Langzeitverhalten von netzgekoppelten Photovoltaikanlagen 2, Forschungsbericht Berner Fachhochschule*, Labor für Photovoltaik, Burgdorf

Häberlin, H, Borgna, L, Kämpfer, M and Zwahlen, U, 2005, 'Totaler Wirkungsgrad – ein neuer Begriff zur besseren Charakterisierung von Netzverbund-Wechselrichtern', Tagungsband des 20. Symposiums Photovoltaische Solarenergie, Hrsg, OTTI-Kolleg

Haferland, F, 1987, *Außenwandentwicklungen, Sonderdruck aus Deutsche Bauzeitung 11/87 und 12/87*, Deutsche Verlags-Anstalt, Stuttgart

Haselhuhn, R, 2004, *Auswertung von Betriebsdaten der PV-Fassadenanlage am Energieforum Berlin*, DGS, Berlin

Haselhuhn, R, 2005, 'Photovoltaik – Gebäude liefern Strom, Hrsg', BINE-Fachinformationsdienst Karlsruhe, TÜV-Verlag, Cologne

Henze, A and Hillebrand, W, 1999, *Strom von der Sonne*, Ökobuch Verlag Staufen bei Freiburg, 1 Auflage

Herrmann, W, 2005, 'Missmatchverluste bei Verschaltung von Solarmodulen - Ertragsgewinn durch Vorsortierung?', Tagungsband des 20. Symposiums Photovoltaische Solarenergie, Hrsg. OTTI-Kolleg

Herzog, T, Krippner, R and Lang, W, 2004, *Fassaden Atlas*, Birkhäuser Basel, Berlin, 1. Aufl.

Hoffmann, W et al, 2004, 'Towards an effective European industrial policy for PV solar electricity', in *Proceedings of the 19th European Photovoltaic Solar Energy Conference, 7–11 June*, Paris, France

Holz, V, 2000, 'Ein neuartiges Konzept für transformatorlose PV-Wechselrichter mit besonderen Vorteilen, Beitrag zum 15', Symposium Photovoltaische Solarenergie, OTTI-Kolleg

Hotopp, R, 1991, 'Private PV-Stromerzeugungsanlagen im Netzparallelbetrieb', RWE-Broschüre, Essen

Jungbluth, N, Frischknecht, R, 2000, 'Literaturstudie ökobilanz photovoltaikstrom und update der ökobilanz für das Jahr 2000', Bericht im Auftrag des Bundesamtes für Energie, Projekt-Nr. 39489

Kaltschmitt, M, 2002, 'Erneuerbare energieträger in Deutschland. Potentiale und kosten', *BwK - das Energie-Fachmagazin*, April

Kern, A, 2004, Blitz- und Überspannungsschutz für Anlagen der regenerativen Energietechnik – Praxisorientierte Schutzkonzepte und –maßnahmen für Photovoltaik- und Windenergie-Anlagen, FH Aachen Abteilung July

KfW, 2002, *Jahresbericht 2000 zum 100.000 Dächer-Solarstrom-Programm*, KfW, Frankfurt

Kiefer, K and Hoffmann, V, 2004, 'Betriebserfahrungen und Messergebnisse von netzgekoppelten Photovoltaik-Anlagen in Deutschland – eine Langzeitstudie', Beitrag zum 19. Symposium Photovoltaische Solarenergie, OTTI-Kolleg

Knapp, K E and Jester, T L, 2000, 'An empirical perspective on the energy payback time for photovoltaic modules', Solar 2000 Conference, Madison, WI

Krampitz, I and Epp, B, 2004, 'Insel-Lösungen', in *Sonne Wind & Wärme*, Ausgabe 10, 28. Jahrgang, BVA Bielefelder Verlag GmbH & Co. KG

Kremer, P and Fuhrmann, 2005, *Entwurf einer Norm zum MPP-Wirkungsgrad*, DKE – Document 373_2004-0080, DKE, Frankfurt

Ladener, H, 1996, *Solare Stromversorgung*, Ökobuch Verlag Staufen bei Freiburg, 2. Auflage

Lafarge Dachsysteme GmbH, 2004, *Handbuch geneigte Dächer, 6.* Auflage, Oberursel

Lang, J and Kiefer, K, 1998, 'Kleine netzgekoppelte PV-Anlagen im Breitentest', BINE Informationsdienst FIZ Karlsruhe, profi-info 1/98

Laupkamp, H, Beitrag zum 2, 1998, *Projektworkshop: Qualifizierung von PV-Fassaden*, Tagungsunterlagen, Cologne

Mandl, W and Lubinski, F, 2001, *Datenblatt Marktübersicht Unterkonstruktionen VHF*, Beitrag in Fassadentechnik, 7. Jahrgang, Charles Coleman Verlag, Lübeck

Meyer, F, 2002, *Modulare Systemtechnik für dezentrale Energieversorgung*, BINE projektinfo 02/02, Fachinformationszentrum Karlsruhe (Hrsg.), Bonn

Mulligan, W P, 2004, 'Manufacture of solar cells with 21% efficiency', in *Proceedings of the 19th European Photovoltaic Solar Energy Conference*, 7–11 June, Paris, France

Nitsch, J et al, 2000, *Klimaschutz durch Nutzung erneuerbarer Energien, Studie im Auftrag des BMU und des UBA*, Berichte des UBA 2/00, Erich Schmidt Verlag, Berlin

Pehnt, M, Bubenzer, A and Räuber, A, 2003, 'Life cycle assessment of photovoltaic systems – Trying to fight deep-seated prejudices', *Photovoltaics Guidebook for Decision Makers. Status and Potential Role in Energy Technological Economy*; Achim Bubenzer, Joachim Luther (eds) Springer-Verlag, pp179–213

Quaschning, V, 1996, *Simulation der Abschattungsverluste bei solarelektrischen Systemen*, Verlag Dr. Köster, Berlin

Quaschning, V, 2000, *Systemtechnik einer klimaverträglichen Elektrizitätsversorgung in Deutschland für das 21. Jahrhundert*, VDI Verlag, pp40–50

Quaschning, V, Grochowski, A and Hanitsch, R, 1998, *Untersuchungen von Alterungserscheinungen bei Photovoltaik-Modulen an der TU-Berlin*, Technische Universität Berlin, Berlin

Renzig, S, 2005, 'Schlankheitskur für zellen, fachartikel neue energie', *Ausgabe*, 15, Jahrgang, Bundesverband WindEnergie e. V. (BWE)

Rexroth, S, 2002, *Gestalten mit Solarzellen*, C.F.Müller Verlag, Heidelberg

Sander, K et al, 2004, 'Stoffbezogene Anforderungen an Photovoltaikprodukte und deren Entsorgung', Kurzfassung, UFOPlan BMU, Forschungsprojekt FKZ 202 33 304, January

Sauer, D U, 2002, *Beitrag zum Fachseminar Dezentrale Stromversorgung mit Photovoltaik*, OTTI-Kolleg

Scharmer, K and Greif, J (eds), 2000, *The European Solar Radiation Atlas*, Les Presses de l'Ecole des Mines, Paris

Schaupp, W, 1993, *Außenwandbekleidungen. Kommentar zu DIN 18515 und DIN 18516*, 1. Auflage, Hrsg.: DIN e. V., Beuth Bauverlag

Scheer, H, 1999, *Solare Weltwirtschaft*, 2. Auflage 1999, Verlag Antje Kunstmann GmbH

Schittich, C, Staib, G, Balkow, D, Schuler, M and Sobek, W, 1998, *Glasbau-Atlas 1*, Auflage, Birkhäuser Verlag, Basel, Boston, Berlin

Schlenker, S and Wambach, K, 2005, *Recycling von Solarzellen und Modulen – Entwurf eines freiwilligen Rücknahmesystems*, Deutsche Solar, Vortrag

Schneider, S, 2000, 'Rechtliche vorschriften an gebäudeintegrierte photovoltaikmodule', Beitrag in Photovoltaik – Architektonische Gebäudeintegration, Informationszentrum Energie des Landesgewerbeamtes Baden-Württemberg [Hrsg.], 1. Auflage

Schneider, S, 2002, *Welche Vorschriften gelten für Solaranlagen?* Beitrag in Erneuerbare Energien, SunMedia Verlag Hanover, 12. Jahrgang, 4

Siegfriedt, U, 1999, *Optimierung von abgeschatteten Solargeneratoren bei Netzeinspeisung*, Diplomarbeit TU-Berlin Fachbereich Elektrotechnik, Berlin

Siegfriedt, U, 2000, 'Solarmodule richtig verschalten – Leistungsverluste bei verschatteten Solarstromanlagen verringern', *Fachartikel Sonnenenergie* 4

Siegfriedt, U and Slickers, G, 2001, Vergleich des solaren Ertrages von nachgeführten und feststehenden PV-Anlagen, Fachbeitrag zum DGS-Workshop – Betriebsergebnisse von Photovoltaikanlagen, DGS, Berlin

Solarenergieförderverein Bayern e.V., 2003, *Genehmigung von Photovoltaik-Anlagen. Ein Leitfaden zum Baurecht*, SeV Bayern, Munich, 1. Auflage

Solarenergieförderverein Bayern e.V., 2004, *Solarstrom aus der Gebäudehülle*, SeV Bayern, Munich, 1. Auflage

Stark, T, 1999 *Strom und Architektur – Die Integration von Photovoltaik*, Diplomarbeit Universität Stuttgart, Stuttgart

Stark, T, 2002, Persönliche Informationen zur Auswertung des PV-Architektur-Wettbewerbs des Landes Baden Württemberg

Steinberger, H, 1995, *Umwelt- und Gesundheitswirkungen der Herstellung und Anwendung sowie Entsorgung von Dünnschichtsolarzellen und Modulen*, BMBF – Forschungsvorhaben 0329205 A, Bericht

Stryi-Hipp, G, 2005, 'Photovoltaik produktion in Deutschland – kapazitäten, lieferfähigkeit, engpässe und wettbewerbsfähigkeit für PV 'made in Germany, 20', Symposium Photovoltaische Solarenergie, 9–11 March, Kloster Banz, Staffelstein

Technische Regeln für die Verwendung von linienförmig gelagerten Verglasungen, Fassung September 1998, DIBt Mitteilungen, Heft 6

TÜV Rheinland, 1984, *Atlas über Solarstrahlung in Europa*, TÜV-Verlag

TÜV Technische Überwachung Hessen GmbH, 2000, *PV-Anlagen-Technische Anforderungen. Planungs- und Installationshinweise*, Hessisches Umweltministerium, 3. Auflage

Vaaßen, W, 2004, 'Degradation von photovoltaik-modulen', *Erneuerbare Energien* 6, Verlag: Sun Media GmbH, Hanover

Vaaßen, W, Althaus, J and Hermann, W, 2003, 'Leistungsangaben und Degradation bei kristallinen Photovoltaik-Modulen, Beitrag zum 18', Symposium Photovoltaische Solarenergie, OTTI-Kolleg

Varta Batterie AG, 1986, 'Bleiakkumulatoren', VDI-Verlag Dusseldorf, 11. Auflage

VDEW, 2001, *Richtlinie für den Anschluss und Parallelbetrieb von Eigenerzeugungsanlagen am Niederspannungsnetz*, 4. Ausg. 2001, VWEW-Verlag, Frankfurt am Main

VDN-Merkblatt zur VDEW, 2004, *Richtlinie für den Anschluss und Parallelbetrieb von Eigenerzeugungsanlagen am Niederspannungsnetz*, VDN, Berlin

Vogdt, F U, 2002, 'Hinterlüftete außenwandbekleidungen', in *Mauerwerk-Kalender*, Ernst & Sohn Verlag, Berlin

von Dohlen, K, 1999, 'Normkonforme Einbindung einer Solarstromanlage mit und ohne äußeren Blitzschutz', Beitrag zum OTTI-Fachseminar EMV und Blitzschutz in PV-Anlagen

von Fabeck, W, 2000, 'Messung und abrechnung einer volleinspeisung unter nutzung des schulnetzes', *Solarbrief* 3

Wagner, A, 1999, *Photovoltaik Engineering- die Methode der effektiven Solarzellenkennlinie*, Springer Verlag, 1. Auflage

Wagner, A, 2001, 'Mathematische lösungen für die praxis – neues solarzellen-ersatzschaltbild ermöglicht qualitätskontrolle', *Fachartikel Sonnenenergie* 5

Wetzell, O W, 2000, *Wendehorst Bautechnische Zahlentafeln*, 29. Auflage, B.G. Teubner Verlag, Wiesbaden

Wildecker, R, 1987, *Praktisches Wetterlexikon*, BeckVerlag, Munich

Zehner, M, Hartung, A, Karg, N, Maier, M and Zettl, M, 2005, 'PV-Datenblätter von herstellern im internet- erfahrungen beim aufbau einer datenbank, tagungsband des 20', Symposiums Photovoltaische Solarenergie, Hrsg. OTTI-Kolleg

Zimmermann, W, 2001, 'Kristalline silizium-dünnschichtsolarzellen auf ssp substraten', Dissertation from Albert-Ludwigs-Universität Freiburg i. Br.

Zweibel, K, von Roedern, B and Ullal, H, 2004, 'Finally: Thin-film PV', *Photon International*, 8 (10)

Index

Note: page numbers in *italics* refer to figures/illustrations and tables

absorption of solar radiation 9, 12
AC (alternating current)
   cabling 5
   connection cable 126
   sizing 171–2
      electrical parameters for *171*
   power 5
   switch disconnector 127
      earth leakage circuit breakers 127
      MCBs (miniature circuit breakers) 127
AC-coupled systems 329–30, *329–30*
acid stratification 307
active tracking 17–18
   astronomical 17
   sensor-controlled 17
additional wind load 231–2, *213-2*
aerospace applications 44
air mass (AM) factor 12, *13*
air pollution 13
albedo value 13, *13*
Altener program 186
alternating current *see* AC
aluminium 49
aluminium-doped zinc oxide 43
aluminium sheets 38
amber warning lantern 74
amorphous silicon cells 15, *23*, 34, 40, 42–3, 48, 51, 52, 53, 61, 76
   cell manufacturers 43
   fabrication 42
   flexible modules *43*
amorphous thin-film modules 78
amorphous triple cells 62–3
anchoring
   fixed systems 228–9, *228–9*
   parts 240
angle definition 11
   in solar technology *11*
angle of elevation, dependence of irradiance on *13*
annual distribution of solar irradiance at module plane of system *159*
anti-reflective coating (AR) on crystalline silicon cells 5, 26, *27*, 28, 30–1, 48, 51
anti-reflective glass 68
APex cells 27, 48
   polycrystalline 29–30
   production process *30*
applications, PV 1–8
arena event hall, Treptow, Berlin *82*
array systems, PV 1–8
astronomical tracking 17
Atomic Institute of the Austrian Universities (AIAU) 33
ATM tracking system 18, *19*
automated cell-stringing 65
automotive engineering 44

back-contacted solar cells 35–6
   cell manufacturers 36
   structure *35*

back contacts 33–4
   rays and circles *34*
back grid contact *34*
back passivation 61
back point contacts *34*
back strip contacts *34*
back surface field (BSF) 33
   and back point contacts *34*
   and back strip contacts *34*
backpack technique 359
ballast-mounted systems (freestanding installations) 225–7, *225–7*
batteries *see* lead-acid batteries
battery chargers 2, 5
battery sizing 325–6
   determining battery capacity *326*
Baumgartner, Franz 112
BAY 45 silicon diode 53
Bayerische Landesbanke, Munich, Germany 79
Berger, Bruno 160
Berliner Bank *228, 229*
bipolar transistors 106
BIMODE international research project 33
bird droppings which soil modules *132*
bituminous sheeting, fitting *203*
body-tinted glass 81
   backing *81*
boron 20
   doping (p-doped) 20, 21, 33, 35, 37, 39, 42
BP Solar 34
breakdowns, typical faults and maintenance for PV systems 284–6, *284–6*
   maintenance 285
   and upkeep checklist *286*
Brösicke, Wolfgang 13
brown polycrystalline cell with special AR coating *31*
building integration and mounting systems 199–271
buoys 2, *4*
busbars 32
bypass diodes
   hot spots and shading 89–92
   module characteristic *I-V* curves with and without *91*
   module junction box with *92*
   shaded PV module with *91*
   shaded PV module without *90*

cable exit
   along glass edge *84*
   on back with junction box *84*
   on side with module lead plug connector system *84*
cable connection systems *125*
cable current carrying capacity 166–7, *166*
cable laying 275
cable losses 318–19
   and voltage drops, minimizing 167
cable requirements 124
cable voltage ratings 166

cables, selecting and sizing for grid-tied PV systems 165–72
   cable current carrying capacity 166–7, *166*
   cable voltage ratings 166
   minimizing cable losses/voltage drops 167
   sizing AC connection cable 171–2
   sizing DC main cable 170–1
   sizing module and string cabling 167–70
     electrical parameters for *167*
cabling, wiring and connection systems 123–6
   AC connection cable 126
   connection systems 124–6
   DC main cable 126
   module and string cables 123–4
CAD 184
cadmium 44, 45
cadmium sulphide 43
cadmium telluride (CdTe) cells 15, *23*, 40, 45–6
   cell manufacturers 45
   fabrication 45
   layered structure *45*
   module *46*
   solar cells *76*
calculation programs 184–5
calculators 2
carbon monoxide 23
cassette fixing 256, *256*
cast polycrystalline silicon blocks *26*
cathode sputtering 40, 43
cell arrangement and transparency 77–8
cell background 79
cell efficiency 41
cell encapsulation 67–72
   with EVA 68
   new module concepts 72–2
   in PVB 69
   in resin 69
   special features when encapsulating thin-film raw modules 70–1
   in Teflon 69
cell parameters 56–7, 59
   and solar cell characteristic *I–V* curves 58–60
cell shapes 78–9
cell stringing 65–6
   automated *65*
   integrated series interconnection with thin-film cells 66–7, *67*
cell thickness of thin-film cells *40*
cell types 23–53, *23*, *76*
central inverter concept 153–5
   concept with higher voltages 154, *154*
   low-voltage concept 153–4, *153–4*
   master–slave concept 154, *155*
central inverter with high power output range (3 phase) 117
central inverter with low power output range (single phase) 118
Centre for Solar Energy and Hydrogen Research 16
ceramic 48
certification and approval testing 98–100
   IEC 61215 98–100
   IEC 61646 98–9
characteristic curve equations *54*, 55
characteristic *I–V* curves
   for inverters *111*
   for modules 85–7, 93–4
     amorphous thin-film modules *94*
     data sheet information (DIN EN 50380) *86*
     for mono-crystalline 50W module *85*
     for three solar cells connected in parallel *86*
     for three solar cells connected in series *85*
charge carriers 21, 33, 55
charge controller cable 325
charge controllers 5, 310–14, *311*
   deep discharge protection 312
   manufacturers 311
   MPP charge controllers 313–14, *313–14*
   series controllers 312, *312*
   shunt controllers (parallel controllers) 312, *312*
charge factor, charge efficiency and energy efficiency of batteries 307, *307*
charge, state of 306, *306*
charging and discharging batteries 305–6, *305–6*
checklists for building survey 146–50
   PV generator, inverter and meter 148–9
   PV system checklist 147–8
     relevant measurements *147*
     sample drawings *147*
   shading checklist 149, *150*
chlorophyll 46
CIS (copper indium diselinide) cells 15, *23*, 40, 43–4
   cell manufacturers 44
   fabrication 43–4
   on flexible metallic foils *44*
   layered structure *43*
   modules based on copper indium disulphide *44*
CIS solar cells *76*
CIS Solartechnik 44
clocks 2
closed questions 21
coefficients for crystalline modules *89*colours of polycrystalline cells with AR coating 31
cold façades 237, *237*
cold roof 207, *207*
collar beam
   roof 201
   truss *202*
combined systems for photovoltaics and solar thermal 223–4, *224*
communication stations 2
concentration principle *50*
concentrating systems 49–50
   Flatcon module prototype *50*
   Fresnel lenses *50*
connection
   comparison of concepts 143–4
     power losses *144*
   and fixing elements 239
   in parallel 142, *143*
   in series (string concept) 142, *142*
connection systems 124–6
   plug connectors 125
   post terminals 125
   screw terminals 125
   spring clamp terminals 125
consulting with customer 130–1
contact fingers 32
contact form *351*
contact grid lines *27*, 28
conversion frequency 109–10
conversion losses in battery 320
copper 43
copper indium diselinide *see* CIS
corrosion 307

cost trends 331–2, *332*
crystal lattice 19–20
crystalline cells *76*
crystalline silicon 23–4
crystalline silicon on glass *see* CSG
crystalline silicon solar cell 19, 23, *23*
   design and functioning 21–2, *22*
      thin-film solar cells made from crystalline silicon 48–9
crystalline silicon sensors 14
crystalline structure of silicon *20*
CSG (crystalline silicon on glass) solar cells 48–9
   layered structure *49*
   prototype module *49*
CSG Solar AG 48
current, battery 304
current equations 54–8
current/voltage characteristic curves
   comparison of crystalline and amorphous silicon solar cells *59*
   for silicon diode BAY-45 *54*
   with different irradiance and constant temperature for amorphous triple cell modules *95*
curved glass custom module *75*
custom-made modules 74
   from acrylic plastic or Makrolon *82*
   with LED *83*
customer orientation 341
Czochralski process (crucible drawing process) 24–5
Czochralski silicon 25

dark equivalent circuit diagrams and characteristic curve *54*
DASTPVPS 185–6, *186*
database management 357
Daystar 44
DC (direct current)
   cabling 5
   installation 273
   load switch (main switch) 126–7, *126*
   main cable 126
      sizing 170–1
   main disconnect/isolator switch 5
      selecting and sizing 172–3
   motor 17
DC-coupled systems 328–9, *329*
deep discharge protection 312
DEGERconecter control system *19*
DEGERtrakers 18, *19*
design and service programs 197
design options 75–83
   cell arrangement and transparency 77–8
   cell background 79
   cell shapes 78–9
   cell types 76
   custom-made modules
      from acrylic plastic or Makrolon 82
      for building integration with connection and bypass diodes 92
      with LED 83
   front contacts 80
   glass format 80
   glass size 80
   glass type: multifunctional modules for building envelopes 81–2
   noise protection modules 83
designer module with diagonal power buses *33*

diffuse radiation 10–11, 13
diffusion 20
diffusion length 21
diode 53
direct coupling of PV array, battery and loads 316–17
direct current *see* DC
direct marketing 347–50
   attention 347–8
   brochures, sales documentation 349
   checklist 349
   desire 348
   face-to-face contact 347
   interest 348
   mail shot 347
   marketing tip 349
   organizing mail shot 347, 349
   publication pyramid *350*
   telephone 343
direct radiation 10–11, 50
   typical daily development (Berlin) 11, *11*
distribution of solar radiation 9
doping atoms 20
double-skin façades (double façades) 238, *238*
drinking water 2, 4
drying out 308
dual-axis tracking 15–16
dye cells *23*
dye-sensitized nano-crystalline cells 46–7, 52
   first commercial module *47*
   layered structure *46*
   prototype module *47*

earth leakage circuit breakers 127
earthing/grounding, lightning and surge protection 173–8, *174*
eaves and ridge of roof with substructure *206*
echo technique 363–4
economics and environmental issues 331–39
   cost trends 331–2, *332*
   economic assessment 333–6
      power production costs 333–6, *334*
   environmental impact 335–9
   technological trends 332, *333*
edge clamps and clips for metal roofs 213, *213*
Edge-Defined Film-Fed Growth *see* EFG Technique
effective solar cell model 56–8
   ideal model *56*
   simple model *56*
   standard model (single-diode model) *56*
   two-diode model *56*
efficiency
   of inverters 109–14, *113*, *114*
   of solar cells and PV modules 62–3
      maximum 52, *52*
EFG Technique 27, *28*
   ribbon-pulling machine *28*
   square cells *28*
electrical characteristics of thin-film modules 93–8
electricity consumption, measuring 317–18
electrical properties of solar cells 53–63
   cell parameters and solar cell characteristic I–V curves 58–60
      standard test conditions (STC) 59, 62–3
   efficiency of solar cells and PV modules 62–3
   equivalent circuit diagrams 53–8
      additional solar cell modules 58

effective solar cell model 58–9
spectral sensitivity 60–1
electricity meter 149
electrolytic baths 40
electron pair bond 19
electronic grade silicon 23
energy balance sheet 22
energy content of annual solar radiation 9
energy flow diagram of grid-connected PV system 179
energy meter with pulse output *290*
energy payback and harvest factor 335–6, *335–6*
energy source, sun as 8–9
energy storage system 5
energy yield 15, 16
environmental impact 335–9
　energy payback and harvest factor 335–6, *335–6*
　module recycling concepts 337–9, *338–9*
　pollutants in production process 336–7
environmental risks 45
equipotential bonding and earthing/grounding of array support structure 279, *279*
equivalent circuit diagrams 53–8
　extended 55, *55*
　for solar cells and their characteristic curve equations *54*, *56*
ethylene vinyl acetate *see* EVA
Euro efficiency 111–13
European Commission Joint Research Centre 98, 99
European Union 186
EVA (ethylene vinyl acetate) 41
　encapsulation 68, *68*
　film module *68*
　glass-film module *68*
　glass-glass module *68*
　metal-film module *68*
extended equivalent circuit diagram 55, *55*
external series connection of crystalline solar cells *65*
external wall structure 235–7
　framed structures 236
　function of individual layers in *235*
　load-bearing wall construction 236
extrinsic conduction in n- and p-doped silicon 20, *20*

façade basics 235–44
　external wall structure 235–7
　façade structures and construction methods 238–42
　fastenings 242–3
　joints and joint sealing 243–4, *243*
façade integration 31
façade systems 52
　with CIS modules *44*
façade structures and construction methods 238–42
　anchoring parts 240
　connection and fixing elements 239
　façade cladding 239, *239*
　lightweight structural glazing systems 241–2, *241*
　mullion-transom stick system 240, *240*
　spandrel panel construction 241
　supporting structure 240
　thermal insulation 239
　unitized façades 241, *241*

ventilated rain-screen façades 238–9, *239*
façade types 237–8
　cold façades 237, *237*
　double-skin façades (double façades) 238, *238*
　warm façades 237–8, *238*
façades with integrated modules 246–56
　cassette fixing 256, *256*
　glazing beads 247, *247*
　module fixing 247, *247*
　point fixing
　　along edges 254, *254*
　　on module back 254–5, *254–5*
　　through boreholes 255, *255*
　pressure plates 248–51, *248–51*
　structural sealant glazing 252–3, *253*
　two-sided linear-supported fixing 253–4, *253*
Fachhochschule für Technik und Wirtschaft (FHTW) Berlin 13
fastenings 242–3
　linear-supported fixing 242
　point fixings 242, *243*
fault types and checks and measures to detect them *288*
filling factor of solar cells 60, *60*
film module (EVA) *68*
fixed solar shading 264–7, *264–7*
　canopies 264–5, *265*
　other solar shadings devices 266, *266–7*
fixing rails, fitting 278–9, *278–9*
flashlights 2
flat roofs 207–8, *207–8*, 224–34
　constructions 202
　on-roof systems for flat roofs 224–34
　reverse roof 208, *208*
　roof-integrated systems 234–5, *235*
　　manufacturers 235
　ventilated roof: cold roof 207, *207*
　unventilated roof: warm roof 208, *208*
Flatcon module prototype *50*
flexible amorphous modules *43*
float-zone method 34, 35, 39
fluidized bed reactors 23
flush-fitted systems 218, *218*
fossil fuels 9
four pillars of marketing concept 342–5
　company 345
　marketing implementation 345
　products 345
framed structures 236
freestanding installations, mounting systems for 270–1, *270–1*
Fresnel lenses 50
front contacts 31–3, 79–80
　along grain boundaries of polycrystalline silicon *33*
　decorative pattern designs *33*
　laser-formed, in mono-crystalline cell *32*
　screen-printed, in polycrystalline cell *32*
front doping 61
fuses 101–3, *102*

gallium 43
gallium arsenide (GaAs) 52
garden light *2*
gardening applications 2
gaseous silane 42
germanium 42, 52
glass-film module (EVA) *68*
　with black front contacts *79*
glass format 80

glass-glass module
  effect on interior spaces 77
  EVA *68*, *70*
  PVB *69*
  resin *70*
  with transparency solar cells 77
glass-free CIS cells on plastic films *44*
glass roofs 257–63
  overhead glazing
    above heated spaces 257–9, *257–9*
      above open spaces *260–62*
      above unheated spaces *259–60*
    skylights on listed buildings 262, *262–3*
glass size 80
glass type: multifunctional modules for building envelopes 81–2
glazing beads 247, *247*
global radiation *10*
golden polycrystalline cell with special AR coating *31*
GOMBIS 183
gradient of current/voltage characteristic curve of solar cell *58*
graphite 28, 48
Grätzel, Michael 46
Greatcell Solar 47
green polycrystalline cell with special AR coating *31*
Greenius 183, 186, *187*
grid-connected inverters 103–23
  characteristics, characteristic curves and properties 109–17
    addition properties 115–17, *116–17*
    conversion frequency 109–10
    Euro efficiency 111–13
    overload behaviour 113–14
    recording operating data 114–15
    static efficiency 110–11
    tracking efficiency 110
  further developments in technology 118–23
    developments relating to large-scale grid-connected inverters 120–1
    inverter with multiple MPP trackers (multi-string concept) 118–19, *118*, *119*
    inverter with separate MPP trackers (string converter concept) 119–20, *119*
    master-slave concept in low power ranges 120, *120*
    three-phase concept in low power ranges 120
  grid-controlled inverters 105–6
    manufacturers 105
    principle *104*
      with single-phase and three-phase inverter *104*
  self-commutated inverters 106–9
  types and construction sizes in various power classes 117–18
    central inverter with high power output range (3 phase) 117
    central inverter with low power output range (single phase) 118
    module inverter 118
    string inverter 118
  wiring symbol and method of operation 103–5
grid-connected PV systems 5–8, 52
  at chicken farm 7
  example installation 276–83
  on noise barrier, Switzerland 8

  on roof of family house 6
  principle 6
  PV cube at Discovery Science Center 7
  PV modules and other components of 65–127
  on urban commercial estate 6
grid-controlled inverters 105–6
  principle *105*
grid-tied PV systems, selecting and sizing cables for 165–72
ground mounting of PV systems 8, *8*, *46*
ground reflection 13
Group III–V semiconductors 49, 52
GTOs (gate turn-off thyristors) 106
guarantee 283

Häberlin, Heinrich 112
hanger bolts for eternite corrugated roofing and trapezoidal sheet roofs 213–14, *213–14*
hanger systems 218
heat-strengthened glass *see* HSG
hermetic isolation of cooling system and electronics for inverters in outside applications *123*
heterojunction with intrinsic thin-layer *see* HIT
HF (high-frequency) transformer 107
  inverters with *108*
high-performance cells 34–40, 61
  back-contacted solar cells 35–6
  new solar cell concepts 37–40
    silver cells 39–40
    spherical solar cells 37–8
    transparent solar cells 36–7
higher-voltages concept 154, *154*
HIT (heterojunction with intrinsic thin-layer) hybrid technology 34, 52
holiday homes 2
HORIcatcher software 139–40, 183
horizON software 140, *140*, 183
hot spot(s)
  bypass diodes and shading 89–92
  solar cell with *90*
HSG (heat-strengthened glass) 71
hybrid HIT solar cells *23*, 51, 52
  cell manufacturer 51
  fabrication 51
  layered structure *51*
hybrid modules 48
hydrogen 23, 24
hydrogen chloride 23

iceberg principle 342, *342*
IGBTs (insulated gate bipolar transistors) 106
illuminated equivalent circuit diagram and characteristic curve *55*
impurity atoms 20, 21
impurity conduction 20, *20*
indium 43
indium gallium arsenide (InGaAs) 49
indium gallium phosphide (InGaPh) 49
indium tin oxide 41, 42, 45
in-roof systems 220–4, *220–1*
  combined systems for photovoltaics and solar thermal 223–4, *224*
  manufacturers of combined systems 224
  manufacturers of in-roof section systems 222
  manufacturers of solar roof elements 223
    roof coverings with integrated PV module 223, *223*

section systems for standard modules 221–2, *221*
solar roof elements 222–3
INSEL 194, *194*
insolation 9, 11
installing, commissioning and operating grid-connected PV systems 273–95
   breakdowns, typical faults and maintenance for PV systems 284–6
   example installation of grid-connected PV system 276–83
      preparation 276–7, *276*
   guarantee 283
   installation notes 273–5
      cable laying 275
      DC installation 273
      module interconnection 275
      module mounting 273–4
   long-term experience and quality 293–5
   monitoring operating data and presentation 288–92
   system installation, step by step 277–83
      attaching roof hooks 277, *277*
      equipotential bonding and earthing/grounding of array support structure 279, *279*
      fitting fixing rails 278–9, *278–9*
      installing mains connection 283, *283*
      inverter installation 282, *282*
      mounting modules 279–80, *280*
      running string cables through roof 280–91, *281*
      string wiring installation inside building 281–2, *282*
      tile cutting 278, *278*
   troubleshooting 286–8
integrated series interconnection with thin-film cells 66–7, *67*
interconnection of PV modules 100–1, *100–1*
   parallel 101, *101*
   series 100, *100*
International Electrotechnical Commission (IEC) 98
International Motor Show (IAA), Frankfurt, Germany 82
internet addresses for researching weather data 197
internet-based system evaluation 291, *291*
internet-based user interface for system visualization *121*
intrinsic conductivity 19–20, *20*
inverter(s) 5
   choosing number and power rating of 159–61
   installation 282, *282*
   with multiple MPP trackers (multi-string concept) 118–19, *118*, *119*
   plug-in module *119*
   with separate MPP trackers (string converter concept) 119–20, *119*
   sizing 159–65
      determining number of strings 164
      sizing using simulation programs 164, *164–5*
      voltage selection 161–4
   see also grid-connected inverters
irradiance 9, 12, 16, 55, 62–3
   dependence
      on angle of elevation *13*
      and temperature characteristics 87–9

differences on horizontal and solar-tracking surfaces *15*
frequency and energy *111*
irradiance meter with PV sensor *14*, 14, *290*
   manufacturers and providers 15
irradiation sensors 15
irrigation 2
ISC (short circuit current) 59
Isofoton 83
Italian National Agency for New Technologies, Energy and the Environment (ENFA) 16

Joint Research Centre (EC) 98, 99
joints and joint sealing 243, *243*
   silicone jointing with PV modules 244, *244*
Jubilee Campus, Nottingham University 78
junction boxes 84, *84*

Kaneka 48
Konarka Technologies, inc. 47
Kyosemi Corporation 37–8

laminates 73
landscaping applications 2
lasers 53
lead-acid batteries
   ageing effects 307–8
   construction and operating principles 299–300, *299–300*
   conversion losses 320
   operating behaviour and characteristics 303–7
      battery capacity 303–4, *304*
      charge factor, charge efficiency and energy efficiency 307, *307*
      charging and discharging 305–6, *305–6*
      current 304
      safety and maintenance 309–10
      state of charge 306–7, *307*
      voltage 304–5
   recycling 310
   selection criteria 308–9, *309*
   types and design 300–3
      block batteries with positive flat plates (OGi block) 303, *303*
      lead-acid gel batteries 301–2, *302*
      lead-acid grid-plate batteries with fluid electrolyte (wet cells) 300–1, *300*
      stationary tubular plate batteries (types OPzS and OPzV) 302–3, *302*
   for use in PV systems, manufacturers 300
leakage currents 55
LEDS, PV modules with integrated 83, *83*
LF (low-frequency) transformer 107
   circuit design for inverter with *107*
light-emitting diodes *see* LEDs
lightning protection
   direct strikes 175–6, *175–6*
   earthing/grounding and surge protection 173–8, *174*
   indirect lightning effects 176–8
   internal 176–8
lightweight structural glazing systems 241–2, *241*
linear clamping 217, *217*
linear-supported fixing 242, 264
load-bearing wall construction 235–6
long-term experience and quality 293–5
   long-term behaviour of PV modules 293–4, *294*

quality and reliability of inverters 294–5, *295*
long wavelength solar radiation 61
loop formation in module wiring *177*
low-cost lenses 49
low light conditions of thin-film modules 95
low-voltage concept 153–4, *153–4*

mains connection, installing 283, *283*
maintenance 285
   and upkeep checklist *286*
Makrolon 82
manufacturing solar cells 23–4, *24*
market overview and classification 183
marketing and promotion 341–65
   customer orientation 341
   database management 357
   four pillars of marketing concept 344–7
   greater success through systematic marketing 343–58
   iceberg principle 342, *342*
   marketing PV: basics 341–3
   marketing reach 346
   pull concept 342–3
   questioning technique 361
   range of marketing options 347–53
      direct marketing 347–50
      events 350
      mass media 350–3
   sales talk 357–65
   six steps to target 353–8
   successful selling 357–9
mass media, marketing via 350–3
master–slave
   concept 154, *155*
      in low power ranges 120
   efficiency curve of master-slave unit with three inverters *120*
maximum power point *see* MPP
Maxis BC+ back-contacted cells 36
MCBs (miniature circuit breakers) 127
measurement
   of electricity consumption 317–18
   of solar radiation 14–15
medium wavelength light 61
metal-film module (EVA) 68
metal roofs
   edge clamps and clips for 213, *213*
   penetrating 214, *214*
metallic contacts 31, 33
metallurgical silicon 23, 24
METEONORM 192, 195, *196*
meter cupboard 5
microcrystalline cells *23*, 48
micromorphous cells *23*, 48, 53
Mie scattering 12
milk frother *2*
milled polycrystalline cells with different AR coatings *37*
miniature circuit breakers *see* MCBs
mini-module with anti-fog coated Fresnel lenses *50*
mismatching losses 318
mobile ice-cream stand with solar freezer system *3*
mobile systems 2
modified sine-wave inverters 313
module and string cables 123–4
   module cable types and characteristics *124*
module cable outlets and junction boxes 84
module efficiency 41, 264

under open air conditions with amorphous triple cells 95
module fixing 216–17, 264
   linear clamping 217, *217*
   linear support 264
   point clamping 216–17, *216–17*
   point-fixing through boreholes 264
module *I–V* curves
   at different module temperatures *88*
   for varying irradiance and constant temperature *87*
   with and without bypass diodes *91*
module fixing 247, *247*
module interconnection 275
module inverter 118
   with operating data recorder and PC interface *115*
module mounting 273–4
module plug connectors, manufacturers of 126
module power at different module temperatures *88*
module recycling concepts 337–9, *338–9*
module structure and prototype module with moulded PU frame *71*
module wiring variants *165*
modules
   standard 73
      frame *73*
      frameless *73*
   types 72–4
molecular scattering 12
molybdenum 43
monitoring operating data and presentation 288–92, *288–92*
   internet-based system evaluation 291, *291*
   presentation and visualization 292, *292–3*
   web-based data transmission and evaluation 292
mono-crystalline silicon cells 15, 23–4, *23*, 24–5
   cell manufacturers 25
   fabrication *24*, 24–5
   with laser-formed front contacts *32*
mono-crystalline silicon wafers 23
Mont Ceris Training Academy, Herne, Germany 156, *156*
MOSFETs (metal-oxide semiconductor power field effect transistors) 106
mountain cabins 2, *3*
mounting modules 279–80, *280*
   on existing façades 245–6, *245–6*
mounting solutions for large-scale and lightweight roofs 233, *233*
mounting systems and building integration 199–271
   façade basics 235–44
   flat roofs 224–34
      variations 231–2, *231–2*
   for freestanding installations 270–1, *270–1*
   glass roofs 257–63
   photovoltaic façades 244–56
   roof basics 200–8
   solar protection devices 263–9
   sloping roofs 209–24
mounting tiles for tiles, concrete roof tiles or plain tiles 212, *212*
moveable solar shading 267–9, *267*
   solar shading louvers 267, *268–9*
Mover tracking systems 18
MPP (maximum power point) 59, 87, 104, 141–2, 144

charge controllers 311–12, *311–12*
controllers 58
current 57
tracker/string converter *119*
voltage 57
mullion-transom stick system 240, *240*
multifunctional PV system, sports hall, Burgweinting, Regensburg, Germany
multi-junction cells 49
multi-string inverter *119*

nail plate truss *202*
National Aeronautics and Space Administration (NASA) 35
negative contact 36
Neue Messe exhibition centre, Munich, Germany 155, *155*
new industrial solar cell encapsulation *see* NICE
new module concepts 72–2
New Sloten housing estate, Amsterdam *199*
new solar cell concepts 37–40
    silver cells 39–40
    spherical solar cells 37–8
NICE (new industrial solar cell encapsulation) 72

observation systems 2
on-roof system *46*, 209–20
    corrosion 210–11
    edge clamps and clips for metal roofs 213, *213*
    flush-fitted systems 218, *218*
    hanger bolts for eternite corrugated roofing and trapezoidal sheet roofs 213–14, *213–14*
    hanger systems 218
    loads 210
    manufacturers of 220
    module fixing 216–17, *216–17*
    mounting *209*
    mounting tiles for tiles, concrete roof tiles or plain tiles 212, *212*
on-roof mounting systems for sloping roofs 219–20, *219–20*
    penetrating metal roofs 214, *214*
    rail system 214–16, *215–16*
    roof fastening 211
    roof hooks for tiles, concrete roof tiles, plain tiles or slates 211, *211*
    stability and structural requirements 210
    structural specifications 210
    theft protection 218, *219*
on-roof systems for flat roofs 224–34
    additional wind load 231–2, *213–2*
    anchoring (fixed systems) 227–9, *228–9*
    ballast-mounted systems (freestanding installations) 225–7, *225–7*
    fixed to roof covering 229, *229*
    manufacturers of systems for large-scale roofs 233
    module fixing 232
    mounting 224, *224*, 230, *230*
    mounting solutions for large-scale and lightweight roofs 233, *233*
    mounting systems
        manufacturers 232–3
        variations 231–2, *231–2*
    roof fastening 224

tracking systems for flat roofs 233–4, *233*
    manufacturers 235
on-site visit and site survey 129–30
open-circuit voltage 57
    and short-circuit current ISC depending upon irradiance *60*
open questions 21
operating data
    capture devices *288–9*
    and presentation, monitoring 288–92, *288–92*
Origin 39
overhead glazing
    above heated spaces 257–9, *257–9*
    above open spaces *260–2*
    above unheated spaces *259–60*
overload behaviour 113–14

p-n junction 24, 35
Panorama Master 140, *140*
parking ticket machines 2
parallel connection
    concept *154*
    of modules with four connection cables *153*
passivation 30, 33, 35, 61
passive thermodynamic tracking system *16*
passive tracking 16–17
peak power meter 57
Peccell 47
penetrating metal roofs 214, *214*
phosphorus 20, 24
    doping (n-doped) 20, 21, 26, 28, 35, 37, 39, 42
photoelectric current (photocurrent) 55
photons 55
photovoltaic effect *see* PV effect
photovoltaic façades *see* PV façades
photovoltaic systems *see* PV systems
photovoltaics in decentral electricity grids/mini-grids 328–31, *328*
    AC-coupled systems 329–30, *329–30*
    DC-coupled systems 328–9, *329*
Photovoltech 36
plain tile *203*
planning and sizing grid-connected PV systems 151–80
    lightning protection, earthing/grounding and surge protection 173–8
    selecting and sizing cables for grid-tied PV systems 165–72
    selecting and sizing PV array combiner/junction box and DC main disconnect/isolator switch 172–3
    sizing inverter 159–65
    system concepts 152–8
    system size and module choice 151–2
    yield forecast 178–80
plastic roof sheeting *203*
platinum 46
plug connectors 125
point clamping 216–17, *216–17*
point fixings 242, *243*
    through boreholes 255, *255*, 264
    along edges 254, *254*
    on module back 254, *254*
    with module clamps 264
pollutants in production process 336–7
pollution 13
polycrystalline APex cells 29–30
    cell manufacturer 30
polycrystalline band cells 23

polycrystalline EFG silicon cells 28–9
    cell manufacturer 29
    fabrication 28–9
polycrystalline POWER cells 23
polycrystalline silicon bars, sawn 26
polycrystalline silicon blocks, cast 26
polycrystalline silicon cells 15, 23, 24–30
    with AR and contact grid lines 27
    cell manufacturers 26
    fabrication 24, 26–7
    with screen-printed front contacts 32
    six-inch and eight-inch 27
    with three busbars 32
polycrystalline silicon wafers 23
    with AR coating 27
    without AR coating 27
polycrystalline string ribbon silicon cells 29
    cell manufacturer 29
polycrystalline thin-line cells 23
polysilicon 23
    manufacturing solar cells from 24
polyvinyl butyral see PVB
polyurethane see PU
positive contact 36
post terminals 125
POWER cell 36
power consumption table 318
power of silence 362
power restriction on day with strong cloud/sun alternation 114
power station towers, Stadtwerke Duisburg, Germany 7
power temperature coefficient
    for amorphous tandem module 97
    comparison of typical temperature coefficients 98
    for polycrystalline module 97
Power Tracker 18
PowerLight 18
pressure plates 248–51, 248–51
price question 364–5
promotion see marketing and promotion 341–65
Protection Class II test 100, 102, 172
prototype Lumiwall module 83
PU (polyurethane) 71
pull concept 342–3
purlin
    and rafters 201
    truss 201, 201
PV array(s) 16, 89
    combiner/junction boxes
        selecting and sizing 172–3
        and string diodes and fuses 101–3, 102–3
    configuration and system concept 141–4
    and inverter's operating range 161
    mismatch 152
    sizing 318–23
    tracking 15–18
PV-DesignPro (Solar Studio Suite) 187, 187
PV effect 19–22
    design and functioning of crystalline silicon solar cell 21–2, 22
    how solar cells work 19–21
PV F-chart 184–5, 185
PV façades 244–56, 245
    façades with integrated modules 246–56
    mounting modules on existing façades 245–6, 245–6

PV generator
    general parameters 320
    inverter and meter 148–9
PV installations, Olympic Village, Sydney 80
PV irradiation sensors 15
PV modules and other components of grid-connected systems 65–127
    AC switch disconnector 127
    cabling, wiring and connection systems 123–6
    cell encapsulation 67–72
    cell stringing 65–6
    characteristic I–V curves for modules 85–7
    comparison of typical filling factors of PV modules 93
    design options 75–83
    direct current load switch (DC main switch) 126–7, 126
    electrical characteristics of thin-film modules 93–8
    grid-connected inverters 103–23
    hot spots, bypass diodes and shading 89–92
    incorporated within insulating glass 82
    interconnection 100–1, 100–1
    irradiance dependence and temperature characteristics 87–9
    junction box with one bypass diode 92
    with laminated 'strip' bypass diodes 93
    with load 90
    module cable outlets and junction boxes 84
    quality certification for modules 98–100
    types of modules 72–4
        custom-made modules 74
        laminates 73
        special modules 73–4
        standard modules 73
    wiring symbols 84
PV resistors 58
PV sensors 14, 14
    manufacturers and providers 15
PV soundless noise barrier 83
PV system(s)
    applications 1–8
    array systems 1–8
        overview 1
        stand-alone systems 1–5, 3
    checklist 147–8
    concepts 152–8
        central inverter concept 153–5
        module inverter concept 156–7, 157
            sub-array and string inverter concept 155–6
    flowers 4
    grid-connected systems 5–8
    potential locations for installing 199
    size and module choice 151–2, 151
    types of 1
    with modules using various solar cell technologies 53
PVB (polyvinyl butyral) encapsulation 68, 69
    glass-glass module 69
PVcad 141, 183
PVS 140, 183, 184, 188, 188–9
PV*SOL 140, 164, 184, 190–2
PVSYST 141, 183, 192, 193
PVWATTS 198
pyranometer 14, 14
    manufacturers and providers 15

quality and reliability of inverters 294–5, *295*
quality certification for modules 98–100
    certification and approval testing 98–100
    manufacturers' warranty periods 100
    Protection Class II test 100
    TÜV-PROOF certification mark 100
quartz
    crucible 24, 26
    sand 23
questioning technique 361
    closed questions 21
    open questions 21

rack-mounted PV arrays 144–6, *145, 209*
    reducing mutual shading losses of 145–6
radiation, solar 8–18
    average daily global radiation on horizontal surfaces *318*
        deviation from horizontal *319*
    intensity 12
rail system 214–16, *215–16*
raised tie or collar beam roof 201
rafter(s)
    and purlin 201
    and raised tie or collar beam roof 201
    tied 201
    truss 202, *202*
Rayleigh scattering 12
RCDs (residual current devices) 127
rechargeable batteries 5
recombination 21, 22
recording operating data 114–15
recycling concepts, module 337–9, *338–9*
reflection 9, 22
reflectivity 13
residual current devices (RCDs) 127
resin encapsulation 68, 69–70
resistors 55, 58
reverse roof 208, *208*
ribbon-pulled silicon cells 27–30, *29*
    polycrystalline EFG silicon cells 28
        cell manufacturers 28
    polycrystalline string ribbon silicon cells 29
        cell manufacturer 29
    polycrystalline APex cells 29–30
        cell manufacturer 30
ridge tiles *205*
roof basics 200–8
    flat roof 207–8, *207–8*
    roof constructions 201–2, *201–2*
    roof shapes 200
    roof skin 202–5
    roof tasks 200, *200*
    sloping roof 206–7, *206–7*, 209–24
roof constructions 201–2, *201–2*
    flat roof constructions 202
    purlin and rafters 201
    rafter and raised tie or collar beam roof 201
    tied rafters 201
    truss constructions 202
roof coverings with integrated PV module 223, *223*
roof fastening 211, 224
roof hooks for tiles *205*, 211, *211*
    attaching 277, *277*
roof installations and superstructures 207, *207*
roof mounting of PV systems 8
roof overhang *206*

roof skin 202–5
    roof covering: drainage covering used in sloping roof applications 202, *203*
    roof sealing: sealing covering used in flat roof applications 203, *203*
    types *204*
round mono-crystalline cell 25, *25*
round timber holder *205*
running string cables through roof 280–1, *281*
ruthenium-base dye 46

safety and maintenance, battery 309–10
Sanyo 34
Saturn process 32, 34
    cf screen printing method *31*
sawn polycrystalline silicon bars *26*
scattering of solar radiation 9
Schott administration building, Barcelona, Spain 81
screen printing method 32, *32*
    cf Saturn technology *31*
screw terminals 125
section systems for standard modules 221–2, *221*
selenium 43, 44
self-commutated inverters 106–9
    with HF transformer 107, *108*
    with LF transformer 107
        circuit design for *107*
    principle *106*
    transformerless 108–9, *108*
self-shading 134, *135*
semiconductors 23
    III–V 49
    photoactive 40
semi-round mono-crystalline cell 25, *25*
semi-transparent glazing 52
semi-transparent sunshade louvres *79*
sensor-controlled tracking 17
series controllers 311–312, *311–312*
series resistor 55
7C strategy 346, *355*, 355–6
shaded PV module without bypass diodes *90*
shading 22
analysis 137–9
    tools, using software 139–41
    using sun path diagram
        on acetate 137–9, *138*
        and site plan 137
    of cell in 50W standard module with two bypass diodes *91*
    characteristic I–V curves
        in case of connection in parallel *143*
        in case of connection in series *142*
checklist 149, *150*
    direct shading 135–6, *135–6*
    with freestanding/rack-mounted PV arrays 144–6, *145*
        reducing mutual shading losses of rack-mounted PV modules 145–6
    and hot spot and bypass diodes 89–92
    losses in relation to shading angle and tilt angle *146*
    PV-array configuration and system concept 141–4
        comparison of connection concepts 143–4
        connection
            in parallel 142, *143*
            in series (string concept) 142, *142*
    resulting from building 133–4, *134*

resulting from location 133, *133*
self-shading 134, *135*
simulation of 183
temporary 131–2
of thin-film modules 94
and crystalline silicon modules *94*
shadow types 131–6
direct shading 135–6, *135–6*
self-shading 134, *135*
shading resulting from building 133–4, *134*
shading resulting from location 133, *133*
temporary shading 131–2
sheet metal roofs *203*
Shell solar cell factory 77
Shell Solar Path 197
short-circuit current (ISC) 57, 59
and open-circuit voltage VOC depending upon irradiance *60*
short wavelength light 61
shunt controllers (parallel controllers) 310, *310*
silane 23–4, 42
silicon 19, 48
crystalline structure *20*
silicon carbide 24
silicon cells *see* mono-crystalline silicon cells; polycrystalline silicon cells; ribbon-pulled silicon cells
silicon diode 53
silicon dioxide 23
silicon granulate 24
silicon nitride 21, 30, 48
silicon oxide 34, 35, 61
silicon tetrachloride 24
silicone jointing with PV modules 244, *244*
silver cells 39–40
as strip cells and modules *39*
silver polycrystalline cell with special AR coating *31*
simulated annual efficiencies of inverters *160*
simulation
checking results 182–3
classification of sizing and design software for PV systems *184*
program descriptions 184–98
calculation programs 184–5
design and service programs 197
simulation systems 194–5
supplementary programs and data sources 195–7
time-step simulation programs 185–93
web-based simulation programs 198
of shading 183
sine-wave inverters 313, *313*
single-diode model of solar cell 55, *58*
single-axis tracking 15
single-wire solar cables 124
sintering 33
site plan and sun path diagram, using 137
site surveys and shading analysis 129–50
checklists for building survey 146–50
consulting with customer 130–1
on-site visit and site survey 129–30
shading analysis 137–9
tools, using software 139–41
shading with freestanding/rack-mounted PV arrays 144–6, *145*
shading, PV-array configuration and system concept 141–4
shadow types 131–6
SITOP solar select 197

six-stage marketing cycle *354*, 354–8
sizing PV array 318–23
battery sizing 325–6
cable, conversion and adjustment losses 320–21
calculating method for designing PV array 322
determining summer excess and winter reserve 321–23
model for calculating yield 318–20
photovoltaics in decentral electricity grids/mini-grids 328–30
sizing cable cross-sections 323–5, *323–5*
charge controller cable 325
summary of design outcome 321–3
use of inverter 326–7, *328*
skylights on listed buildings 262, *262–3*
slate, natural *203*
sloping roof 206–7, *206–7*
in-roof systems 220–4
on-roof mounting systems for 219–20, *219–20*
on-roof systems 209–20
roof installations and superstructures 207, *207*
sludging 308
SMA inverter with integrated electronic DC circuit breaker *123*
SMILE 194–5
snow
arrangement of tilted PV modules in case of *132*
on PV system *132*
snow trap grid *205*
solar altitude 12, *12*
and solar spectrum 11–13
solar array 18
solar boat *3*
solar bus-stop lighting *4*
solar car *3*
solar cell(s) 2, 14
characteristic curve 58
electrical properties 53–63
with hot spot *90*
how they work 19–21
manufacturing 23–4
types 23–53
amorphous silicon cells 42–3
anti-reflective coating (AR) on crystalline silicon cells 30–1
back contacts 33–4
cadmium telluride (CdTe) cells 45–6
CIS (copper indium diselinide) cells 43–4
comparison and trends 52–3
concentrating systems 49–50
crystalline silicon 23–4
dye-sensitized nano-crystalline cells 46–7
front contacts 31–3
high-performance cells 34–40
hybrid HIT solar cells 51
mono-crystalline silicon cells 24–5
polycrystalline silicon cells 26–7
ribbon-pulled silicon cells 27–30
thin-film cell technology 40–1
thin-film solar cells made from crystalline silicon 48–9
solar charger *2*
solar constant 9
solar energy 5
solar farms 18

Germany 17
solar flags with transparent crystalline solar
 cells and integrated LEDs 83
solar freezer system 3
solar-generated electricity 16
solar-grade siliucon 23
solar lighting system 5
solar modules 17
solar opening tile for routing solar wiring 205
Solar Pathfinder 139, 139
solar power 2
Solar Pro 183
solar protection devices 263–9, 263
 fixed solar shading 264–7, 264–7
 module fixing 264
 moveable solar shading 267–9, 267
solar pump systems 2, 4
solar radiation 8–18
 angle definition 11
 direct and diffuse radiation 10–11
 distribution of solar radiation 9
 ground reflection 13
 how solar radiation is measured 14–15
 solar attitude and solar spectrum 11–13
 sun as energy source 8–9
 tracking PV arrays 15–18
 worldwide distribution of annual solar
  radiation 10
solar radios 2
solar roof elements 222–3
 special modules with roof covering
  properties 222–3
solar roof tile 74
solar shadings devices 266, 266–7
solar silicon 23
solar spectrum 13
 and solar attitude 11–13
solar tracker with electric motor system 16
solar water desalination 2, 5
solar water disinfection 2
Solarion 44
soldering cell strings 65
SOLDIM 183, 193, 193
SolEm 140, 164, 183
SOLON Mover 18
Sony Europe 47
SOS telephones 2
space charge region 20, 21, 21
space travel 35
spandrel panel construction 241
special modules 73–4
spectral sensitivity 60–1, 60
Spheral spherical solar cells 38, 38
spherical solar cells 37–8, 38–9
 prototype modules 39
Spherical Solar Power 37
spring clamp terminals 125
sputter processes 40
square EFG cells 28
square mono-crystalline cell 25, 25
square-wave inverters 313
stacked solar cells 42, 61
Staebler-Wronski effect 42
standalone inverters 313–15
 application criteria 315
 manufacturers 314
 modified sine-wave inverters 315
 sine-wave inverters 315, 314
 square-wave inverters 315
standalone PV systems 1–5, 3, 297–330

AC through inverter 298
batteries 298–310
charge controllers 310–14
DC only 297
measuring electricity consumption 317–18
modules 298
planning and designing 316–17
 direct coupling of PV array, battery and
  loads 316–17
 sizing PV array 318–23
 standalone inverters 313–15
 PV-wind hybrid system 298
standard modules 73
 frame 73
 frameless 73
standard test conditions (STC) 59, 88, 99
static efficiency 110–11
stationary tubular plate batteries (types OPzS
 and OPzV) 302–3, 302
STI 47
string cables, running through roof 280–1, 281
string cabling, recommended wiring lengths
 168–70
string diodes 101–3, 102
string fuses 102, 102
string inverter 118
 concept 155–6, 156
stringing machine 66
string ribbon process 27, 29
string wiring installation inside building 281–2,
 282
structural sealant glazing 252–3, 252–3
sub-array and string inverter concept 155–6,
 156
Sulfurcell Solartechnik 44
sulphation 307
sulphur 43
sun
 as energy source 8–9
 path
  diagram 137
   on acetate 137–9, 138
  at particular times of year 12
SUNDI 196, 196
sunlight
 energy in 9
 passing through atmosphere 10
 and spectral sensitivity 60
SunPower A-300 cell 35, 52, 61
 back with contact lines and front 35
 modules without front contacts 35
Sunways 36
surge and lightning protection and
 earthing/grounding 173–8, 174
surge arresters 178
 types and rates voltages for 177
system planning with thin-film modules and
 shading 94
system sizing, design and simulation software
 181–98
 checking simulation results 182–3
 market overview and classification 183
 program descriptions 184–98
 simulation of shading 183
 use of 181–2

Teflon encapsulation 68
 module 69
temperature
 behaviour of thin-film modules 96–8

characteristics 87–9
comparison of typical temperature coefficients 98
dependence
of amorphous modules 96
of CdTe modules 97
and irradiance dependence 98
of CIS modules 96
of electrical characteristics of PV modules 88
increase and reduction in annual energy yield for various PV array installation methods 89
and irradiance sensor 290
temporary shading 131–2
Test Specification No.503 98
Texas Instruments 37
theft protection 218, 219
thermal insulation 239
thermo-hydraulic control system 16, 16
thermoplastic polyurethane see TPU
thermoplastic spacer see TPS
thin-film modules, electrical characteristics of 93–8
characteristic I–V curves for modules 93–4
low light conditions 95
shading 94
thin-film cell technology 40–1, 53, 61
cell thickness, material consumption and energy expenditure 40
semi-transparent modules 41
integrated series interconnection 66–7, 67
typical strip appearance 41
thin-film solar cells made from crystalline silicon 48–9
crystalline silicon on glass (CSG) solar cells 48–9
microcrystalline and micromorphous solar cells 48
thin layer cells 23
3DSolarWelt 141, 183
three-phase concept in low power ranges 120
tied rafters 201
tile cutting 278, 278
time-step simulation programs 185–93
DASTPVPS 185–6, 186
Greenius 186, 187
tin oxide 41, 42
titanium dioxide 46–7
titanium oxide 21, 30
touch-proof plug connector 125
Town Hall, Monthey, Switzerland 77
TPS (thermoplastic spacer) 72
TPU (thermoplastic polyurethane)
tracking PV arrays 15–18
active 17–18
efficiency 110
passive 16–17
traffic signals 2
tracking systems 15–18
ATM 18
DEGERtrakers 18, 19
for flat roofs 235, 235
manufacturers 235
manufacturers 18
passive thermodynamic tracking system 16
Power Tracker 18
on roof of ufaBarbrik, Berlin 16
solar tracker with electric motor system 16
SOLON Mover 18

transformerless inverter 108–9, 108
comparison of inverters with and without 109
principle of 108
transistors 106
transparent conductive oxide (TCO) 41, 42, 48, 66
transparent solar cells 36–7
cell manufacturers 37
Transparent Sunways Solar Cell 37
isometry 37
trapezoidal custom-made module in production 74
tread step 205
trichlorsilane 23–4
Trier-Birkenfeld University of Applied Sciences, Germany 43
triple solar cells 61
amorphous 62–3
layered structure 62
TRNSYS 195
troubleshooting 286–8, 288
truss constructions 202
nail plate truss 202
rafter truss 202
TÜV-PROOF certification mark 100
two-diode model 56, 58
two-sided linear-supported fixing 253–4, 253

unitized façades 241, 241
unventilated roof 208, 208

valence electrons 19
vapour deposition 40
vent pipe 205
vent tiles 205
ventilated rain-screen façades 238–9, 239
ventilated roof 207
village electrification 2
violet polycrystalline cell with special AR coating 31
voltage
battery 304–5
drop 55
equations 54–8
optimization 163–4, 163
selection 161–4
maximum number of modules in string 162
minimum number of modules in string 162–3

wafers cut by laser 53
from octagonal tubes 28
warm façades 237–8, 238
warm roof 208, 208
watering system 4, 53
web-based data transmission and evaluation 292
providers of systems for operational data monitoring and yield checks 292
wind load, additional 231–2, 213–2
wind power 3
wireless operating data capture device with display 289
wiring diagram for 2kW system with central inverter 165
wiring symbol and method of operation 103–5
wood shingles 203

worldwide distribution of annual solar
    radiation *10*

yield checking, principle of (PVSAT project)
    290

yield forecast 178–80
    and simulation program *180*

zinc 24, 45
zinc oxide 41, 42, 43